PRINCIPLES AND APPLICATIONS OF MASS TRANSFER

PRINCIPLES AND APPLICATIONS OF MASS TRANSFER

The Design of Separation Processes for Chemical and Biochemical Engineering

Fourth Edition

Jaime Benitez

WILEY

This edition first published 2023
© 2023 John Wiley & Sons, Inc.

Edition History
3e © 2017 by John Wiley & Sons, Inc.

All rights reserved. No part of this publication may be reproduced, stored in a retrieval system, or transmitted, in any form or by any means, electronic, mechanical, photocopying, recording or otherwise, except as permitted by law. Advice on how to obtain permission to reuse material from this title is available at http://www.wiley.com/go/permissions.

The right of Jaime Benitez to be identified as the author of this work has been asserted in accordance with law.

Registered Office
John Wiley & Sons, Inc., 111 River Street, Hoboken, NJ 07030, USA

Editorial Office
111 River Street, Hoboken, NJ 07030, USA

For details of our global editorial offices, customer services, and more information about Wiley products visit us at www.wiley.com.

Wiley also publishes its books in a variety of electronic formats and by print-on-demand. Some content that appears in standard print versions of this book may not be available in other formats.

Limit of Liability/Disclaimer of Warranty
In view of ongoing research, equipment modifications, changes in governmental regulations, and the constant flow of information relating to the use of experimental reagents, equipment, and devices, the reader is urged to review and evaluate the information provided in the package insert or instructions for each chemical, piece of equipment, reagent, or device for, among other things, any changes in the instructions or indication of usage and for added warnings and precautions. While the publisher and authors have used their best efforts in preparing this work, they make no representations or warranties with respect to the accuracy or completeness of the contents of this work and specifically disclaim all warranties, including without limitation any implied warranties of merchantability or fitness for a particular purpose. No warranty may be created or extended by sales representatives, written sales materials or promotional statements for this work. The fact that an organization, website, or product is referred to in this work as a citation and/or potential source of further information does not mean that the publisher and authors endorse the information or services the organization, website, or product may provide or recommendations it may make. This work is sold with the understanding that the publisher is not engaged in rendering professional services. The advice and strategies contained herein may not be suitable for your situation. You should consult with a specialist where appropriate. Further, readers should be aware that websites listed in this work may have changed or disappeared between when this work was written and when it is read. Neither the publisher nor authors shall be liable for any loss of profit or any other commercial damages, including but not limited to special, incidental, consequential, or other damages.

Library of Congress Cataloging-in-Publication Data
Names: Benitez, Jaime, 1948- author.
Title: Principles and applications of mass transfer : the design of separation processes for chemical and biochemical engineering / Jaime Benitez.
Other titles: Principles and modern applications of mass transfer operations
Description: Fourth edition. | Hoboken, NJ : John Wiley & Sons Inc., 2023. | Revised edition of: Principles and modern applications of mass transfer operations / Jamie Benitez. Third edition. 2017. | Includes bibliographical references and index.
Identifiers: LCCN 2022034854 (print) | LCCN 2022034855 (ebook) | ISBN 9781119785248 (hardback) | ISBN 9781119785255 (pdf) | ISBN 9781119785262 (epub)
Subjects: LCSH: Mass transfer. | Chemical engineering.
Classification: LCC TP156.M3 B44 2023 (print) | LCC TP156.M3 (ebook) | DDC 660/.28423--dc23/ eng/20220726
LC record available at https://lccn.loc.gov/2022034854
LC ebook record available at https://lccn.loc.gov/2022034855

Cover image: © Baac3nes/Getty Images
Cover design: Wiley

Set in 9/12pt TexGyreSchola by Integra Software Services Pvt. Ltd, Pondicherry, India

SKY10036809_101422

A nuestro hijo, Jaime, un verdadero milagro de supervivencia. A nuestros nietos Teresita, Oliver y Diego; que la vida les sonría.

Contents

Preface to the Fourth Edition xvii

Preface to the Third Edition.. xix

Preface to the Second Edition... xxi

Preface to the First Edition xxiii

Nomenclature xxv

1. FUNDAMENTALS OF MASS TRANSFER................................... 1

1.1 INTRODUCTION.. 1

1.2 MOLECULAR MASS TRANSFER 3

1.2.1 Concentrations 4
1.2.2 Velocities and Fluxes... 10
1.2.3 The Maxwell–Stefan Relations.. 12
1.2.4 Fick's First Law for Binary Mixtures................................... 15

1.3 THE DIFFUSION COEFFICIENT 16

1.3.1 Diffusion Coefficients for Binary Ideal Gas Systems 17
1.3.2 Diffusion Coefficients for Dilute Liquids 22
1.3.3 Diffusion Coefficients for Concentrated Liquids 26
1.3.4 Effective Diffusivities in Multicomponent Mixtures 28

1.4 STEADY-STATE MOLECULAR DIFFUSION IN FLUIDS 34

1.4.1 Molar Flux and the Equation of Continuity 34
1.4.2 Steady-State Molecular Diffusion in Gases.......................... 35
1.4.3 Steady-State Molecular Diffusion in Liquids 47

1.5 STEADY-STATE DIFFUSION IN SOLIDS................................. 50

1.5.1 Steady-State Binary Molecular Diffusion in Porous Solids 51
1.5.2 Knudsen Diffusion in Porous Solids..................................... 52

vii

viii   Contents

1.5.3 Hydrodynamic Flow of Gases in Porous Solids 55
1.5.4 "Dusty Gas" Model for Multicomponent Diffusion 57

1.6 TRANSIENT MOLECULAR DIFFUSION IN SOLIDS........................... 58

1.7 DIFFUSION WITH HOMOGENEOUS CHEMICAL REACTION 62

1.8 ANALOGIES AMONG MOLECULAR TRANSFER PHENOMENA...... 66

PROBLEMS 68

REFERENCES.. 83

APPENDIX 1.1 84

APPENDIX 1.2.. 85

APPENDIX 1.3.. 86

APPENDIX 1.4.. 89

2. CONVECTIVE MASS TRANSFER...91

2.1 INTRODUCTION... 91

2.2 MASS-TRANSFER COEFFICIENTS... 92

2.2.1 Diffusion of A Through Stagnant B ($N_B = 0$, $\Psi_A = 1$) 92
2.2.2 Equimolar Counterdiffusion ($N_B = -N_A$, Ψ_A = undefined) 95

2.3 DIMENSIONAL ANALYSIS 96

2.3.1 The Buckingham Method... 97

**2.4 FLOW PAST FLAT PLATE IN LAMINAR FLOW; BOUNDARY
LAYER THEORY** 101

2.5 MASS- AND HEAT-TRANSFER ANALOGIES 108

Contents

ix

2.6 CONVECTIVE MASS-TRANSFER CORRELATIONS 116

2.6.1 Mass-Transfer Coefficients for Flat Plates 116
2.6.2 Mass-Transfer Coefficients for a Single Sphere 118
2.6.3 Mass-Transfer Coefficients for Single Cylinders 122
2.6.4 Turbulent Flow in Circular Pipes 122
2.6.5 Mass Transfer in Packed and Fluidized Beds 128
2.6.6 Mass Transfer in Hollow-Fiber Membrane Modules 130

2.7 MULTICOMPONENT MASS-TRANSFER COEFFICIENTS 133

PROBLEMS 135

REFERENCES 149

APPENDIX 2.1 152

APPENDIX 2.2 153

3. INTERPHASE MASS TRANSFER 155

3.1 INTRODUCTION 155

3.2 EQUILIBRIUM CONSIDERATIONS IN CHEMICAL AND BIOCHEMICAL SYSTEMS 155

3.2.1 Chemical Phase Equilibria 156
3.2.2 Biochemical Equilibrium Concepts (Seader et al., 2011) 160

3.3 DIFFUSION BETWEEN PHASES 166

3.3.1 Two-Resistance Theory 166
3.3.2 Overall Mass-Transfer Coefficients 168
3.3.3 Local Mass-Transfer Coefficients: General Case 172

3.4 MATERIAL BALANCES 180

3.4.1 Countercurrent Flow 180
3.4.2 Cocurrent Flow 194
3.4.3 Batch Processes 195

x   Contents

3.5 EQUILIBRIUM-STAGE OPERATIONS 196

PROBLEMS 204

REFERENCES 216

APPENDIX 3.1 217

APPENDIX 3.2 218

APPENDIX 3.3 219

APPENDIX 3.4 220

APPENDIX 3.5 221

4. EQUIPMENT FOR GAS–LIQUID MASS-TRANSFER OPERATIONS 225

4.1 INTRODUCTION 225

4.2 GAS–LIQUID OPERATIONS: LIQUID DISPERSED 225

4.2.1 Types of Packing 226
4.2.2 Liquid Distribution 229
4.2.3 Liquid Holdup 230
4.2.4 Pressure Drop 237
4.2.5 Mass-Transfer Coefficients 239

4.3 GAS–LIQUID OPERATIONS: GAS DISPERSED 243

4.3.1 Sparged Vessels (Bubble Columns) 244
4.3.2 Tray Towers 249
4.3.3 Tray Diameter 252
4.3.4 Tray Gas-Pressure Drop 255
4.3.5 Weeping and Entrainment 257
4.3.6 Tray Efficiency 258

Contents

PROBLEMS 264

REFERENCES 274

5. ABSORPTION AND STRIPPING 277

5.1 INTRODUCTION 277

5.2 COUNTERCURRENT MULTISTAGE EQUIPMENT 278

5.2.1 Graphical Determination of the Number of Ideal Trays 278
5.2.2 Tray Efficiencies and Real Trays by Graphical Methods 279
5.2.3 Dilute Mixtures 279

5.3 COUNTERCURRENT CONTINUOUS-CONTACT EQUIPMENT 285

5.3.1 Dilute Solutions; Henry's Law 290

5.4 THERMAL EFFECTS DURING ABSORPTION AND STRIPPING ... 292

5.4.1 Adiabatic Operation of a Packed-Bed Absorber 296

PROBLEMS 300

REFERENCES 311

APPENDIX 5.1 312

6. DISTILLATION 315

6.1 INTRODUCTION 315

6.2 SINGLE-STAGE OPERATION—FLASH VAPORIZATION 316

6.3 DIFFERENTIAL DISTILLATION 320

xii   Contents

6.4 CONTINUOUS RECTIFICATION—BINARY SYSTEMS ... 322

6.5 McCABE–THIELE METHOD FOR TRAYED TOWERS ... 324

6.5.1 Rectifying Section ... 325
6.5.2 Stripping Section ... 326
6.5.3 Feed Stage ... 328
6.5.4 Number of Equilibrium Stages and Feed-Stage Location ... 330
6.5.5 Limiting Conditions ... 332
6.5.6 Optimum Reflux Ratio ... 333
6.5.7 Large Number of Stages ... 339
6.5.8 Use of Open Steam ... 342
6.5.9 Tray Efficiencies ... 343

6.6 BINARY DISTILLATION IN PACKED TOWERS ... 350

6.7 MULTICOMPONENT DISTILLATION ... 354

6.8 FENSKE–UNDERWOOD–GILLILAND METHOD ... 357

6.8.1 Total Reflux: Fenske Equation ... 357
6.8.2 Minimum Reflux: Underwood Equations ... 361
6.8.3 Gilliland Correlation for Number of Stages at Finite Reflux ... 366

6.9 RIGOROUS CALCULATION PROCEDURES FOR MULTICOMPONENT DISTILLATION ... 368

6.9.1 Equilibrium Stage Model ... 368
6.9.2 Nonequilibrium, Rate-Based Model ... 370

6.10 BATCH DISTILLATION ... 371

6.10.1 Binary Batch Distillation with Constant Reflux ... 372
6.10.2 Batch Distillation with Constant Distillate Composition ... 375
6.10.3 Multicomponent Batch Distillation ... 377

PROBLEMS ... 378

REFERENCES ... 389

APPENDIX 6.1 ... 390

APPENDIX 6.2 ... 391

APPENDIX 6.3 ... 392

Contents　　　　　　　　　　　　　　　　　　　　　　**xiii**

7. LIQUID–LIQUID EXTRACTION...........393

7.1 INTRODUCTION.................393

7.2 LIQUID EQUILIBRIA.................394

7.3 STAGEWISE LIQUID–LIQUID EXTRACTION.................399

7.3.1 Single-Stage Extraction.................400
7.3.2 Multistage Crosscurrent Extraction.................403
7.3.3 Countercurrent Extraction Cascades.................404
7.3.4 Insoluble Liquids.................409
7.3.5 Continuous Countercurrent Extraction with Reflux.................412

7.4 EQUIPMENT FOR LIQUID–LIQUID EXTRACTION.................419

7.4.1 Mixer-Settler Cascades.................419
7.4.2 Multicompartment Columns.................428

7.5 LIQUID–LIQUID EXTRACTION OF BIOPRODUCTS.................430

PROBLEMS.................437

REFERENCES.................446

8. HUMIDIFICATION OPERATIONS.................447

8.1 INTRODUCTION.................447

8.2 EQUILIBRIUM CONSIDERATIONS.................448

8.2.1 Saturated Gas–Vapor Mixtures.................448
8.2.2 Unsaturated Gas–Vapor Mixtures.................451
8.2.3 Adiabatic-Saturation Curves.................452
8.2.4 Wet-Bulb Temperature.................454

8.3 ADIABATIC GAS–LIQUID CONTACT OPERATIONS.................457

8.3.1 Fundamental Relationships.................458
8.3.2 Water Cooling with Air.................460
8.3.3 Dehumidification of Air–Water Vapor.................466

xiv   Contents

PROBLEMS 468

REFERENCES.. 472

APPENDIX 8.1 473

APPENDIX 8.2.. 474

9. MEMBRANES AND OTHER SOLID: SORPTION AGENTS.....477

9.1 INTRODUCTION.. 477

9.2 MASS TRANSFER IN MEMBRANES 478

9.2.1 Solution-Diffusion for Liquid Mixtures.. 479
9.2.2 Solution-Diffusion for Gas Mixtures 481
9.2.3 Module Flow Patterns.. 484

9.3 EQUILIBRIUM CONSIDERATIONS IN POROUS SORBENTS......... 489

9.3.1 Adsorption and Chromatography Equilibria 489
9.3.2 Ion-Exchange Equilibria 494

9.4 MASS TRANSFER IN FIXED BEDS OF POROUS SORBENTS......... 497

9.4.1 Basic Equations for Adsorption... 499
9.4.2 Linear Isotherm 500
9.4.3 Langmuir Isotherm.. 501
9.4.4 Length of Unused Bed 505
9.4.5 Mass-Transfer Rates in Ion Exchangers... 506
9.4.6 Mass-Transfer Rates in Chromatographic Separations.............................. 507
9.4.7 Electrophoresis.. 510

9.5 APPLICATIONS OF MEMBRANE-SEPARATION PROCESSES....... 512

9.5.1 Dialysis 513
9.5.2 Reverse Osmosis.. 515
9.5.3 Gas Permeation... 518
9.5.4 Ultrafiltration and Microfiltration... 518
9.5.5 Bioseparations... 522

Contents xv

9.6 APPLICATIONS OF SORPTION-SEPARATION PROCESSES 524

PROBLEMS 529

REFERENCES 535

APPENDIX 9.1 536

APPENDIX 9.2 538

APPENDIX 9.3 540

APPENDIX 9.4 542

APPENDIX 9.5 544

APPENDIX 9.6 546

APPENDIX 9.7 548

Appendix A Binary Diffusion Coefficients 551

Appendix B Lennard-Jones Constants 555

Appendix C-1 Maxwell-Stefan Equations (Mathcad) 557

Appendix C-2 Maxwell-Stefan Equations (Python) 559

Appendix D-1 Packed-Column Design (Mathcad) 563

Appendix D-2 Packed-Column Design (Python) 569

Appendix E-1 Sieve-Tray Design (Mathcad) 573

xvi **Contents**

Appendix E-2 Sieve-Tray Design (Python) 579

Appendix F-1 McCabe-Thiele Method: Saturated Liquid Feed (Mathcad).......... 583

Appendix F-2 McCabe-Thiele Method: Saturated Liquid Feed (Python) 587

Appendix G-1 Single-Stage Extraction (Mathcad)....................................... 591

Appendix G-2 Single-Stage Extraction (Python) 593

Appendix G-3 Multistage Crosscurrent Extraction (Mathcad) 595

Appendix G-4 Multistage Crosscurrent Extraction (Python) 598

Appendix H Constants and Unit Conversions 601

Index.. 603

Preface to the Fourth Edition

The new title of this edition has the objective of increasing the discoverability of the book by college instructors. When the book was first published two decades ago, most chemical engineering departments had a required course titled Mass-Transfer Operations. However, now the name of the course has changed in most US universities to Separation Processes; and the name of most of the departments have changed to Chemical and Biochemical Engineering. The content of the book is basically unchanged (highlighting the biochemical applications already in the previous edition and some added biochemical content), just the name has changed. The material covered in this book is still perfectly appropriate for the course. I suggested eliminating from the title of the book the word "modern" because it is really aimed to be used in a first course of separation processes and the emphasis is on classical processes such as distillation and gas absorption.

This edition incorporates, for the first time, the use of Python, a very useful computational tool. Python is a free, open-source language and environment that has tremendous potential for use within the domain of scientific and technical computing. It is a modern language (object oriented) that is concise, easy to read, and quick to learn. It is full of freely available libraries, in particular scientific ones such as linear algebra, visualization tools, plotting, differential equations solving, statistics, and many more. It is widely used in industrial applications. Learning to use Python for engineering calculations is a very desirable and valuable skill for chemical engineers.

A distinguishing feature of the first three editions of this book was the incorporation of Mathcad for both example problems and homework questions. The fourth edition maintains the use of Mathcad but supplemented with the corresponding Python versions of the programs. Appendices were added to each chapter to include in them the corresponding Python versions of the example problems of the given chapter solved by Mathcad. The library of Mathcad programs in the appendices at the end of the book were also supplemented with the corresponding Python versions. This constitutes the most significant modification of the book content. Other differences in this edition are listed in the following paragraph.

Chapter 1 includes a new section on transient molecular diffusion in solids. Chapter 3 includes an analysis of equilibrium thermodynamics of biochemical systems. A new section on extraction of bioproducts was added to Chapter 7. Electrophoresis and the use of membranes for bioseparations were included in Chapter 9. Additional end-of-chapter problems related to environmental and biochemical applications were added throughout the book.

xvii

xviii   **Preface to the Fourth Edition**

I want to acknowledge all the help provided to me by Katrina Maceda, Kanimozhi Ramamurthy, and Summers Scholl from Wiley. It was a pleasure working with you. I want to also acknowledge the help from Prof. Bruce A. Finlayson. He is Professor Emeritus of Chemical Engineering at the University of Washington and has published extensively in computational methods for chemical engineering, including Python. I have followed and admired his distinguished career ever since I was a graduate student. I wish him health and a long life to enjoy his retirement and many hobbies. Thanks again to my wife Teresa for her unconditional love and support, especially during the dark times of the COVID-19 pandemic.

Jaime Benítez
Gainesville, Florida

Preface to the Third Edition

The most significant difference between the first two editions and the third edition is the adoption in the latter of PTC Mathcad Prime most recent version (version 3.1 as of this writing). PTC Mathcad Prime—one of the world's leading tools for technical computing in the context of engineering, science, and math applications—is a significant departure from the previous versions of Mathcad. There is a definite learning curve associated with making the switch from Mathcad to Mathcad Prime. However, the new features included in Mathcad Prime make switching from Mathcad worthwhile. Besides, programs written for the previous versions of Mathcad will not run in Mathcad Prime. Other differences in this edition are listed in the following paragraphs.

In Chapter 3 of the third edition, the material covered in Problems 3.14 and 3.15 of the second edition to determine analytically minimum flow rates in absorbers and strippers is incorporated in the theoretical presentation of Section 3.4 (Material Balances), and the corresponding Mathcad Prime code for solving these problems is given. In Section 3.5 of the third edition, Mathcad Prime code is given to determine analytically the number of ideal stages required for absorbers and strippers.

Section 4.2 of the third edition use, exclusively, the updated Billet and Schultes correlations for estimating the loading and flooding points in packed beds, and the corresponding gas-pressure drop for operation between these limits. The Generalized Pressure Drop Correlation (GPDC) is not included. The updated Billet and Schultes correlations are also used to estimate the volumetric mass-transfer coefficients in both liquid and gas phases. New end-of-chapter problems have been added throughout this third edition.

I want to acknowledge the extraordinarily thorough editing job that Katrina Maceda, Production Editor at Wiley, and Baljinder Kaur, Project Manager at Aptara did on this edition. The book is much better now because of them. It was a pleasure working with both of you. Ludo de Wolf, a physical therapist with gifted hands and a delightful sense of humor, literally removed from my shoulders the heavy load of completing this edition. Thanks to my wife Teresa for her unconditional love and support. I know it is not easy!

Jaime Benítez
Gainesville, Florida

Preface to the Second Edition

The idea for the first edition of this book was born out of my experience teaching a course on mass-transfer operations at the Chemical Engineering Department of the University of Puerto Rico during the previous 25 years. This course is the third in a three-course unit operations sequence. The first course covers momentum transfer (fluid mechanics), and the second course covers heat transfer. Besides these two courses, another prerequisite of the mass-transfer course is a two-semester sequence of chemical engineering thermodynamics.

I decided to write a textbook for a first course on mass-transfer operations with a level of presentation that was easy to follow by the reader, but with enough depth of coverage to guarantee that students using the book will, upon successful completion of the course, be able to specify preliminary designs of the most common mass-transfer equipment (such as absorbers, strippers, distillation columns, liquid extractors, etc.). I decided also to incorporate, from the very beginning of the book, the use of Mathcad, a computational tool that is, in my opinion, very helpful and friendly. The first edition of this book was the result of that effort.

Part of my objective was achieved, as evidenced by the following excerpt from a very thorough review of the first edition of my book, written by Professor Mark J. McCready, a well-known expert in chemical engineering education: "If the topics that are needed for a given course are included in this text, I would expect the educational experience to go smoothly for both student and instructor. I think that students will like this book, because the explanations are clear, the level of difficulty is appropriate, and the examples and included data give the book very much of a 'handbook' flavor. Instructors will find that, overall, the topics are presented in a logical order and the discussion makes sense; there are many examples and lots of homework problems" (McCready, M. J., *AIChE J.*, Vol. 49, No. 1, January 2003).

"Each major section of the book has learning objectives which certainly benefit the students and perhaps the instructor. A key feature of the book, which separates it from the other texts mentioned above, is the incorporation of Mathcad for both example problems and homework questions. A library of Mathcad programs for solving the Maxwell-Stefan equations, packed column calculations, sieve-tray design, binary distillation problems by McCabe-Thiele method, and multistage crosscurrent extraction is given in the appendices. These programs enable students to obtain useful solutions with less effort, as well as allow them to explore the different variables or parameters. The wide availability, low cost, and ease of use of Mathcad allow it to be the modern equivalent of "back of the envelope' calculations, which can be refined, if necessary, using full-scale process simulators" (McCready, 2003).

xxii   **Preface to the Second Edition**

However, the same reviewer also points out some limitations of the book. One of the main objectives of this second edition is to remedy those shortcomings of the first edition to make it more attractive as a textbook to a broader audience. Another important objective of the second edition is to incorporate material related to mass-transfer phenomena in biological systems. Many chemical engineering departments all over the world are changing their names and curricula to include the area of biochemical engineering in their offerings. The second edition includes pertinent examples such as convection and diffusion of oxygen through the body's circulatory system, bio-artificial kidneys, separation of sugars by chromatography, and purification of monoclonal antibodies by affinity adsorption.

As with the first edition, the first four chapters of the book present a basic framework for analysis that is applicable to most mass-transfer operations. Chapters 5 to 7 apply this common methodology to the analysis and design of some of the most popular types of mass-transfer operations. Chapter 5 covers gas absorption and stripping; Chapter 6 covers distillation; and Chapter 7 covers liquid extraction. Chapter 8, new to the second edition, covers humidification operations in general, and detailed design of packed cooling towers specifically. These operations—in particular, cooling towers—are very common in industry. Also, from the didactic point of view, their analysis and design involve simultaneous mass- and heat-transfer considerations. Therefore, the reader is exposed in detail to the similarities and differences between these two transport phenomena. Chapter 9, also new, covers mass-transfer processes using barriers (membranes) and solid sorption agents (adsorption, ion exchange, and chromatography).

In response to suggestions by Professor McCready and other reviewers, some other revisions and additions to the second edition are:

- In Chapter 1, the Maxwell-Stefan equations (augmented by the steady-state continuity equation for each component) are solved numerically using a combination of a Runge-Kutta-based differential equation solver (*Rkfixed*) and an algebraic equation solver (*Given-Find*), both included in Mathcad. This methodology is much more flexible than the one presented in the first edition (orthogonal collocation), and its theoretical justification is well within the scope of the mathematical background required for a first course in mass-transfer operations.
- Chapter 1 includes a section on diffusion in solids.
- Chapter 2 includes a section on boundary-layer theory and an example on simultaneous mass and heat transfer during air humidification.
- Chapter 6 includes a section on multistage batch distillation.

I wish to acknowledge gratefully the contribution of the University of Puerto Rico at Mayagüez to this project. My students in the course INQU 4002 reviewed the material in the book, found quite a few errors, and gave excellent suggestions on ways to improve its content and presentation. My students are my source of motivation; they make all the effort to prepare this book worthwhile!

Jaime Benítez
Mayagüez, Puerto Rico

Preface to the First Edition

The importance of the mass-transfer operations in chemical processes is profound. There is scarcely any industrial process that does not require a preliminary purification of raw materials or final separation of products. This is the realm of mass-transfer operations. Frequently, the major part of the cost of a process is that for the separations accomplished in the mass-transfer operations, a good reason for process engineers and designers to master this subject. The mass-transfer operations are largely the responsibility of chemical engineers, but increasingly practitioners of other engineering disciplines are finding them necessary for their work. This is especially true for those engaged in environmental engineering, where separation processes predominate.

My objective in writing this book is to provide a means to teach undergraduate chemical engineering students the basic principles of mass transfer and to apply these principles, aided by modern computational tools, to the design of equipment used in separation processes. The idea for it was born out of my experiences during the last 25 years teaching mass-transfer operations courses at the University of Puerto Rico.

The material treated in the book can be covered in a one-semester course. Chapters are divided into sections with clearly stated objectives at the beginning. Numerous detailed examples follow each brief section of text. Abundant end-of-chapter problems are included, and problem degree of difficulty is clearly labeled for each. Most of the problems are accompanied by their answers. Computer solution is emphasized, both in the examples and in the end- of-chapter problems. The book uses mostly SI units, which virtually eliminates the tedious task of unit conversions and makes it "readable" to the international scientific and technical community.

Following the lead of other authors in the chemical engineering field and related technical disciplines, I decided to incorporate the use of Mathcad into this book. Most readers will probably have a working knowledge of Mathcad. (Even if they don't, my experience is that the basic knowledge needed to begin using Mathcad effectively can be easily taught in a two-hour workshop.) The use of Mathcad simplifies mass-transfer calculations to a point that it allows the instructor and the student to readily try many different combinations of the design variables, a vital experience for the amateur designer.

The Mathcad environment can be used as a sophisticated scientific calculator, can be easily programed to perform a complicated sequence of calculations

xxiii

xxiv **Preface to the First Edition**

(for example, to check the design of a sieve-plate column for flooding, pressure drop, entrainment, weeping, and calculating Murphree plate efficiencies), can be used to plot results, and as a word processor to neatly present homework problems. Mathcad can perform calculations using a variety of unit systems, and will give a warning signal when calculations that are not dimensionally consistent are tried. This is a most powerful didactic tool, since dimensional consistency in calculations is one of the most fundamental concepts in chemical engineering education.

The first four chapters of the book present a basic framework of analysis that is applicable to any mass-transfer operation. Chapters 5 to 7 apply this common methodology to the analysis and design of the most popular types of mass-transfer operations. Chapter 5 covers gas absorption and stripping, chapter 6 distillation columns, and chapter 7 liquid extraction. This choice is somewhat arbitrary, and based on my own perception of the relevance of these operations. However, application of the general framework of analysis developed in the first four chapters should allow the reader to master, with relative ease, the peculiarities of any other type of mass-transfer operation.

I wish to acknowledge gratefully the contribution of the University of Puerto Rico at Mayagüez to this project. My students in the course INQU 4002 reviewed the material presented in the book, found quite a few errors, and gave excellent suggestions on ways to improve it. My special gratitude goes to Teresa, my wife, and my four children who were always around lifting my spirits during the long, arduous hours of work devoted to this volume. They make it all worth-while!

Jaime Benítez
Mayagüez, Puerto Rico

Nomenclature

LATIN LETTERS

A	absorption factor; dimensionless.
A	mass flow rate of species A; kg/s.
A_a	active area of a sieve tray; m^2.
A_d	area taken by the downspout in a sieve tray; m^2.
A_h	area taken by the perforations on a sieve tray; m^2.
A_M	membrane area; m^2.
A_n	net cross-section area between trays inside a tray column; m^2.
A_t	total cross-section area, m^2.
a	mass-transfer surface area per unit volume; m^{-1}.
a_h	hydraulic, or effective, specific surface area of packing; m^{-1}.
B	mass flow rate of species B; kg/s.
B_0	viscous flow parameter; m^2.
c	total molar concentration; mol/m^3.
c_i, C_i	molar concentration of species i; mol/m^3.
C	total number of components in multicomponent distillation.
C_p	specific heat at constant pressure; J/kg-K.
C_S	humid heat; J/kg-K.
C_D	drag coefficient; dimensionless.
Da	Damkohler number for first-order reaction; dimensionless.
D_{ij}	Maxwell-Stefan diffusivity for pair i-j; m^2/s.
D_{ij}	Fick diffusivity or diffusion coefficient for pair i-j; m^2/s.
$\mathrm{D}_{K,i}$	Knudsen diffusivity for component i; m^2/s.
d_e	equivalent diameter; m.
d_i	driving force for mass diffusion of species i; m^{-1}.
d_i	inside diameter; m.
d_o	outside diameter; m.
d_o	perforation diameter in a sieve plate; m.
d_p	particle size; m.
d_{vs}	Sauter mean drop diameter defined in equation (7-48); m.
DM	dimensional matrix.
D	tube diameter; m.
D	distillate flow rate; moles/s.
E	fractional entrainment; liquid mass flow rate/gas mass flow rate.
E	extract mass flow rate, kg/s.
E_m	mechanical efficiency of a motor-fan system; dimensionless.
Eo	Eotvos number defined in equation (7-53); dimensionless.
EF	extraction factor defined in equation (7-19); dimensionless.
$\mathbf{E}_{ME}$	Murphree stage efficiency in terms of extract composition.
$\mathbf{E}_{MG}$	Murphree gas-phase tray efficiency; dimensionless.

xxv

$\mathbf{E}_{MGE}$	Murphree gas-phase tray efficiency corrected for entrainment.
$\mathbf{E}_O$	overall tray efficiency of a cascade; equilibrium trays/real trays.
$\mathbf{E}_{OG}$	point gas-phase tray efficiency; dimensionless.
f_{12}	proportionality coefficient in equation (1-21).
f	friction factor; dimensionless.
f	fractional approach to flooding velocity; dimensionless.
f_{ext}	fractional extraction; dimensionless.
F	mass-transfer coefficient; mol/m^2-s.
F	molar flow rate of the feed to a distillation column; mol/s.
F	mass flow rate of the feed to a liquid extraction process; kg/s.
$FR_{i,D}$	fractional recovery of component i in the distillate; dimensionless.
$FR_{i,w}$	fractional recovery of component i in the residue; dimensionless.
Fr_L	liquid Froude number; dimensionless.
Ga	Galileo number; dimensionless.
G_M	superficial molar velocity; mol/m^2-s.
G_{Mx}	superficial liquid-phase molar velocity; mol/m^2-s.
G_{My}	superficial gas-phase molar velocity; mol/m^2-s.
G_x	superficial liquid-phase mass velocity; kg/m^2-s.
G_y	superficial gas-phase mass velocity; kg/m^2-s.
Gr_D	Grashof number for mass transfer; dimensionless.
Gr_H	Grashof number for heat transfer; dimensionless.
Gz	Graetz number; dimensionless.
g	acceleration due to gravity; 9.8 m/s^2.
g_c	dimensional conversion factor; 1 kg-m/N-s^2.
H	Henry's law constant; atm, kPa, Pa.
H	molar enthalpy; J/mol.
H	height of mixing vessel; m.
H'	enthalpy of gas-vapor mixture; J/kg.
HETS	height equivalent to a theoretical stage in staged liquid extraction columns; m.
HK	heavy-key component in multicomponent distillation.
ΔHs	heat of solution; J/mol of solution.
H_{tL}	height of a liquid-phase transfer unit; m.
H_{tG}	height of a gas-phase transfer unit; m.
H_{toG}	overall height of a gas-phase transfer unit; m.
H_{toL}	overall height of a liquid-phase transfer unit; m.
h	convective heat-transfer coefficient, W/m^2-K.
h_d	dry-tray head loss; cm of liquid.
h_i	equivalent head of clear liquid on tray; cm of liquid.
h_L	specific liquid holdup; m^3 holdup/m^3 packed bed.
h_t	total head loss/tray; cm of liquid.
h_w	weir height; m.
h_σ	head loss due to surface tension; cm of liquid.
$\mathrm{h}_{2\phi}$	height of two-phase region on a tray; m.
i	number of dimensionless groups needed to describe a situation.
j_D	Chilton-Colburn j-factor for mass transfer; dimensionless.
j_H	Chilton-Colburn j-factor for heat transfer; dimensionless.

Nomenclature xxvii

$\mathbf{j}_i$ mass diffusion flux of species i with respect to the mass-average velocity; kg/m^2-s.

$\mathbf{J}_i$ molar diffusion flux of species i with respect to the molar-average velocity; mol/m^2-s.

J_0 Bessel function of the first kind and order zero; dimensionless.

J_1 Bessel function of the first kind and order one; dimensionless.

K distribution coefficient; dimensionless.

K Krogh diffusion coefficient; cm^3 O_2/cm-s-torr.

K parameter in Langmuir adsorption isotherm; Pa^{-1}.

K_{AB} molar selectivity parameter in ion exchange; dimensionless.

K_W wall factor in Billet-Schultes pressure-drop correlations; dimensionless.

k thermal conductivity, W/m-K.

k_c convective mass-transfer coefficient for diffusion of A through stagnant B in dilute gas-phase solution with driving force in terms of molar concentrations; m/s.

k'_c convective mass-transfer coefficient for equimolar counterdiffusion in gas-phase solution with driving force in molar concentrations; m/s.

k_G convective mass-transfer coefficient for diffusion of A through stagnant B in dilute gas-phase solution with driving force in terms of partial pressure; mol/m^2-s-Pa.

K_G convective mass-transfer coefficient for equimolar counterdiffusion in gas-phase solution with driving force in terms of partial pressure; mol/m^2-s-Pa.

k'_G overall convective mass-transfer coefficient for diffusion of A through stagnant B in dilute solutions with driving force in terms of partial pressures; mol/m^2-s-Pa.

k_L convective mass-transfer coefficient for diffusion of A through stagnant B in dilute liquid-phase solution with driving force in terms of molar concentrations; m/s.

k'_L convective mass-transfer coefficient for equimolar counterdiffusion in liquid-phase solution with driving force in terms of molar concentrations; m/s.

Kn   Knudsen number, dimensionless.

k_r reaction rate constant; mol/m^2-s-mol fraction.

K_r restrictive factor for diffusion of liquids in porous solids; dimensionless.

k_x convective mass-transfer coefficient for diffusion of A through stagnant B in dilute liquid-phase solution with driving force in terms of mol fractions; mol/m^2-s.

K_x overall convective mass-transfer coefficient for diffusion of A through stagnant B in dilute solutions with driving force in terms of liquid-phase mol fractions; mol/m^2-s.

k'_x convective mass-transfer coefficient for equimolar counterdiffusion in liquid-phase solution with driving force in terms of mol fractions; mol/m^2-s.

k_y convective mass-transfer coefficient for diffusion of A through stagnant B in dilute gas-phase solution with driving force in terms of mol fractions; mol/m^2-s.

K_y overall convective mass-transfer coefficient for diffusion of A through stagnant B in dilute solutions with driving force in terms of gas-phase mol fractions; mol/m^2-s.

k'_y	convective mass-transfer coefficient for equimolar counterdiffusion in gas-phase solution with driving force in terms of mol fractions; mol/m^2-s.
L	characteristic length, m.
L	molar flow rate of the L-phase; mol/s.
L	length of settling vessel; m.
LK	light-key component in multicomponent distillation.
L_s	molar flow rate of the nondiffusing solvent in the L-phase; mol/s.
L'	mass flow rate of the L-phase; kg/s.
L'_S	mass flow rate of the nondiffusing solvent in the L-phase; kg/s.
L_e	entrainment mass flow rate, kg/s.
L_w	weir length; m.
l	characteristic length, m.
l	tray thickness; m.
l_M	membrane thickness; m.
Le	Lewis number; dimensionless.
M_i	molecular weight of species i.
M_0	oxygen demand; cm^3 O$_2$/cm^3-min.
MTZ	width of the mass-transfer zone in fixed-bed adsorption; m.
m	amount of mass; kg.
m	slope of the equilibrium distribution curve; dimensionless.
$\mathbf{n}$	total mass flux with respect to fixed coordinates; kg/m^2-s.
$\mathbf{n}_i$	mass flux of species i with respect to fixed coordinates; kg/m^2-s.
n	number of variables significant to dimensional analysis of a given problem.
n	rate of mass transfer from the dispersed to the continuous phase in liquid extraction; kg/s.
n	number of species in a mixture.
$\mathbf{N}$	total molar flux with respect to fixed coordinates; mol/m^2-s.
$\mathbf{N}_i$	molar flux of species i with respect to fixed coordinates; mol/m^2-s.
N	number of equilibrium stages in a cascade; dimensionless.
N_E	mass of B/(mass of A + mass of C) in the extract liquids.
N_R	number of stages in rectifying section; dimensionless.
N_R	mass of B/(mass of A + mass of C) in the raffinate liquids.
N_S	number of stages in stripping section; dimensionless.
N_{tL}	number of liquid-phase transfer units; dimensionless.
N_{tG}	number of gas-phase transfer units; dimensionless.
N_{tOD}	overall number of dispersed-phase transfer units; dimensionless.
Nto_G	overall number of gas-phase transfer units; dimensionless.
N_{tOL}	overall number of liquid-phase transfer units; dimensionless.
Nu	Nusselt number; dimensionless.
O_t	molar oxygen concentration in the air leaving an aeration tank; percent.
O_{eff}	oxygen transfer efficiency; mass of oxygen absorbed by water/total mass of oxygen supplied.
p'	pitch, distance between centers of perforations in a sieve plate; m.
p_i	partial pressure of species i; atm, Pa, kPa, bar.
$P_{B,M}$	logarithmic mean partial pressure of component B; atm, Pa, kPa, bar.
P	total pressure; atm, Pa, kPa, bar.
P	permeate flow through a membrane; mol/s.

Nomenclature   xxix

P	Impeller power; kW.
P_c	critical pressure, Pa, kPa, bar.
Pe_D	Peclet number for mass transfer.
Pe_H	Peclet number for heat transfer.
P_i	vapor pressure of species i; atm, Pa, kPa, bar.
Po	power number defined in equation (7-37); dimensionless.
Pr	Prandtl number; dimensionless.
Q	volumetric flow rate; m^3/s.
Q	net rate of heating; J/s.
Q	membrane permeance; m/s.
q	membrane permeability; barrer, m^2/s.
q	parameter defined by equation (6-27); dimensionless.
q_m	parameter in Langmuir adsorption isotherm; g/g.
r	rank of the dimensional matrix, **DM**; dimensionless.
r_A	solute particle radius; m.
R	radius; m.
R	ideal gas constant; J/mol-K.
R	reflux ratio; mol of reflux/mol of distillate.
R	raffinate mass flow rate; kg/s.
R_A	volumetric rate of formation of A; mol per unit volume per unit time.
R_m	retentate flow in a membrane; mol/s.
Re	Reynolds number; dimensionless.
R_i	volumetric rate of formation of component i; mol/m^3-s.
S	surface area, cross-sectional area; m^2.
S	stripping factor, reciprocal of absorption factor (A); dimensionless.
S	mass flow rate of the solvent entering a liquid extraction process; kg/s.
Sc	Schmidt number; dimensionless.
Sh	Sherwood number; dimensionless.
SR	salt rejection; dimensionless.
St_D	Stanton number for mass transfer; dimensionless.
St_H	Stanton number for heat transfer; dimensionless.
t	tray spacing; m.
t	time; s, h.
t_b	breakthrough time in fixed-bed adsorption; s.
t_{res}	residence time; min.
T	temperature; K.
T_{as}	adiabatic saturation temperature; K.
T_b	normal boiling point temperature; K.
T_c	critical temperature, K.
T_w	wet-bulb temperature; K.
u	fluid velocity past a stationary flat plate, parallel to the surface; m/s.
v	mass-average velocity for multicomponent mixture; m/s.
v$_i$	velocity of species i; m/s.
v_t	terminal velocity of a particle; m/s.
V	molar-average velocity for multicomponent mixture; m/s.
$\underline{V}$	volume; m^3.
V	molar flow rate of the V-phase; mol/s.

V_s	molar flow rate of the nondiffusing solvent in the V-phase; mol/s.
V'	mass flow rate of the V-phase; kg/s.
V_S	mass flow rate of the nondiffusing solvent in the V-phase; kg/s.
V_A	molar volume of a solute as liquid at its normal boiling point; cm^3/mol.
V_B	boilup ratio; mol of boilup/mol of residue.
V_b	molar volume of a substance as liquid at its normal boiling point; cm^3/mol.
V_c	critical volume; cm^3/mol.
w	mass-flow rate; kg/s.
W	work per unit mass; J/kg.
W	molar flow rate of the residue from a distillation column; mol/s.
W_e	Weber number defined in equation (7-49); dimensionless.
x_i	mol fraction of species i in either liquid or solid phase.
x_i	mass fraction of species i in raffinate (liquid extraction).
$x_{B,M}$	logarithmic mean mol fraction of component B in liquid or solid phase.
x	rectangular coordinate.
x'	mass of C/mass of A in raffinate liquids.
X	mol ratio in phase L; mol of A/mol of A-free L.
X	flow parameter; dimensionless.
X	parameter in Gilliland's correlation, see equation (6-87); dimensionless.
X	mass of C/(mass of A + mass of C) in the raffinate liquids.
X'	mass ratio in phase L; kg of A/kg of A-free L.
y	rectangular coordinate.
y'	mass of C/mass of B in extract liquids.
$y_{B,M}$	logarithmic mean mol fraction of component B in gas phase.
y_i	mol fraction of species i in the gas phase.
y_i	mass fraction of species i in extract (liquid extraction).
Y	mol ratio in phase V; mol of A/mol of A-free V.
Y	parameter in Gilliland's correlation, see equation (6-86); dimensionless.
Y	mass of C/(mass of A + mass of C) in the extract liquids.
Y	molal absolute humidity; mol A/mol B.
Y'	absolute humidity; kg A/kg B$'$.
Y'	mass ratio in phase V; kg of A/kg of A-free V.
z	rectangular coordinate.
z_i	average mol fraction of component i in a solution or multiphase mixture.
Z	total height; m.
Z_c	compressibility factor at critical conditions; dimensionless.
Z_r	total height of the rectifying section of a packed fractionator; m.
Z_s	total height of the stripping section of a packed fractionator; m.

GREEK LETTERS

α	thermal diffusivity; m^2/s.
α	relative volatility; dimensionless.
α_m, α_{AB}	membrane separation factor; dimensionless.
β	volume coefficient of thermal expansion; K^{-1}.
Γ	matrix of thermodynamic factors defined by equation (1-32).

Nomenclature

Γ	concentration polarization factor; dimensionless.
γ_i	activity coefficient of species i in solution.
δ	length of the diffusion path; m.
δ	velocity boundary-layer thickness; m.
δ_{ij}	Kronecker delta; 1 if $i = k$, 0 otherwise.
Δ_R	difference in flow rate, equation (7-12); kg/s.
ε	porosity or void fraction; dimensionless.
ε_{AB}	Lennard-Jones parameter; erg.
θ	membrane cut; mol of permeate/mol of feed.
κ	Boltzmann constant; 1.38×10^{-16} erg / K.
κ	constant in equation (4-50), defined in equation (4-52); dimensionless.
k_i	molar latent heat of vaporization of component i; J/mol.
λ_i	similar to the stripping factor, S, in equations (4-61) to (4-66).
λ	mean free path in gases; m.
μ_i	chemical potential of species i; J/mol.
μ_B	solvent viscosity; cP.
υ	momentum diffusivity, or kinematic viscosity; m^2/s.
υ_i	stoichiometric number of species i.
ξ	reduced inverse viscosity in Lucas method; $(\mu P)^{-1}$.
π	constant; 3.1416.
π	Pi groups in dimensional analysis.
π	osmotic pressure; Pa.
ρ	mass density; kg/m^3.
ρ_i	mass density of species i; kg/m^3.
σ_{AB}	Lennard-Jones parameter; Å.
σ	surface tension, dyn/cm, N/m.
τ	shear stress; N/m^2.
τ	pore-path tortuosity; dimensionless.
Φ_B	association factor of solvent B; dimensionless.
ϕ	packing fraction in hollow-fiber membrane module; dimensionless.
ϕ	root of equation (6-82); dimensionless.
ϕ_e	effective relative froth density; height of clear liquid/froth height.
ϕ_C	fractional holdup of the continuous liquid phase.
ϕ_D	fractional holdup of the dispersed liquid phase.
ϕ_G	specific gas holdup; m^3 holdup/m^3 total volume.
ω_i	mass fraction of species i.
Ω_D	diffusion collision integral; dimensionless.
Ω	impeller rate of rotation; rpm.
Ψ	stream function; m^2/s.
Ψ_A	molar flux fraction of component A; dimensionless.
Ψ_0	dry-packing resistance coefficient in Billet-Schultes pressure-drop correlations; dimensionless.

1

Fundamentals of Mass Transfer

1.1 INTRODUCTION

When a system contains two or more components whose concentrations vary from point to point, there is a natural tendency for mass to be transferred, minimizing the concentration differences within the system, and moving it toward equilibrium. The transport of one component from a region of higher concentration to that of a lower concentration is called *mass transfer*.

Many of our daily experiences involve mass-transfer phenomena. The invigorating aroma of a cup of freshly brewed coffee and the sensuous scent of a delicate perfume both reach our nostrils from the source by diffusion through air. A lump of sugar added to the cup of coffee eventually dissolves and then diffuses uniformly throughout the beverage. Laundry hanging under the sun during a breezy day dries fast because the moisture evaporates and diffuses easily into the relatively dry moving air.

Mass transfer plays an important role in many industrial processes. A group of operations for separating the components of mixtures is based on the transfer of material from one homogeneous phase to another. These methods—covered by the term mass-transfer operations—include such techniques as distillation, gas absorption, humidification, liquid extraction, adsorption, membrane separations, and others. The driving force for transfer in these operations is a concentration gradient, much as a temperature gradient provides the driving force for heat transfer.

Distillation separates, by partial vaporization, a liquid mixture of miscible and volatile substances into individual components or, in some cases, into groups of components. The separation of a mixture of methanol and water into its components; of liquid air into oxygen, nitrogen, and argon; and of crude petroleum into gasoline, kerosene, fuel oil, and lubricating stock are examples of distillation.

Principles and Applications of Mass Transfer: The Design of Separation Processes for Chemical and Biochemical Engineering, Fourth Edition. Jaime Benitez.
© 2023 John Wiley & Sons, Inc. Published 2023 by John Wiley & Sons, Inc.

In *gas absorption*, a soluble vapor is absorbed by means of a liquid in which the solute gas is more or less soluble, from its mixture with an inert gas. The washing of ammonia from a mixture of ammonia and air by means of liquid water is a typical example. The solute is subsequently recovered from the liquid by distillation, and the absorbing liquid can be either discarded or reused. When a solute is transferred from the solvent liquid to the gas phase, the operation is known as *desorption or stripping*.

In *humidification* or *dehumidification* (depending upon the direction of transfer), the liquid phase is a pure liquid containing but one component while the gas phase contains two or more substances. Usually, the inert or carrier gas is virtually insoluble in the liquid. Removal of water vapor from air by condensation on a cold surface and the condensation of an organic vapor, such as carbon tetrachloride out of a stream of nitrogen, are examples of dehumidification. In humidification operations, the direction of transfer is from the liquid to the gas phase.

The *adsorption* operations exploit the ability of certain solids preferentially to concentrate specific substances from solution onto their surfaces. In this manner, the components of either gaseous or liquid solutions can be separated from each other. A few examples will illustrate the great variety of practical applications of adsorption. It is used to dehumidify air and other gases, to remove objectionable odors and impurities from industrial gases, to recover valuable solvent vapors from dilute mixtures with air and other gases, to remove objectionable taste and odor from drinking water, and many other applications.

Liquid extraction is the separation of the constituents of a liquid solution by contact with another insoluble liquid. If the substances constituting the original solution distribute themselves differently between the two liquid phases, a certain degree of separation will result. The solution which is to be extracted is called the *feed*, and the liquid with which the feed is contacted is called the *solvent*. The solvent-rich product of the operation is called the *extract*, and the residual liquid from which the solute has been removed is called the *raffinate*.

Membrane separations are rapidly increasing in importance. In general, the membranes serve to prevent intermingling of two miscible phases. They also prevent ordinary hydrodynamic flow, and movement of substances through them is by diffusion. Separation of the components of the original solution takes place by selectively controlling their passage from one side of the membrane to the other. An example of a membrane-mediated, liquid–liquid separation process is *dialysis*. In this process, a colloid is removed from a liquid solution by contacting the solution with a solvent through an intervening membrane which is permeable to the solution, but not to the larger colloidal particles. For example, aqueous beet–sugar solutions containing undesired colloidal material are freed of the latter by contact with water through a semipermeable membrane. Sugar and water diffuse through the membrane, but not the colloid.

Returning to the lump of sugar added to the cup of coffee, it is evident that the time required for the sugar to distribute uniformly depends upon whether the liquid is quiescent, or whether it is mechanically agitated by a spoon. In general, the mechanism of mass transfer depends upon the dynamics of the system in which it occurs. Mass can be transferred by random molecular motion in quiescent fluids,

1.2 Molecular Mass Transfer

or it can be transferred from a surface into a moving fluid, aided by the dynamic characteristics of the flow. These two distinct modes of transport, *molecular mass transfer* and *convective mass transfer*, are analogous to conduction heat transfer and convective heat transfer. Each of these modes of mass transfer will be described and analyzed. The two mechanisms often act simultaneously. Frequently, when this happens, one mechanism can dominate quantitatively so that approximate solutions involving only the dominant mode can be used.

1.2 MOLECULAR MASS TRANSFER

As early as 1815, it was observed qualitatively that whenever a gas mixture contains two or more molecular species, whose relative concentrations vary from point to point, an apparently natural process results which tends to diminish any inequalities in composition. This macroscopic transport of mass, independent of any convection effects within the system, is defined as *molecular diffusion*.

In the specific case of a gaseous mixtures, a logical explanation of this transport phenomenon can be deduced from the kinetic theory of gases. At any temperature above absolute zero, individual molecules are in a state of continual yet random motion. Within dilute gas mixtures, each solute molecule behaves independently of the other solute molecules since it seldom encounters them. Collisions between the solute and the solvent molecules are continually occurring. As a result of the collisions, the solute molecules move along a zigzag path, sometimes toward a region of higher concentration, sometimes toward a region of lower concentration.

Consider a hypothetical section passing normal to the concentration gradient within an isothermal, isobaric gaseous mixture containing solute and solvent molecules. The two thin, equal elements of volume above and below the section will contain the same number of molecules, as stipulated by Avogadro's law (Welty et al., 1984).

Although it is not possible to state which way any molecule will travel in a given interval of time, a definite number of the molecules in the lower element of volume will cross the hypothetical section from below, and the same number of molecules will leave the upper element and cross the section from above. With the existence of a concentration gradient, there are more solute molecules in one of the elements of volume than in the other; accordingly, an overall net transfer from a region of higher concentration to one of lower concentration will result. The net flow of each molecular species occurs in the direction of a negative concentration gradient.

The laws of mass transfer show the relation between the flux of the diffusing substance and the concentration gradient responsible for this mass transfer. Since diffusion occurs only in mixtures, its evaluation must involve an examination of the effect of each component. For example, it is often desired to know the diffusion rate of a specific component relative to the velocity of the mixture in which it is moving. Since each component may possess a different mobility, the mixture velocity must be evaluated by averaging the velocities of all the components present.

To establish a common basis for future discussions, definitions and relations which are often used to explain the role of components within a mixture are considered next.

4   **Fundamentals of Mass Transfer**

1.2.1 Concentrations

Your objectives in studying this section are to be able to:

1. Convert a composition given in mass fraction to mole fraction, and the reverse.
2. Transform a material from one measure of concentration to another, including mass/volume and moles/volume.

In a multicomponent mixture, the concentration of species can be expressed in many ways. A mass concentration for each species, as well as for the mixture, can be defined. For species A, the *mass concentration*, ρ_A, is defined as the mass of A per unit volume of the mixture. The total mass concentration, or *density*, ρ, is the total mass of the mixture contained in a unit volume; that is,

$$\rho = \sum_{i=1}^{n} \rho_i \tag{1-1}$$

where n is the number of species in the mixture. The *mass fraction*, ω_A, is the mass concentration of species A divided by the total mass density,

$$\omega_A = \frac{\rho_A}{\sum_{i=1}^{n} \rho_i} = \frac{\rho_A}{\rho} \tag{1-2}$$

The sum of the mass fractions, by definition, must be 1:

$$\sum_{i=1}^{n} \omega_i = 1 \tag{1-3}$$

The *molar concentration* of species A, c_A, is defined as the number of moles of A present per unit volume of the mixture. One mol of any species contains a mass equivalent to its molecular weight; therefore, the mass concentration and the molar concentration are related by

$$c_A = \frac{\rho_A}{M_A} \tag{1-4}$$

where M_A is the molecular weight of species A. When dealing with a gas phase under conditions in which the ideal gas law applies, the molar concentration is given by

$$c_A = \frac{p_A}{RT} \tag{1-5}$$

1.2 Molecular Mass Transfer 5

where p_A is the partial pressure of the species A in the mixture, T is the absolute temperature, and R is the gas constant. The total molar concentration, c, is the total moles of mixture contained in a unit volume; that is,

$$c = \sum_{i=1}^{n} c_i \tag{1-6}$$

For a gaseous mixture that obeys the ideal gas law,

$$c = \frac{P}{RT} \tag{1-7}$$

where P is the total pressure. The *mol fraction* for liquid or solid mixtures, x_A, and for gaseous mixtures, y_A, are the molar concentrations of species A divided by the total molar concentration:

$$x_A = \frac{c_A}{c} \quad \text{(liquids and solids)}$$
$$y_A = \frac{c_A}{c} \quad \text{(gases)} \tag{1-8}$$

For a gaseous mixture that obeys the ideal gas law, the mol fraction, y_A, can be written in terms of pressures:

$$y_A = \frac{c_A}{c} = \frac{p_A}{P} \tag{1-9}$$

Equation (1-9) is an algebraic representation of Dalton's law for gas mixtures. The sum of the mol fractions, by definition, must be 1.

$$\sum_{i=1}^{n} y_i = \sum_{i=1}^{n} x_i = 1.0 \tag{1-10}$$

Example 1.1 Concentration of Feed to a Gas Absorber

A gas containing 88% (by volume) CH_4, 4% C_2H_6, 5% $n\text{-}C_3H_8$, and 3% $n\text{-}C_4H_{10}$ at 300 K and 500 kPa will be scrubbed by contact with a nonvolatile oil in a gas absorber. The objective of the process is to recover in the liquid effluent as much as possible of the heavier hydrocarbons in the feed (see Figure 1.1). Calculate

(a) Total molar concentration in the gas feed.

(b) Density of the gas feed.

(c) Composition of the gas feed expressed in terms of mass fractions.

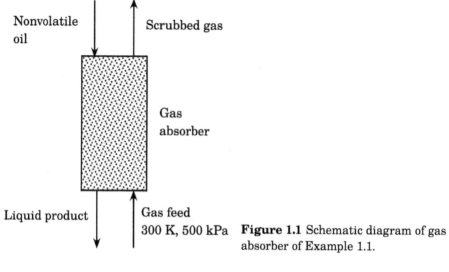

Figure 1.1 Schematic diagram of gas absorber of Example 1.1.

Solution

(a) Use Equation (1-7) to calculate the total molar concentration:

$$c = \frac{P}{RT} = \frac{500}{8.314 \times 300} = 0.20 \frac{\text{kmol}}{\text{m}^3}$$

(b) Calculate the gas average molecular weight, M_{av}:

Basis: 100 kmol of gas mixture

Component	kmol	Molecular weight	Mass, kg
CH_4	88	16.04	1411.52
C_2H_6	4	30.07	120.28
$n\text{-}C_3H_8$	5	44.09	220.45
$n\text{-}C_4H_{10}$	3	58.12	174.36
Total	100		1926.61

$$M_{av} = \frac{1926.61}{100} = 19.26 \frac{\text{kg}}{\text{kmol}}$$

To calculate the mixture mass density:

$$\rho = c M_{av} = 0.20 \times 19.26 = 3.85 \frac{\text{kg}}{\text{m}^3}$$

1.2 Molecular Mass Transfer

(c) Calculate the mass fraction of each component in the gas mixture from the intermediate results of part (b):

Component	Mass, kg	Mass fraction
CH_4	1411.52	(1411.52/1926.61) = 0.733
C_2H_6	120.28	0.062
$n\text{-}C_3H_8$	220.45	0.114
$n\text{-}C_4H_{10}$	174.36	0.091
Total	1926.61	1.000

Example 1.2 Concentration of Potassium Nitrate Wash Solution

In the manufacture of potassium nitrate, potassium chloride reacts with a hot aqueous solution of sodium nitrate according to

$$KCl + NaNO_3 \rightarrow KNO_3 + NaCl$$

The reaction mixture is cooled down to 293 K and pure KNO_3 crystallizes. The resulting slurry contains the KNO_3 crystals and an aqueous solution of both KNO_3 and NaCl. The crystals in the slurry are washed in a multistage process with a saturated KNO_3 solution to free them of NaCl (see Figure 1.2). The equilibrium solubility of KNO_3 in water at 293 K is 24% (by weight); the density of the saturated solution is 1162 kg/m^3 (Perry and Chilton, 1973). Calculate

(a) Total molar density of the fresh wash solution.

(b) Composition of the fresh wash solution expressed in terms of molar fractions.

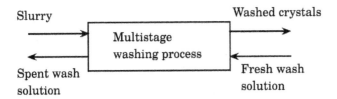

Figure 1.2 Schematic diagram of the washing process in Example 1.2.

Solution
(a) Calculate the average molecular weight, M_{av}, of the wash solution, and then its total molar density.

Basis: 100 kg of fresh wash solution

Component	Mass, kg	Molecular weight	kmol
KNO_3	24	101.10	0.237
H_2O	76	18.02	4.218
Total	100		4.455

Therefore,

$$M_{av} = \frac{100}{4.455} = 22.45 \frac{kg}{kmol}$$

$$c = \frac{\rho}{M_{av}} = \frac{1162}{22.45} = 51.77 \frac{kmol}{m^3}$$

(b) Calculate the mol fractions from intermediate results in part (a):

Component	kmol	Mol fraction
KNO_3	0.237	(0.237/4.455) = 0.053
H_2O	4.218	0.947
Total	4.455	1.000

Example 1.3 Material Balances on a Bio-Artificial Kidney (Montgomery et al., 1998)

The primary functions of the kidneys are to remove waste products (such as urea, uric acid, and creatinine), and to maintain the fluid and salt balance in the blood. Blood consists of two parts: blood cells, mostly red (45% by volume), and plasma (55% by volume). Urea, uric acid, creatinine, and water are all found in the plasma. If the kidneys fail, wastes start to accumulate, and the body becomes overloaded with fluid. Fortunately, patients with renal failure can use an external dialysis machine, also known as an artificial kidney, to clean the blood. The cleaning of the blood in the artificial kidney is due to the difference in toxin concentrations between the blood and the dialysis fluid. Semipermeable membranes in the machine selectively allow toxins to pass from the blood to the dialysis fluid.

During a dialysis procedure, a patient was connected to the machine for 4 h. The blood was pumped through the artificial kidney at the rate of 1200 mL/min. The partially cleansed blood was returned to the patient's body, and the wastes removed were collected in the used dialysis fluid. During the procedure, the patient's kidneys were completely inactive. A total of 1540 g of urine was collected with an urea concentration of 1.3% by weight. A sample of the blood plasma was analyzed before the dialysis and found to contain 155.3 mg/dL of urea. The specific gravity of the plasma was measured at 1.0245. Calculate:

1.2 Molecular Mass Transfer

9

(a) The urea removal efficiency by the artificial kidney.

(b) The urea concentration in the plasma of the cleansed blood, in mg/dL.

Solution

(a) Write a material balance for urea on the artificial kidney.

$$\text{Basis: 4 h}$$

Assuming that the rate of formation and decomposition of urea during the procedure is negligible, and that no urea is removed by the patient's kidneys:

$$\text{urea in ``clean'' blood = urea in ``dirty'' blood – urea in urine}$$

The mass of urea in the urine is simply $1540 \times 0.013 = 20.0$ g. Calculate the mass of urea in the "dirty" blood by the following procedure:

- Calculate the total volume of plasma that flows through the artificial kidney in 4 h:

$$\frac{1200\,\text{mL of blood}}{\text{min}}\left|\frac{60\,\text{min}}{\text{h}}\right|\frac{0.55\,\text{mL of plasma}}{\text{mL of blood}}\left|\frac{1\,\text{dL}}{100\,\text{mL}}\right|\frac{4\text{h}}{} = 1584\,\text{dL}$$

- Calculate the urea in the "dirty" blood from the given plasma concentration:

$$\frac{155.3\,\text{mg of urea}}{\text{dL of plasma}}\left|\frac{1.0\,\text{g}}{1000\,\text{mg}}\right|\frac{1584\,\text{dL of plasma}}{1} = 246\,\text{g of urea}$$

The urea removal efficiency is then $(20/246) \times 100 = 8.1\%$.

(b) Substituting in the material balance, the mass of urea in the clean blood is found to be $246 - 20 = 226$ g. To calculate the corresponding concentration, the volume of plasma remaining after dialysis must be calculated. Assuming that no cells are removed by the machine during the procedure, the mass of plasma remaining is the difference between the mass of plasma entering the artificial kidney and the mass of urine removed. The mass of plasma entering is given by

$$\frac{1.0245\,\text{g of plasma}}{\text{mL of plasma}}\left|\frac{100\,\text{mL}}{1.0\,\text{dL}}\right|\frac{1584\,\text{dL of plasma}}{} = 162,280\,\text{g of plasma}$$

The mass of plasma remaining is $162,280 - 1540 = 160,740$ g. The volume of plasma remaining is $160,740/(1.0245 \times 100) = 1569$ dL of plasma. Therefore, the urea concentration in the remaining plasma is $(226 \times 1000)/1569 = 144$ mg/dL.

Notice that in the artificial kidney the removal efficiency of all the wastes must be kept relatively low because of the need to always maintain homeostasis in the patient's body. If too many wastes are removed at one time, death may occur.

10   **Fundamentals of Mass Transfer**

1.2.2 Velocities and Fluxes

Your objectives in studying this section are to be able to:

1. Define the following terms: mass-average velocity, molar-average velocity, mass (or molar) flux, and diffusion mass (or molar) flux.
2. Write down an expression to calculate the mass (or molar) flux relative to a fixed coordinate system in terms of the diffusion mass (or molar) flux and the bulk motion contribution.

The basic empirical relation to estimate the rate of molecular diffusion, first postulated by Fick (1855) and, accordingly, often referred to as Fick's first law, quantifies the diffusion of component A in an isothermal, isobaric system. According to Fick's law, a species can have a velocity relative to the mass or molar average velocity (called *diffusion velocity*) only if gradients in the concentration exist. In a multicomponent system, the various species will normally move at different velocities; for that reason, an evaluation of a characteristic velocity for the gas mixture requires the averaging of the velocities of each species present.

The *mass-average velocity* for a multicomponent mixture is defined in terms of the mass densities:

$$\mathbf{v} = \frac{\sum_{i=1}^{n} \rho_i \mathbf{v}_i}{\sum_{i=1}^{n} \rho_i} = \frac{\sum_{i=1}^{n} \rho_j \mathbf{v}_i}{\rho} = \sum_{i=1}^{n} \omega_i \mathbf{v}_i \qquad (1\text{-}11)$$

where $\mathbf{v}_i$ denotes the absolute velocity of species i relative to stationary coordinate axis. The mass-average velocity is the velocity that would be measured by a pitot tube. On the other hand, the *molar-average velocity* for a multicomponent mixture is defined in terms of the molar concentrations of all components by

$$\mathbf{V} = \frac{\sum_{i=1}^{n} c_i \mathbf{v}_i}{\sum_{i=1}^{n} c_i} = \frac{\sum_{i=1}^{n} c_i \mathbf{v}_i}{c} = \sum_{i=1}^{n} x_i \mathbf{v}_i \qquad (1\text{-}12)$$

Diffusion rates are most conveniently described in terms of fluxes. The *mass (or molar) flux* of a given species is a vector quantity denoting the amount of the species, in either mass or molar units, that passes per given unit time through a unit area normal to the vector. The flux may be defined with reference to coordinates that are fixed in space, coordinates which are moving with the mass-average velocity, or coordinates which are moving with the molar-average velocity.

The *mass flux* of species i with respect to coordinates that are fixed in space is defined by

$$\mathbf{n}_i = \rho_i \mathbf{v}_i \qquad (1\text{-}13)$$

1.2 Molecular Mass Transfer

If we sum the component fluxes, we obtain the *total mass flux* given by

$$\boldsymbol{n} = \rho \mathrm{v} \tag{1-14}$$

The *molar flux* of species i with respect to coordinates that are fixed in space is given by

$$\mathbf{N}_i = c_i \mathbf{v}_i \tag{1-15}$$

The *total molar flux* is the sum of these quantities:

$$\mathbf{N} = c\mathrm{V} \tag{1-16}$$

The *mass diffusion flux* of species i with respect to the mass-average velocity is given by

$$\mathbf{j}_i = \rho_i \left(\mathbf{v}_i - \mathbf{v} \right) \text{ and } \sum_{i=1}^{n} \mathbf{j}_i = 0 \tag{1-17}$$

The *molar diffusion flux* of species i with respect to the molar-average velocity is given by

$$\mathbf{J}_i = c_i \left(\mathbf{v}_i - \mathbf{V} \right) \quad \text{and} \quad \sum_{i=1}^{n} \mathbf{J}_i = 0 \tag{1-18}$$

The mass flux $\mathbf{n}_i$ is related to the mass diffusion flux as

$$\boldsymbol{n}_i = \boldsymbol{j}_i + \rho_i \boldsymbol{v} = \boldsymbol{j}_i + \omega_i \boldsymbol{n} \tag{1-19}$$

The molar flux $\mathbf{N}_i$ is related to the molar diffusion flux as

$$\mathbf{N}_i = \mathbf{J}_i + c_i \mathbf{V} = \mathbf{J}_i + y_i \mathbf{N} \tag{1-20}$$

It is important to note that the molar flux, $\mathbf{N}_i$, described by equation (1-20) is a resultant of the two vector quantities:

$$\mathbf{J}_i$$

the molar diffusion flux, $\mathbf{J}_i$, resulting from the concentration gradient; this term is referred to as the *concentration gradient contribution*,

and

$$y_i \mathbf{N} = c_i \mathbf{V}$$

the molar flux resulting as component i is carried in the bulk flow of the fluid; this flux term is designated the *bulk motion contribution*.

Either or both quantities can be a significant part of the total molar flux, $\mathbf{N}_i$. Whenever equation (1-20) is applied to describe molar diffusion, the vector nature of the

12   Fundamentals of Mass Transfer

individual fluxes, $\mathbf{N}_i$, must be considered. The same applies for the mass fluxes in equation (1-19). Example 1.4 applies these concepts to mass transport of oxygen in the blood.

Example 1.4 Oxygen Transport in the Human Body

Oxygen is important for living organisms since metabolic processes in the body involve oxygen. The transport of oxygen in the body is therefore crucial to life. The blood in the human circulatory system transports oxygen from the lungs to every single cell in the body. In the larger blood vessels, most of the oxygen transport is due to the bulk motion contribution, while the concentration gradient contribution is more important in smaller blood vessels and capillaries. In the larger vessels, blood flow is rapid and unimpeded, and the oxygen concentration in the fluid remains relatively constant. On the other hand, in smaller vessels oxygen diffuses out of the veins into the surrounding tissue, resulting in decreasing concentrations in the blood.

When referring to the oxygen concentration in the blood, both oxygen bound to hemoglobin and aqueous oxygen dissolved in the plasma are included. Each molecule of hemoglobin, a protein in red blood cells, can bind four oxygen molecules. Most of the blood oxygen is coupled to hemoglobin. In the larger blood vessels, any dissolved oxygen lost to the surrounding tissues by diffusion is replaced by oxygen dissociating from the hemoglobin into the plasma to maintain its constant concentration.

1.2.3 The Maxwell–Stefan Relations

Your objectives in studying this section are to be able to:

1. Write down the Maxwell–Stefan (MS) equations for a binary system, and for multicomponent systems.
2. Define the concepts MS diffusivity and thermodynamic factor.
3. Express the driving force for mass transfer in terms of mole fraction gradients for ideal and nonideal systems.

Consider a binary mixture of ideal gases 1 and 2 at constant temperature and pressure. From a momentum balance describing collisions between molecules of species 1 and molecules of species 2, we obtain (Taylor and Krishna, 1993)

$$\nabla p_1 = -f_{12} y_1 y_2 \left(\mathbf{v}_1 - \mathbf{v}_2 \right) \tag{1-21}$$

where f_{12} is an empirical parameter analogous to a friction factor or a drag coefficient. For convenience, we define an inverse drag coefficient $D_{12} = P/f_{12}$ and rewrite equation (1-21) as

1.2 Molecular Mass Transfer

$$\mathbf{d}_1 = \frac{\nabla p_1}{P} = -\frac{y_1 y_2 \left(\mathbf{v}_1 - \mathbf{v}_2\right)}{D_{12}} \tag{1-22}$$

where $\mathbf{d}_1$ is the driving force for diffusion of species 1 in an ideal gas mixture at constant temperature and pressure.

Equation (1-22) is the MS equation for the diffusion of species 1 in a binary ideal gas mixture. The symbol D_{12} is the MS diffusivity. From a similar analysis for species 2,

$$\mathbf{d}_2 = \frac{\nabla p_2}{P} = -\frac{y_1 y_2 \left(\mathbf{v}_2 - \mathbf{v}_1\right)}{D_{21}} \tag{1-23}$$

For all the applications considered in this book the system pressure is constant across the diffusion path. Then, equations (1-22) and (1-23) simplify to

$$\begin{aligned}
\frac{\nabla y_1}{P} &= -\frac{y_1 y_2 \left(\mathbf{v}_1 - \mathbf{v}_2\right)}{D_{12}} \\
\frac{\nabla y_2}{P} &= -\frac{y_1 y_2 \left(\mathbf{v}_2 - \mathbf{v}_1\right)}{D_{21}}
\end{aligned} \tag{1-24}$$

Example 1.5 Diffusivities in Binary Mixtures

Show that, for a binary mixture, $D_{12} = D_{21}$.

Solution

Since for a binary mixture $(y_1 + y_2) = 1.0$,

$$\nabla y_1 = -\nabla y_2$$

Then, from equation (1-24)

$$-\frac{y_1 y_2 \left(\mathbf{v}_1 - \mathbf{v}_2\right)}{D_{12}} = \frac{y_1 y_2 \left(\mathbf{v}_2 - \mathbf{v}_1\right)}{D_{21}}$$

which can be true only if $D_{12} = D_{21}$.

For multicomponent mixtures, equation (1-22) can be generalized to (Taylor and Krishna, 1993)

$$\mathbf{d}_i = -\sum_{j=1}^{n} \frac{y_i y_j \left(\mathbf{v}_i - \mathbf{v}_j\right)}{D_{ij}} \quad i = 1, 2, \ldots, n-1 \tag{1-25}$$

Equation (1-25) can be written in terms of the molar fluxes $\mathbf{N}_i = c_i\mathbf{v}_i$ to get

$$\mathbf{d}_i = -\sum_{j=1}^{n}\frac{y_i\mathbf{N}_j - y_j\mathbf{N}_i}{cD_{ij}} \quad i-1,2,\ldots,n-1 \tag{1-26}$$

or, in terms of the diffusion fluxes, $\mathbf{J}_i$

$$\mathbf{d}_i = -\sum_{j=1}^{n}\frac{y_j\mathbf{J}_j - y_j\mathbf{J}_i}{cD_{ij}} \quad i=1,2,\ldots,n-1 \tag{1-27}$$

These are the MS diffusion equations for multicomponent systems. They are named after the Scottish physicist James Clerk Maxwell and the Austrian scientist Josef Stefan who were primarily responsible for their development around 1870. It is important to point out that only $(n-1)$ of the MS equations are independent because the $\mathbf{d}_i$, must sum to zero. Also, for a multicomponent ideal gas mixture, a more elaborate analysis than that of Example 1.5 is needed to show that (Taylor and Krishna, 1993)

$$D_{ij} = D_{ji} \tag{1-28}$$

It is easy to show, for a binary mixture of ideal gases where the driving force for diffusion is the mol fraction gradient, equation (1-27) reduces to

$$\mathbf{J}_1 = -cD_{12}\mathbf{d}_1 = -cD_{12}\nabla y_1 \tag{1-29}$$

For nonideal fluids the driving force for diffusion must be defined in terms of chemical potential gradients as

$$\mathbf{d}_i = \frac{x_i}{RT}\nabla_{T,P}\mu_i \tag{1-30}$$

The subscripts T, P are to emphasize that the gradient is to be calculated under constant temperature and pressure conditions. We may express equation (1-30) in terms of activity coefficients, γ_i, and mol fraction gradients as

$$\mathbf{d}_i = \sum_{j=1}^{n-1}\Gamma_{ij}\nabla x_j = \Gamma\nabla x \tag{1-31}$$

where the *thermodynamic factor matrix* Γ is given by

$$\Gamma_{ij} = \delta_{ij} + \frac{\partial \ln \gamma_i}{\partial x_j}\bigg|_{T,P,\Sigma} \tag{1-32}$$

where δ_{ij} is the *Kronecker delta* defined as

$$\delta_{ij} = \begin{cases} 1 \text{ if } i = j \\ 0 \text{ if } i \neq j \end{cases} \tag{1-33}$$

1.2 Molecular Mass Transfer

The symbol Σ is used in equation (1-32) to indicate that the differentiation with respect to mol fraction is to be carried out in such a manner that the mol fractions always add up to 1.0. Combining equations (1-26), (1-27), and (1-31) we obtain

$$\Gamma \nabla x_i = \sum_{j=1}^{n} \frac{x_i \mathbf{J}_j - x_j \mathbf{J}_i}{c D_{ij}} = \sum_{j=1}^{n} \frac{x_i \mathbf{N}_j - x_j \mathbf{N}_i}{c D_{ij}} \quad i = 1, 2, \ldots, n-1 \quad (1\text{-}34)$$

For a multicomponent mixture of ideal gases, equation (1-34) becomes

$$\nabla y_i = \sum_{j=1}^{n} \frac{y_i \mathbf{J}_j - y_j \mathbf{J}_i}{c D_{ij}} = \sum_{j=1}^{n} \frac{y_i \mathbf{N}_j - y_j \mathbf{N}_i}{c D_{ij}} \quad i = 1, 2, \ldots, n-1 \quad (1\text{-}35)$$

For a binary mixture, equation (1-34) reduces to

$$\mathbf{J}_1 = -c D_{12} \Gamma \nabla x_1 \quad (1\text{-}36)$$

where the *thermodynamic factor* Γ is given by

$$\Gamma = 1 + x_1 \frac{\partial \ln \gamma_1}{\partial x_1} \quad (1\text{-}37)$$

The thermodynamic factor is evaluated for liquid mixtures from activity coefficient models. For a *regular solution*, for example,

$$\ln \gamma_1 = A \left(1 - x_1\right)^2 \quad (1\text{-}38)$$

therefore, equation (1-37) yields

$$\Gamma = 1 - 2A x_1 x_2 \quad (1\text{-}39)$$

1.2.4 Fick's First Law for Binary Mixtures

Your objectives in studying this section are to be able to:

1. Write down Fick's first law for diffusion in a binary, isothermal, isobaric mixture.
2. Define the Fick diffusivity and establish the relation between the Fick diffusivity and the MS diffusivity for a binary mixture.

At about the same time that Maxwell and Stefan were developing their ideas of diffusion in multicomponent mixtures, Adolf Fick and others were attempting to uncover the basic diffusion equations through experimental studies involving

16   **Fundamentals of Mass Transfer**

binary mixtures (Fick, 1855). The result of Fick's work was the "law" that bears his name. The Fick equation for a binary mixture in an isothermal, isobaric system is

$$\mathbf{J}_1 = -cD_{12}\nabla x_1 \tag{1-40}$$

where D_{12} is the *Fick diffusivity or diffusion coefficient*. Comparing equations (1-36) and (1-40) we see that, for a binary system, the Fick diffusivity D and the MS diffusivity D are related by

$$D_{12} = D_{12}\Gamma \tag{1-41}$$

For ideal systems, Γ is unity and the diffusivities are identical.

$$D_{12} = D_{12} \text{ for ideal systems} \tag{1-42}$$

The correlation and prediction of Fick and MS diffusion coefficients are discussed in the section that follows. The Fick diffusivity incorporates two aspects: (1) the significance of an inverse drag coefficient (D) and (2) thermodynamic nonideality (Γ). Consequently, the physical interpretation of the Fick diffusion coefficient is less transparent than for the MS diffusivity.

1.3 THE DIFFUSION COEFFICIENT

Fick's law proportionality factor, D_{12}, is known as the diffusion coefficient or diffusivity. Its fundamental dimensions, which are obtained from equation (1-40),

$$D_{12} = -\frac{\mathbf{J}_1}{c\nabla x_1} = \left(\frac{M}{L^2 t}\right)\left(\frac{1}{M/L^3 \times 1/L}\right) = \frac{L^2}{t}$$

are identical to the fundamental dimensions of the other transport properties: kinematic viscosity, ν, and thermal diffusivity, α. The mass diffusivity is usually reported in units of cm^2/s; the SI units are m^2/s, which is a factor 10^{-4} smaller.

The diffusion coefficient depends upon the pressure, temperature, and composition of the system. Experimental values for the diffusivities of gases, liquids, and solids are tabulated in Appendix A. As one might expect from consideration of the mobility of the molecules, the diffusivities are generally higher for gases (in the range of 0.5×10^{-5} to 1.0×10^{-5} m^2/s) than for liquids (in the range of 10^{-10} to 10^{-9} m^2/s) which are higher than the values reported for solids (in the range of 10^{-14} to 10^{-10} m^2/s). In the absence of experimental data, semitheoretical expressions have been developed which give approximations, sometimes as valid as experimental values due to the difficulties encountered in their measurement.

1.3 The Diffusion Coefficient

1.3.1 Diffusion Coefficients for Binary Ideal Gas Systems

Your objectives in studying this section are to be able to:

1. Estimate diffusion coefficients for binary gas systems using the Wilke-Lee equation with tabulated values of the Lennard-Jones parameters.
2. Estimate diffusion coefficients for binary gas systems using the Wilke-Lee equation with values of the Lennard-Jones parameters estimated from empirical correlations.
3. Use a Mathcad® routine to implement calculation of diffusion coefficients for binary gas systems using the Wilke-Lee equation.

The theory describing diffusion in binary gas mixtures at low to moderate pressures has been well developed. Modern versions of the kinetic theory of gases have attempted to account for the forces of attraction and repulsion between molecules. Hirschfelder et al. (1949), using the Lennard-Jones potential to evaluate the influence of intermolecular forces, presented the following equation to estimate the diffusion coefficient for gas pairs of nonpolar, nonreacting molecules:

$$D_{AB} = \frac{0.00266T^{1.5}}{PM_{AB}^{0.5}\sigma_{AB}^2\Omega_D} \tag{1-43}$$

where

$$M_{AB} = \frac{2M_A M_B}{M_A + M_B}$$

D_{AB} = diffusion coefficient, cm^2/s
M_A, M_B = molecular weights of A and B
T = temperature, K
P = pressure, bar
σ_{AB} = "collision diameter," a Lennard-Jones parameter, A
Ω_D = diffusion collision integral, dimensionless

The collision integral, Ω_D, is a function of the temperature and of the intermolecular potential field for one molecule of A and one molecule of B. It is usually tabulated as a function of $T^* = \kappa T/\epsilon_{AB}$, where κ is the Boltzmann constant $(1.38 \times 10^{-16}$ erg/K) and ϵ_{AB} is the energy of molecular interaction for the binary system A and B—a Lennard-Jones parameter—in erg. An accurate approximation of Ω_D can be obtained from Neufield et al. (1972)

where $T^* = \kappa T/\epsilon_{AB}$, $a = 1.06036$ $b = 0.15610$

$c = 0.19300$ $d = 0.47635$ $el = 1.03587$ (1-44)

$f = 1.52996$ $g = 1.76474$ $h = 3.89411$

For a binary system composed of nonpolar molecular pairs, the Lennard-Jones parameters of the pure components may be combined empirically by the following relations:

$$\sigma_{AB} = \frac{\sigma_A + \sigma_B}{2} \qquad \varepsilon_{AB} = \sqrt{\varepsilon_A \varepsilon_B} \qquad (1\text{-}45)$$

These relations must be modified for polar–polar and polar–nonpolar molecular pairs; the proposed modifications are discussed by Hirschfelder et al. (1954).

The Lennard-Jones parameters for the pure components are usually obtained from viscosity data. Appendix B tabulates some of the data available. In the absence of experimental data, the values of the parameters for pure components may be estimated from the following empirical correlations:

$$\sigma = 1.18 V_b^{1/3} \qquad (1\text{-}46)$$

$$\varepsilon / \kappa = 1.15 T_b \qquad (1\text{-}47)$$

where V_b is the molar volume of the substance as liquid at its normal boiling point, in cm^3/gmol, and T_b is the normal boiling point temperature. Molar volumes at normal boiling point for some commonly encountered compounds are listed in Table 1.1. For other compounds not listed in Table 1.1, if a reliable value of the critical volume (V_c) is available, the Tyn and Calus (1975) method is recommended:

$$V_b = 0.285 V_c^{1.048} \qquad (1\text{-}48)$$

Otherwise, the atomic volume of each element present are added together as per the molecular formula of the compound. Table 1.2 lists the contributions for each of the constituent atoms.

Several proposed methods for estimating D_{AB} in low-pressure binary gas systems retain the general form of equation (1-43), with empirical constants based on experimental data. One of the most widely used methods, shown to be quite general and reliable, was proposed by Wilke and Lee (1955):

$$D_{AB} = \frac{\left[3.03 - \left(\dfrac{0.98}{M_{AB}^{0.5}}\right)\right] \left(10^{-3}\right) T^{1.5}}{P M_{AB}^{0.5} \sigma_{AB}^2 \Omega_D} \qquad (1\text{-}49)$$

where all the symbols are as defined under equation (1-43). The use of the Wilke-Lee equation is illustrated in the Examples 1.6 and 1.7.

Example 1.6 Calculation of Diffusivity by the Wilke-Lee Equation with Known Values of the Lennard-Jones Parameters

Estimate the diffusivity of carbon disulfide vapor in air at 273 K and 1 bar using the Wilke-Lee equation (1-49). Compare this estimate with the experimental value reported in Appendix A.

1.3 The Diffusion Coefficient

Table 1.1 Molar Volumes at Normal Boiling Point

Compound	Volume $(cm^3/gmol)$	Compound	Volume $(cm^3/gmol)$
Hydrogen, H_2	14.3	Nitric oxide, NO	23.6
Oxygen, O_2	25.6	Nitrous oxide, N_2O	36.4
Nitrogen, N_2	31.2	Ammonia, NH_3	25.8
Air	29.9	Water, H_2O	18.9
Carbon monoxide, CO	30.7	Hydrogen sulfide, H_2S	32.9
Carbon dioxide, CO_2	34.0	Bromine, Br_2	53.2
Carbonyl sulfide, COS	51.5	Chlorine, Cl_2	48.4
Sulfur dioxide, SO_2	44.8	Iodine, I_2	71.5

Source: Data from Welty et al. (1984).

Table 1.2 Atomic Volume Contributions of the Elements

Element	Volume $(cm^3/gmol)$	Element	Volume $(cm^3/gmol)$
Bromine	27.0	Oxygen, except as noted below	7.4
Carbon	14.8		
Chlorine	24.6	Oxygen, in methyl esters	9.1
Hydrogen	3.7		
Iodine	37.0	Oxygen, in methyl ethers	9.9
Nitrogen	15.6		
Nitrogen, in primary amines	10.5	Oxygen, in higher ethers and other esters	11.0
Nitrogen, in secondary amines	12.0	Oxygen, in acids	12.0
		Sulfur	25.6

For three-membered ring, such as ethylene oxide, subtract	6.0
For four-membered ring, such as cyclobutane, subtract	8.5
For five-membered ring, such as furan, subtract	11.5
For pyridine, subtract	15.0
F or benzene ring, subtract	15.0
For naphthalene ring, subtract	30.0
For anthracene ring, subtract	47.5

Source: Data from Welty et al. (1984).

Solution

Values of the Lennard-Jones parameters (σ and ε/k) are obtained from Appendix B:

	σ, in Å	ε/κ, in K	M, g/mol
CS_2	4.483	467	76
Air	3.620	97	29

20   **Fundamentals of Mass Transfer**

Evaluate the various parameters of equation (1-49) as follows:

$$\sigma_{AB} = \frac{\sigma_A + \sigma_B}{2} = 4.052\,\text{Å}$$

$$\frac{\varepsilon_{AB}}{\kappa} = \sqrt{\frac{\varepsilon_A}{\kappa} \times \frac{\varepsilon_B}{\kappa}} = \sqrt{467 \times 97} = 212.8\,\text{K}$$

$$\frac{\kappa T}{\varepsilon_{AB}} = \frac{273.0}{212.8} = 1.283 \quad \Omega_D = 1.282 \;\; [\text{from equation}(1-44)]$$

$$M_{AB} = 2\left[\frac{M_A M_B}{M_A + M_B}\right] = \frac{2 \times 76 \times 29}{76 + 29} = 41.981$$

Substituting these values into the Wilke-Lee equation yields

$$D_{AB} = \frac{0.001\left(3.03 - \dfrac{0.98}{\sqrt{41.981}}\right)(273)^{1.5}}{(1.0)(4.052)^2(1.282)\sqrt{41.981}} = 0.0952\,\frac{\text{cm}^2}{\text{s}}$$

$$D_{AB} = 9.52 \times 10^{-6}\,\frac{\text{m}^2}{\text{s}}$$

As evidenced by this example, estimation of binary diffusivities can be quite tedious. Most mass-transfer problems involve the calculation of one or more values of diffusivities. Convenience therefore suggests the use of a computer software package for technical calculations, such as Mathcad, for that purpose. Figure 1.3 shows a Mathcad routine to estimate the gas phase binary diffusivity using the Wilke-Lee equation and the conditions for this example. For any other set of conditions, only the two lines of the program following the "Enter Data" expression must be modified accordingly. Appendix 1.1 includes a Python version of this program. The experimental value for this example is obtained from Appendix A:

$$D_{AB}P = 0.894\,\frac{\text{m}^2 \times \text{Pa}}{\text{s}}$$

$$D_{AB} = 0.894\,\frac{\text{m}^2 \times \text{Pa} \times 1\text{bar}}{\text{s} \times 1\text{bar} \times 10^5\,\text{Pa}}$$

$$D_{AB} = 8.94 \times 10^{-6}\,\frac{\text{m}^2}{\text{S}}$$

The error of the estimate, compared to the experimental value, is 6.5%.

Example 1.7 Calculation of Diffusivity by the Wilke-Lee Equation with Estimated Values of the Lennard-Jones Parameters

Estimate the diffusivity of allyl chloride (C_3H_5Cl) in air at 298 K and 1 bar using the Wilke-Lee equation (1-49). The experimental value reported by Lugg (1968) is 0.098 cm^2/s.

1.3 The Diffusion Coefficient

21

Solution

Values of the Lennard-Jones parameters for allyl chloride (A) are not available in Appendix B. Therefore, they must be estimated from equations (1-46) and (1-47). From Table 1.2,

$$V_b = (3)(14.8) + (5)(3.7) + 24.6 = 87.5 \text{ cm}^3/\text{mol}$$

An alternate method to estimate V_b for allyl chloride is using equation (1-48) with a value of $V_c = 234$ cm³/mol (Reid et al., 1987).

$$V_b = (0.285)(234)1.048 = 86.7 \text{ cm}^3/\text{mol}$$

The two estimates are virtually identical. From equation (1-46), $\sigma_A = 5.24$ Å. The normal boiling point temperature for allyl chloride is $T_b = 318.3$ K (Reid et al., 1987). From equation (1-47), $\varepsilon_A/\kappa = 366$ K. The molecular weight of allyl chloride is 76.5 g/mol. The corresponding values for air are $\sigma_B = 3.62$ Å, $\varepsilon_B/\kappa = 97$ K, and $M_B = 29$ g/mol. Substituting these values into the Mathcad routine of Figure 1.3 yields

In the Wilke-Lee equation, T is temperature in K, P is absolute pressure in bar, M is molecular weight in g/mol, the Lennard-Jones parameters are from Appendix B with the units specified there. The data presented here are from Example 1.6.

Enter Data: $\quad T := 273 \qquad P := 1.0 \qquad M_A := 76 \qquad M_B := 29$

$$\sigma_A := 4.483 \qquad \sigma_B := 3.620 \qquad \varepsilon_{Ak} := 467 \qquad \varepsilon_{Bk} := 97$$

$$\sigma_{AB} := \frac{(\sigma_A + \sigma_B)}{2} = 4.052$$

$$\varepsilon_{ABk} := \sqrt{\varepsilon_{Ak} \cdot \varepsilon_{Bk}} = 212.836 \qquad x := \frac{T}{\varepsilon_{ABk}} = 1.283$$

$$a := 1.06026 \qquad b := 0.15610 \qquad c := 0.19300$$
$$d := 0.47635 \qquad e1 := 1.03587 \qquad f := 1.52996$$
$$g := 1.76474 \qquad h := 3.89411$$

$$M_{AB} := 2 \cdot \left(\frac{1}{M_A} + \frac{1}{M_B}\right)^{-1} = 41.981$$

$$\Omega := \frac{a}{x^b} + \frac{c}{e^{d \cdot x}} + \frac{e1}{e^{f \cdot x}} + \frac{g}{e^{h \cdot x}} = 1.282$$

$$D_{AB} := \frac{0.001 \cdot \left(3.03 - \dfrac{0.98}{\sqrt{M_{AB}}}\right) \cdot T^{1.5}}{P \cdot \sigma_{AB}^2 \cdot \Omega \cdot \sqrt{M_{AB}}} \cdot \frac{\text{cm}^2}{\text{s}} = 0.0952 \frac{\text{cm}^2}{\text{s}}$$

Figure 1.3 Mathcad routine to estimate gas-phase mass diffusivities using the Wilke-Lee equation.

$$D_{AB} = 0.0992 \text{ cm}^2/\text{s}$$

The error of this estimate, compared to the experimental value, is 1.2%.

It must be emphasized that, for low-pressure binary gas systems, the diffusivity does not depend on the composition of the mixture. Either component can be chosen as component A or component B. As the next section will show, the estimation of diffusivities in liquid mixtures is much more complex.

1.3.2 Diffusion Coefficients for Dilute Liquids

Your objectives in studying this section are to be able to:

1. Estimate diffusion coefficients for binary dilute liquid systems using the Wilke and Chang equation.
2. Estimate diffusion coefficients for binary dilute liquid systems using the Hayduk and Minhas correlations.
3. Use a Mathcad routine to implement calculation of diffusion coefficients for binary dilute liquid systems using the Hayduk and Minhas equation.

In contrast to the case for gases, where an advanced kinetic theory to explain molecular motion is available, theories of the structure of liquids and their transport characteristics are still inadequate to allow a rigorous treatment. Liquid diffusion coefficients are several orders of magnitude smaller than gas diffusivities and depend on concentration due to the changes in viscosity with concentration and changes in the degree of ideality of the solution. As the mol fraction of either component in a binary mixture approaches unity, the thermodynamic factor Γ approaches unity and the Fick diffusivity and the MS diffusivity are equal. The diffusion coefficients obtained under these conditions are the infinite dilution diffusion coefficients and are given the symbol D^0.

The Stokes-Einstein equation is a theoretical method of estimating D^0,

$$D_{AB}^0 = \frac{\kappa T}{6\pi r_A \mu_B} \tag{1-50}$$

where r_A is the solute particle radius, and μ_B is the solvent viscosity. This equation has been successful in describing diffusion of colloidal particles or large round molecules through a solvent which behaves as a continuum relative to the diffusing species.

Equation (1-50) has provided a useful starting point for several semiempirical correlations arranged into the general form:

$$\frac{D_{AB}^0 \mu_B}{\kappa T} = f\left(V_{bA}\right) \tag{1-51}$$

1.3 The Diffusion Coefficient

in which $f(V_{bA})$ is a function of the molecular volume of the diffusing solute. Empirical correlations using the general form of equation (1-51) have been developed which attempt to predict the liquid diffusion coefficient in terms of the solute and solvent properties. Wilke and Chang (1955) have proposed the following still widely used correlation for nonelectrolytes in an infinitely dilute solution:

$$\frac{D^0_{AB}\mu_B}{T} = \frac{7.4 \times 10^{-8}\left(\Phi_B M_B\right)^{0.5}}{V_{bA}^{0.6}} \tag{1-52}$$

where

D^0_{AB} = diffusivity of A in very dilute solution in solvent B, cm^2/s

M_B = molecular weight of solvent B

T = temperature, K

μ_B = viscosity of solvent B, cP

V_{bA} = solute molar volume at its normal boiling point, cm^3/mol

     = 75.6 cm^3/mol for water as solute

Φ_B = association factor of solvent B, dimensionless

     = 2.26 for water as solvent

     = 1.9 for methanol as solvent

     = 1.5 for ethanol as solvent

     = 1.0 for unassociated solvents, e.g., benzene, ether, heptane

The value of V_{bA} may be the true value or, if necessary, estimated from equation (1-48), or from the data of Table 1.2, except when water is the diffusing solute, as noted above. The association factor for a solvent can be estimated only when diffusivities in that solvent have been experimentally measured. There is also some doubt about the ability of the Wilke-Chang equation to handle solvents of extremely high viscosity, say 100 cP or more.

Hayduk and Minhas (1982) considered many correlations for the infinite dilution binary diffusion coefficient. By regression analysis, they proposed several correlations depending on the type of solute–solvent system:

(a) For *solutes in aqueous solutions:*

$$D^0_{AB} = 1.25 \times 10^{-8}\left(V_{bA}^{-0.19} - 0.292\right)T^{1.52}\mu_B^{\varepsilon}$$

$$\varepsilon = \frac{9.58}{V_{bA}} - 1.12 \tag{1-53}$$

where

D^0_{AB} = diffusivity of A in very dilute aqueous solution, cm^2/s

T = temperature, K

μ_B = viscosity of water, cP

V_{bA} = solute molar volume at its normal boiling point, cm^3/mol

(b) For *nonaqueous (nonelectrolyte) solutions*:

$$D_{AB}^0 = 1.55 \times 10^{-8} \frac{V_{bB}^{0.27} T^{1.29} \sigma_B^{1.25}}{V_{bA}^{0.42} \mu_B^{0.92} \sigma_A^{0.105}} \tag{1-54}$$

where σ is surface tension at the normal boiling point temperature, in dyn/cm. If values of the surface tension are not known, they may be estimated by the Brock and Bird (1955) corresponding states method (limited to nonpolar liquids):

$$\sigma\left(T_b\right) = P_c^{2/3} T_c^{1/3} \left(0.132\alpha_c - 0.278\right)\left(1 - T_{br}\right)$$

$$T_{br} = \frac{T_b}{T_c} \tag{1-55}$$

$$\alpha_c = 0.9076 \left[1 + \frac{T_{br} \ln\left(P_c / 1.013\right)}{1 - T_{br}}\right]$$

where

$\sigma(T_b)$ = surface tension at the normal boiling point temperature, dyn/cm
P_c = critical pressure, bar
T_c = critical temperature, K

When using the correlation shown in equation (1-54), the authors note several restrictions:

1. The method should not be used for diffusion in viscous solvents. Values of μ_B above about 20 cP would classify the solvent as viscous.
2. If the solute is water, a dimer value of V_{bA} should be used (V_{bA} = 37.4 cm^3/mol).
3. If the solute is an organic acid and the solvent is other than water, methanol, or butanol, the acid should be considered a dimer with twice the expected value of V_{bA}.
4. For nonpolar solutes diffusing into monohydroxy alcohols, the values of V_{bB} should be multiplied by a factor equal to $8\mu_B$ where μ_B is the solvent viscosity in cP.

Examples 1.8 and 1.9 illustrate the use of these correlations.

Example 1.8 Calculation of Liquid Diffusivity in Aqueous Solution

Estimate the diffusivity of ethanol (C_2H_6O) in a dilute solution in water at 288 K. Compare your estimate with the experimental value reported in Appendix A.

Solution
(a) Use the Wilke-Chang correlation, equation (1-52). Calculate the molar volume of ethanol using equation (1-48) with V_c = 167.1 cm^3/mol (Reid et al., 1987):

$$V_{bA} = 0.285(167.1)^{1.048} = 60.9 \frac{\text{cm}^3}{\text{mol}}$$

1.3 The Diffusion Coefficient

The viscosity of liquid water at 288 K is $\mu_B = 1.153$ cP (Reid et al., 1987). Substituting in equation (1-52) gives

$$D_{AB}^0 = \frac{\left(7.4\times10^{-8}\right)(288)[(2.26)(18)]^{1/2}}{(1.153)(60.9)^{0.6}} = 1.002\times10^{-5}\,\frac{cm^2}{s}$$

The experimental value reported in Appendix A is 1.0×10^{-5} cm^2/s. Therefore, the error of the estimate is only 0.2%. (This is not typical!)

(b) Using the Hayduk and Minhas correlation for aqueous solutions:

$$\varepsilon = \frac{9.58}{60.9} - 1.12 = -0.963$$

$$D_{AB}^0 = 1.25\times10^{-8}\left[60.9^{-0.19} - 0.292\right](288)^{1.52}(1.153)^{-0.963}$$

$$= 0.991\times10^{-5}\,\frac{cm^2}{s}$$

The error of the estimate in this case is –0.9%.

Example 1.9 Calculation of Liquid Diffusivity in Dilute Nonaqueous Solution

Estimate the diffusivity of acetic acid ($C_2H_4O_2$) in a dilute solution in acetone (C_3H_6O) at 313 K. Compare your estimate with the experimental value reported in Appendix A. The following data are available (Reid et al., 1987):

Parameter	Acetic acid	Acetone
T_b, K	390.4	329.2
T_c, K	594.8	508.0
P_c, bars	57.9	47.0
μ_c, cm^3/mol	171	209
μ, cP	—	0.264
M	60	58

Solution

(a) Use the Wilke-Chang correlation, equation (1-52). Calculate the molar volume of acetic acid using equation (1-48) with $V_c = 171$ cm^3/mol: $V_{bA} = 62.4$ cm^3/mol. The association factor for acetone is $\Phi = 1.0$. From equation (1-52),

$$D_{AB}^0 = \frac{\left(7.4\times10^{-8}\right)(313)[(1)(58)]^{1/2}}{(0.264)(62.4)^{0.6}} = 5.595\times10^{-5}\,\frac{cm^2}{s}$$

From Appendix A, the experimental value is 4.04×10^{-5} cm^2/s. Therefore, the error of the estimate is 38.5%.

26   **Fundamentals of Mass Transfer**

(b) Use the Hayduk and Minhas correlation for nonaqueous solutions. According to restriction 3 mentioned above, the molar volume of the acetic acid to be used in equation (1-54) should be $V_{bA} = 2 \times 62.4 = 124.8$ cm^3/mol. The molar volume of acetone is calculated from equation (1-48): $V_{bB} = 77.0$ cm^3/mol. Estimates of the surface tension at the normal boiling point of each component are calculated from equation (1-55) as follows:

For acetone (B):

$$T_{br} = \frac{329.2}{508} = 0.648$$
$$\alpha_c = 7.319$$
$$\sigma_B = 20.0 \, \text{dyn} / \text{cm}$$

For acetic acid (A):

$$T_{br} = \frac{390.4}{594.8} = 0.656$$
$$\alpha_c = 7.910$$
$$\sigma_B = 26.2 \, \text{dyn} / \text{cm}$$

Substituting numerical values in equation (1-54):

$$D_{AB}^0 = 1.55 \times 10^{-8} \frac{(77)^{0.27}(313)^{1.29}(20)^{0.125}}{(124.8)^{0.42}(0.264)^{0.92}(26.2)^{0.105}}$$
$$D_{AB}^0 = 3.84 \times 10^{-5} \frac{\text{cm}^2}{\text{s}}$$

From Appendix A, the experimental value is 4.04×10^{-5} cm^2/s. Therefore, the error of the estimate is -5.0%.

Figure 1.4 shows a Mathcad routine to estimate liquid-phase binary diffusivities for nonaqueous solutions using the Hayduk and Minhas equation and the conditions for this example. For any other set of conditions, only the four lines of the program following the "Enter Data" expression must be modified accordingly. Appendix 1.2 presents a Python routine to estimate liquid-phase mass diffusivities.

1.3.3 Diffusion Coefficients for Concentrated Liquids

Most methods for predicting D in concentrated liquid solutions attempt to combine the infinite dilution diffusion coefficients $(D_{12})^0$ and $(D_{21})^0$ in a single function of composition as shown in Example 1.10. The Vignes formula is recommended by Reid et al. (1987):

$$D_{12} = \left(D_{12}^0\right)^{x_2} \left(D_{21}^0\right)^{x_1}$$
$$D_{12} = D_{12}\Gamma$$

$$(1\text{-}56)$$

1.3 The Diffusion Coefficient

> In this equation, temperatures are in K, pressures in bar, molar volumes in cc/mol, and viscosity of the solvent in cP.
>
> Enter Data:
>
> $$T_{bA} := 390.4 \qquad T_{cA} := 594.8 \qquad T := 313$$
>
> $$P_{cA} := 57.9 \qquad V_{bA} := 124.8 \qquad \mu_B := 0.264$$
>
> $$T_{bB} := 329.2 \qquad T_{cB} := 508.0$$
>
> $$P_{cB} := 47.0 \qquad V_{bB} := 77.0$$
>
> $$T_{brA} := \frac{T_{bA}}{T_{cA}} = 0.656 \qquad\qquad T_{brB} := \frac{T_{bB}}{T_{cB}} = 0.648$$
>
> $$\alpha_{cA} := 0.9076 \cdot \left[1 + \frac{T_{brA} \cdot \ln\left(\dfrac{P_{cA}}{1.013}\right)}{1 - T_{brA}} \right] = 7.921$$
>
> $$\alpha_{cB} := 0.9076 \cdot \left[1 + \frac{T_{brB} \cdot \ln\left(\dfrac{P_{cB}}{1.013}\right)}{1 - T_{brB}} \right] = 7.32$$
>
> $$\sigma_A := P_{cA}^{\frac{2}{3}} \cdot T_{cA}^{\frac{1}{3}} \cdot \left[0.132 \cdot \alpha_{cA} - 0.278 \right] \cdot \left[1 - T_{brA} \right]^{\frac{11}{9}} = 26.185$$
>
> $$\sigma_B := P_{cB}^{\frac{2}{3}} \cdot T_{cB}^{\frac{1}{3}} \cdot \left[0.132 \cdot \alpha_{cB} - 0.278 \right] \cdot \left[1 - T_{brB} \right]^{\frac{11}{9}} = 19.959$$
>
> $$D_{L0} := \frac{1.55 \cdot 10^{-8} \cdot V_{bB}^{0.27} \cdot T^{1.29} \cdot \sigma_B^{0.125}}{V_{bA}^{0.42} \cdot \mu_B^{0.92} \cdot \sigma_A^{0.105}} \cdot \frac{cm^2}{s} = \left(3.839 \cdot 10^{-5} \right) \frac{cm^2}{s}$$

Figure 1.4 Mathcad routine to estimate liquid-phase mass diffusivities in dilute nonaqueous solutions using the Hayduk and Minhas equation.

Example 1.10 Diffusion Coefficients for Acetone-Benzene (Taylor and Krishna, 1993)

Estimate the MS and Fick diffusion coefficients for an acetone (1)–benzene (2) mixture of composition $x_1 = 0.7808$ at 298 K. The infinite dilution diffusivities are

$$D_{12}^0 = 2.75 \times 10^{-9} \frac{m^2}{s}$$

$$D_{21}^0 = 4.15 \times 10^{-9} \frac{m^2}{s}$$

28   **Fundamentals of Mass Transfer**

From the nonrandom-two-liquid (NRTL) equation, for this system at the given temperature and concentration, the thermodynamic correction factor Γ = 0.871. The experimental value of D_{12} at this concentration is 3.35×10^{-9} m²/s.

Solution
Substituting in equation (1-56):

$$D_{12} = \left(2.75\times10^{-9}\right)^{0.2192} \times \left(4.155\times10^{-9}\right)^{0.7808}$$

$$D_{12} = 3.792\times10^{-9}\,\frac{\text{m}^2}{\text{s}}$$

$$D_{12} = 3.792\times10^{-9}\times0.871$$

$$D_{12} = 3.30\times10^{-9}\,\frac{\text{m}^2}{\text{s}}$$

The predicted value of the Fick diffusivity is in excellent agreement with the experimental result.

1.3.4 Effective Diffusivities in Multicomponent Mixtures

Your objectives in studying this section are to be able to:

1. Estimate the effective diffusivity of a component in a multicomponent mixture of gases from its binary diffusion coefficients with each of the other constituents of the mixture.
2. Estimate the effective diffusivity of a dilute solute in a homogeneous solution of mixed solvents from its infinite dilution binary diffusion coefficients into each of the individual solvents.

The MS equations for diffusion in multicomponent systems can become extremely complicated, and they are sometimes handled by using an *effective diffusivity or pseudobinary approach*. The effective diffusivity is defined by assuming that the rate of diffusion of component i depends only on its own composition gradient; that is,

$$\mathbf{J}_i = -c\mathrm{D}_{i,eff}\nabla x_i \quad i = 1,2,\ldots,n \tag{1-57}$$

where $\mathrm{D}_{i,eff}$ is some characteristic diffusion coefficient of species i in the mixture which must be synthesized from the binary MS diffusivities.

In gases, the binary MS diffusivities are normally assumed independent of composition. With this approximation, the effective diffusivity of component i in a multicomponent gas mixture can be derived from the MS equation to give (Treybal, 1980)

$$\mathrm{D}_{i,eff} = \frac{\mathbf{N}_i - y_i\sum_{j=1}^{n}\mathbf{N}_j}{\sum_{\substack{j=1\\j\neq i}}^{n}\frac{1}{D_{ij}}\left(y_j\mathbf{N}_i - y_i\mathbf{N}_j\right)} \tag{1-58}$$

1.3 The Diffusion Coefficient

In order to use this equation, the relation between the values of the N's must be known. This relation, called a *bootstrap condition* (Taylor and Krishna, 1993), is usually fixed by other considerations, such as reaction stoichiometry or energy balances. Consider, for instance, the general chemical reaction, written as

$$\sum_i v_i A_i = 0$$

where the v_i are called stoichiometric numbers and A_i stands for a chemical formula. The sign convention for v_i is positive (+) for products and negative (−) for reactants (Smith et al., 1996). For diffusion with heterogeneous chemical reaction, the relations between the fluxes are specified by the stoichiometry of the reaction as described by the following equations:

$$\frac{\mathbf{N}_1}{v_1} = \frac{\mathbf{N}_2}{v_2} = \frac{\mathbf{N}_3}{v_3} = \cdots = \frac{\mathbf{N}_i}{v_i} \tag{1-59}$$

For example, if ammonia is being cracked on a solid catalyst (see Example 1-11) according to

$$NH_3 \rightarrow \frac{1}{2}N_2 + \frac{3}{2}H_2$$

under circumstances such that NH_3 (1) diffuses to the catalyst surface, and N_2 (2) and H_2 (3) diffuse back, applying equation (1-59) gives

$$N_2 = -\tfrac{1}{2}N_1$$
$$N_3 = -\tfrac{3}{2}N_1$$

Equation (1-58) suggests that the effective diffusivity may vary considerably along the diffusion path, but a linear variation with distance can usually be assumed (Bird et al., 1960). Therefore, if $D_{i,eff}$ changes significantly from one end of the diffusion path to the other, the arithmetic average of the two values should be used.

A common situation in multicomponent diffusion, illustrated in Example 1.12, is when all the $\mathbf{N}$'s except $\mathbf{N}_1$ are zero (i.e., all but component 1 is stagnant). Equation (1-58) then becomes

$$D_{1,eff} = \frac{1 - y_1}{\displaystyle\sum_{i=2}^{n} \frac{y_i}{D_{1,i}}} = \frac{1}{\displaystyle\sum_{i=2}^{n} \frac{y_i'}{D_{1,i}}} \tag{1-60}$$

where y_i' is the mole fraction of component i on a component 1-free base, that is

$$y_i' = \frac{y_i}{1 - y_i} \quad i = 2,3,\ldots,n \tag{1-61}$$

Example 1.11 Calculation of Effective Diffusivity in a Multicomponent Gas Mixture

Ammonia is being cracked on a solid catalyst according to the reaction

$$NH_3 \rightarrow \tfrac{1}{2}N_2 + \tfrac{3}{2}H_2$$

At one place in the apparatus where the pressure is 1 bar and the temperature 300 K, the analysis of the gas is 40% NH_3 (1), 20% N_2 (2), and 40% H_2 (3) by volume. Estimate the effective diffusivity of ammonia in the gaseous mixture.

Solution

Calculate the binary diffusivities of ammonia in nitrogen (D_{12}) and ammonia in hydrogen (D_{13}) using the Mathcad routine developed in Example 1.6 to implement the Wilke-Lee equation. Values of the Lennard-Jones parameters (σ and ε/κ) are obtained from Appendix B. The data needed are

	σ, in Å	ε/κ, in K	M, g/mol
NH_3	2.900	558.3	17
N_2	3.798	71.4	28
H_2	2.827	59.7	2

Therefore, the following results were obtained:

$$D_{12} = 0.237 \text{ cm}^2/\text{s}$$
$$D_{13} = 0.728 \text{ cm}^2/\text{s}$$

Figure 1.5 shows the flux of ammonia (1) toward the catalyst surface where it is consumed by chemical reaction, and the fluxes of nitrogen (2) and hydrogen (3) produced by the reaction migrating away from the same surface.

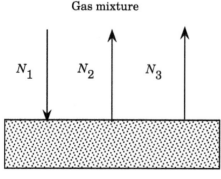

Figure 1.5 Catalytic cracking of ammonia (1) into N_2 (2) and H_2 (3).

1.3 The Diffusion Coefficient

As was shown before, the stoichiometry of the reaction fixes the relationship between the individual fluxes, which, in this case, becomes $N_2 = -(1/2)N_1$, $N_3 = -(3/2)N_1$, $(N_1 + N_2 + N_3) = -N_1$. Substituting in equation (1-58), we obtain

$$
\begin{aligned}
D_{1,eff} &= \frac{N_1\left(1+y_1\right)}{\dfrac{y_2 N_1 + \dfrac{1}{2}y_1 N_1}{D_{12}} + \dfrac{y_3 N_1 + \dfrac{3}{2}y_1 N_1}{D_{13}}} \\[2em]
&= \frac{\left(1+y_1\right)}{\dfrac{y_2 + \dfrac{1}{2}y_1}{D_{12}} + \dfrac{y_3 + \dfrac{3}{2}y_1}{D_{13}}} \\[2em]
&= \frac{1.4}{\dfrac{0.2+0.5\times0.4}{0.237} + \dfrac{0.4+1.5\times0.4}{0.728}} \\[2em]
&= 0.457\,\frac{cm^2}{s}
\end{aligned}
$$

Example 1.12 Calculation of Effective Diffusivity in a Multicomponent Stagnant Gas Mixture

Ammonia is being absorbed from a stagnant mixture of nitrogen and hydrogen by contact with a 2 N sulfuric acid solution. At one place in the apparatus where the pressure is 1 bar and the temperature 300 K, the analysis of the gas is 40% NH_3 (1), 20% N_2 (2), and 40% H_2 (3) by volume. Estimate the effective diffusivity of ammonia in the gaseous mixture.

Solution
The binary diffusivities are the same as for the previous example. However, in this case, only the ammonia diffuses toward the liquid interphase since neither nitrogen nor hydrogen is soluble in the sulfuric acid solution. Therefore, equation (1-60) applies. Calculate the mole fractions of nitrogen (2) and hydrogen (3) on an ammonia (1)-free base from equation (1-61):

$$
y_2' = \frac{y_2}{1-y_1} = \frac{0.2}{1-0.4} = 0.333
$$

$$
y_3' = \frac{y_3}{1-y_1} = \frac{0.4}{1-0.4} = 0.667
$$

32   **Fundamentals of Mass Transfer**

Substituting in equation (1-60) gives us

$$D_{1,eff} = \cfrac{1}{\cfrac{y_2'}{D_{12}} + \cfrac{y_3'}{D_{13}}}$$

$$= \cfrac{1}{\cfrac{0.333}{0.237} + \cfrac{0.667}{0.728}}$$

$$= 0.431\, \frac{cm^2}{s}$$

For multicomponent liquid mixtures, it is usually difficult to obtain numerical values of the diffusion coefficients relating fluxes to concentration gradients. One important case of multicomponent diffusion for which simplified empirical correlations have been proposed is when a dilute solute diffuses through a homogeneous solution of mixed solvents. Perkins and Geankoplis (1969) evaluated several methods and suggested

$$D_{1,eff}^0 \mu_M^{0.8} = \sum_{j=2}^{n} x_j D_{1,j}^0 \mu_j^{0.8} \tag{1-62}$$

where

$D_{1,eff}^0$ = effective diffusivity for a dilute solute A into the mixture, cm^2/s

$D_{1,j}^0$ = infinite dilution binary diffusivity of solute A into solvent j, cm^2/s

x_j = mol fraction of component j

μ_M = mixture viscosity, cP

μ_j = pure component viscosity, cP

When tested with data for eight ternary systems, errors were normally less than 20%, except for cases where CO_2 was the solute. For CO_2 as a dilute solute diffusing into mixed solvents, Takahashi et al. (1982) recommend

$$D_{CO_2,eff}^0 \left(\frac{\mu_M}{V_M}\right)^{1/3} = \sum_{\substack{j=1 \\ j \neq CO_2}}^{n} x_j D_{CO_2,j}^0 \left(\frac{\mu_j}{V_j}\right)^{1/3} \tag{1-63}$$

where

V_M = molar volume of the mixture at T, cm^3/mol

V_j = molar volume of the pure component j at T, cm^3/mol

Tests with several ternary systems involving CO_2 led to deviations from experimental values usually less than 4%, as shown in Example 1.13 (Reid et al., 1987).

1.3 The Diffusion Coefficient

33

Example 1.13 Calculation of Effective Diffusivity of a Dilute Solute in a Homogeneous Mixture of Solvents

Estimate the diffusion coefficient of acetic acid at extremely low concentrations diffusing into a mixed solvent containing 40.0 wt% ethyl alcohol in water at a temperature of 298 K.

Solution

Let 1 = acetic acid, 2 = water, and 3 = ethyl alcohol. Estimate the infinite dilution diffusion coefficient of acetic acid in water at 298 K using equation (1-53). The required data are (Reid et al., 1987): (1) at 298 K, μ_B = 0.894 cP; (2) V_{cA} = 171.0 cm^3/mol. From equation (1-48), V_{bA} = 62.4 cm^3/mol. Substituting in equation (1-53),

$$D_{12}^0 = 1.32 \times 10^{-5} \, \text{cm}^2 \, / \, \text{s}$$

Estimate the infinite dilution diffusion coefficient of acetic acid in ethanol at 298 K. The following data are available (Reid et al., 1987):

Parameter	Acetic acid	Ethanol
T_b, K	390.4	351.4
Tc, K	594.8	513.9
Pc, bar	57.9	61.4
Vc, cm^3/mol	171	167
μ @ 298 K, cP	—	1.043

Use the Hayduk and Minhas correlation for nonaqueous solutions. According to restriction 3 mentioned above, the molar volume of the acetic acid to be used in equation (1-54) should be V_{bA} = 2 × 62.4 = 124.8 cm^3/mol. The molar volume of ethanol is calculated from equation (1-48): V_{bB} = 60.9 cm^3/mol. Using the Mathcad routine developed in Example 1.9, we have

$$D_{13}^0 = 1.0 \times 10^{-5} \, \text{cm}^2 \, / \, \text{s}$$

The viscosity of a 40 wt% aqueous ethanol solution at 298 K is μ_M = 2.35 cP (Perkins and Geankoplis, 1969). Before substituting in equation (1-62), the solution composition must be changed from mass to molar fractions following a procedure like that illustrated in Example equation (1.2. Accordingly, a 40 wt% aqueous ethanol solution converts to 20.7 mol%. Substituting in equation (1-62) yields

$$D_{1,\,\text{eff}}^0 = \frac{10^{-5}}{(2.35)^{0.8}} \left[(0.207)(1.0)(1.043)^{0.8} + (0.793)(1.32)(0.894)^{0.8} \right]$$
$$= 0.591 \times 10^{-5} \, \text{cm}^2 \, / \, \text{s}$$

The experimental value reported by Perkins and Geankoplis (1969) is 0.571 × 10^5 cm^2/s. Therefore, the error of the estimate is 3.5%.

1.4 STEADY-STATE MOLECULAR DIFFUSION IN FLUIDS

1.4.1 Molar Flux and the Equation of Continuity

Your objectives in studying this section are to be able to:

1. State the equation of continuity for a system of variable composition.
2. Simplify the equation of continuity for steady-state diffusion in one direction in rectangular coordinates and without chemical reaction.

The *equation of continuity* for component A in a mixture describes the change of concentration of A with respect to time at a fixed point in space resulting from mass transfer of A and chemical reactions producing A. In vector symbolism, the resulting equation is (Bird et al., 1960)

$$\nabla \cdot \mathbf{N}_A + \frac{\partial c_A}{\partial t} - R_A = 0 \tag{1-64}$$

The first term in equation (1-64), the divergence of the molar flux vector, represents the net rate of molar efflux per unit volume. In rectangular coordinates, this term can be expanded as

$$\nabla \cdot \mathbf{N}_A = \frac{\partial N_{A,x}}{\partial x} + \frac{\partial N_{A,y}}{\partial y} + \frac{\partial N_{A,z}}{\partial z} \tag{1-65}$$

The second term in equation (1-64) represents the molar rate of accumulation of A per unit volume. The term R_A represents the volumetric rate of formation of A by chemical reaction (mol of A formed/volume-time). An expression like equation (1-64) can be written for each component of the mixture:

$$\nabla \cdot \mathbf{N}_i + \frac{\partial c_i}{\partial t} - R_i = 0 \qquad i = 1, 2, ..., n \tag{1-66}$$

Consider, now, steady-state diffusion (no accumulation) without chemical reaction. In that case, equation (1-66) reduces to

$$\nabla \cdot \mathbf{N}_i = 0 \qquad i = 1, 2, ..., n \tag{1-67}$$

In rectangular coordinates, with diffusion only in the z direction, equation (1-67) simplifies to

$$\frac{dN_{i,z}}{dz} = 0 \qquad i = 1, 2, ..., n \tag{1-68}$$

This relation stipulates a constant molar flux of each component in the mixture, and a constant total molar flux. Therefore, for steady-state, one-dimensional diffusion without chemical reaction:

1.4 Steady-State Molecular Diffusion in Fluids

$$N_i = N_{i,z} = \text{constant}_i \quad i = 1,2...,n$$

$$\sum_{i=1}^{n} N_i = \text{constant} \tag{1-69}$$

1.4.2 Steady-State Molecular Diffusion in Gases

Your objectives in studying this section are to be able to:

1. Calculate molar fluxes during steady-state molecular, one-dimensional diffusion in gases.
2. Calculate molar fluxes for steady-state gaseous diffusion of A through stagnant B, and for equimolar counterdiffusion.

Consider steady-state diffusion only in the z direction without chemical reaction in a binary gaseous mixture. For the case of one-dimensional diffusion, equation (1-20) becomes

$$N_A = -c\mathrm{D}_{AB}\frac{dy_A}{dz} + y_A N_A = -c\mathrm{D}_{AB}\frac{dy_A}{dz} + y_A\left(N_A + N_B\right) \tag{1-70}$$

For a gaseous mixture, if the temperature and pressure are constant, c and D_{AB} are constant, independent of position and composition. According to equation (1-69), all the molar fluxes are also constant. Equation (1-70) may then be integrated between the two boundary conditions:

$$\text{at } z = z_1 \quad y_A = y_{A1}$$
$$\text{at } z = z_2 \quad y_A = y_{A2}$$

where 1 indicates the beginning of the diffusion path (y_A high) and 2 the end of the diffusion path (y_A low).

However, before equation (1-70) can be integrated, the bootstrap relation between the molar flux of component A and the total molar flux (N) must be established a priori based on stoichiometric or energy balance considerations as mentioned before. The next example will illustrate the simplest case: that when the total molar flux vanishes and there is no bulk motion contribution to the flux of either component.

Example 1.14 Steady-State Equimolar Counterdiffusion

This is a situation frequently encountered in binary distillation operations when the molar latent heat of vaporization of both components is similar. In this case, $N_B = -N_A = \text{const.}$, $\Sigma N_i = N = 0$. For an ideal gas mixture, equation (1-70) simplifies to

$$N_A = -c\mathrm{D}_{AB}\frac{dy_A}{dz} = -\frac{\mathrm{D}_{AB}P}{RT}\frac{dy_A}{dz} \tag{1-71}$$

36   **Fundamentals of Mass Transfer**

Separating variables and integrating with constant N_A, we have

$$\begin{aligned} N_A &= \frac{D_{AB}P}{RT\delta}\left(y_{A1} - y_{A2}\right) \\ &= \frac{D_{AB}}{RT\delta}\left(p_{A1} - p_{A2}\right) \end{aligned} \tag{1-72}$$

where we let $z_2-z_1 = \delta$. Notice that in equation (1-72), derived for equimolar counterdiffusion, the driving force for mass transfer is linear, that is, the flux of component A is directly proportional to the difference in concentrations or partial pressures between the extremes of the diffusion path. This result is a direct consequence of the fact that the process described is of a purely diffusional nature, with no bulk motion contribution. A plot of y_A versus position would be a straight line.

Consider the following numerical example. A binary gaseous mixture of components A and B at a pressure of 1 bar and temperature of 300 K undergoes steady-state equimolar counterdiffusion along a 1-mm-thick diffusion path. At one end of the path the mole fraction of component A is 70%, while at the other end it is 20%. Under these conditions, $D_{AB} = 0.1$ cm^2/s. Calculate the molar flux of component A.

Solution
Direct substitution into equation (1-72) yields

$$N_A = \frac{\left(10^{-5}\,\mathrm{m}^2/\mathrm{s}\right)\left(10^5\,\mathrm{Pa}\right)(0.7 - 0.2)}{\left(8.314\,\dfrac{\mathrm{m}^3\cdot\mathrm{Pa}}{\mathrm{mol}\cdot\mathrm{K}}\right)(300\mathrm{K})\left(10^{-3}\,\mathrm{m}\right)}$$

$$= 0.20\ \mathrm{mol/m}^2\cdot\mathrm{s}$$

In the general case, the bulk motion contribution to the flux must be included, as shown by equation (1-70). Separating the variables in equation (1-70):

$$\frac{-dy_A}{N_A - y_A\left(N_A + N_B\right)} = \frac{dz}{cD_{AB}} \tag{1-73}$$

Equation (1-73) may then be integrated between the two boundary conditions:

$$\frac{1}{N_A + N_B}\ln\left[\frac{N_A - y_{A2}\left(N_A + N_B\right)}{N_A - y_{A1}\left(N_A + N_B\right)}\right] = \frac{\delta}{cD_{AB}} \tag{1-74}$$

Multiplying both sides of equation (1-74) by N_A and rearranging:

$$N_A = \frac{N_A}{N_A + N_B}\frac{cD_{AB}}{\delta}\ln\left|\frac{\dfrac{N_A}{N_A + N_B} - y_{A2}}{\dfrac{N_A}{N_A + N_B} - y_{A1}}\right| \tag{1-75}$$

1.4 Steady-State Molecular Diffusion in Fluids

Define the molar flux fraction, Ψ_A:

$$\Psi_A = \frac{N_A}{N_A + N_B} \tag{1-76}$$

In this expression, N_A gives the positive direction of z. Any flux in the opposite direction will carry a negative sign. Notice that the molar flux fraction Ψ_A depends exclusively on the bootstrap relation between N_A and the other fluxes, a relation that is known a priori based on other considerations, as was shown before. Therefore, equation (1-75) becomes

$$N_A = \Psi_A \frac{c\mathrm{D}_{AB}}{\delta} \ln\left[\frac{\Psi_A - y_{A2}}{\Psi_A - y_{A1}}\right] \tag{1-77}$$

Equation (1-77) is one of the most fundamental relations in the analysis of mass-transfer phenomena. It was derived by integration assuming that all the molar fluxes were constant, independent of position. Integration under conditions where the fluxes are not constant is also possible. Consider steady-state radial diffusion from the surface of a solid sphere into a fluid. Equation (1-70) can be applied, but the fluxes are now a function of position owing to the geometry. Most practical problems which deal with such matters, however, are concerned with diffusion under turbulent conditions, and the transfer coefficients which are then used are based upon a flux expressed in terms of some arbitrarily chosen area, such as the surface of the sphere. These matters are discussed in detail in Chapter 2.

Most practical problems of mass transfer in the gas phase are at conditions such that the mixture behaves like an ideal gas as illustrated in Example 1.15. Then, combining equations (1-7) and (1-77):

$$N_A = \Psi_A \frac{D_{AB}P}{RT\delta} \ln\left[\frac{\Psi_A - y_{A2}}{\Psi_A - y_{A1}}\right] \tag{1-78}$$

Example 1.15 Steady-State Diffusion of A Through Stagnant B

Consider steady-state, one-dimensional diffusion of A through nondiffusing B. This might occur, for example, if ammonia (A) is absorbed from air (B) into water. In the gas phase, since air does not dissolve appreciably in water, and neglecting the evaporation of water, only the ammonia diffuses. Thus, $N_B = 0$, N_A = constant, and

$$\Psi_A = \frac{N_A}{N_A + N_B} = 1$$

Equation (1-78) then becomes

$$N_A = \frac{D_{AB}P}{RT\delta} \ln\left[\frac{1 - y_{A2}}{1 - y_{A1}}\right] \tag{1-79}$$

Substituting equation (1-9) into equation (1-79), the driving force for mass transfer is expressed in terms of partial pressures:

$$N_A = \frac{D_{AB}P}{RT\delta} \ln\left[\frac{P - p_{A2}}{P - p_{A1}}\right] \qquad (1\text{-}80)$$

Notice that in equation (1-80), as in all the relations derived so far to calculate the molar flux when the bulk motion contribution is significant, the driving force is of a logarithmic nature. The resulting concentration profiles for components A and B are no longer linear. The concentration profile of component A can be derived integrating equation (1-73) from point 1 to any arbitrary position z, substituting the expression for N_A obtained in equation (1-79), and rearranging,

$$y_A(z) = 1 - (1 - y_{A1})\left[\frac{1 - y_{A2}}{1 - y_{A1}}\right]^{\frac{z}{\delta}} \qquad (1\text{-}81)$$

$$y_B(z) = 1 - y_A(z)$$

Figure 1.6 shows schematically the concentration profiles for diffusion of A through stagnant B. Substance A diffuses by virtue of its concentration gradient, $-dy_A/dz$. Substance B is also diffusing relative to the average molar velocity at a flux J_B which depends upon $-dy_B/dz$, but like a fish which swims upstream at the same velocity as the water flows downstream, $N_B = 0$ relative to a fixed place in space.

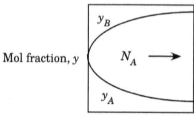

Figure 1.6 Diffusion of A through stagnant B.

It will be shown in Chapter 2 that, for diffusion under turbulent conditions, the flux is usually calculated as the product between an empirical mass-transfer coefficient and a linear driving force. For comparison purposes, we will now manipulate equation (1-80) to rewrite it in terms of a linear driving force. Since

$$P - p_{A2} = p_{B2} \quad P - p_{A1} = p_{B1}$$
$$p_{B2} - p_{B1} = p_{A1} - p_{A2}$$

then

$$N_A = \frac{D_{AB}P}{RT\delta}\left[\frac{p_{A1} - p_{A2}}{p_{B2} - p_{B1}}\right]\ln\left[\frac{p_{B2}}{p_{B1}}\right] \qquad (1\text{-}82)$$

1.4 Steady-State Molecular Diffusion in Fluids

Define the logarithmic mean partial pressure of B, $p_{B,M}$:

$$p_{B,M} = \frac{p_{B2} - p_{B1}}{\ln\left[\dfrac{p_{B2}}{p_{B1}}\right]} \tag{1-83}$$

then

$$N_A = \frac{D_{AD}P}{RT\delta p_{B,M}}\left(p_{A1} - p_{A2}\right) \tag{1-84}$$

Comparing equations (1-72) and (1-84), when all other conditions are equal, the ratio of the molar flux of component A under conditions of stagnant B ($N_B = 0$) to the corresponding value under conditions of equimolar counterdiffusion ($N_B = -N_A$) is given by $P/P_{B,M}$. For very dilute mixtures of component A in B, this ratio approaches a unit value. However, for concentrated mixtures, this ratio can be significantly higher than 1.0.

For illustration purposes, calculate the molar flux of component A for the conditions of Example 1.14, but for diffusion of A through stagnant B.

Solution
Calculate the partial pressures of component B at the ends of the diffusion path: $P_{B2} = 0.8$ bar, $P_{B1} = 0.3$ bar. Substituting in equation (1-83), $P_{B,M} = 0.51$ bar. Therefore, using the result of Example 1.14,

$$N_A = 0.2\,\text{mol}/\text{m}^2 \cdot \text{s} \times \frac{1.0\,\text{bar}}{0.51\,\text{bar}} = 0.39\,\text{mol}/\text{m}^2 \cdot \text{s}$$

Example 1.16 Production of Nickel Carbonyl: Steady-State, One-Dimensional Binary Flux Calculation

Nickel carbonyl (A) is produced by passing carbon monoxide (B) at 323 K and 1 atm over a nickel slab. The following reaction takes place at the solid surface:

$$Ni(s) + 4CO(g) \rightarrow Ni(CO)_4(g)$$

The reaction is very rapid, so that the partial pressure of CO at the metal surface is essentially zero. The gases diffuse through a film 0.625 mm thick. At steady state, estimate the rate of production of nickel carbonyl, in mol/m^2 of solid surface per second. The composition of the bulk gas phase is 50 mol% CO. The binary gas diffusivity under these conditions is $D_{AB} = 20.0$ mm^2/s.

Solution
Figure 1.7 is a schematic diagram of the diffusion process. The stoichiometry of the reaction determines the relation between the fluxes: from equation (1-59),

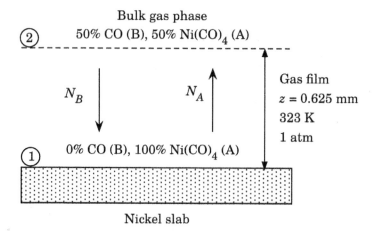

Figure 1.7 Schematic diagram for Example 1.16.

$N_B = -4N_A$, $N_A + N_B = -3N_A$. Calculate the molar flux fraction, Ψ_A. From equation (1-76):

$$\Psi_A = \frac{N_A}{N_A + N_B} = \frac{N_A}{-3N_A} = -\frac{1}{3}$$

Combining this result with equation (1-78)

$$N_A = -\frac{D_{AB}P}{3RT\delta} \ln\left[\frac{-\frac{1}{3} - y_{A2}}{-\frac{1}{3} - y_{A1}}\right]$$

$$= \frac{D_{AB}P}{3RT\delta} \ln\left[\frac{\frac{1}{3} + y_{A1}}{\frac{1}{3} + y_{A2}}\right]$$

Substituting numerical values:

$$N_A = \frac{(2.0 \times 10^{-5})(1.013 \times 10^5)}{3 \times (8.314)(323)(0.625 \times 10^{-3})} \ln\left[\frac{\frac{1}{3} + 1.0}{\frac{1}{3} + 0.5}\right]$$

$$= 0.189 \text{ mol/m}^2 \cdot \text{s}$$

Example 1.17 Multicomponent, Steady-State, Gaseous Diffusion in a Stefan Tube (Taylor and Krishna, 1993)

The Stefan tube, depicted schematically in Figure 1.8, is a simple device sometimes used for measuring diffusion coefficients in binary vapor mixtures. In the bottom of the tube is a pool of quiescent liquid. The vapor that evaporates from this pool

1.4 Steady-State Molecular Diffusion in Fluids

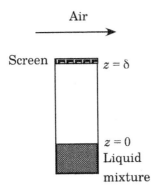

Figure 1.8 Schematic diagram of a Stefan tube.

diffuses to the top of the tube. A stream of gas across the top of the tube keeps the mole fraction of the diffusing vapors there to essentially zero. The composition of the vapor at the vapor–liquid interface is its equilibrium value. Carty and Schrodt (1975) evaporated a binary liquid mixture of acetone (1) and methanol (2) in a Stefan tube. Air (3) was used as the carrier gas. In one of their experiments the composition of the vapor at the liquid interface was $y_1 = 0.319$, $y_2 = 0.528$, and $y_3 = 0.153$. The pressure and temperature in the gas phase were 99.4 kPa and 328.5 K, respectively. The length of the diffusion path was 0.238 m. The MS diffusion coefficients of the three binary pairs are

$$D_{12} = 8.48 \text{ mm}^2/\text{s}$$
$$D_{13} = 13.72 \text{ mm}^2/\text{s}$$
$$D_{23} = 19.91 \text{ mm}^2/\text{s}$$

Calculate the molar fluxes of acetone and methanol, and the composition profiles predicted by the MS equations. Compare your results with the experimental data.

Solution

For a multicomponent mixture of ideal gases such as this, the MS equation (1-35) must be solved simultaneously for the fluxes and the composition profiles. At constant temperature and pressure, the total molar density c and the binary diffusion coefficients are constant. Furthermore, diffusion in the Stefan tube takes place in only one direction, up the tube. Therefore, the continuity equation simplifies to equation (1-69). Air (3) diffuses down the tube as the evaporating mixture diffuses up, but because air does not dissolve in the liquid its flux N_3 is zero (i.e., the diffusion flux J_3 of the gas down the tube is exactly balanced by the diffusion-induced bulk flux $y_3 \times N$ up the tube).

With the above assumptions, the MS relations given by equation (1-35) reduce to a system of two first-order differential equations

$$\frac{dy_1}{dz} = \frac{y_1 N_2 - y_2 N_1}{cD_{12}} + \frac{y_1 N_3 - (1 - y_1 - y_2)N_1}{cD_{13}}$$
$$\frac{dy_2}{dz} = \frac{y_2 N_1 - y_1 N_2}{cD_{12}} + \frac{y_2 N_3 - (1 - y_1 - y_2)N_2}{cD_{23}} \quad (1\text{-}85)$$

42   **Fundamentals of Mass Transfer**

subject to the boundary conditions

$$z = 0, \quad y_i = y_{i0} \quad z = \delta, \quad y_i = y_{i\delta} \quad i = 1,2 \tag{1-86}$$

Define

$$\eta = \frac{z}{\delta} \quad F'_{12} = \frac{cD_{12}}{\delta} \quad F'_{13} = \frac{cD_{13}}{\delta} \quad F'_{23} = \frac{cD_{23}}{\delta}$$

The MS relations then become

$$\frac{dy_1}{d\eta} = \frac{y_1 N_2 - y_2 N_1}{F_{12}} + \frac{y_1 N_3 - \left(1 - y_1 - y_2\right) N_1}{F_{13}}$$

$$\frac{dy_2}{d\eta} = \frac{y_2 N_1 - y_1 N_2}{F_{12}} + \frac{y_2 N_3 - \left(1 - y_1 - y_2\right) N_2}{F_{23}} \tag{1-87}$$

Equation (1-87) involves not only two variable concentrations (y_1 and y_2), but also three unknown fluxes (N_1, N_2, and N_3). From equation (1-68) we obtain three additional constraints:

$$\frac{dN_1}{d\eta} = 0 \quad \frac{dN_2}{d\eta} = 0 \quad \frac{dN_3}{d\eta} = 0 \tag{1-88}$$

Equation (1-87) and (1-88) with boundary conditions given by equation (1-86) constitute a *boundary-value problem*. Notice that there are five first-order ordinary differential equations (ODEs) and only four boundary conditions specified. Therefore, an additional piece of information, the bootstrap condition, must be specified a priori to completely define the problem. Before attempting to solve this set of equations, we will reformulate them in dimensionless form. Define a vector of dimensionless dependent variables, **u**, such that

$$u_1 \equiv y_1 \quad u_2 \equiv y_2 \quad u_3 \equiv \frac{N_1}{F_{12}} \quad u_4 \equiv \frac{N_2}{F_{12}} \quad u_5 \equiv \frac{N_3}{F_{12}} \tag{1-89}$$

Equations (1-87) and (1-88) can be written in terms of the components of vector **u** as

$$\frac{du_1}{d\eta} = u_1 u_4 + \frac{u_1 u_5 - \left(1 - u_1 - u_2\right) u_3}{r_1}$$

$$\frac{du_2}{d\eta} = u_2 u_3 + \frac{u_2 u_5 - \left(1 - u_1 - u_2\right) u_4}{r_2} \tag{1-90}$$

$$\frac{du_3}{d\eta} = 0, \quad \frac{du_4}{d\eta} = 0, \quad \frac{du_5}{d\eta} = 0$$

where $r_1 = F_{13}/F_{12}$ and $r_2 = F_{23}/F_{12}$.

An analytical solution of these equations for Stefan diffusion ($N_3 = 0$) in a ternary mixture is available (Carty and Schrodt, 1975). However—with the objective of presenting a general method of solution that can be used for any number of

1.4 Steady-State Molecular Diffusion in Fluids 43

components, and any relation between the fluxes—we solved these equations numerically using the shooting method (Fausett, 2002). This method reduces the solution of a boundary-value problem to the iterative solution of an *initial-value problem*. The usual approach involves a trial-and-error procedure. That boundary point having the most known conditions is selected as the initial point. Any other missing initial conditions are assumed, and the initial-value problem is solved using a step-by-step procedure, such as the *fourth-order Runge-Kutta method* (Fausett, 2002). Unless the computed solution agrees with the known boundary conditions (unlikely on the first try!), the assumed initial conditions are adjusted, and the problem is solved again. The process is repeated until the assumed initial conditions yield, within specified tolerances, a solution that agrees with the known boundary conditions.

An initial-value problem, consisting of a system of first-order ODEs and the corresponding set of initial conditions, can be expressed compactly in vector notation as

$$\mathbf{u}' = \mathbf{f}(x, \mathbf{u}) \quad \mathbf{u}(a) = \mathbf{u}^{(0)} \quad a \leq x \leq b \tag{1-91}$$

The fourth-order Runge-Kutta method approximates the solution $\mathbf{u}(x)$ of the system of ODEs using n steps of size $h = (b - a)/n$ and grid points located at $x_i = a + h \times i$ according to the following algorithm (Fausett, 2002):

$$\mathbf{k}1 = h\mathbf{f}\left[x_i, \mathbf{u}^{(i)}\right]$$

$$\mathbf{k}2 = h\mathbf{f}\left[x_i + \frac{h}{2}, \mathbf{u}^{(i)} + \frac{\mathbf{k}1}{2}\right]$$

$$\mathbf{k}3 = h\mathbf{f}\left[x_i + \frac{h}{2}, \mathbf{u}^{(i)} + \frac{\mathbf{k}2}{2}\right] \tag{1-92}$$

$$\mathbf{k}4 = h\mathbf{f}\left[x_i + h, \mathbf{u}^{(i)} + \mathbf{k}3\right]$$

$$\mathbf{u}^{(i+1)} = \mathbf{u}^{(i)} + \frac{\mathbf{k}1}{6} + \frac{\mathbf{k}2}{3} + \frac{\mathbf{k}3}{3} + \frac{\mathbf{k}4}{6}$$

The Mathcad intrinsic function *rkfixed* provides the basic approach for solving numerically a system of first-order ODEs implementing the fourth-order Runge-Kutta algorithm. The call to this function has the form

$$rkfixed\left(u^{(0)}, a, b, n, \mathbf{D}\right)$$

where $\mathbf{D}$ is a vector containing the first derivatives of the dependent variables (the right-hand side of the system of ODEs, $\mathbf{u}' = \mathbf{f}(x, \mathbf{u})$). The function *rkfixed* returns the solution in the form of a matrix with $(n + 1)$ rows and $(j + 1)$ columns, where j is the number of dependent variables. The first column contains the values of the independent variable at which the solution is given; the remaining columns contain the corresponding solution values for the dependent variables.

To implement the shooting method in Mathcad, define a function explicitly in terms of the solution matrix returned by the function *rkfixed* and the missing initial conditions (in this case, $u_3^{(0)}$, $u_4^{(0)}$, and $u_5^{(0)}$). After specifying the bootstrap condition (in this case, $u_5^{(0)} = 0$), use the "solve block" feature of Mathcad to solve for the values of the missing initial conditions that yield a solution in agreement

44   **Fundamentals of Mass Transfer**

with the known boundary conditions at the end of the diffusion path. Once convergence is achieved by the "solve block," calculate the fluxes from the corresponding initial conditions. Remember that, according to the equation of continuity (equation 1-69), the fluxes are constant, independent of position; therefore,

$$u_3 = u_3^{(0)} \quad N_1 = u_3^{(0)} F_{12}$$
$$u_4 = u_4^{(0)} \quad N_2 = u_4^{(0)} F_{12} \tag{1-93}$$
$$u_5 = u_5^{(0)} \quad N_3 = u_5^{(0)} F_{12}$$

The concentration profiles are obtained directly from the converged solution.

Figure 1.9 shows the Mathcad implementation of the shooting method for Stefan flow, and compares the results predicted by solving the MS equations to the experimental values obtained by Carty and Schrodt (1975). The predicted fluxes and concentration profiles are in excellent agreement with the observed experimental results. Appendix 1.3 presents a Python solution to this problem.

Example 1.18 Multicomponent, Steady-State, Gaseous Diffusion with Chemical Reaction

Consider the vapor phase dehydrogenation of ethanol (1) to produce acetaldehyde (2) and hydrogen (3):

$$C_2H_6O(g) \rightarrow C_2H_4O(g) + H_2(g)$$

carried out at a temperature of 548 K and 101.3 kPa (Taylor and Krishna, 1993). The reaction takes place at a catalyst surface with a reaction rate that is first order in the ethanol mole fraction, with a rate constant $k_r = 10$ mol/(m^2-s-mol fraction). Reactant and products diffuse through a gas film 1 mm thick. The bulk gas-phase composition is 60 mol% ethanol (1), 20% acetaldehyde (2), and 20% hydrogen (3). Estimate the overall rate of reaction at steady state. The MS diffusion coefficients are

$$D_{12} = 72 \text{ mm}^2/\text{s} \qquad D_{13} = 230 \text{ mm}^2/\text{s} \qquad D_{23} = 230 \text{ mm}^2/\text{s}$$

Solution

Equation (1-85) also applies to this problem; however, there are three additional unknown quantities: N_3, $y_{1\delta}$, and $y_{2\delta}$. The mol fractions at the interface ($z = \delta$) cannot be specified arbitrarily but are fixed by the reaction stoichiometry. Thus, the interface mol fractions must be determined simultaneously with the molar fluxes. The flux of ethanol is determined by the rate of chemical reaction at the catalyst surface. Furthermore, for every mol of ethanol that diffuses to the catalyst surface, one mol of acetaldehyde and one mol of hydrogen diffuse in the opposite direction. That is, the boundary condition and bootstrap relations are, respectively:

$$N_1 = k_r y_{1\delta}$$
$$N_2 = N_3 = -N_1 \tag{1-94}$$

1.4 Steady-State Molecular Diffusion in Fluids

Enter data:

$$\text{ORIGIN} := 1 \qquad P := 0.994 \text{ bar} \qquad T := 328.5 \text{ K} \qquad \delta := 0.238 \text{ m} \qquad num := 100$$

$$R := 8.314 \ \frac{\text{J}}{\text{mol} \cdot \text{K}} \qquad D12 := 8.48 \ \frac{\text{mm}^2}{\text{s}} \qquad D13 := 13.72 \ \frac{\text{mm}^2}{\text{s}} \qquad D23 := 19.91 \ \frac{\text{mm}^2}{\text{s}}$$

Preliminary calculations:

$$c := \frac{P}{R \cdot T} = 36.395 \ \frac{\text{mol}}{\text{m}^3} \qquad F12 := \frac{c \cdot D12}{\delta} = 0.0013 \ \frac{\text{mol}}{\text{m}^2 \cdot \text{s}}$$

$$F13 := \frac{c \cdot D13}{\delta} = 0.0021 \ \frac{\text{mol}}{\text{m}^2 \cdot \text{s}} \qquad r_1 := \frac{F13}{F12} = 1.618$$

$$F23 := \frac{c \cdot D23}{\delta} = 0.00304 \ \frac{\text{mol}}{\text{m}^2 \cdot \text{s}} \qquad r_2 := \frac{F23}{F12} = 2.348$$

Define dimensionless dependent variables:

$$u_1 = y_1 \qquad u_2 = y_2 \qquad u_3 = \frac{N1}{F12} \qquad u_4 = \frac{N2}{F12} \qquad u_5 = \frac{N3}{F12}$$

Fourth-order Runge--Kutta:

$$D(\eta, u) := \begin{bmatrix} \dfrac{u_1 \cdot u_4 - u_2 \cdot u_3}{1} + \dfrac{u_1 \cdot u_5 - \left(1 - u_1 - u_2\right) \cdot u_3}{r_1} \\[2mm] \dfrac{u_2 \cdot u_3 - u_1 \cdot u_4}{1} + \dfrac{u_2 \cdot u_5 - \left(1 - u_1 - u_2\right) \cdot u_4}{r_2} \\[2mm] 0 \\ 0 \\ 0 \end{bmatrix} \qquad \text{Matrix of first derivatives}$$

$$S1(u30, u40, u50) := \text{rkfixed}\left(\begin{bmatrix} 0.319 \\ 0.528 \\ u30 \\ u40 \\ u50 \end{bmatrix}, 0, 1, num, D\right) \qquad \begin{array}{l}\text{Function in terms of unknown}\\ \text{initial values and Runge--Kutta}\\ \text{solution}\end{array}$$

$$u1(u30, u40, u50) := S1(u30, u40, u50)^{\langle 2 \rangle} \qquad \begin{array}{l}\text{The solutions u1 and u2 are the second}\\ \text{and third columns of S1 and depend on}\\ u2(u30, u40, u50) := S1(u30, u40, u50)^{\langle 3 \rangle} \qquad \text{the guess values of u30, u40, and u50.}\end{array}$$

Figure 1.9 Mathcad solution of Example 1.17.

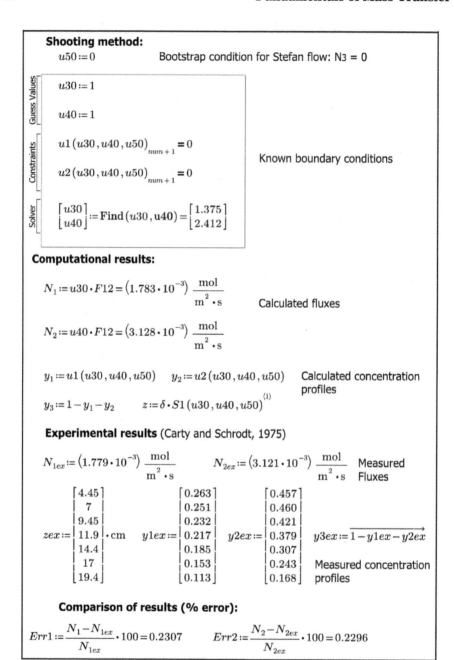

Figure 1.9 Mathcad solution of Example 1.17 (continuation).

1.4 Steady-State Molecular Diffusion in Fluids

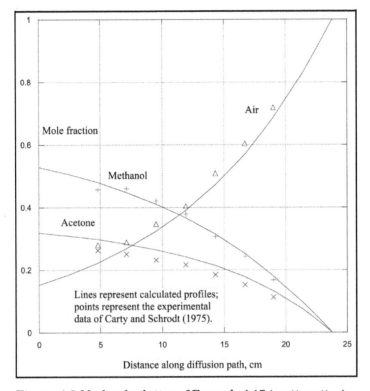

Figure 1.9 Mathcad solution of Example 1.17 (continuation).

Equation (1-94) can be expressed in terms of the components of the dimensionless vector of dependent variables **u**—defined in equation (1-89)—as

$$u_3^{(0)} = \kappa u_{1\delta}$$
$$u_4^{(0)} = u_5^{(0)} = -u_3^{(0)}$$
(1-95)

where $\kappa = k_r/F_{12}$. The shooting method is again implemented using Mathcad to solve the boundary-value problem consisting of equations (1-90) and (1-95). The converged solution predicts an overall rate of reaction $N_1 = 0.888$ mol of ethanol/m² surface area-s, and an interface composition of 8.8 mol% ethanol, 58.3% acetaldehyde, and 32.9% hydrogen. More details of the solution of Example 1.18, including the complete concentration profiles, are provided in Appendix C.

1.4.3 Steady-State Molecular Diffusion in Liquids

Your objectives in studying this section are to be able to:

1. Calculate molar fluxes during steady-state molecular, one-dimensional diffusion in liquids.
2. Calculate molar fluxes for steady-state diffusion of A through stagnant B, and for equimolar counterdiffusion, both in the liquid phase.

48   **Fundamentals of Mass Transfer**

The integration of equation (1-70), to put it in the form of equation (1-77), requires the assumption that and c are constant. This is satisfactory for gas mixtures but not in the case of liquids, where both may vary considerably with concentration. Nevertheless, it is customary to use equation (1-77) to predict molecular diffusion in liquids with an average c and the best average value of D_{AB} available. equation (1-77) is conveniently written for dilute solutions as

$$N_A = \Psi_A \frac{D^0_{AB}}{\delta} \left(\frac{\rho}{M} \right)_{av} \ln \left[\frac{\Psi_A - x_{A2}}{\Psi_A - x_{A1}} \right] \tag{1-96}$$

where ρ and M are the solution density and molecular weight, respectively. As for gases, the value of Ψ_A must be established by the prevailing circumstances. For the most commonly occurring cases, we have, as for gases:

1. Steady-state diffusion of A through stagnant B

For this case, N_A = const, N_B = 0, Ψ_A = 1, then equation (1-96) becomes

$$N_A = \frac{D^0_{AB}}{\delta x_{B,M}} \left(\frac{\rho}{M} \right)_{av} \left(x_{A1} - x_{A2} \right) \tag{1-97}$$

where

$$x_{B,M} = \frac{x_{B2} - x_{B1}}{\ln \left(x_{B2} / x_{B1} \right)} \tag{1-98}$$

2. Steady-state equimolar counterdiffusion

For this case, $N_A = -N_B$ and

$$N_A = \frac{D^0_{AB}}{\delta} \left(\frac{\rho}{M} \right)_{av} \left(x_{A1} - x_{A2} \right) \tag{1-99}$$

Example 1.19 Steady-State Molecular Diffusion in Liquids

A crystal of chalcanthite ($CuSO_4 \cdot 5H_2O$) dissolves in a large tank of pure water at 273 K. Estimate the rate at which the crystal dissolves by calculating the flux of $CuSO_4$ from the crystal surface to the bulk solution. Assume that molecular diffusion occurs through a liquid film uniformly 0.01 mm thick surrounding the crystal. At the inner side of the film—adjacent to the crystal surface—the solution is saturated with $CuSO_4$, while at the outer side of the film the solution is virtually pure water. The solubility of chalcanthite in water at 273 K is 24.3 g of crystal/100 g of water and the density of the corresponding saturated solution is 1140 kg/m^3 (Perry and Chilton, 1973). The diffusivity of $CuSO_4$ in dilute aqueous solution at 273 K can be estimated as 3.6 × 10^{-10} m^2/s (see Problem 1.20). The density of pure liquid water at 273 K is 999.8 kg/m^3.

Solution

The crystal dissolution process may be represented by the equation

$$CuSO_4 \cdot 5H_2O(s) \rightarrow CuSO_4(aq) + 5H_2O(l)$$

1.4 Steady-State Molecular Diffusion in Fluids

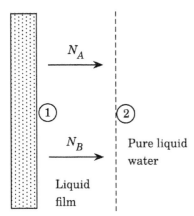

Figure 1.10 Schematic diagram for the crystal dissolution process of Example 1.19.

Accordingly, for each mol of chalcanthite (molecular weight 249.71) that dissolves, one mol of $CuSO_4$ (molecular weight 159.63) and five moles of hydration water diffuse through the liquid film from the surface of the crystal to the bulk of the liquid phase. This is shown schematically in Figure equation (1.10 (substance A is $CuSO_4$ and substance B is water). Notice that both fluxes are in the same direction; therefore, they are both positive and the bootstrap relation is $N_B = 5N_A$. From equation (1-76), $\Psi_A = 1/6 = 0.167$.

Next, we calculate the mol fraction of component A at the inner side of the film (x_{A1}), a saturated solution of chalcanthite in water at 273 K. The solubility of the salt under these conditions is 24.3 g/100 g H_2O:

Basis: 100 g of H_2O (24.3 g of $CuSO_4 \bullet 5H_2O$)

The mass of $CuSO_4$ in 24.3 g of the crystal is $(24.3)(159.63)/(249.71) = 15.53$ g. The mass of hydration water in the crystal is $24.3 - 15.53 = 8.77$ g. Then, the total mass of water is $100 + 8.77 = 108.77$ g. Therefore,

$$x_{A1} = \frac{\frac{15.53}{159.63}}{\frac{15.53}{159.63} + \frac{108.77}{18}} = 0.0158$$

The other end of the film is virtually pure water; therefore,

$$x_{A2} \approx 0$$

Next, we calculate the film average molar density. At point 1, the average molecular weight is $M_1 = (0.0158)(159.63) + (1 - 0.0158)(18) = 20.24$ kg/kmol. The corresponding molar density is $\rho_1/M_1 = 1140/20.24 = 56.32$ kmol/m^3. At point 2, the molar density is $\rho_2/M_2 = 999.8/18 = 55.54$ kmol/m^3. Then

$$\left(\frac{\rho}{M}\right)_{av} = \frac{56.32 + 55.54}{2} = 55.93 \, \text{kmol/m}^3$$

50   **Fundamentals of Mass Transfer**

Substituting in equation (1-96):

$$N_A = \frac{0.167 \times 3.6 \times 10^{-10} \times 55.93 \times \ln\left[\dfrac{0.167 - 0}{0.167 - 0.0158}\right]}{0.01 \times 10^{-3}}$$

$$= 3.342 \times 10^{-5} \frac{\text{kmol}}{\text{m}^2 \cdot \text{s}}$$

The rate of dissolution of the crystal is $(3.342 \times 10^{-5})(249.71)(3600) = 30$ kg/m^2 of crystal surface area per hour.

1.5 STEADY-STATE DIFFUSION IN SOLIDS

Your objectives in studying this section are to be able to:

1. Describe the mechanisms of diffusion through polymeric, crystalline, and porous solids.
2. Calculate the mean free path for gases, and the Knudsen number for gas flow through porous solids.
3. Calculate effective molecular and Knudsen diffusion coefficients in porous solids.
4. Calculate fluxes through porous solids when both molecular and Knudsen diffusion are important.
5. Calculate the hydrodynamic flow of gases through porous solids under a total pressure gradient.
6. Use the "dusty gas" model to calculate multicomponent fluxes in porous solids.

Some of the mass-transfer operations, such as adsorption and membrane separations, involve contact of fluids with solids. In these cases, some of the diffusion occurs in the solid phase and may proceed according to any of several mechanisms. The structure of the solid and its interaction with the diffusing substance determine how diffusion occurs and the rate of mass transport. The solids may be *polymeric, crystalline,* or *porous*.

Thin, dense, nonporous polymeric membranes are widely used to separate gas and liquid mixtures. Diffusion of solutes through certain types of polymeric solids is more like diffusion through liquid solutions than any of the other solid-diffusion phenomena, at least for the permanent gases as solutes. Consider, for example, two portions of a gas at different pressures separated by a polymeric membrane. The gas dissolves in the solid at the faces exposed to the gas to an extent usually directly proportional to the pressure. The dissolved gas then diffuses from the high- to the low-pressure side in a manner describable by Fick's first law.

1.5 Steady-State Diffusion in Solids

Diffusion through nonporous crystalline solids depends markedly on the crystal lattice structure and the diffusing entity. The mechanisms of diffusion in crystalline solids include (Seader and Henley, 2006):

1. Direct exchange of lattice position by two atoms or ions.
2. Migration of small solutes through interlattice spaces called *interstitial sites*.
3. Migration to a vacant site in the lattice.
4. Migration along lattice imperfections or grain boundaries.

Diffusion coefficients associated with the first three mechanisms can vary widely and are almost always at least one order of magnitude smaller than diffusion coefficients in low-viscosity liquids. As might be expected, diffusion by the fourth mechanism can be much faster than by the other three mechanisms.

When solids are porous, it is possible to predict the diffusivity of gaseous and liquid solute species in the pores. In this case, any of the following mass-transfer mechanisms or combinations thereof may take place (Seader and Henley, 2006):

1. Ordinary molecular diffusion through pores, which present tortuous paths and hinder the movement of large molecules when their diameter is more than 10% of the pore diameter.
2. Knudsen diffusion, which involves collisions of diffusing gaseous molecules with the pore walls when the pore diameter and gas pressure are such that the molecular mean free path is large compared to the pore diameter.
3. Surface diffusion involving the jumping of molecules adsorbed on the pore walls from one adsorption site to another based on surface concentration-driving force.
4. Hydrodynamic flow through or into the pores driven by a total pressure gradient.

Most of the solids used for mass-transfer operations are porous. Porous solids are also frequently used in catalytic chemical reactors. Because of their importance, the following sections explore with greater detail the four mass-transfer mechanisms in porous solids mentioned above.

1.5.1 Steady-State Binary Molecular Diffusion in Porous Solids

When treating diffusion of solutes in porous materials where diffusion is considered to occur only in the fluid inside the pores, it is common to refer to an effective diffusivity, $D_{AB, eff}$, which is based on (1) the total cross-sectional area of the porous solid rather than the cross-sectional area of the pore and (2) on a straight path, rather than the pore actual path, which is usually quite tortuous. In a binary system, if pore diffusion occurs only by ordinary molecular diffusion, Fick's law can be used with an effective diffusivity that can be expressed in terms of the ordinary diffusion coefficient, D_{AB}, as

$$D_{AB,eff} = \frac{\varepsilon D_{AB}}{\tau} \tag{1-100}$$

where ε is the fractional porosity (typically 0.5) of the solid and t is the pore-path tortuosity (typically 2 to 4). The *tortuosity* is defined as the ratio of the pore actual length to the length if the pore were straight in the direction of diffusion.

For binary mixtures diffusing inside porous solids, the applicability of equation (1-100) depends upon the value of a dimensionless ratio called the *Knudsen number*, Kn, defined as

$$Kn = \frac{\lambda}{d} \qquad (1\text{-}101)$$

where d is the pore diameter and λ is the mean free path of the diffusing molecules. Equation (1-100) applies when Kn is less than about 0.05 for all the diffusing species (Treybal, 1980).

For liquids, the mean free path is commonly a few angstroms, so the Knudsen number is almost always small and diffusion inside the pores is usually only by ordinary molecular diffusion. In gases, the mean free path can be estimated, as shown in Example 1.20, from the following (Cussler, 1997):

$$\lambda = \frac{\kappa T}{\sqrt{2}\pi\sigma_{AB}^2 P} \qquad (1\text{-}102)$$

where κ is Boltzmann constant and σ_{AB} is the collision diameter of the diffusing species.

Example 1.20 Steady-State Molecular Diffusion in Porous Solid

Estimate the effective diffusivity of hydrogen in ethane in a porous solid with an average pore size of 4000 Å, 40% porosity, and tortuosity of 2.5. The gas mixture is at a pressure of 10 atm and temperature of 373 K. For this system, the ordinary diffusion coefficient is given by $D_{AB} = 0.86/P$ in cm^2/s with total pressure P in atm (Seader and Henley, 2006).

Solution
Calculate the mean free path from equation (1-102). Using data from Appendix B for hydrogen and ethane, and equation (1-45), $\sigma_{AB} = 3.635$ Å. Then, from equation (1-102), $\lambda = 86.54$ Å. From equation (1-101), Kn = 0.022 < 0.05. Therefore, diffusion inside the pores occurs only by ordinary molecular diffusion. At a total pressure of 10 atm, $D_{AB} = 0.086$ cm^2/s. Substituting the given values of porosity and tortuosity in equation (1-100), $D_{AB,eff} = 0.014$ cm^2/s.

1.5.2 Knudsen Diffusion in Porous Solids

When the Knudsen number is large (Kn > 5.0), the molecular mean free path is much larger than the diameter of the channel in which the diffusing molecules reside. When this happens, the molecules bounce from wall to wall rather than colliding with other molecules. Since molecular collisions are unimportant under these

1.5 Steady-State Diffusion in Solids 53

conditions, each gas diffuses independently. This mode of transport is known as *Knudsen diffusion*. From the kinetic theory of gases, in a straight cylindrical pore of diameter d and length l, the Knudsen diffusivity, $D_{K,i}$, is given by (Treybal, 1980)

$$D_{K,i} = \frac{d}{3}\left(\frac{8RT}{\pi M_i}\right)^{0.5} \tag{1-103}$$

and the Knudsen flux is given by

$$N_i = \frac{D_{K,i}\left(p_{i1} - p_{i2}\right)}{RTl} \tag{1-104}$$

For Knudsen diffusion at constant total pressure, it can be shown that there is a bootstrap condition given by (Do, 1998)

$$\sum_{j=1}^{n} N_j \sqrt{M_j} = 0 \tag{1-105}$$

This is known as *Graham's law of effusion* for Knudsen diffusion of a multicomponent system at constant total pressure. For binary gas mixtures, equation (1-105) simplifies to

$$\frac{N_B}{N_A} = -\sqrt{\frac{M_A}{M_B}} \tag{1-106}$$

For Knudsen diffusion in porous solids of porosity ε and tortuosity τ, and

$$D_{K,i,eff} = \frac{\varepsilon D_{K,i}}{\tau} \tag{1-107}$$

and

$$N_i = \frac{D_{K,i,eff}\left(p_{i1} - p_{i2}\right)}{RTl} \tag{1-108}$$

Example 1.21 Knudsen Diffusion in Porous Solid

A mixture of O_2 (A) and N_2 (B) diffuses through the pores of a 2-mm-thick piece of unglazed porcelain at a total pressure of 0.1 atm and a temperature of 293 K. The average pore diameter is 0.1 μm, the porosity is 30.5%, and the tortuosity is 4.39. Estimate the diffusion fluxes of both components when the mole fractions of O_2 are 80 and 20% on either side of the porcelain.

Solution

Calculate the mean free path from equation (1-102). Using data from Appendix B for oxygen and nitrogen, and equation (1-45), σ_{AB} = 3.632 Å. Then, from equation (1-102), λ = 6807 Å. From equation (1-101), Kn = 6.807 > 5.0. Therefore, transport inside

54   **Fundamentals of Mass Transfer**

the pores is mainly by Knudsen diffusion. From equation (1-103), D_{KA} = 0.147 cm^2/s. From equation (1-107), $D_{KA,eff}$ = 0.0102 cm^2/s; from equation (1-108), Na = 1.27 × 10^{-3} mol/m^2-s. From equation (1-106), $N_B = -N_A \times (32/28)^{0.5} = -1.07Na = -1.36 \times 10^{-3}$ mol/m^2 s.

In the range of Kn from roughly 0.02 to 5.0, a transition range, both molecular and Knudsen diffusion have influence, and the flux for binary mixtures is given by (Treybal, 1980)

$$N_A = \Psi_A \frac{PD_{ABeff}}{RTl} \ln \frac{\Psi_A \left(1 + \dfrac{D_{ABeff}}{D_{K,A,eff}}\right) - y_{A2}}{\Psi_A \left(1 + \dfrac{D_{AB,eff}}{D_{K,A,eff}}\right) - y_{A1}} \qquad (1\text{-}109)$$

Further, for open-ended pores, the bootstrap relation given by equation (1-106) for constant total pressure applies throughout the transition range for solids whose pore diameters are of the order of 10 μm or less (Treybal, 1980).

Example 1.22 Combined Molecular and Knudsen Diffusion in Porous Solid

Repeat Example 1.21 for an average pore diameter of 0.3 μm, while the porosity and tortuosity remain unchanged. At the temperature of 293 K and pressure of 0.1 atm, D_{AB} = 2.01 cm^2/s.

Solution
For a pore diameter of 0.3 μm, Kn = 2.27, which means that both molecular and Knudsen diffusion are important, and equation (1-109) must be used to calculate N_A. Since the pore diameter is less than 10 μm, the bootstrap relation of equation (1-106) applies

$$\Psi_A = \frac{N_A}{N_A + N_B} = \frac{1}{1 + N_B / N_A} = \frac{1}{1 - 1.069} = -14.49$$

From equation (1-100),

$$D_{AB,\ eff} = \frac{2.01 \times 0.305}{4.39} = 0.14 \frac{cm^2}{s}$$

From equation (1-103), $D_{K,A}$ = 0.440 cm^2/s; from equation (1-107), $D_{K,A,eff}$ = 0.031 cm^2/s. Then, $D_{AB,eff}/D_{K,A,eff}$ = 0.14/0.031 = 4.57. From equation (1-109):

$$N_A = \frac{-14.49 \left(1.4 \times 10^{-5}\right)(10,130)}{8.314(293)(0.002)} \ln \left[\frac{-14.49(1 + 4.57) - 0.2}{-14.49(1 + 4.57) - 0.8}\right]$$

$$= 3.11 \times 10^{-3} \frac{mol}{m^2 \cdot s}$$

$$N_B = -1.069 N_A = -3.32 \times 10^{-3} \frac{mol}{m^2 \cdot s}$$

1.5 Steady-State Diffusion in Solids
55

Knudsen diffusion is not known for liquids, but important reductions in diffusion rates occur when the molecular size of the diffusing solute, d_m, becomes significant relative to the pore size of the solid as shown in Example 1.23. In that case, the effective diffusivity is given by (Seader and Henley, 2006)

$$D_{AB,eff} = \frac{\varepsilon D_{AB}}{\tau} K_r \qquad (1\text{-}110)$$

where K_r is a restrictive factor that accounts for interfering collisions of the diffusing solute with the pore wall, and is given by

$$K_r = \left[1 - \frac{d_m}{d}\right]^4 \quad \frac{d_m}{d} \leq 1.0 \qquad (1\text{-}111)$$

Example 1.23 Dextrin Diffusion in a Porous Membrane

Beck and Schultz (1972) measured the effective diffusivity of several sugars in aqueous solutions through microporous membranes of mica. One of the sugars studied was β-dextrin. Dextrins are a group of low-molecular-weight carbohydrates produced by the hydrolysis of starch. They have the same general formula as carbohydrates but are of shorter chain length. Dextrins find widespread use in industry, due to their nontoxicity and low price. They are used as water-soluble glues, as thickening agents in food processing, and as binding agent in pharmaceuticals. The membrane used by Beck and Schultz (1972) had an average pore diameter of 88.8 Å, tortuosity of 1.1, and a porosity of 2.33%. The molecular diameter of β-dextrin is 17.96 Å, and its diffusivity in dilute aqueous solution at 298 K is 3.22×10^{-6} cm^2/s. Estimate the effective diffusivity of β-dextrin through this membrane at 298 K.

Solution
Calculate the restrictive factor and effective diffusivity from equations (1-111) and (1-110), respectively:

$$K_r = \left[1 - \frac{17.96}{88.8}\right]^4 = 0.405$$

$$D_{AB,eff} = \frac{0.0233 \left(3.22 \times 10^{-6}\right)(0.405)}{1.1} = 2.78 \times 10^{-8} \, \text{cm}^2 \, / \, \text{s}$$

1.5.3 Hydrodynamic Flow of Gases in Porous Solids

If there is a difference in total pressure across a porous solid, a hydrodynamic flow of gas through the solid will occur. Consider a solid consisting of uniform straight capillary tubes of diameter d and length l reaching from the high- to low-pressure side. In most practical applications, the flow inside the capillaries will

56   Fundamentals of Mass Transfer

be laminar. For a single gas, this can be described by Poiseuille's law for compressible fluids obeying the ideal gas law:

$$N_A = \frac{d^2 P_{av} \left(P_1 - P_2 \right)}{32 \mu RTl}$$
(1-112)

where $P_{av} = (P_1 + P_2)/2$. This assumes that the entire pressure difference is the result of friction in the pores and ignores entrance and exit losses and kinetic-energy effects, which is satisfactory for present purposes. Since usually the pores are neither straight nor of constant diameter, it is best to base N_A on the gross external cross-sectional area of the solid and write equation (1-112) as

$$N_A = \frac{B_0 P_{av} \left(P_1 - P_2 \right)}{\mu RTl}$$
(1-113)

where B_0 is an empirical factor characterizing the porous solid called the *viscous flow parameter* (Do, 1998).

When the gas is a mixture, the hydrodynamic flux given by equation (1-112) or (1-113) is the flux of the mixture, that is, the mixture moves as a whole under the total pressure gradient. All species move at the same speed (that is, no separation is achieved) and the individual flux of each component is the product of its mol fraction times the total flux. In that case, the viscosity in equations (1-112) and (1-113) is the viscosity of the mixture. If the gas is a mixture with different concentrations and different total pressure on either side of the porous solid, the flow may be a combination of hydrodynamic, Knudsen, and diffusive.

Example 1.24 Hydrodynamic Flow in a Porous Diaphragm

A porous carbon diaphragm 25.4 mm thick of average pore diameter 0.01 cm allowed the flow of nitrogen at the rate of 0.05 m^3 (measured at 300 K and 1 atm)/m^2-s with a pressure difference across the diaphragm of 500 Pa. The temperature was 300 K and the downstream pressure 0.1 atm. The viscosity of nitrogen at 300 K is 180 μP.

(a) Estimate the value of the viscous flow parameter, B_0, for this solid.
(b) Calculate the nitrogen flow to be expected at 400 K with the same pressure difference. At 400 K, the viscosity of nitrogen is 220 μP.

Solution
(a) From the information given, P_2 = 10,130 Pa, P_1 = 10,630 Pa, P_{av} = 10,380 Pa. At the average pressure and the temperature of 300 K, the mean free path for nitrogen is estimated from equation (1-102) to be λ = 0.622 μm. For a pore diameter of 0.01 cm, this corresponds to a Knudsen number, Kn = 6.22 × 10^{-3}. Therefore, Knudsen diffusion will not occur, and all the flow observed is of a hydrodynamic nature.

1.5 Steady-State Diffusion in Solids

From the ideal gas law, the nitrogen flux corresponding to the volumetric flow rate of 0.05 m^3/m^2-s at 300 K and 1 atm is 2.031 mol/m^2-s. equation (1-113) is solved for B_0 and numerical values are substituted:

$$B_0 = \frac{1.8 \times 10^{-5}(8.314)(300)(2.031(0.0254))}{(10,380)(500)}$$
$$= 4.462 \times 10^{-10} \, m^2$$

(b) At 393 K, the viscosity of nitrogen is 220 μP. Substituting in equation (1-113) the new values of temperature and viscosity and the value of B_0 obtained in part (a), while maintaining the pressure conditions unchanged, $N_A = 1.269$ mol/m^2-s. This corresponds to a volumetric flow rate of 0.041 m^3 (measured at 300 K and 1 atm)/m^2.s.

1.5.4 "Dusty Gas" Model for Multicomponent Diffusion

Modeling multicomponent gaseous diffusion in porous media depends upon the value of the Knudsen number. For large pores, corresponding to very small values of Kn, the MS equation (1-34) can be used with effective binary diffusivities given by

$$D_{ij,eff} = \frac{\varepsilon D_{ij}}{\tau} \tag{1-114}$$

For very small pores, corresponding to very large values of Kn, Knudsen diffusion prevails, and the flux of each individual component is calculated using equations (1-103) to (1-108). In either case, if there is a significant difference in total pressure across the solid, the hydrodynamic flow contribution must be added.

Modeling multicomponent diffusion in porous media is complicated when the mean free path of the molecules is of the order of magnitude of the pore diameter. The difficulties posed by this intermediate case may be circumvented by a method originally proposed by Maxwell in 1866. He suggested that the porous material be described as a supplementary "dust" species consisting of large molecules that are kept motionless by some unspecified external force. The Chapman-Enskog kinetic theory is then applied to the new pseudo gas mixture, whereby the interaction between the dust and gas molecules simulates the interaction between the solid matrix and the gas species. In this way, one is no longer faced with the problem of flux and composition variations across a pore and problems related to the solid geometry.

The price paid for this simplification is that certain physical features of the porous medium are "lost." One of the more important issues is that the model is no longer dealing with channels of finite size corresponding to the pores in the real medium. Thus, there is no introduction of the hydrodynamic fluxes in the formal development of the equations. These must be added empirically. The working form of the "dusty ideal gas" model equations is (Higler et al., 2000):

$$\frac{P}{RT} \nabla y_i + \frac{y_i}{RT} \left(1 + \frac{B_0 P}{\mu D_{K,i,eff}} \right) \nabla P = \sum_{j=1}^{n} \frac{y_i N_j - y_j N_i}{D_{ij,ff}} - \frac{N_i}{D_{K,i,eff}} \quad i = 1, \ldots, n \tag{1-115}$$

58   **Fundamentals of Mass Transfer**

Notice that equation (1-115) can be written for all the n components, unlike the original MS equations which are valid for only $(n - 1)$ components. However, another variable, the total pressure, has been added. Therefore, a bootstrap relation between the fluxes is still needed to complete the problem formulation.

An explicit expression for the total pressure gradient can be obtained from equation (1-115) by summing with respect to all components:

$$\frac{\nabla P}{RT} = -\frac{\sum_{i=1}^{n} \dfrac{N_i}{D_{K,i,eff}}}{1 + \dfrac{B_0 P}{\mu} \sum_{i=1}^{n} \dfrac{1}{D_{K,i,eff}}} \qquad (1\text{-}116)$$

Equation (1-116) can be substituted into (1-115), written for $(n - 1)$ components, to eliminate the total pressure gradient from those equations. The resulting $(n - 1)$ equations are then augmented with equation (1-116) and the bootstrap condition to solve for the n fluxes and the total pressure gradient. The resulting boundary-value problem can be solved by modifying the Mathcad program developed in Examples 1.17 and 1.18.

1.6 TRANSIENT MOLECULAR DIFFUSION IN SOLIDS

Your objectives in studying this section are to be able to:

1. Write down Fick's second law for transient molecular diffusion in solids.
2. Use the results of the method of separation of variables to obtain the series solution of Fick's second law equation for one-dimensional transient diffusion in solids of different geometries.
3. Integrate the series solution over the whole solid to evaluate the average concentration in the solid as a function of time.

Since solids are not as easily transported through equipment as fluids, application of batch and semibatch processes—and consequently—transient diffusional conditions arise much more frequently than with fluids. For diffusion through solids where there is no bulk flow, and in the absence of chemical reactions, the equation of continuity for component A (equation 1-64) reduces to (assuming constant diffusivity):

$$\frac{\partial c_A}{\partial t} = D_{AB} \nabla^2 c_A \qquad (1\text{-}117)$$

This is known as Fick's second law. For rectangular coordinates, equation (1-117) becomes

1.6 Transient Molecular Diffusion in Solids 59

$$\frac{\partial c_A}{\partial t} = D_{AB}\left[\frac{\partial^2 c_A}{\partial x^2} + \frac{\partial^2 c_A}{\partial y^2} + \frac{\partial^2 c_A}{\partial z^2}\right] \tag{1-118}$$

Fick's second law can be used to solve problems of transient diffusion by integration with appropriate boundary conditions as shown in Examples 1.25 and 1.26.

Example 1.25 Diffusion from a Slab with Sealed Edges

Consider a solid slab with thickness $2l$ in the x direction with sealed edges on four sides so that diffusion can take place only toward and from the exposed flat parallel faces located at $x = -l$ and $x = +l$. The initial concentration of solute A throughout the slab is constant at c_{A0}. The slab is immersed in a medium so that the solute will diffuse out of the slab. Let the solute concentration at the surfaces be c_{Al} invariant with the passage of time. If Fick's second law applies with diffusion only along the x direction, calculate the concentration of A as a function of time and position along the slab.

Solution

Write Fick's second law for one-dimensional transient diffusion:

$$\frac{\partial c_A}{\partial t} = D_{AB}\frac{\partial^2 c_A}{\partial x^2} \tag{1-119}$$

Equation (1-119) must be solved subject to the following initial and boundary conditions:

Initial condition: $t = 0$, $c_A = c_{A0}$
Boundary conditions: $x = \pm l$, $c_A = c_{Al}$

This problem can be solved by applying the method of separation of variables (Crank, 1975). The first step is to write the problem in dimensionless terms. Define a dimensionless concentration, position, and time:

$$\Theta = \frac{c_A - c_{Al}}{c_{A0} - c_{Al}} \qquad \xi = \frac{x}{l} \qquad \tau = \frac{D_{AB}t}{l^2} \tag{1-120}$$

Equation (1-119) can be rewritten in terms of these dimensionless variables as

$$\frac{\partial \Theta}{\partial \tau} = \frac{\partial^2 \Theta}{\partial \xi^2} \tag{1-121}$$

The initial and boundary conditions become

$$\begin{aligned} \tau = 0 \quad &\Theta = 1 \quad \text{initial condition} \\ \xi = \pm 1 \quad &\Theta = 0 \quad \text{boundary conditions} \end{aligned} \tag{1-122}$$

Applying the method of separation of variables to the solution of equation (1-121) subject to the conditions of equation (1-122) gives the series solution:

$$\frac{c_A(x,t) - c_{Al}}{c_{A0} - c_{Al}} = \frac{4}{\pi} \sum_{n=0}^{\infty} \frac{(-1)^n}{2n+1} \exp\left[-\frac{(2n+1)^2 \pi^2 t D_{AB}}{4l^2}\right] \cos\frac{(2n+1)\pi x}{2l} \qquad (1\text{-}123)$$

Even though the solution given by equation (1-123) is an infinite series, it converges rapidly and only 3 or 4 terms are enough to give a very precise estimate of the true solution.

If the diffusion could continue indefinitely, the concentration would fall to the uniform value c_{Al}. On the other hand, if the diffusion from the slab were stopped at time θ, the distribution of solute would be symmetrical with the maximum located at the center of the slab. Eventually, by internal diffusion, the solute concentration would level off to the uniform concentration $c_{A\theta}$. Here $c_{A\theta}$ is the average concentration at time θ obtained by integrating equation (1-123) over the whole slab $(-l \leq x \leq l)$. The quantity $c_{A\theta} - c_{Al}$ is a measure of the amount of solute still unremoved. The fraction unremoved, $E(\theta)$, is given by

$$E(\theta) = \frac{c_{A\theta} - c_{Al}}{c_{A0} - c_{Al}} = \frac{8}{\pi^2} \sum_{n=0}^{\infty} \frac{1}{(2n+1)^2} \exp\left[-\frac{(2n+1)^2 \pi^2 \theta D_{AB}}{4l^2}\right] \qquad (1\text{-}124)$$

It is useful to note that equation (1-119) is of the same form as Fourier's equation for transient unidirectional heat conduction, with molecular rather than thermal diffusivity and concentration rather than temperature. Consequently, the lengthy catalog of solutions to the problems of heat transfer of Carslaw and Jaeger (1959) can be made applicable to diffusion by appropriate substitution. Crank (1975) deals particularly with problems of this sort for diffusion in rectangular, cylindrical, and spherical coordinates and different initial and boundary conditions.

Example 1.26 Diffusion of Urea in an Agar Gel

An agar gel containing a uniform urea concentration of 50 g/L is molded in the form of a 3-cm cube. Two opposite faces of the cube are exposed to a running supply of fresh water unto which urea diffuses. The other faces are protected by the mold. The temperature is 280 K. It was observed that at the end of 16.6 h the average urea concentration in the gel had fallen to 30 g/L. The resistance to diffusion may be considered as residing wholly within the gel. (a) Calculate the diffusivity of urea in the gel at this temperature. (b) Plot the urea concentration in the gel as a function of position on the slab currently. (c) How long would it take for the average concentration to fall to 10 g/L?

Solution

(a) In this case, c_A can be expressed in terms of g/L. The initial uniform concentration is $c_{A0} = 50$ g/L. Since pure water is used as the leaching agent and it is assumed that there is no resistance to diffusion in the liquid phase, $c_{Al} = 0$. Then, at $\theta = 16.6$ h = 59,760 s:

$$E(\theta) = \frac{c_{A\theta} - c_{Al}}{c_{A0} - c_{Al}} = \frac{30 - 0.0}{50 - 0.0} = 0.6$$

1.6 Transient Molecular Diffusion in Solids

From the slab dimensions, $l = 3/2$ cm = 1.5 cm. Using the first three terms ($n = 0, 1, 2$) of the series solution, equation (1-124):

$$0.6 \cong \frac{8}{\pi^2}\left[\exp\left(-\frac{59{,}760 \times \pi^2 \times D_{AB}}{4 \times 1.5^2}\right) + \frac{1}{9}\exp\left(-\frac{9 \times 59{,}760 \times \pi^2 \times D_{AB}}{4 \times 1.5^2}\right) + \right.$$
$$\left. \frac{1}{25}\exp\left(-\frac{25 \times 59{,}760 \times \pi^2 \times D_{AB}}{4 \times 1.5^2}\right)\right]$$

Solving this equation by trial-and-error, $D_{AB} = 4.7 \times 10^{-6}$ cm²/s.

(b) Figure 1.11 shows the concentration profile at 16.6 h, again using only the first three terms in the series, obtained using Mathcad. The maximum concentration,

Example 1-26b

$$c_{A0} := 50 \,\frac{gm}{L} \qquad D_{AB} := 4.7 \cdot 10^{-6} \,\frac{cm^2}{s} \qquad l := 1.5 \; cm$$

$$\theta := 16.6 \; hr \qquad i := 0..150 \qquad x_i := (-1.5 + 0.02 \cdot i) \; cm$$

$$c_{A_i} := (c_{A0}) \cdot \frac{4}{\pi} \cdot \sum_{n=0}^{2} \frac{(-1)^n}{2 \cdot n + 1} \cdot \exp\left(-\frac{(2 \cdot n + 1)^2 \cdot \pi^2 \cdot D_{AB} \cdot \theta}{4 \cdot l^2}\right) \cdot \cos\left(\frac{(2 \cdot n + 1) \cdot \pi \cdot x_i}{2 \cdot l}\right)$$

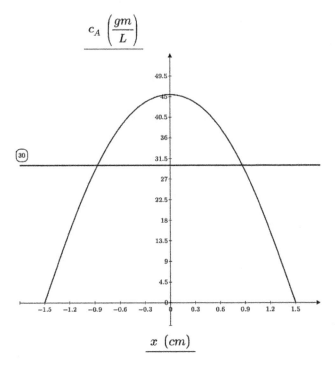

Figure 1.11 Urea concentration profile in the agar gel slab of Example 1.26.

62 — Fundamentals of Mass Transfer

found at the center of the slab, is 45.4 g/L. The horizontal marker shows the current average concentration of 30 g/L. Appendix 1.4 presents a Python solution to this problem.

c) When the average urea concentration is 10 g/L, $E(\theta) = 10/50 = 0.2$. Using the first three terms of equation (1-124) and solving by trial-and-error, $\theta = 74.92$ h.

1.7 DIFFUSION WITH HOMOGENEOUS CHEMICAL REACTION

Your objectives in studying this section are to be able to:

1. Define the concept of homogeneous chemical reaction and learn to incorporate it into the continuity equation.
2. Define the Damkohler number for first-order chemical reaction and ascertain its effect on mass-transfer rates.
3. Derive the Krogh model for oxygen transport in skeletal muscle with Michaelis-Menten chemical reaction kinetics.

In many processes, diffusion rates are altered by homogeneous chemical reactions. A homogeneous reaction is one that takes place in solution. In that case, the R_A term in equation (1-64), representing the volumetric rate of formation of component A by chemical reaction, is different from zero. Examples 1.27 and 1.28 illustrate some of the effects of homogeneous chemical reactions on mass-transfer rates.

Example 1.27 Mass Transfer with First-Order Chemical Reaction

A liquid is in contact with a well-mixed gas containing substance A to be absorbed. Near the surface of the liquid there is a film of thickness δ across which A diffuses steadily while being consumed by a first-order homogeneous chemical reaction with a rate constant k_1. At the gas–liquid interface, the liquid solution is in equilibrium with the gas and its concentration is c_{Ai}; at the other side of the film, its concentration is virtually zero. Assuming dilute solutions, derive an expression for the ratio of the absorption flux with chemical reaction to the corresponding flux without chemical reaction.

Solution

Let us first determine the one-dimensional steady-state absorption flux of component A (that is, at the gas–liquid interface, $z = 0$) without chemical reaction. From the equation of continuity, assuming that the bulk-motion contribution to the flux is negligible for dilute solutions:

$$\frac{d^2 c_A}{dz^2} = 0 \qquad c_A(z) = c_{Ai}\left(1 - \frac{z}{\delta}\right)$$

$$N_A \,|_{z=0} = N_A = -D_{AB}\frac{dc_A}{dz} = \frac{c_{Ai}D_{AB}}{\delta}$$

1.7 Diffusion with Homogeneous Chemical Reaction

For one-dimensional steady-state diffusion including a first-order homogeneous reaction, neglecting the bulk-motion contribution, the equation of continuity becomes

$$\frac{dN_A^1}{dz} = -k_1 c_A \qquad N_A^1 = -D_{AB} \frac{dc_A}{dz}$$

$$\frac{d^2 c_A}{dz^2} - \frac{k_1}{D_{AB}} c_A = 0$$

where $N_A{}^1$ is the flux in the presence of a first-order reaction. The second-order differential equation describing the concentration profile is easily solved substituting in it the trial solution emz, solving the resulting characteristic equation for the values of m, and applying the boundary conditions (Perry and Chilton, 1973). The complete solution is

$$c_A(z) = c_{Ai} \frac{\sinh\left(\sqrt{\frac{k_1}{D_{AB}}}(\delta - z)\right)}{\sinh\left(\sqrt{\frac{k_1}{D_{AB}}}\delta\right)}$$

Evaluate $N_A{}^1$ from the expression for the concentration profile:

$$N_A^1 = -D_{AB} \frac{dc_A}{dz} = c_{Ai} \sqrt{k_1 D_{AB}} \frac{\cosh\left[\sqrt{k_1/D_{AB}}(\delta - z)\right]}{\sinh\left(\sqrt{k_1/D_{AB}}\delta\right)}$$

The absorption flux at the gas–liquid interface ($z = 0$) is, then

$$N_A^1 \big|_{z=0} = c_{Ai} \sqrt{k_1 D_{AB}} \coth\left[\sqrt{k_1/D_{AB}}\delta\right]$$

Therefore, the ratio of absorption flux with chemical reaction to absorption flux without reaction is given by

$$\frac{N_A^1 \big|_{z=0}}{N_A \big|_{z=0}} = \frac{c_{Ai} \sqrt{k_1 D_{AB}} \coth\left[\sqrt{k_1/D_{AB}}\delta\right]}{c_{Ai} D_{AB}/\delta} = \sqrt{\frac{k_1 \delta^2}{D_{AB}}} \coth\left(\sqrt{\frac{k_1 \delta^2}{D_{AB}}}\right)$$

Define the *Damkohler number for first-order reaction*, $Da = k_1{}^2/D_{AB}$. Then,

$$\frac{N_A^1 \big|_{z=0}}{N_A \big|_{z=0}} = \sqrt{Da} \coth\left[\sqrt{Da}\right]$$

This is the desired result. Two limits are instructive. First, when the reaction is slow, Da is small and the series expansion for coth (x) can be truncated to include only the first two terms, giving

$$\frac{N_A^1 \big|_{z=0}}{N_A \big|_{z=0}} = \sqrt{Da}\left[\frac{1}{\sqrt{Da}} + \frac{\sqrt{Da}}{3} - \frac{(\sqrt{Da})^3}{15} + \cdots\right] \approx 1 + \frac{Da}{3}$$

Second, when the chemical reaction is extremely fast,[3] Da is large and, since coth (x) approaches 1.0 for large values of x,

$$\frac{N_A^1 \big|_{z=0}}{N_A \big|_{z=0}} = \sqrt{Da}$$

Example 1.28 Oxygen Transport in Skeletal Muscle: Krogh Model

In 1922, August Krogh proposed that oxygen was transported from capillaries to surrounding tissue by passive diffusion. He formulated a simple geometric model to describe this phenomenon. The Krogh model is a simplification of a capillary bed; a group of parallel identical units, each containing a central capillary and a surrounding cylinder of tissue, as shown in Figure 1.12. The goal of the Krogh model is to predict the concentration of oxygen as a function of position within the tissue cylinder. Each tissue cylinder is assumed to be supplied with oxygen exclusively by the capillary within it. Oxygen diffusion in the axial direction is neglected, as evidenced by the fact that the oxygen concentration gradient is much steeper in the radial direction than in the axial direction.

Applying Fick's law of diffusion in the radial direction ($R_c \leq r \leq R_t$) and conservation of mass lead to the following equation for oxygen diffusion in muscle tissue (McGuire and Secomb, 2001):

$$D_{AB,eff}\left[\frac{1}{r}\frac{d}{dr}\left(r\frac{dc_A}{dr}\right)\right] = R_A(c_A) \quad (1\text{-}125)$$

where c_A is the oxygen concentration in tissue. In studies of oxygen transport, it is customary to express the concentration in tissue in terms of an equivalent oxygen partial pressure, $p_A = Hc_A$, where H is a modified form of Henry's law constant. Then, equation (1-125) can be written in terms of the oxygen partial pressure in tissue as

$$K\left[\frac{1}{r}\frac{d}{dr}\left(r\frac{dp_A}{dr}\right)\right] = M(p_A) \quad R_c \leq r \leq R \quad (1\text{-}126)$$

where K (known as the *Krogh diffusion coefficient*) = $D_{AB,eff}/H$, and $M(p_A)$ is the oxygen consumption rate per unit volume of the tissue cylinder. Oxygen consumption in skeletal muscle is usually assumed to follow *Michaelis-Menten kinetics*, a model that describes the kinetics of many enzymatic reactions, given by

$$M(p_A) = \frac{M_0 p_A}{p_0 + p_A} \quad (1\text{-}127)$$

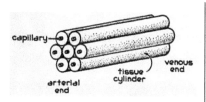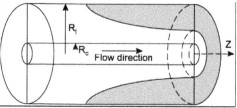

Figure 1.12 Geometry of the Krogh cylinder-type model. Inner cylinder represents the capillary. Outer cylinder corresponds to tissue cylinder. Shaded area: example of hypoxic region under conditions of high demand. R_t, tissue cylinder radius; R_c, capillary radius; z, distance along the capillary.(McGuire and Secomb, 2001. Reproduced with permission of the American Physiological Society.)

1.7 Diffusion with Homogeneous Chemical Reaction 65

where M_0 is the oxygen demand (i.e., the consumption when the oxygen supply is not limiting) and p_0 is the partial pressure of oxygen when consumption is half of the demand. The Michaelis-Menten equation reduces to first-order kinetics for values of p_A much smaller than p_0, and to zero-order kinetics for values of p_A much larger than p_0. For this example, we will assume zero-order kinetics; therefore, $M(p_A) = M_0$.

The two boundary conditions needed to solve equation (1-126) are

1. Neglecting the oxygen-diffusion resistance of the capillary wall, at the capillary–tissue interface, the oxygen partial pressure in the tissue is approximately equal to the average partial pressure of oxygen within the blood, $P_{A\,b}$; that is, $P_A^{(R_c)} = P_{A,b}$.

2. It is assumed that no oxygen is exchanged across the outer boundary of the tissue cylinder, so that

$$\frac{dp_A}{dr}\Big|_{R_t} = 0$$

The complete solution for zero-order kinetics is

$$p_A(r) = p_{A,b} + \frac{M_0}{4K}\left(r^2 - R_c^2\right) - \frac{M_0 R_t^2}{2K}\ln\left(\frac{r}{R_c}\right) \quad R_c \le r \le R_t \quad (1\text{-}128)$$

Equation (1-128) can be used to predict whether, under a given set of parameters, there are *hypoxic regions*—where the partial pressure of oxygen is less than 1 torr—in the tissue cylinder. Table 1.3 summarizes parameter values used in the Krogh model.

To illustrate the use of equation (1-128), we use the parameters of Table 1.3, with an oxygen consumption of 0.4 cm^3 O^2/cm^3-min, and estimate that the lowest oxygen partial pressure in the tissue surrounding the capillary entrance is 49.9 torr at $r = R_t$. Therefore, all the tissue at this position is exposed to a healthy oxygen concentration with no hypoxic regions. Moreover, since p_A is everywhere much higher than p_0, the assumption of zero-order kinetics is validated.

Table 1.3 Parameter Values in Krogh Model

Parameter	Value
Partial pressure of oxygen within the blood at capillary entrance, $p_{A,b}$	100 torr
Krogh diffusion constant in tissue, K	9.4×10^{-10} cm²/s-torr
Oxygen demand, M_0	0–0.80 cm^3 O_2/cm^3-min
Half-maximal oxygen consumption pressure, p_0	1 torr
Capillary radius, R_c	2.5 μm
Tissue cylinder radius, R_t	26 μm
Capillary length, L	500 μm

Source: Data from McGuire and Secomb (2001).

66   **Fundamentals of Mass Transfer**

However, as blood flows along the capillary, oxygen is extracted by the tissue and the oxygen partial pressure in the blood declines rapidly. The oxygen partial pressure in the tissue also declines rapidly and a point is reached where zero-order kinetics no longer applies. The complete form of the Michaelis-Menten equation must be used, requiring a numerical solution of equation (1-126). McGuire and Secomb (2001) showed numerically that, under the set of conditions considered in this example, as much as 37% of the tissue was hypoxic.

1.8 ANALOGIES AMONG MOLECULAR TRANSFER PHENOMENA

Your objectives in studying this section are to be able to:

1. Identify the analogies among molecular momentum, heat, and mass transfer for the simple case of fluid flow past a flat plate.
2. Define and interpret the following dimensionless numbers: Schmidt, Prandtl, and Lewis.

In the flow of a fluid past a phase boundary, there will be a velocity gradient within the fluid which results in a transfer of momentum through it. In some cases, there is also a transfer of heat by virtue of a temperature gradient. The processes of momentum, heat, and mass transfer under these conditions are intimately related, and it is useful to consider at this point the analogies among them.

Consider the velocity profile for the case of a fluid flowing past a stationary flat plate. Since the velocity at the solid surface is zero, there must necessarily be a sublayer adjacent to the surface where the flow is predominantly laminar. Within this region, the shearing stress τ required to maintain the velocity gradient is given by

$$\tau = -\mu \frac{du}{dz} \tag{1-129}$$

where u is the fluid velocity parallel to the surface, and z is measured as increasing toward the surface. This can be written as

$$\tau = -\frac{\mu}{\rho} \frac{d(u\rho)}{dz} = -\nu \frac{d(u\rho)}{dz} \tag{1-130}$$

where ν is the *kinematic viscosity*, μ/ρ, also known as the *momentum diffusivity*.

The kinematic viscosity has the same dimensions as the mass diffusivity, length^2/time, while the quantity u_ρ can be interpreted as a volumetric momentum

1.8 Analogies Among Molecular Transfer Phenomena 67

concentration. The shearing stress τ may also be interpreted as a viscous momentum flux toward the solid surface. equation (1-130) is, therefore, a rate equation analogous to Fick's first law, equation (1-40). The *Schmidt number* is defined as the dimensionless ratio of the momentum and mass diffusivities:

$$\text{Sc} = \frac{\mu}{\rho D_{AB}} \qquad (1\text{-}131)$$

When a temperature gradient exists between the fluid and the plate, the rate of heat transfer in the laminar region is

$$q = -k\frac{dT}{dz} \qquad (1\text{-}132)$$

where k is the thermal conductivity of the fluid. This can also be written as

$$q = -\frac{k}{\rho C_p}\frac{d\left(\rho C_p T\right)}{dz} = -\alpha\frac{d\left(\rho C_p T\right)}{dz} \qquad (1\text{-}133)$$

where C_p is the specific heat at constant pressure. The quantity $TC_p\rho$ may be looked upon as a volumetric thermal concentration, and $\alpha = k/C_p\rho$ is the thermal diffusivity, which, like the momentum and mass diffusivities, has dimensions of length^2/time. equation (1-133) is therefore a rate equation analogous to the corresponding equations for momentum and mass transfer.

The *Prandtl number* is defined as the dimensionless ratio of the momentum and thermal diffusivities:

$$\text{Pr} = \frac{v}{\alpha} = \frac{C\mu}{k} \qquad (1\text{-}134)$$

A third dimensionless group, the *Lewis number*, is formed by dividing the thermal by the mass diffusivity:

$$\text{Le} = \frac{\alpha}{D_{AB}} = \frac{\text{Sc}}{\text{Pr}} \qquad (1\text{-}135)$$

The Lewis number plays an important role in problems of simultaneous heat and mass transfer, such as humidification operations.

An elementary consideration of the processes of momentum, heat, and mass transfer leads to the conclusion that in certain simplified situations there is a direct analogy between them. In general, however, modification of the simple analogy is necessary when mass and momentum transfer occur simultaneously. Nevertheless, even the limited analogies which exist are put to important practical use, as will be seen in Chapter 2.

68   **Fundamentals of Mass Transfer**

PROBLEMS

The problems at the end of each chapter have been grouped into four classes (designated by a superscript after the problem number).

Class a: Illustrates direct numerical application of the formulas in the text.
Class b: Requires elementary analysis of physical situations, based on the subject material in the chapter.
Class c: Requires somewhat more mature analysis.
Class d: Requires computer solution.

1.1[a]. Concentration of a gas mixture

A mixture of noble gases (helium, argon, krypton, and xenon) is at a total pressure of 150 kPa and a temperature of 500 K. If the mixture has equal mole fractions of each of the gases, determine:

(a) The composition of the mixture in terms of mass fractions.

(b) The average molecular weight of the mixture.

(c) The total molar concentration.

(d) The mass density.

Answer: 2.34 kg/m^3

1.2[a]. Concentration of liquid solution fed to a distillation column

A solution of carbon tetrachloride and carbon disulfide containing 50% by weight of each is to be continuously distilled at the rate of 5000 kg/h.

(a) Determine the concentration of the mixture in terms of mole fractions.

(b) Determine the average molecular weight of the mixture.

(c) Calculate the feed rate in kmol/h.

Answer: 49.10 kmol/h

1.3[a]. Concentration of liquefied natural gas

A sample of liquefied natural gas, LNG, from Alaska has the following molar composition: 93.5% CH_4, 4.6% C_2H_6, 1.2% C_3H_8, and 0.7% CO_2. Calculate:

(a) Average molecular weight of the LNG mixture.

(b) Weight fraction of CH_4 in the mixture.

Answer: 87.11%

Problems   **69**

(c) The LNG is heated to 320 K and 140 kPa and vaporizes completely. Estimate the density of the gas mixture under these conditions.

Answer: 0.904 kg/m^3

1.4^b. Concentration of a flue gas

A flue gas consists of carbon dioxide, oxygen, water vapor, and nitrogen. The *molar* fractions of CO_2 and O_2 in a sample of the gas are 12% and 6%, respectively. The *weight* fraction of H_2O in the gas is 6.17%. Estimate the density of this gas at 500 K and 110 kPa.

Answer: 0.772 kg/m^3

1.5^b. Material balances around an ammonia gas absorber

A gas stream flows at the rate of 10.0 m^3/s at 300 K and 102 kPa. It consists of an equimolar mixture of ammonia and air. The gas enters through the bottom of a packed-bed gas absorber, where it flows countercurrent to a stream of pure liquid water that absorbs 95% of all the ammonia, and virtually no air. The absorber vessel is cylindrical with an internal diameter of 2.2 m.

(a) Neglecting the evaporation of water, calculate the ammonia mol fraction in the gas leaving the absorber.

(b) Calculate the outlet gas mass velocity (defined as mass flow rate per unit empty tube cross-sectional area).

Answer: 1.61 kg/m^2-s

1.6^b. Velocities and fluxes in a gas mixture

A gas mixture at a total pressure of 150 kPa and 295 K contains 20% H_2, 40% O_2, and 40% H_2O by volume. The absolute velocities of each species are -10 m/s, -2 m/s, and 12 m/s, respectively, all in the direction of the z-axis.

(a) Determine the mass average velocity, $\mathbf{v}$, and the molar average velocity, $\mathbf{V}$, for the mixture.

(b) Evaluate the four fluxes: j_{O2}, n_{O2}, J_{O2}, $\mathbf{N}_{O2}$.

Answer: $j_{O2} = -3.74$ kg/m^2 s

1.7^b. Properties of air saturated with water vapor

Air, stored in a 30-m^3 container at 340 K and 150 kPa, is saturated with water vapor. Determine the following properties of the gas mixture:

(a) Mol fraction of water vapor.

(b) Average molecular weight of the mixture.

70   **Fundamentals of Mass Transfer**

(c) Total mass contained in the tank.

(d) Mass of water vapor in the tank.

Answer: 5.22 kg

1.8^c. Water balance around an industrial cooling tower

The cooling water flow rate to the condensers of a big coal-fired power plant is 8970 kg/s. The water enters the condensers at 29 °C and leaves at 45 °C. From the condensers, the water flows to a cooling tower, where it is cooled down back to 29 °C by countercurrent contact with air (see Figure 1.13). The air enters the cooling tower at the rate of 6500 kg/s of dry air, at a dry-bulb temperature of 30 °C, pressure of 1 atm, and a humidity of 0.016 kg of water/kg of dry air. It leaves the cooling tower saturated with water vapor at 38 °C.

(a) Calculate the water losses by evaporation in the cooling tower.

(b) To account for water losses in the cooling tower, part of the effluent from a nearby municipal wastewater treatment plant will be used as makeup water. This makeup water contains 500 mg/L of dissolved solids. To avoid fouling of the condenser heat-transfer surfaces, the circulating water is to contain no more than 2000 mg/L of dissolved solids. Therefore, a small amount of the circulating water must be deliberately discarded (blowdown). Windage losses from the tower are estimated at 0.2% of the recirculation rate. Estimate the makeup-water requirement.

Answer: 238 kg/s

1.9^b. Water balance around a soap dryer

It is desired to dry 10 kg/min of soap continuously from 20% moisture by weight to 4% moisture in a countercurrent stream of hot air. The air enters the dryer at the rate of 30.0 m^3/min at 350 K, 101.3 kPa, and initial water-vapor partial pressure of 2 kPa. The dryer operates at constant temperature and pressure.

(a) Calculate the moisture content of the entering air, in kg of water/kg of dry air.

(b) Calculate the flow rate of dry air, in kg/min.

(c) Calculate the water-vapor partial pressure and relative humidity in the air leaving the dryer.

Answer: 24.7% relative humidity

1.10^b. Activated carbon adsorption; material balances

A waste gas contains 0.5% toluene in air and occupies a volume of 2500 m^3 at 298 K and 101.3 kPa. To reduce the toluene content of this gas, it is exposed to 100 kg of activated carbon, initially free of toluene. The system is allowed to reach equilibrium at constant temperature and pressure.

Problems

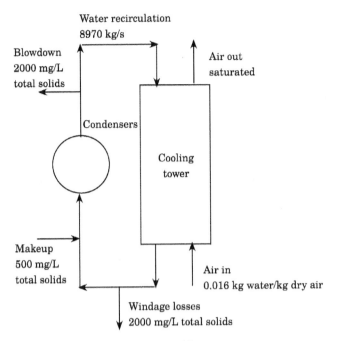

Figure 1.13 Flowsheet for Problem 1.8.

If the air does not adsorb on the carbon, calculate the equilibrium concentration of toluene in the gaseous phase, and the amount of toluene adsorbed by the carbon. The adsorption equilibrium for this system is given by the Freundlich isotherm (USEPA, 1987):

$$W = 0.208 \left(p^*\right)^{0.11}$$

where W is the carbon equilibrium adsorptivity, in kg of toluene/kg of carbon, and p^* is the equilibrium toluene partial pressure, in Pa, and must be between 0.7 and 345 Pa.

1.11[b]. Activated carbon adsorption; material balances

It is desired to adsorb 99.8% of the toluene originally present in the waste gas of Problem 1.10. Estimate how much activated carbon should be used if the system is allowed to reach equilibrium at constant temperature and pressure.

Answer: 225 kg

1.12[a, d]. Estimation of gas diffusivity by the Wilke-Lee equation

Larson (1964) measured the diffusivity of chloroform in air at 298 K and a pressure of 1 atm and reported its value as 0.093 cm²/s. Estimate the diffusion coefficient by the Wilke-Lee equation and compare it with the experimental value.

72 Fundamentals of Mass Transfer

1.13[a, d]. Estimation of gas diffusivity by the Wilke-Lee equation

(a) Estimate the diffusivity of naphthalene ($C_{10}H_8$) in air at 303 K and 1 bar. Compare it with the experimental value of 0.087 cm^2/s reported in Appendix A. The normal boiling point of naphthalene is 491.1 K, and its critical volume is 413 cm^3/mol.

Answer: Error = −20.1%

(b) Estimate the diffusivity of pyridine (C_5H_5N) in hydrogen at 318 K and 1 atm. Compare it with the experimental value of 0.437 cm^2/s reported in Appendix A. The normal boiling point of pyridine is 388.4 K, and its critical volume is 254 cm^3/mol.

Answer: Error = −9.9%

(c) Estimate the diffusivity of aniline (C_6H_7N) in air at 273 K and 1 atm. Compare it with the experimental value of 0.061 cm^2/s (Guilliland, 1934). The normal boiling point of aniline is 457.6 K, and its critical volume is 274 cm^3/mol.

Answer: Error = 15.1%

1.14[d]. Diffusivity of polar gases

If one or both components of a binary gas mixture are polar, a modified Lennard-Jones relation is often used. Brokaw (equation (1969) has suggested an alternative method for this case. 1-49) is used, but the collision integral is now given by

$$\Omega_D = \Omega_D[\text{Eq.}(1-44)] + \frac{0.19\delta_{AB}^2}{T^*} \tag{1-136}$$

where

$$\delta = \frac{1.94 \times 10^{-3} \mu_p^2}{V_b T_b} \tag{1-137}$$

μ_p = dipole moment, debye [1 debye = 3.162 × 10^{-25} (J m^3)$^{1/2}$]

$$\frac{\varepsilon}{k} = 1.18\left(1 + 1.3\delta^2\right)T_b \tag{1-138}$$

$$\sigma = \left[\frac{1.585V_b}{1+1.3\delta^2}\right]^{1/3} \tag{1-139}$$

$$\delta_{AB} = \sqrt{\delta_A\delta_B} \quad \sigma_{AB} = \sqrt{\sigma_A\sigma_B} \quad \frac{\varepsilon_{AB}}{K} = \sqrt{\frac{\varepsilon_A}{K}\frac{\varepsilon_B}{K}} \tag{1-140}$$

Problems

(a) Modify the Mathcad routine of Figure 1.3 to implement Brokaw's method.

(b) Estimate the diffusion coefficient for a mixture of methyl chloride and sulfur dioxide at 1 bar and 323 K, and compare it to the experimental value of 0.078 cm^2/s. The data required to use Brokaw's relation are shown below (Reid et al., 1987):

Parameter	Methyl chloride	Sulfur dioxide
T_b, K	249.1	263.2
V_b, cm^3/mol	50.6	43.8
μ_p, debye	1.9	1.6
M	50.5	64.1

Answer: Error = 7.8%

1.15^d. Diffusivity of polar gases

Evaluate the diffusion coefficient of hydrogen chloride in water at 420 K and 1.2 bar. The data required to use Brokaw's relation (see Problem 1.14) are shown below (Reid et al., 1987):

Parameter	Hydrogen chloride	Water
T_b, K	188.1	373.2
V_b, cm^3/mol	30.6	18.9
μ_ρ, debye	1.1	1.8
M	36.5	18.0

Answer: 0.283 cm^2/s

1.16^d. Diffusivity of polar gases

Evaluate the diffusion coefficient of hydrogen sulfide in sulfur dioxide at a temperature of 298 K and a pressure of 1.5 bar. The data required to use Brokaw's relation (see Problem 1.14) are shown below (Reid et al., 1987):

Parameter	Hydrogen sulfide	Sulfur dioxide
T_b, K	189.6	263.2
V_b, cm^3/mol	35.03	43.8
μ_ρ, debye	0.9	1.6
M	34.08	64.06

Answer: 0.065 cm /s

74 Fundamentals of Mass Transfer

1.17[a,d]. Effective diffusivity in a multicomponent stagnant gas mixture

Calculate the effective diffusivity of nitrogen through a stagnant gas mixture at 400 K and 1.5 bar. The mixture composition is

O_2	15 mol%
CO	30%
CO_2	35%
N_2	20%

Answer: 0.216 cm^2/s

1.18[a,d]. Mercury removal from flue gases by sorbent injection

Mercury is considered for possible regulation in the electric power industry under Title III of the 1990 Clean Air Act Amendments. One promising approach for removing mercury from fossil-fired flue gas involves the direct injection of activated carbon into the gas. Meserole et al. (1999) describe a theoretical model for estimating mercury removal by the sorbent injection process. An important parameter of the model is the effective diffusivity of mercuric chloride vapor traces in the flue gas. If the flue gas is at 1.013 bar and 408 K, and its composition (on a mercuric chloride-free basis) is 6% O_2, 12% CO_2, 7% H_2O, and 75% N_2, estimate the effective diffusivity of mercuric chloride in the flue gas. Assume that only the $HgCl_2$ is adsorbed by the activated carbon. Meserole et al. reported an effective diffusivity value of 0.22 cm^2/s.

Answer: 0.156 cm^2/s

1.19[b]. Diffusion in electrolyte solutions

When a salt dissociates in solution, ions rather than molecules diffuse. In the absence of an electric potential, the diffusion of a single salt may be treated as molecular diffusion. For dilute solutions of a *single salt*, the diffusion coefficient is given by the Nernst-Haskell equation (Harned and Owen, 1950):

$$D_{AB}^0 = \frac{RT\left[\left(1/n_+\right)+\left(1/n_-\right)\right]}{F^2\left[\left(1/\lambda_+^0\right)+\left(1/\lambda_-^0\right)\right]}$$

(1-141)

Where

D_{AB}^0	=	diffusivity at infinite dilute solution, based on molecular concentration, cm^2/s
R	=	gas constant, 8.314 J/mol-K
T	=	temperature, K
λ_+^0, λ_-^0	=	limiting (zero concentration) ionic conductance, (A/cm^2)(V/cm)(g-equiv/cm^3)
n_+, n_-	=	valences of cation and anion, respectively
F	=	Faraday's constant = 96,500 C/g-equiv

Problems   **75**

Values of limiting ionic conductances for some ionic species at 298 K are included in Table 1.4. If conductance values at other temperatures are needed, an approximate correction factor is $T / \left(334 \mu_W \right)$, where μ_W is the viscosity of water at T, in cP.

(a) Estimate the diffusion coefficient at 298 K for a very dilute solution of HCl in water.

Answer: $3.33 \times 10^{-5} \text{ cm}^2/\text{s}$

(b) Estimate the diffusion coefficient at 273 K for a very dilute solution of $CuSO_4$ in water. The viscosity of liquid water at 273 K is 1.79 cP.

Answer: $3.59 \times 10^{-6} \text{ cm}^2/\text{s}$

1.20[a]. Oxygen diffusion in water: Hayduk and Minhas correlation

Estimate the diffusion coefficient of oxygen in liquid water at 298 K. Use the Hayduk and Minhas correlation for solutes in aqueous solutions. At this temperature, the viscosity of water is 0.9 cP. The critical volume of oxygen is 73.4 cm^3/mol. The experimental value of this diffusivity was reported as $2.1 \times 10^{-5} \text{ cm}^2/\text{s}$ (Cussler, 1997).

Answer: Error $= -8.1\%$

1.21[a, d]. Liquid diffusivity: Hayduk and Minhas correlation

Estimate the diffusivity of carbon tetrachloride in a dilute solution in n-hexane at 298 K using the Hayduk and Minhas correlation for nonaqueous solutions. Compare the estimate to the reported value of $3.7 \times 10^{-5} \text{ cm}^2/\text{s}$. The following data are available (Reid et al., 1987):

Table 1.4 Limiting Ionic Conductance in Water at 298

Anion	λ^0_-	Cation	λ^0_+
OH^-	197.6	H^+	349.8
Cl^-	76.3	Li^+	38.7
Br^-	78.3	Na^+	50.1
NO_3^-	71.4	NH_4^+	73.4
HCO_3^-	44.5	Mg^{2+}	53.1
$CH_3CO_2^-$	40.9	Ca^{2+}	59.5
SO_4^{2-}	80.0	Cu^{2+}	54.0
$C_2O_4^{2-}$	74.2	Zn^{2+}	53.0

Units: $(A \times cm^2)/(V \times \text{g-equiv})$
Source: Data from Harned and Owen (1950).

76   **Fundamentals of Mass Transfer**

Parameter	Carbon tetrachloride	n-Hexane
T_b, K	349.9	341.9
T_c, K	556.4	507.5
P_c, bar	45.6	30.1
V_c, cm³/mol	275.9	370
μ, cP	—	0.3
M	153.8	86.2

Answer:Error =8.4%

1.22[b]. Estimating molar volumes from liquid diffusion data

The diffusivity of allyl alcohol (C_3H_6O) in a very dilute aqueous solution at 288 K is 0.9×10^{-5} cm²/s (Reid et al., 1987). Based on this result, and the Hayduk and Minhas correlation for aqueous solutions, estimate the molar volume of allyl alcohol at its normal boiling point. Compare to the result obtained using Table 1.2. The viscosity of water at 288 K is 1.15 cP.

1.23[b, d]. Concentration dependence of binary liquid diffusivities

(a) Estimate the diffusivity of ethanol in water at 298 K when the mol fraction of ethanol in solution is 40%. Under these conditions (Hammond and Stokes, 1953),

$$\left[1 + \frac{\partial \ln \gamma_A}{\partial \ln x_A}\right]\bigg|_{T,P} = 0.355$$

The experimental value reported by Hammond and Stokes (1953) is 0.42×10^{-5} cm²/s. The critical volume of ethanol is 167.1 cm³/mol; the viscosity of water at 298 K is 0.91 cP.

Answer: Error = 8.3%

(b) Estimate the diffusivity of acetone in water at 298 K when the mol fraction of acetone in solution is 65%. For this system at 298 K, the activity coefficient for acetone is given by the Wilson equation (Smith et al., 1996):

$$\ln \gamma_A = -\ln\left(x_A + x_B \Lambda_{AB}\right) + x_B \left[\frac{\Lambda_{AB}}{x_A + x_B \Lambda_{AB}} - \frac{\Lambda_{BA}}{x_B + x_A \Lambda_{BA}}\right]$$

$$\Lambda_{AB} = 0.149 \quad \Lambda_{BA} = 0.355 \quad @298 \text{ K}$$

The critical volume for acetone is 209 cm³/mol.

Answer: 0.73×10^{-5} cm²/s

1.24[b, d]. Steady-state, multicomponent, gas-phase flux calculation

A flat plate of solid carbon is being burned in the presence of pure oxygen according to the reaction

$$2C(s) + \frac{3}{2}O_2(g) \rightarrow CO(g) + CO_2(g)$$

Problems

Molecular diffusion of gaseous reactant and products takes place through a gas film adjacent to the carbon surface; the thickness of this film is 1.0 mm. On the outside of the film, the gas concentration is 40% CO, 20% O_2, and 40% CO_2. The reaction at the surface may be assumed to be instantaneous; therefore, next to the carbon surface, there is virtually no oxygen. The temperature of the gas film is 600 K, and the pressure is 1 bar. Estimate the rate of combustion of the carbon, in kg/m²-min, and the interface concentration.

Answer: 0.241 kg / m² · min

1.25[b]. Steady-state, one-dimensional, liquid-phase flux calculation

A crystal of Glauber's salt $(Na_2SO_4 \cdot 10H_2O)$ dissolves in a large tank of pure water at 288 K. Estimate the rate at which the crystal dissolves by calculating the flux of Na_2SO_4 from the crystal surface to the bulk solution. Assume that molecular diffusion occurs through a liquid film 0.085 mm thick surrounding the crystal. At the inner side of the film—adjacent to the crystal surface—the solution is saturated with Na_2SO_4, while at the outer side of the film the solution is virtually pure water. The solubility of Glauber's salt in water at 288 K is 36 g of crystal/100 g of water and the density of the corresponding saturated solution is 1240 kg/m³ (Perry and Chilton, 1973). The diffusivity of Na_2SO_4 in dilute aqueous solution at 288 K can be estimated as suggested in Problem 1.19. The density of pure liquid water at 288 K is 999.8 kg/m³; the viscosity is 1.153 cP.

Answer: 12.9 kg of crystal / m² · h

1.26[c, d]. Molecular diffusion through a gas–liquid interface

Ammonia, NH_3, is being selectively removed from an air–NH_3 mixture by absorption into water. In this steady-state process, ammonia is transferred by molecular diffusion through a stagnant gas layer 5 mm thick and then through a stagnant water layer 0.1 mm thick. The concentration of ammonia at the outer boundary of the gas layer is 3.42 mol% and the concentration at the lower boundary of the water layer is essentially zero.

The temperature of the system is 288 K, and the total pressure is 1 atm. The diffusivity of ammonia in air under these conditions is 0.215 cm²/s and in liquid water is 1.77×10^{-5} cm² / s. Neglecting water evaporation, determine the rate of diffusion of ammonia, in kg/m²-h. Assume that the gas and liquid are in equilibrium at the interface.

The equilibrium data for ammonia over very dilute aqueous solution at 288 K and 1 atm can be represented by (Welty et al., 1984)

$$y_A = 0.121 + 0.013 \ln x_A$$

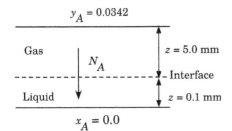

Hint: At steady state, the ammonia rate of diffusion through the gas layer must equal the rate of diffusion through the liquid layer.

Answer: 0.053 kg / m² · h

1.27[c]. Steady-state molecular diffusion in gases

A mixture of ethanol and water vapor is being rectified in an adiabatic distillation column. The alcohol is vaporized and transferred from the liquid to the vapor phase. Water vapor condenses—enough to supply the latent heat of vaporization needed by the alcohol being evaporated—and is transferred from the vapor to the liquid phase. Both components diffuse through a gas film 0.1 mm thick. The temperature is 368 K and the pressure is 1 atm. The mol fraction of ethanol is 0.8 on one side of the film and 0.2 on the other side of the film. Calculate the rate of diffusion of ethanol and of water, in $kg/m^2 \cdot s$. The latent heat of vaporization of the alcohol and water at 368 K can be estimated by the Pitzer acentric factor correlation (Reid et al., 1987)

$$\frac{\Delta H_v}{RT_c} = 7.08\left(1 - T_r\right)^{0.354} + 10.95\omega\left(1 - T_r\right)^{0.456}$$

$$0.5 \leq \left(T_r = T/T_c\right) \leq 1.0$$

where ω is the acentric factor.

Answer: 0.17 kg ethanol / $m^2 \cdot s$

1.28[a, d]. Analogy between molecular heat and mass transfer

It has been observed that for the system air-water vapor at near-ambient conditions, Le = 1.0 (Treybal, 1980). This observation, called the *Lewis relation*, has profound implications in humidification operations, as will be seen later. Based on the Lewis relation, estimate the diffusivity of water vapor in air at 300 K and 1 atm. Compare your result with the value predicted by the Wilke-Lee equation. For air at 300 K and 1 atm: $C_p = 1.01$ kJ/kg·K, $k = 0.0262$ W/m·K, $\mu = 1.846 \times 10^{-5}$ kg/m·s, and $\rho = 1.18\,kg/m^3$.

1.29[b, d]. Steady-state molecular diffusion in gases

Water evaporating from a pond at 298 K does so by molecular diffusion across an air film 1.5 mm thick. If the relative humidity of the air at the outer edge of the film is 20%, and the total pressure is 1 bar, estimate the drop in the water level per day if conditions in the film remain constant. The vapor pressure of water as a function of temperature can be accurately estimated from the Wagner equation (Reid et al., 1987)

$$\ln\left(\frac{P_W}{P_c}\right) = \frac{-7.7645\tau + 1.4584\tau^{1.5} - 2.7758\tau^3 - 1.2330\tau^6}{T_r}$$

$$\tau = 1 - T_r$$

$$T_r = \frac{T}{T_c}$$

$$P_W = \text{water vapor pressure}$$

Answer: 2.81 cm

Problems

1.30[b, d]. Steady-state molecular diffusion in a ternary gas system

Calculate the fluxes and concentration profiles for the ternary system hydrogen (1), nitrogen (2), and carbon dioxide (3) under the following conditions. The temperature is 308 K and the pressure is 1 atm. The diffusion path length is 86 mm. At one end of the diffusion path, the concentration is 20 mol% H_2, 40% N_2, 40% CO_2; at the other end, the concentration is 50% H_2, 20% N_2, 30% CO_2. The total molar flux is zero, $N = 0$. The MS diffusion coefficients are $D_{12} = 83.8$ mm^2 / s, $D_{13} = 68.0$ mm^2 / s, $D_{23} = 16.8$ mm^2 / s.

Answer: $N_1 = -0.0104 \, \text{mol} / \text{m}^2 \cdot \text{s}$

1.31[b]. Membrane area required for dialysis

One of the oldest membrane materials used for dialysis is porous cellophane, a thin transparent sheet made of regenerated cellulose. Typical values of parameters for commercial cellophane membranes are as follows: thickness = 80 m, porosity = 0.45, tortuosity = 5.0, and pore diameter = 40 Å (Seader and Henley, 2006).

(a) Estimate the effective diffusivity of urea through this membrane at 298 K. The diffusivity of urea in dilute aqueous solution at this temperature is 1.38×10^{-5} cm^2 / s. The molecular diameter of urea is 5.28 Å.

(b) The artificial kidney of Example 1.3 uses a cellophane membrane. If the urea concentration difference across the membrane is 25 mg/dL, estimate the membrane area required.

Answer: 6.3 m^2

1.32[a]. Knudsen diffusion in a porous solid

Porous silica gel is used to adsorb propane from helium at 373 K and 1 atm. Typical values of parameters for porous silica gel are as follows: porosity = 0.486, tortuosity = 3.35, and pore diameter = 22 Å.

(a) Calculate the Knudsen number for diffusion inside the pores.

(b) Calculate the effective diffusivity of propane under these conditions.

Answer: 0.045 mm^2/s

1.33[b]. Combined molecular and Knudsen diffusion in a porous solid

It is desired to increase by 50% the oxygen flux in the process described in Example 1.22 by changing the pore size while maintaining all the other conditions unchanged. Calculate the required pore size. Remember to calculate the new value of the Knudsen number to corroborate that you have used the correct set of equations to relate flux to pore size.

80   **Fundamentals of Mass Transfer**

1.34[b]. Flow of pure gases through a porous porcelain plate

An unglazed porcelain plate 5 mm thick has an average pore diameter of 0.2 μm. Pure oxygen gas at an absolute pressure of 20 mm Hg, 400 K, on one side of the plate passed through at a rate of 0.10 cm^3 (at 20 mm Hg, 400 K)/cm^2-s when the pressure on the downstream side was so low as to be considered negligible.

(a) Calculate the Knudsen number for flow inside the pores.

(b) Estimate the rate of passage of hydrogen gas through the plate at 300 K and an absolute pressure of 10 mm Hg, with negligible downstream pressure.

$$\text{Answer: } 1.85 \times 10^{-3} \, \text{mol} / \text{m}^2 \cdot \text{s}$$

1.35[b]. Hydrodynamic flow through a porous diaphragm

Consider the hydrodynamic flow of nitrogen at 400 K through the porous diaphragm described in Example 1.24. It is desired to achieve a superficial gas velocity through the diaphragm of 0.28 kg of N_2/m^2·s by increasing the upstream pressure while maintaining the downstream pressure constant at 0.1 atm. Calculate the upstream pressure required.

$$\text{Answer: } 13.64 \text{ kPa}$$

1.36[b]. Mass transfer with first-order chemical reaction

Oxygen uptake by the blood is faster than oxygen uptake by water because of the reaction of oxygen with hemoglobin. Many chemists have dreamed of inventing a new compound capable of fast, selective reaction with oxygen. Aqueous solutions of this compound could then be used as the basis of a process for oxygen separation from air. Such a compound would complex with oxygen at low temperatures but would give up the oxygen at high temperatures. In an experiment with one potential compound, a researcher observed that the oxygen absorption flux in dilute aqueous solution of this compound at 298 K was 2.5 higher than the absorption flux with pure water at the same temperature, for a film thickness of 0.5 mm.

(a) Calculate the corresponding value of the Damkohler number.

(b) Estimate the reaction rate constant, assuming first-order kinetics.

$$\text{Answer: } 3.06 \text{ min}^{-1}$$

1.37[b]. Krogh model of oxygen transport in muscles

Consider the Krogh cylinder model for oxygen transport in skeletal muscles described in Example 1.28.

(a) At what radial position in the tissue surrounding the capillary entrance is the oxygen partial pressure down to 60.0 torr?

(b) At a position along the capillary where the oxygen partial pressure in the blood

Problems
81

is 25 torr, the lowest oxygen partial pressure in the surrounding tissue is 10.0 torr. Estimate the corresponding value of the oxygen demand, M_0.

(c) For the conditions described in part (b), estimate the oxygen flux leaving the capillary wall.

Answer: 0.109 cm^3 O$_2$/cm^2·h

1.38^d. Krogh model of oxygen transport using Michaelis-Menten kinetics

As discussed in Example equation (1.28, zero-order kinetics is a good approximation to the Michaelis-Menten model if p_A is much larger than p_0. However, when p_A and p_0 are of the same order of magnitude, the complete Michaelis-Menten equation must be used to describe the reaction kinetics. In that case, 1-126) must be solved numerically.

(a) Show that equations (1-126) and (1-127) can be written in dimensionless form as

$$\frac{d^2 u_1}{d\eta^2} + \frac{1}{\eta}\frac{du_1}{d\eta} = \frac{u_1 \text{Da}}{\theta + u_1} \tag{1-142}$$

where

$$u_1 = p_A / p_{A,b} \qquad \eta = r / R_t$$
$$\theta = p_0 / p_{A,b} \qquad \text{Da} = M_0 R_t^2 / K p_{A,b}$$

with boundary conditions

$$(1)\ \eta = R_c / R_t \quad u_1 = 1.0$$
$$(2)\ \eta = 1.0\, du_1 / d\eta = 0$$

(b) Transform the second-order differential equation (1-142) into a system of two simultaneous first-order equations by defining $u_2 = du_1 / d\eta$. The resulting boundary-value problem can be solved numerically by the shooting method using the *rkfixed* intrinsic function of Mathcad as illustrated in Example 1.17.

(c) For the conditions of Example equation (1.28, McGuire and Secomb (2001) showed that at the venous end of the capillary the oxygen partial pressure in blood had dropped to 25 torr. Solve 1-142) numerically and show that for these conditions, hypoxic conditions prevail all the way from a radial position of 20 μm to the end of the tissue cylinder at 26 μM.

1.39^d. Transient diffusion in a solid sphere

Consider the diffusion of urea in an agar gel discussed in Example 1.26. In this case the gel, containing a uniform initial urea concentration of 50 g/L, is molded in the form of a 3-cm diameter sphere. All the sphere surface is exposed to a running supply of fresh water into which the urea diffuses. The temperature is 278 K and the resistance to diffusion is assumed to reside wholly within the gel. The diffusivity of urea in this gel at this temperature was found in Example 1.26 to be 4.7×10^{-6} cm^2/s. How long will it take for the average urea concentration in the gel to be 30 g/L?

Hint: From Crank (1975), for spherical geometry:

$$E(\theta) = \frac{c_{A\theta} - c_{Al}}{c_{A0} - c_{Al}} = \frac{6}{\pi^2} \sum_{n=1}^{\infty} \frac{1}{n^2} \exp\left[-\frac{n^2 \pi^2 \theta D_{AB}}{a^2}\right] \qquad (1\text{-}143)$$

Here, a is the radius of the sphere. This series converges rapidly, and the first three terms are enough to give accurate results.

1.40^d. Transient diffusion in a slab of clay

A slab of clay 50 mm thick was dried from both flat surfaces—with the four thin edges sealed—by exposure to dry air. The initial uniform moisture content was 15%. The drying took place by internal diffusion of the liquid water to the surface, followed by evaporation at the surface. The surface moisture content remained constant at 3%. In 5 h, the average moisture content dropped to 10.2%. The resistance to diffusion is assumed to reside wholly within the solid. The diffusivity is assumed to be constant independent of the water concentration and position.

(a) Calculate the effective diffusivity.

(b) Under the same drying conditions, how long would it take to reduce the average water content to 6%?

1.41^c. Symmetric molecular diffusion from an initial sharp concentration peak

To illustrate the slowness of molecular diffusion, an environmental engineering professor claims that one could place a droplet of water containing 100 µg of n-hexadecane into the middle of a stagnant water-filled pipe (4-cm diameter), and that it would take at least a week until the maximum concentration anywhere in the pipe drops to 1.0 mg/L. How long does it really take? The water temperature is 298 K. At this temperature, the molecular diffusivity of n-hexadecane in water is 4.13×10^{-6} cm^2/s (Yaws, 2009).

Hint: Consider the symmetric propagation by molecular diffusion of a chemical with total mass m'' per unit area into an infinite space ($x = \pm\infty$). At time $t = 0$, this compound is all concentrated at $x = 0$. It can be shown that application of Fick's second law to this case gives the following expression for concentration of the chemical as function of time and position (Scwarzenbach et al., 2017):

$$c_A(x,t) = \frac{m''}{2(\pi D_{AB}t)^{1/2}} \exp\left[-\frac{x^2}{4D_{AB}t}\right] \qquad (1\text{-}144)$$

REFERENCES

Beck, R. E., and J. S. Schultz, *Biochim. Biophys. Acta*, **255**, 273 (1972).

Bird, R. B., W. E. Stewart, and E. N. Lightfoot, *Transport Phenomena*, Wiley, New York (1960).

Brock, J. R., and R. B. Bird, *AIChE J*, **1**, 174 (1955).

Brokaw, E., *Ind. Eng. Chem. Process Design Develop.*, **8**, 240 (1969).

Carslaw, H. S., and J. C. Jaeger, *Conduction of Heat in Solids*, 2nd ed., Oxford University Press, Fair Lawn, NJ (1959).

Carty, R., and T. Schrodt, *Ind. Eng. Chem., Fundam.*, **14**, 276 (1975).

Crank, J., *The Mathematics of Diffusion*, 2nd ed., Oxford University Press, UK (1975).

Cussler, E. L., *Diffusion*, 2nd ed., Cambridge University Press, Cambridge, UK (1997).

Do, D. D., *Adsorption Analysis: Equilibria and Kinetics*, Imperial College Press, London, UK (1998).

Fausett, L., *Numerical Methods Using MathCad*, Prentice Hall, Upper Saddle River, NJ (2002).

Fick, A., *Ann. Physik.*, **94**, 59 (1855).

Guilliland, E. R., *Ind. Eng. Chem.*, **26**, 681 (1934).

Hammond, B. R., and R. H. Stokes, *Trans. Faraday Soc.*, **49**, 890 (1953).

Harned, H. S., and B. B. Owen, The Physical Chemistry of Electrolytic Solutions, *ACS Monogr.* **95** (1950).

Hayduk, W., and B. S. Minhas, *Can. J. Chem. Eng.*, **60**, 295 (1982).

Higler, A., R. Krishna, and R. Taylor, *Ind. Eng. Chem. Res.*, **39**, 1596–1607 (2000).

Hirschfelder, J. O., R. B. Bird, and E. L. Spotz, *Chem. Revs.*, **44**, 205–231 (1949).

Hirschfelder, J. O., C. F. Curtiss, and R. B. Bird, *Molecular Theory of Gases and Liquids*, Wiley, New York (1954).

Larson, E. M., MS thesis, Oregon State University (1964).

Lugg, G. A., *Anal. Chem.*, **40**, 1072 (1968).

McGuire, B. J., and T. W. Secomb, *J. Appl. Physiol.*, **91**, 2255 (2001).

Meserole, F. B. et al., *J. Air & Waste Manage. Assoc.*, **49**, 694 (1999).

Montgomery, S. et al., *Chemical Engineering Fundamentals in Biological Systems*, University of Michigan, Distributed by CACHE Corp., Austin, TX (1998).

Neufield, P. D., A. R. Janzen, and R. A. Aziz, *J. Chem. Phys.*, **57**, 1100 (1972).

Perkins, L. R., and C. J. Geankoplis, *Chem. Eng. Sci.*, **24**, 1035 (1969).

Perry, R. H., and C. H. Chilton (eds.), *Chemical Engineers' Handbook*, 5th ed., McGraw-Hill, New York (1973).

Reid, R. C., J. M. Prausnitz, and B. E. Poling, *The Properties of Gases and Liquids*, 4th ed., McGraw-Hill, Boston (1987).

Scwarzenbach, R. P. et al., *Environmental Organic Chemistry*, 3rd ed., Wiley, New York (2017).

84 **Fundamentals of Mass Transfer**

Seader, J. D., and E. J. Henley, *Separation Process Principles*, 2nd ed., Wiley, New York (2006).

Smith, J. M. et al., *Introduction to Chemical Engineering Thermodynamics*, 5th ed., McGraw-Hill, New York (1996).

Takahashi, M., Y. Kobayashi, and H. Takeuchi, *J. Chem. Eng. Data*, **27**, 328 (1982).

Taylor, R., and R. Krishna, *Multicomponent Mass Transfer*, Wiley, New York (1993).

Treybal, R. E., *Mass-Transfer Operations*, 3rd ed., McGraw-Hill, New York (1980).

Tyn, M. T., and W. F. Calus, *Processing*, **21** (4), 16 (1975).

USEPA, *EAB Control Cost Manual*, 3rd ed., USEPA, Research Triangle Park, NC (1987).

Welty, J. R. et al., *Fundamentals of Momentum, Heat, and Mass Transfer*, 3rd ed., Wiley, New York (1984).

Wilke, C. R., and P. Chang, *AIChE J.*, **1**, 264 (1955).

Wilke, C. R., and C. Y. Lee, *Ind. Eng. Chem.*, **47**, 1253 (1955).

Yaws, C. L., *Transport Properties of Chemicals and Hydrocarbons*, William Andrew Inc., Norwich, NY, 597 (2009).

APPENDIX 1.1

Python routine to estimate gas-phase mass diffusivities using the Wilke-Lee equation. Data are from Example 1.6.

```
In the Wilke-Lee equation, T is temperature in K, P is absolute pressure in bar,
M is molecular weight in g/mol, the Lennard-Jones parameters are from Appendix B
with the units specified there. The data presented here are from Example 1.6.

import math; import numpy as np
# Enter Data
T = 273;  P = 1.0;  MA = 76;  MB = 29
sigmaA = 4.483; sigmaB = 3.620
epsAk = 467; epsBk = 97


sigmaAB = (sigmaA + sigmaB)/2; epsABk = (epsAk*epsBk)**0.5
x = T/epsABk
print('\u03C3AB = %5.3f' %sigmaAB); print('\u03B5ABk = %5.3f' %epsABk)
print('x = %5.3f'%x)


a = 1.06026; b = 0.15610; c = 0.19300; d = 0.47635
e = 1.03587; f = 1.52996; g = 1.76474; h = 3.89411
omega = a/x**b + c/math.exp(d*x)+ e/math.exp(f*x)+g/math.exp(h*x)
print ('\u03A9 = %5.3f' %omega)
MAB = 2*(1/MA + 1/MB)**-1; print ('MAB = %5.3f' %MAB)


D1 = 0.001*(3.03-0.98/MAB**0.5)*T**1.5
D2 = P*sigmaAB**2*omega*MAB**0.5
DAB = D1/D2; print('DAB (cm^2/s) = %5.3f' %DAB)
```

Appendix 1.2

85

The results from the program for the conditions of Example 1.6 are

$$\sigma_{AB} = 4.051 \quad \varepsilon_{ABk} = 21.84\,\text{K} \quad x = T/\varepsilon_{ABk} = 1.283$$
$$\Omega_D = 1.282 \quad M_{AB} = 41.98 \text{ g/mol} \quad D_{AB} = 0.095 \text{ cm}^2/\text{s}$$

To use the routine for other sets of conditions, you need to modify only lines 3, 4, and 5 of the program.

APPENDIX 1.2

Python routine to estimate liquid-phase mass diffusivities using the Hayduk and Minhas equation. Data are from Example 1.9.

```
In the Hayduk and Minhas equation temperatures are in K, pressures in bar, molar volumes in
cc/mol, and viscosity of the solvent in cP.

import numpy
# Enter data:
TbA = 390.4; TcA = 594.8; T = 313; PcA = 57.9; VbA = 124.8; muB = 0.264
TbB = 329.2;  TcB = 508.0;  PcB = 47.0;  VbB = 77.0

#Calculations:

TbrA = TbA/TcA;  TbrB = TbB/TcB
alphacA = 0.9076*(1+ TbrA*numpy.log(PcA/1.013)/(1-TbrA))
alphacB = 0.9076*(1+ TbrB*numpy.log(PcB/1.013)/(1-TbrB))
print('\u03B1cA = %5.3f' %alphacA); print('\u03B1cB = %5.3f' %alphacB)

alphacA = 0.9076*(1+ TbrA*numpy.log(PcA/1.013)/(1-TbrA))
alphacB = 0.9076*(1+ TbrB*numpy.log(PcB/1.013)/(1-TbrB))
print('\u03B1cA = %5.3f' %alphacA); print('\u03B1cB = %5.3f' %alphacB)
sigA = PcA**(2/3)*TcA**(1/3)*(0.132*alphacA - 0.278)*(1-TbrA)**(11/9)
print('\u03C3A = %5.3f' %sigA)

sigB = PcB**(2/3)*TcB**(1/3)*(0.132*alphacB - 0.278)*(1-TbrB)**(11/9)
print('\u03C3B = %5.3f' %sigB)

D1 = 1.55*10**-8*VbB**0.27*T**1.29*sigB**0.125
D2 = VbA**0.42*muB**0.92*sigA**0.105
DL0 = D1/D2
print('DL0 (cm^2/s) = %10.3e' %DL0)
```

The results from the program for the conditions of Example 1.9(b) are

$$\alpha_{CA} = 7.921 \quad \alpha_{CB} = 7.32 \quad \sigma_A = 26.185$$
$$\sigma_B = 19.959 \quad D_{L0} = 3.839 \times 10^{-5}\,\text{cm}^2/\text{s}$$

To use the routine for other sets of conditions, you need to modify only lines 3 and 4 of the program.

APPENDIX 1.3

Python solution for multicomponent diffusion in a Stefan tube. Data are from Example 1.17.

```
#Enter data:
P = 0.994; T = 328.5; delta = 0.238
# Pressure in bar; temperature in K; length of diffusion path in m
R = 8.314e-05; D12 = 8.48e-06; D13 = 13.72e-06; D23 = 19.91e-06
#Ideal gas constant in bar*m^3/mol-K; diffusivities in m^2/s

#Preliminary calculations:
c = P/(R*T); F12 = c*D12/delta; F13 = c*D13/delta; F23 = c*D23/delta
r1 = F13/F12; r2 = F23/F12
import numpy as np
import scipy as sp
from scipy.optimize import fsolve
import matplotlib.pyplot as plt
from scipy.integrate import odeint

#Define dimensionless dependent variables
#u0 = y1; u1 = y2; u2 = N1/F12; u3 = N2/F12; n4 = N3/F12

#Fourth-Order Runge-Kutta Method; Shooting Method
def D(u,t):
    z = np.zeros(5,dtype=float)
    z[0] = u[0]*u[3] - u[1]*u[2] + (u[0]*u[4] - (1 - u[0] - u[1])*u[2])/r1
    z[1] = u[1]*u[2] - u[0]*u[3] + (u[1]*u[4] - (1 - u[0] - u[1])*u[3])/r2
    z[2] = 0
    z[3] = 0
    z[4] = 0
    return[z[0], z[1], z[2], z[3], z[4]]
def r(u0):
    a = u0[0]
    b = u0[1]
    distance = np.arange(0,1.01,0.01)
    z = odeint(D, [0.319, 0.528, a, b, 0], distance)
#Bootstrap condition for Stefan flow: N3=0; u40=0
    return z[100,0], z[100,1]
u0 = np.zeros(2,dtype=float)
u0[0] = 1
u0[1] = 1
u0 = fsolve(r,u0)
N1 = u0[0]*F12; N2 = u0[1]*F12
```

APPENDIX 1.3 (continuation).

```
print('u20 = %5.3f' %u0[0]); print('N1 (mol/m^2-s) = %5.3e' %N1)
print('u30 = %5.3f' %u0[1]); print('N2 (mol/m^2-s) = %5.3e' %N2)


#Experimental Results (Carty and Schrodt, 1975)
N1exp = 1.779e-03; N2exp = 3.121e-03 #Measured experimental fluxes
zexp = [4.45, 7.0, 9.45, 11.9, 14.4, 17.0, 19.4]
#Position along the diffusion path, cm
y1exp = [0.263, 0.251, 0.232, 0.217, 0.185, 0.153, 0.113]
y2exp = [0.457, 0.460, 0.421, 0.379, 0.307, 0.243, 0.168]
y3exp = np.zeros(7,dtype=float)
for i in range (7):
    y3exp[i] = 1 - y1exp[i] -y2exp[i]
    i+=1


#Comparison of Results (%Error in Predicted Fluxes)
Err1 = (N1 - N1exp)*100/N1exp; Err2 = (N2 - N2exp)*100/N2exp
print('The percent error in the predicted N1 flux = %4.2f' %Err1)
print('The percent error in the predicted N2 flux = %4.2f' %Err2)


#Plot the Concentration Profiles
distance = np.arange(0,1.01,0.01)
u = odeint(D, [0.319, 0.528, u0[0], u0[1], 0], distance)
y1 = np.zeros(101,dtype=float)
y2 = np.zeros(101,dtype=float)
y3 = np.zeros(101,dtype=float)
for i in range(101):
    y1[i] = u[i,0]
    y2[i] = u[i,1]
    y3[i] = 1 - y1[i] - y2[i]
    i+=1
x = distance*delta*100 #Calculate distance along diffusion path in cm.
plt.plot(x, y1,'-.k', label = 'Acetone')
plt.plot(x, y2, '--k',label='Methanol')
plt.plot(x, y3, '-k',label='Air')
plt.plot(zexp, y1exp, '*k', label='Acetone experimental')
plt.plot(zexp, y2exp, '+k', label='Methanol experimental')
plt.plot(zexp, y3exp, 'xk', label='Air experimental')
plt.legend()
plt.xlabel('Distance along diffusion path, cm')
plt.ylabel('Molar fraction')
plt.grid(True)
plt.show()
```

The results from the program for the conditions of Example 1.17 are

$$u20 = 1.375 \quad N_1 = 1.783 \times 10^{-3} \, \text{mol/m}^2\text{-s}$$
$$u30 = 2.412 \quad N_2 = 3.128 \times 10^{-3} \, \text{mol/m}^2\text{-s}$$

Error in the predicted N_1 flux = 0.23%
Error in the predicted N_2 flux = 0.23%
Concentration profiles:

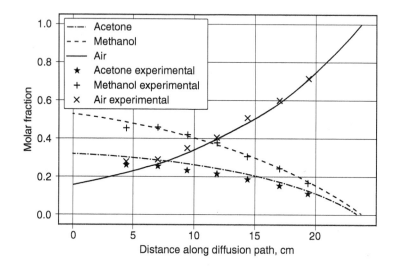

APPENDIX 1.4

Python solution for transient diffusion in solid slab. Data are from Example 1.26.

```python
from math import pi
import numpy as np
import matplotlib.pyplot as plt
from scipy.optimize import fsolve

#(a) Estimate the diffusivity
l = 1.5; theta = 16.6  # Slab half-thickness in cm; time in hours
# Use the first three terms in the series to calculate fraction unremoved, E
def ED(DAB):
    ED1 = np.exp(-DAB*theta*pi**2/(4*l**2))
    ED2 = (1/9)*np.exp(-9*DAB*theta*pi**2/(4*l**2))
    ED3 = (1/25)*np.exp(-25*DAB*theta*pi**2/(4*l**2))
    return (8/pi**2)*(ED1 + ED2 + ED3) - 0.6
# Find the value of diffusivity that will give a value of E = 0.6 in 16.6 hrs
a = fsolve(ED, 0.001) #Initial estimate of diffusivity = 0.001 cm^2/hr
print ('(a) DAB (cm^2/hr) = %5.3f' %a)
b = a/3600 #Convert the diffusivity units
print ('    DAB (cm^2/s) = %5.2e' %b)

#(c) Calculate the time required for E to drop to a value of 0.2

DAB = a
def ET(theta):
    ET1 = np.exp(-DAB*theta*pi**2/(4*l**2))
    ET2 = (1/9)*np.exp(-9*DAB*theta*pi**2/(4*l**2))
    ET3 = (1/25)*np.exp(-25*DAB*theta*pi**2/(4*l**2))
    return (8/pi**2)*(ET1 + ET2 + ET3) - 0.2
c = fsolve(ET, 80)   #Initial estimate of 80 hours
print('(c) Time required (hr) = %5.2f' %c)

#(b) Generate the concentration profile at 16.6 hrs (59760 s)

# Enter data:
cA0 = 50; theta = 59760
DAB = b;
# Concentrations in g/L; position in cm
# time in seconds; diffusivity in cm^2/s
x = np.linspace(-1.5,1.5,150)
def CA(z):
```

```
        CA1 = np.exp(-DAB*theta*pi**2/(4*l**2))*np.cos(pi*z/(2*l))
        CA2 = (1/3)*np.exp(-9*DAB*theta*pi**2/(4*l**2))*np.cos(3*pi*z/(2*l))
        CA3 = (1/5)*np.exp(-25*DAB*theta*pi**2/(4*l**2))*np.cos(5*pi*z/(2*l))
        return (4*cA0/pi)*(CA1-CA2+CA3)
y = CA(x)
x1 = [-1.5 , 0, 1.5]
y1 = [30, 30, 30] #Average concentration
plt.plot(x,y, '-k',label='Local concentration')
plt.plot (x1, y1, '--k', label='Average concentration')
plt.legend()
plt.xlabel('Position along slab, cm')
plt.ylabel('Urea concentration, gm/L')
plt.grid(True)
plt.show()
```

The results from the program for the conditions of Example 1.26 are

(a) D_{AB} = 0.017 cm²/h = 4.73 × 10^{-6} cm²/s

(b) The plot showing the concentration profile is as follows:

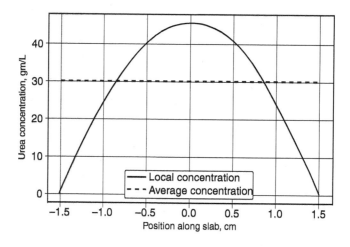

(c) Time θ = 74.92 h

2

Convective
Mass Transfer

2.1 INTRODUCTION

Mass transfer by convection involves the transport of material between a boundary surface and a moving fluid, or between two relatively immiscible moving fluids. This mode of transfer depends both on the transport properties and on the dynamic characteristics of the flowing fluid. When a fluid flows past a solid surface under conditions such that turbulence generally prevails, there is a region close to the surface where the flow is predominantly laminar, and the fluid particles immediately adjacent to the solid boundary are at rest. Molecular diffusion is responsible for mass transfer through the stagnant and laminar flowing fluid layers. The controlling resistance to convective mass transfer is often the result of this "film" of fluid. However, under most convective conditions, this film is extremely thin, and its thickness is virtually impossible to measure or predict theoretically. Therefore, the mass-transfer rates cannot be calculated based on the concepts of molecular diffusion alone.

With increasing distance from the surface, the character of the flow gradually changes, becoming increasingly turbulent, until in the outermost region of the fluid fully turbulent conditions prevail. In the turbulent region, particles of fluid no longer flow in the orderly manner found in the laminar sublayer. Instead, relatively large portions of the fluid, called *eddies*, move rapidly from one position to another with an appreciable component of their velocity in the direction perpendicular to the surface past which the fluid is flowing. These eddies contribute considerably to the mass-transfer process. Because the eddy motion is rapid, mass transfer in the turbulent region is much more rapid than it would be under laminar flow conditions. This situation cannot be modeled in terms of Fick's law. Instead, it is explained in terms of a *mass-transfer coefficient*, an approximate engineering idea that simplifies the analysis of an overly complex problem.

Principles and Applications of Mass Transfer: The Design of Separation Processes for Chemical and Biochemical Engineering, Fourth Edition. Jaime Benitez.
© 2023 John Wiley & Sons, Inc. Published 2023 by John Wiley & Sons, Inc.

92   Convective Mass Transfer

2.2 MASS-TRANSFER COEFFICIENTS

Your objectives in studying this section are to be able to:

1. Explain the concept of a mass-transfer coefficient for turbulent diffusion by analogy with molecular diffusion.
2. Define and use special mass-transfer coefficients for diffusion of A through stagnant B, and for equimolar counterdiffusion.

The mechanism of the flow process involving the movement of the eddies in the turbulent region is not thoroughly understood. On the other hand, the mechanism of molecular diffusion, at least for gases, is well known, since it can be described in terms of a kinetic theory to give results which agree well with experience. It is therefore natural to attempt to describe the rate of mass transfer through the various regions from the surface to the turbulent zone in the same manner found useful for molecular diffusion. Thus, the cD_{AB}/δ of equation (1-77), which is characteristic of molecular diffusion, is replaced by F, a mass-transfer coefficient (Treybal, 1980). Then,

$$N_A = \Psi_A F \ln\left[\frac{\Psi_A - c_{A2}/c}{\Psi_A - c_{A1}/c}\right] \qquad (2\text{-}1)$$

where c_A/c is the mol-fraction concentration, x_A for liquids, y_A for gases. As in molecular diffusion, Ψ_A is ordinarily established by nondiffusional considerations.

Since the surface through which the transfer takes place may not be plane, so that the diffusion path in the fluid may be of variable cross section, N_A is defined as the flux at the phase interface or boundary where substance A enters or leaves the phase for which F has been defined. When c_{A1} is at the beginning of the transfer path and c_{A2} is at the end, N_A is positive. In any case, one of these concentrations will be at the phase boundary. The manner of defining the concentration of A in the fluid will influence the value of F, and it must be clearly established when the mass-transfer coefficient is defined for a particular situation.

The two situations noted in Chapter 1, equimolar counterdiffusion and diffusion of A through stagnant B, occur so frequently that special mass-transfer coefficients are usually defined for them. These are defined by equations of the form

$$\text{flux} = (\text{coefficient}) \times (\text{concentration difference})$$

Since concentration may be defined in several ways and standards have not been established, there are a variety of coefficients that can be defined for each situation.

2.2.1 Diffusion of A Through Stagnant B ($N_B = 0$, $\Psi A = 1$)

In this case, equation (2-1) becomes

$$N_A = F_G \ln\left[\frac{1 - y_{A2}}{1 - y_{A1}}\right] \text{ for gases} \qquad (2\text{-}2)$$

2.2 Mass-Transfer Coefficients

$$N_A = F_L \ln\left[\frac{1-x_{A2}}{1-x_{A1}}\right] \text{ for liquids} \qquad (2\text{-}3)$$

Equations (2-2) and (2-3) can be manipulated algebraically in a manner like that applied to equations (1-80) and (1-82) to yield

$$N_A = \frac{F_G}{y_{B,M}}\left(y_{A1} - y_{A2}\right) = k_y\left(y_{A1} - y_{A2}\right) \qquad (2\text{-}4)$$

$$N_A = \frac{F_L}{x_{B,M}}\left(y_{A1} - y_{A2}\right) = k_x\left(x_{A1} - x_{A2}\right) \qquad (2\text{-}5)$$

where

$$y_{B,M} = \frac{y_{B2} - y_{B1}}{\ln\left[\dfrac{y_{B2}}{y_{B1}}\right]} \quad x_{B,M} = \frac{x_{B2} - x_{B1}}{\ln\left[\dfrac{x_{B2}}{x_{B1}}\right]}$$

$$k_y = \frac{F_G}{y_{B,M}} \qquad k_x = \frac{F_L}{x_{B,M}} \qquad (2\text{-}6)$$

Since concentrations may be defined in several equivalent ways, other definitions of mass-transfer coefficients for this case ($N_B = 0$; dilute solutions) are frequently used, such as

$$N_A = k_G\left(p_{A1} - p_{A2}\right) = k_c\left(c_{A1} - c_{A2}\right) \qquad (2\text{-}7)$$

$$N_A = k_L\left(c_{A1} - c_{A2}\right) \qquad (2\text{-}8)$$

where it is easily shown that

$$F_G = k_G p_{B,M} = k_c \frac{p_{B,M}}{RT} = k_y \frac{p_{B,M}}{P} \qquad (2\text{-}9)$$

$$F_L = k_L x_{B,M} c = k_x x_{B,M} \qquad (2\text{-}10)$$

Notice that in the case of diffusion of A through stagnant B, there is a bulk motion contribution to the total flux that results in a logarithmic form of the concentration driving force. For that reason, the newly defined k-type mass-transfer coefficients are not constant, but usually depend on the concentration driving force. For very dilute solutions, the bulk-motion contribution is negligible, and the driving force becomes approximately linear. In mathematical terms, this follows from the fact that for very dilute solutions $y_{B,M} \cong x_{B,M} \cong 1.0$.

> Therefore, for diffusion of A through stagnant B, equations (2-4), (2-5), (2-7), and (2-8) in terms of the k-type mass-transfer coefficients are recommended only for very dilute solutions, otherwise equations (2-2) and (2-3) should be used.

The relationships between different mass-reansfer coefficients are illustrated in Examples 2.1, 2.2, and 2.3,

Example 2.1 Mass-Transfer Coefficients in a Blood Oxygenator (Cussler, 1997)

Blood oxygenators are used to replace the human lungs during open-heart surgery. To improve oxygenator design, you are studying mass transfer of oxygen into water at 310 K in one specific blood oxygenator. From published correlations of mass-transfer coefficients, you expect that the coefficient based on the oxygen concentration difference in the water is 3.3 × 10^{-5} m/s. Calculate the corresponding mass-transfer coefficient based on the mole fraction of oxygen in the liquid.

Solution

Neglecting water evaporation, in the liquid phase this is a case of diffusion of A (oxygen) through stagnant B (water). Since the solubility of oxygen in water at 310 K is extremely low, we are dealing with dilute solutions. Therefore, the use of k-type coefficients is appropriate. From the information given (coefficient based on the oxygen concentration difference in the water), $k_L = 3.3 \times 10^{-5}$ m/s. You are asked to calculate the coefficient based on the mole fraction of oxygen in the liquid, or k_x. From equation (2-10)

$$k_x = k_L c \tag{2-11}$$

Since we are dealing with very dilute solutions

$$c = \frac{\rho}{M_{av}} \approx \frac{993}{18} = 55.2\,\text{kmol/m}^3$$

Therefore,

$$k_x = 3.3 \times 10^{-5} \times 55.2 = 1.82 \times 10^{-3} \frac{\text{kmol}}{\text{m}^2 \cdot \text{s}}$$

Example 2.2 Mass-Transfer Coefficient in a Gas Absorber

A gas absorber is used to remove ammonia from air by scrubbing the gas mixture with water at 300 K and 1 atm. At a certain point in the absorber, the ammonia mol fraction in the bulk of the gas phase is 0.80, while the interfacial

2.2 Mass-Transfer Coefficients 95

ammonia gas-phase concentration is 0.732. The ammonia flux at that point is measured as 4.3×10^{-4} kmol/m^2·s. Neglecting the evaporation of water, calculate the mass-transfer coefficient in the gas phase at that point in the equipment.

Solution

Neglecting water evaporation and the solubility of air in water, in the gas phase this is a case of diffusion of A (ammonia) through stagnant B (air). However, because of the high ammonia concentration, k-type coefficients should not be used; instead, use equation (2-2) to solve for F_G. In this case, since the ammonia flux is from the bulk of the gas phase to the interface, $y_{A1} = 0.8$, $y_{A2} = 0.732$.

$$F_G = \frac{4.3 \times 10^{-4}}{\ln\left[\left(1 - 0.732\right)/\left(1 - 0.8\right)\right]} = 1.47 \times 10^{-4}\,\frac{\text{kmol}}{\text{m}^2 \cdot \text{s}}$$

2.2.2 Equimolar Counterdiffusion ($N_B = -N_A$, Ψ_A = undefined)

In this case, there is no bulk-motion contribution to the flux, and the flux is related linearly to the concentration difference driving force. Special k'-type mass-transfer coefficients are defined specifically for the case of equimolar counterdiffusion as follows:

$$\begin{aligned} N_A &= k'_G\left(p_{A1} - p_{A2}\right) = k'_c\left(c_{A1} - c_{A2}\right) = k'_y\left(y_{A1} - y_{A2}\right) \\ &= F_G(y_{A1} - y_{A2}) \end{aligned} \tag{2-12}$$

$$\begin{aligned} N_A &= k'_L\left(c_{A1} - c_{A2}\right) = k'_x\left(x_{A1} - x_{A2}\right) \\ &= F_L(x_{A1} - x_{A2}) \end{aligned} \tag{2-13}$$

Equations (2-12) and (2-13) are always valid for equimolar counterdiffusion, regardless of whether the solutions are dilute or concentrated. It is easily shown that the various mass-transfer coefficients defined for this case are related through

$$F_G = k'_G P = k'_c\,\frac{P}{RT} = k'_y \tag{2-14}$$

$$F_L = k'_L c = k'_x \tag{2-15}$$

Example 2.3 Mass-Transfer Coefficient in a Packed-Bed Distillation Column

A packed-bed distillation column is used to adiabatically separate a mixture of methanol and water at a total pressure of 1 atm. Methanol—the more volatile of the two

96 **Convective Mass Transfer**

components—diffuses from the liquid phase toward the vapor phase, while water diffuses in the opposite direction. Assuming that the molar latent heat of vaporization of both components is similar, this process is usually modeled as one of equimolar counterdiffusion. At a point in the column, the mass-transfer coefficient is estimated as 1.62×10^{-5} kmol/m^2·s·kPa. The gas-phase methanol mol fraction at the interface is 0.707, while at the bulk of the gas it is 0.656. Estimate the methanol flux at that point.

Solution

Equimolar counterdiffusion can be assumed in this case (as will be shown in a later chapter, this is the basis of the *McCabe–Thiele method* of analysis of distillation columns). Methanol diffuses from the interface toward the bulk of the gas phase; therefore, $y_{A1} = 0.707$ and $y_{A2} = 0.656$. Since they are not limited to dilute solutions, k'-type mass-transfer coefficients may be used to estimate the methanol flux. From the units given, it can be inferred that the coefficient given in the problem statement is k'_G. From equations (2-14) and (2-12),

$$k'_y = k'_G P = 1.62 \times 10^{-5} \times 101.3 = 1.64 \times 10^{-3} \text{ kmol / m}^2 \cdot \text{s}$$

$$N_A = 1.64 \times 10^{-3} \times (0.707 - 0.656) = 1.36 \times 10^{-5} \text{kmol/m}^2 \cdot \text{s}$$

2.3 DIMENSIONAL ANALYSIS

Your objectives in studying this section are to be able to:

1. Explain the concept and importance of dimensional analysis in correlating experimental data on convective mass-transfer coefficients.
2. Use the Buckingham method to determine the dimensionless groups significant to a given mass-transfer problem.

Most practically useful mass-transfer situations involve turbulent flow, and for these it is generally not possible to compute mass-transfer coefficients from theoretical considerations. Instead, we must rely principally on experimental data. The data are limited in scope, however, with respect to circumstances and situations as well as to range of fluid properties. Therefore, it is important to be able to extend their applicability to conditions not covered experimentally and to draw upon knowledge of other transport processes (of heat, particularly) for help. A particularly useful procedure to this end is *dimensional analysis*.

In dimensional analysis, the significant variables in each situation are grouped into dimensionless parameters which are less numerous than the original variables. Such a procedure is extremely helpful in experimental work in which the very number of significant variables presents an imposing task of correlation. By combining the variables into a smaller number of dimensionless parameters, the work of experimental data reduction is considerably reduced.

2.3 Dimensional Analysis 97

2.3.1 The Buckingham Method

Dimensional analysis predicts the various dimensionless parameters which are helpful in correlating experimental data. Certain dimensions must be established as fundamental, with all others expressible in terms of these. One of these fundamental dimensions is length, symbolized L. Thus, area and volume may dimensionally be expressed as L^2 and L^3, respectively. A second fundamental dimension is time, symbolized t. Velocity and acceleration may be expressed as L/t and L/t^2, respectively. Another fundamental dimension is mass, symbolized M. The mol is included in M. An example of a quantity whose dimensional expression involves mass is the density (mass or molar) which would be expressed as M/L^3.

If the differential equation describing a given situation is known, then dimensional homogeneity requires that each term in the equation have the same units. The ratio of one term in the equation to another must then, of necessity, be dimensionless. With knowledge of the physical meaning of the various terms in the equation we are then able to give some physical interpretation to the dimensionless parameters thus formed. A more general situation in which dimensional analysis may be profitably employed is one in which there is no governing differential equation which clearly applies. In such cases, the *Buckingham method* is used.

The initial step in applying the Buckingham method requires the listing of the variables significant to a given problem. It is then necessary to determine the number of dimensionless parameters into which the variables may be combined. This number may be determined using the *Buckingham pi theorem*, which states (Buckingham, 1914):

> The number of dimensionless groups used to describe a situation, i, involving n variables is equal to $n - r$, where r is the rank of the dimensional matrix of the variables.

Thus,

$$i = n - r \qquad (2\text{-}16)$$

The dimensional matrix is simply the matrix formed by tabulating the exponents of the fundamental dimensions M, L, and t, which appear in each of the variables involved. The rank of a matrix is the number of rows in the largest nonzero determinant which can be formed from it. Examples 2.4 and 2.5 illustratethe evaluation of r and i, as well as the application of the Buckingham method.

Example 2.4 Mass Transfer into a Dilute Stream Flowing under Forced Convection in a Circular Conduit ($N_B = 0$)

Consider the transfer of mass from the walls of a circular conduit to a dilute stream flowing through the conduit. The transfer of A through stagnant B is a result of the

98 **Convective Mass Transfer**

concentration driving force, $(c_{A1} - c_{A2})$. Use the Buckingham method to determine the dimensionless groups formed from the variables significant to this problem.

Solution

The first step is to construct a table of the significant variables in the problem and their dimensions. For this case, the important variables, their symbols, and their dimensional representations are listed below:

Variable	Symbol	Dimensions
Tube diameter	D	L
Fluid density	ρ	M/L^3
Fluid viscosity	μ	M/Lt
Fluid velocity	v	L/t
Mass diffusivity	D_{AB}	L^2/t
Mass-transfer coefficient	k_c	L/t

The above variables include terms descriptive of the system geometry, flow, fluid properties, and the quantity, which is of primary interest, k_c. To determine the number of dimensionless parameters to be formed, we must know the rank, r, of the dimensional matrix. The matrix is formed from the following tabulation:

	k_c	v	ρ	μ	D_{AB}	D
M	0	0	1	1	0	0
L	1	1	-3	-1	2	1
t	-1	-1	0	-1	-1	0

The numbers in the table represent the exponent of M, L, and t in the dimensional expression of each of the six variables involved. For example, the dimensional expression of μ is M/Lt; hence, the exponents 1, -1, and -1 are tabulated versus M, L, and t, respectively, the dimensions with which they are associated. The dimensional matrix, **DM**, is then the array of numbers

$$\text{DM} = \begin{bmatrix} 0 & 0 & 1 & 1 & 0 & 0 \\ 1 & 1 & -3 & -1 & 2 & 1 \\ -1 & -1 & 0 & -1 & -1 & 0 \end{bmatrix}$$

The rank of the matrix is easily obtained using the *rank* (**A**) function of Mathcad. Therefore, $r = rank$ (**DM**) = 3. From equation (2-16), $i = 6 - 3 = 3$, which means that there will be three dimensionless groups.

The three dimensionless parameters will be symbolized π_1, π_2, and π_3 and may be formed in several different ways. Initially, a *core group* of r variables must be chosen which will appear in each of the pi groups and, among them, contain all the fundamental dimensions. One way to choose a core is to exclude from it those variables whose effect one wishes to isolate. In the present problem, it would be desirable to have the mass-transfer coefficient in only one dimensionless group; hence, it will not be in the core. Let

2.3 Dimensional Analysis

us arbitrarily exclude the fluid velocity and viscosity from the core. The core group now consists of D_{AB}, D, and ρ, which include M, L, and t among them.

We now know that all π_1, π_2, and π_3 contain D_{AB}, D, and ρ, that one of them includes k_c, one includes μ, and the other includes v; and that all must be dimensionless. For each to be dimensionless, the variables must be raised to certain exponents. Therefore,

$$\pi_1 = D_{AB}^a \rho^b D^c k_c$$
$$\pi_2 = D_{AB}^d \rho^e D^f v$$
$$\pi_3 = D_{AB}^g \rho^h D^i \mu$$

Writing π_1 in dimensional form,

$$M^0 L^0 t^0 = \left[\frac{L^2}{t}\right]^a \left[\frac{M}{L^3}\right]^b L^c \left[\frac{L}{t}\right]$$

Equating the exponents of the fundamental dimensions on both sides of the equation, we have for

$$L: \quad 0 = 2a - 3b + c + 1$$
$$t: \quad 0 = -a - 1$$
$$M: \quad 0 = b$$

The solution of these equations for the three unknown exponents is $a = -1$, $b = 0$, $c = 1$, thus

$$\pi_1 = \frac{k_c D}{D_{AB}} = \text{Sh}$$

where Sh represents the *Sherwood number*, the mass-transfer analog to the Nusselt number of heat transfer. The other two pi groups are determined in the same manner, yielding

$$\pi_2 = \frac{vD}{D_{AB}} = \text{Pe}_D$$

where Pe_D represents the *Peclet number for mass transfer*, analogous to Pe_H—the Peclet number for heat transfer—and

$$\pi_3 = \frac{\mu}{\rho D_{AB}} = \text{Sc}$$

where Sc represents the Schmidt number, already defined in Chapter 1. Dividing π_2 by π_3, we obtain

$$\frac{\pi_2}{\pi_3} = \frac{\dfrac{Dv}{D_{AB}}}{\dfrac{\mu}{\rho D_{AB}}} = \frac{Dv\rho}{\mu} = \text{Re}$$

100 Convective Mass Transfer

the Reynolds number. The result of the dimensional analysis of forced-convection mass transfer in a circular conduit indicates that a correlating relation could be of the form

$$\mathrm{Sh} = f(\mathrm{Re}, \mathrm{Sc}) \tag{2-17}$$

which is analogous to the heat-transfer correlation

$$\mathrm{Nu} = f(\mathrm{Re}, \mathrm{Pr}) \tag{2-18}$$

Example 2.5 Mass Transfer with Natural Convection

Natural convection currents will develop if there exists a significant variation in density within a liquid or gas phase. The density variations may be due to temperature differences or to relatively large concentration differences. Consider natural convection involving mass transfer from a vertical plane wall to an adjacent fluid. Use the Buckingham method to determine the dimensionless groups formed from the variables significant to this problem.

Solution

The important variables in this case, their symbols, and dimensional representations are listed below.

Variable	Symbol	Dimensions
Characteristic length	L	L
Fluid density	ρ	M/L^3
Fluid viscosity	μ	M/Lt
Buoyant force	$g\Delta\rho_A$	M/L^2t^2
Mass diffusivity	D_{AB}	L^2/t
Mass-transfer coefficient	F	M/L^2t

By the Buckingham theorem, there will be three dimensionless groups. With D_{AB}, L, and ρ as the core variables, the three pi groups to be formed are

$$\pi_1 = D_{AB}^a L^b \rho^c F$$
$$\pi_2 = D_{AB}^d L^e \rho^f \mu$$
$$\pi_3 = D_{AB}^g L^h \rho^i g\Delta\rho_A$$

Solving for the three pi groups, we obtain

$$\pi_1 = \frac{FL}{\rho D_{AB}} = \mathrm{Sh}$$

$$\pi_2 = \frac{\mu}{\rho D_{AB}} = \mathrm{Sc}$$

2.4 Flow Past Flat Plate in Laminar Flow; Boundary Layer Theory   101

and

$$\pi_3 = \frac{g\Delta\rho_A L^3}{\rho D_{AB}^2}$$

Dividing π_3 by the square of π_2, we obtain

$$\frac{\pi_3}{\pi_2^2} = \frac{\dfrac{g\Delta\rho_A L^3}{\rho D_{AB}^2}}{\left[\dfrac{\mu}{\rho D_{AB}}\right]^2} = \frac{g\rho\Delta\rho_A L^3}{\mu^2} = \mathrm{Gr}_D$$

the *Grashof number for mass transfer*, analogous to Gr_H—the Grashof number in natural-convection heat transfer. The result of the dimensional analysis of natural-convection mass transfer indicates that a correlating relation could be of the form

$$\mathrm{Sh} = f\left(\mathrm{Gr}_D, \mathrm{Sc}\right) \tag{2-19}$$

which is analogous to the heat-transfer correlation

$$\mathrm{Nu} = f\left(\mathrm{Gr}_H, \mathrm{Pr}\right) \tag{2-20}$$

For both forced and natural convection, relations have been obtained by dimensional analysis which suggest that a correlation of experimental data may be in terms of three variables instead of the original six. This reduction in variables has aided investigators who have developed correlations for estimating convective mass-transfer coefficients in a variety of situations.

2.4 FLOW PAST FLAT PLATE IN LAMINAR FLOW; BOUNDARY LAYER THEORY

Your objectives in studying this section are to be able to:

1. Define the concepts of three types of boundary layers: velocity, thermal, and concentration.
2. Formulate the equations of continuity, momentum, energy, and continuity of component A for the corresponding boundary layers over a flat plate in laminar flow.
3. Solve the Blasius problem using the shooting method in Mathcad.
4. Derive expressions for the local and average Sherwood numbers for mass transfer over a flat plate in laminar flow.

In a few limited situations, mass-transfer coefficients can be deduced from theoretical principles. One important case in which an analytical solution of the equations of momentum transfer, heat transfer, and mass transfer has been achieved is that for the laminar boundary layer on a flat plate in steady flow.

The observation of a decreasing region of influence of shear stress as the Reynolds number is increased led Ludwig Prandtl to the boundary layer concept in 1904. According to Pandtl's hypothesis, the effects of fluid friction at high Reynolds number are limited to a thin layer near the boundary of a body, hence the term *boundary layer*. Accordingly, the fluid flow is characterized by two distinct regions, a thin fluid layer in which velocity gradients and shear stresses are large, and a region outside the boundary layer in which velocity gradients and shear stresses are negligible. With increasing distance, the effects of viscosity penetrate further into the free stream and the boundary layer grows.

Near the leading edge, flow within the boundary layer is laminar, and this is designated as the laminar boundary-layer region. As the boundary layer grows, eventually a transition zone is reached across which a conversion from laminar to turbulent conditions occurs. In the region in which the boundary layer is always turbulent, there exists a very thin film of fluid, called the *laminar sublayer*, wherein flow is still laminar and large velocity gradients exist.

Because it pertains to the fluid velocity, the boundary layer described above may be referred to more specifically as the *velocity boundary layer*. Just as a velocity boundary layer develops when there is fluid flow over a surface, a *thermal boundary layer* must develop if the fluid free stream and surface temperatures differ. Similarly, a *concentration boundary layer* must develop if the fluid free stream and surface concentrations differ. Figure 2.1 illustrates how the thickness of the concentration boundary layer increases with distance x from the leading edge.

In laminar incompressible flow with low mass-transfer rates and constant physical properties past a flat plate, the equations to be solved for a binary mixture are the following (Welty et al., 1984):

$$\frac{\partial u_x}{\partial x} + \frac{\partial u_y}{\partial y} = 0 \quad \text{continuity} \quad (2\text{-}21)$$

$$u_x \frac{\partial u_x}{\partial x} + u_y \frac{\partial u_x}{\partial y} = \nu \frac{\partial^2 u_x}{\partial y^2} \quad \text{momentum} \quad (2\text{-}22)$$

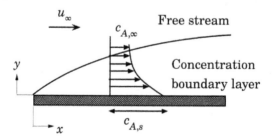

Figure 2.1 Concentration boundary layer over a flat plate.

2.4 Flow Past Flat Plate in Laminar Flow; Boundary Layer Theory 103

$$u_x \frac{\partial T}{\partial x} + u_y \frac{\partial T}{\partial y} = \alpha \frac{\partial^2 T}{\partial y^2} \qquad \text{energy} \qquad (2\text{-}23)$$

$$u_x \frac{\partial c_A}{\partial x} + u_y \frac{\partial c_A}{\partial y} = D_{AB} \frac{\partial^2 c_A}{\partial y^2} \qquad \text{continuity of A} \qquad (2\text{-}24)$$

The following are the boundary conditions for the three boundary layers (Welty et al., 1984):

$$\frac{u_x - u_{x,s}}{u_\infty - u_{x,s}} = 0 \quad \text{at } y = 0 \qquad \frac{u_x - u_{x,s}}{u_\infty - u_{x,s}} = 1 \quad \text{at } y = \infty$$

$$\frac{T - T_s}{T_\infty - T_s} = 0 \quad \text{at } y = 0 \qquad \frac{T - T_s}{T_\infty - T_s} = 1 \quad \text{at } y = \infty$$

$$\frac{c_A - c_{A,s}}{c_{A,\infty} - c_{A,s}} = 0 \quad \text{at } y = 0 \qquad \frac{c_A - c_{A,s}}{c_{A,\infty} - c_{A,s}} = 1 \quad \text{at } y = \infty$$

The similarity of the three differential equations (2-22), (2-23), and (2-24), and of the boundary conditions suggests that similar solutions should be obtained for the three transfer phenomena.

For the case of constant momentum diffusivity, ν, Blasius (1908) obtained a solution for the velocity boundary layer equations (2-21) and (2-22). He first introduced the concept of a *stream function*, $\Psi(x,y)$, such that

$$u_x = \frac{\partial \Psi}{\partial y} \qquad u_y = -\frac{\partial \Psi}{\partial x}$$

The stream function defined this way automatically satisfies the two-dimensional continuity equation (2-21). The original set of two partial differential equations may be reduced to a single ordinary differential equation by transforming the independent variables x and y to η and the dependent variables from $\Psi(x,y)$ to $f(\eta)$, where

$$\eta(x,y) = y\sqrt{\frac{u_\infty}{\nu x}} \qquad f(\eta) = \frac{\Psi(x,y)}{\sqrt{\nu x u_\infty}} \qquad (2\text{-}25)$$

The appropriate terms in equation (2-22) in terms of the dimensionless variables may be determined from equation (2-25), the following expressions result:

$$u_x = \frac{\partial \Psi}{\partial y} = u_\infty f'(\eta) \qquad (2\text{-}26)$$

$$u_y = -\frac{\partial \Psi}{\partial y} = \frac{1}{2}\sqrt{\frac{u_\infty}{\nu x}}\left(\eta f' - f\right) \tag{2-27}$$

$$\frac{\partial u_x}{\partial x} = -\frac{u_\infty \eta}{2x}f'' \qquad \frac{\partial u_x}{\partial y} = u_\infty\sqrt{\frac{u_\infty}{\nu x}}f'' \tag{2-28}$$

$$\frac{\partial^2 u_x}{\partial y^2} = \frac{u_\infty^2}{\nu x}f''' \tag{2-29}$$

Substitution of equations (2-26) to (2-29) into (2-22) and simplification gives a single ordinary differential equation,

$$f''' + \frac{1}{2}ff'' = 0 \tag{2-30}$$

with the appropriate boundary conditions

$$\begin{aligned} f = f' = 0 \quad &\text{at } \eta = 0 \\ f' = 1 \quad &\text{as } \eta \to \infty \end{aligned} \tag{2-31}$$

Observe that equations (2-30) and (2-31) constitute a nonlinear boundary value problem that must be solved numerically. It was first solved by Blasius (1908), and later by Howarth (1938). It is easily solved by the shooting method using the *rkfixed* function of Mathcad, as shown in Figure 2.2 (Appendix 2.1 gives a Python program for this purpose). Before applying the shooting method, however, the upper boundary must be expressed in finite terms. Define the velocity boundary layer thickness, δ, as that value of y at which $u_x/u_\infty = 0.99$. Moreover, to be able to use the shooting method to solve simultaneously for the unknown dimensionless velocity gradient, f''', at the solid–fluid interface and for the boundary layer thickness, another boundary condition must be specified. At the outer edge of the boundary layer, the shear stress must vanish; therefore, the dimensionless velocity gradient must approach zero. We arbitrarily chose a value of $f'' = 0.016$ at $y = \delta$ to reproduce the results obtained by Howarth (1938), who used a series expansion to solve the Blasius problem. Therefore, the boundary conditions of equation (2-31) are replaced in the shooting method by

$$\begin{aligned} f = f' = 0 \quad &\text{at } \eta = 0 \\ f' = 0.99 \quad f'' = 0.016 \qquad &\text{at } \eta = \eta_{\text{max}} = \delta\sqrt{\frac{u_\infty}{\nu x}} \end{aligned} \tag{2-32}$$

From Figure 2.2, $\eta_{\text{max}} = 5$; therefore, the velocity boundary layer thickness is given by

$$\delta = 5\sqrt{\frac{\nu x}{u_\infty}} \quad \text{or} \quad \delta = \frac{5x}{\sqrt{u_\infty x / \nu}} = \frac{5x}{\sqrt{\text{Re}_x}} \tag{2-33}$$

The velocity gradient at the surface is given by equation (2-28), which combined with the results of Figure 2.2 gives

2.4 Flow Past Flat Plate in Laminar Flow; Boundary Layer Theory

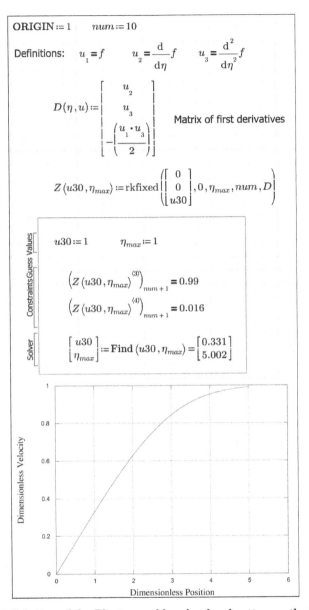

Figure 2.2 Solution of the Blasius problem by the shooting method.

$$\left.\frac{\partial u_x}{\partial y}\right|_{y=0} = u_\infty \sqrt{\frac{u_\infty}{\nu x}} f''(0) = 0.331 u_\infty \sqrt{\frac{u_\infty}{\nu x}} \qquad (2\text{-}34)$$

Substituting equation (2-34) into the expression for shear stress at the surface gives

$$\tau_{s,x} = -\mu \left.\frac{\partial u_x}{\partial y}\right|_{y=0} = 0.331 u_\infty \sqrt{\frac{u_\infty \rho \mu}{x}} \qquad (2\text{-}35)$$

106 **Convective Mass Transfer**

The local friction coefficient is then

$$C_{f,x} = \frac{\tau_{s,x}}{\rho u_\infty^2 / 2} = \frac{0.662}{\sqrt{\mathrm{Re}_x}} \qquad (2\text{-}36)$$

From knowledge of conditions in the velocity boundary layer, the continuity equation for component A may now be solved. To solve equation (2-24) we introduce the dimensionless concentration c^* defined as

$$c^* = \frac{c_A - c_{A,s}}{c_{A,\infty} - c_{A,s}}$$

and assume a solution of the form $c^* = c^*(\eta)$. Making the necessary substitutions and simplifying, equation (2-24) reduces to

$$\frac{d^2 c^*}{d\eta^2} + \frac{\mathrm{Sc}}{2} f \frac{dc^*}{d\eta} = 0 \qquad (2\text{-}37)$$

Observe the dependence of the concentration boundary layer on the hydrodynamic conditions through the appearance of the variable f in equation (2-37). The appropriate boundary conditions are

$$c^*(0) = 0 \quad \text{and} \quad c^*(\infty) = 1.0 \qquad (2\text{-}38)$$

Equation (2-37) may be solved numerically for different values of the Schmidt number. One important consequence of this solution is that, for $\mathrm{Sc} \geq 0.6$, results for the surface concentration gradient may be correlated by

$$\left. \frac{\partial c^*}{\partial \eta} \right|_{\eta=0} = 0.331 \mathrm{Sc}^{1/3} \qquad (2\text{-}39)$$

In terms of the original variables, equation (2-39) can be written as

$$\left. \frac{\partial c_A}{\partial y} \right|_{y=0} = \frac{0.331}{x} \mathrm{Re}_x^{1/2} \mathrm{Sc}^{1/3}(c_{A,\infty} - c_{A,s}) \qquad (2\text{-}40)$$

Recall that the Blasius solution for the velocity boundary layer did not involve a velocity in the y direction at the surface. Accordingly, equation (2-40) involves the important assumption that the rate at which mass enters or leaves the boundary layer at the surface is so small that it does not alter the velocity profile predicted by the Blasius solution.

When the velocity in the y direction at the surface is essentially zero, the bulk-motion contribution to mass transfer is negligible. Then, the flux of component A from the surface into the laminar boundary layer is given by

$$N_{A,y} = -D_{AB} \left. \frac{\partial c_A}{\partial y} \right|_{y=0} \qquad (2\text{-}41)$$

2.4 Flow Past Flat Plate in Laminar Flow; Boundary Layer Theory   107

Substituting equation (2-40) into (2-41) gives

$$N_{A,y} = D_{AB} \left[\frac{0.331}{x} \mathrm{Re}_x^{1/2} \mathrm{Sc}^{1/3} \right] (c_{A,s} - c_{A,\infty}) \qquad (2\text{-}42)$$

The mass-transfer conditions prevailing in the boundary layer described so far correspond to diffusion of A through stagnant B in dilute solutions. Therefore, the flux can be written in terms of a k-type mass-transfer coefficient, as described by equation (2-7), using the difference in surface and bulk concentrations as driving force:

$$N_{A,y} = k_c (c_{A,s} - c_{A,\infty}) \qquad (2\text{-}43)$$

The right-hand sides of equations (2-42) and (2-43) may be equated to give

$$k_c = D_{AB} \left[\frac{0.331}{x} \mathrm{Re}_x^{1/2} \mathrm{Sc}^{1/3} \right] \qquad (2\text{-}44)$$

or, in terms of the local Sherwood number, Sh_x

$$\frac{k_c x}{D_{AB}} \equiv \mathrm{Sh}_x = 0.331 \mathrm{Re}_x^{1/2} \mathrm{Sc}^{1/3} \quad \mathrm{Sc} \geq 0.6 \qquad (2\text{-}45)$$

From the numerical solution of equation (2-37), it also follows that the ratio of the velocity to concentration boundary layer thickness is given by

$$\frac{\delta}{\delta_c} \approx \mathrm{Sc}^{1/3} \qquad \mathrm{Sc} \geq 0.6 \qquad (2\text{-}46)$$

From the foregoing local results, the average boundary layer parameters may be determined. If the flow is laminar over the entire boundary layer length, L, the average mass-transfer coefficient is given by

$$\bar{k}_c = \frac{1}{L} \int_0^L k_c \, dx = 0.331 \frac{D_{AB}}{L} \mathrm{Sc}^{1/3} \sqrt{\frac{u_\infty}{\nu}} \int_0^L \frac{dx}{x^{1/2}} \qquad (2\text{-}47)$$

Integrating and rearranging in terms of the average Sherwood number, Sh_L,

$$\mathrm{Sh}_L = \frac{\bar{k}_c L}{D_{AB}} = 0.662 \mathrm{Re}_L^{1/2} \mathrm{Sc}^{1/3} \qquad \mathrm{Sc} \geq 0.6 \qquad (2\text{-}48)$$

Equation (2-48) has been experimentally verified and shown to give accurate results for values of $\mathrm{Re}_L < 3 \times 10^5$ (Welty et al., 1984).

2.5 MASS- AND HEAT-TRANSFER ANALOGIES

Your objectives in studying this section are to be able to:

1. Define the dimensionless groups of mass transfer corresponding to those for heat transfer.
2. Convert correlations of data on heat transfer to correlations on mass transfer.
3. Use the Chilton–Colburn analogy in problems where the heat-transfer data are correlated exclusively in terms of the Reynolds number.

Most practically useful mass-transfer situations involve turbulent flow, and for these it is generally not possible to compute mass-transfer coefficients from theoretical considerations. Instead, we must rely principally on experimental data. There are many more experimental data available for heat transfer than for mass transfer. In the previous dimensional analyses of mass-transfer situations, we have recognized the similarities between the dimensionless groups for heat and mass transfer suggested by application of the Buckingham method. There are also similarities in the differential equations that govern convective mass and heat transfer and in the boundary conditions when the transport gradients are expressed in terms of dimensionless variables. In this section, we will consider several analogies among transfer phenomena which have been proposed because of the similarity in their mechanisms. The analogies are useful in understanding the underlying transport phenomena and as a satisfactory means of predicting behavior of systems for which limited quantitative data exist.

> Therefore, to convert equations or correlations of data on heat transfer to correlations on mass transfer, the dimensionless groups of heat transfer are replaced by the corresponding groups of mass transfer.

For analogous circumstances, temperature and concentration profiles in dimensionless form and heat- and mass-transfer coefficients in the form of dimensionless groups, respectively, are given by the same functions (Treybal, 1980).

Table 2.1 lists the commonly appearing dimensionless groups. The limitations to the above rule are

1. The flow conditions and geometry must be the same.
2. Most heat-transfer data are based on situations involving no mass transfer. Use of the analogy would then produce mass-transfer coefficients corresponding to no net mass transfer, in turn corresponding most closely to k'_G, k'_c, or k'_y ($= F$). Sherwood numbers are commonly written in terms of any of the coefficients, but when derived by replacement of Nusselt numbers for use where the net mass transfer is not zero, they should be taken as $\text{Sh} = Fl/c\text{D}_{AB}$, and the F used with equation (2-1).

2.5 Mass- and Heat-Transfer Analogies

Table 2.1 Dimensionless Groups for Mass and Heat Transfer[a]

Mass transfer	Heat transfer
Reynolds number $$\mathrm{Re} = \frac{l v \rho}{\mu}$$	Reynolds number $$\mathrm{Re} = \frac{l v \rho}{\mu}$$
Schmidt number $$\mathrm{Sc} = \frac{\mu}{\rho \mathrm{D}_{AB}}$$	Prandtl number $$\mathrm{Pr} = \frac{C_p \mu}{k}$$
Sherwood number $$\mathrm{Sh} = \frac{Fl}{c \mathrm{D}_{AB}}, \frac{k_G p_{B,M} RTl}{P \mathrm{D}_{AB}},$$ $$\frac{k_c p_{B,M} l}{P \mathrm{D}_{AB}}, \frac{k_c' l}{\mathrm{D}_{AB}}, \frac{k_y' RTl}{P \mathrm{D}_{AB}}, \text{ etc.}$$	Nusselt number $$\mathrm{Nu} = \frac{hl}{k}$$
Grashof number $$\mathrm{Gr}_D = \frac{g l^3 \rho \Delta \rho}{\mu^2}$$	Grashof number $$\mathrm{Gr}_H = \frac{g l^3 \rho^2 \beta \Delta T}{\mu^2}$$
Peclet number $$\mathrm{Pe}_D = \mathrm{Re}\,\mathrm{Sc} = \frac{lv}{\mathrm{D}_{AB}}$$	Peclet number $$\mathrm{Pe}_H = \mathrm{Re}\,\mathrm{Pr} = \frac{C_p l v \rho}{k}$$
Stanton number $$\mathrm{St}_D = \frac{\mathrm{Sh}}{\mathrm{Re}\mathrm{Sc}} = \frac{\mathrm{Sh}}{\mathrm{Pe}_D} = \frac{F}{cv}$$	Stanton number $$\mathrm{St}_H = \frac{\mathrm{Nu}}{\mathrm{Re}\mathrm{Pr}} = \frac{\mathrm{Nu}}{\mathrm{Pe}_H} = \frac{h}{C_p v \rho}$$
Chilton–Colburn j-factor $$j_D = \mathrm{St}_D \mathrm{Sc}^{2/3}$$	Chilton–Colburn j-factor $$j_H = \mathrm{St}_H \mathrm{Pr}^{2/3}$$

[a]l = characteristic length β = volume coefficient of expansion
Source: Data from Treybal (1980).

Example 2.6 Mass Transfer to Fluid Flow Normal to a Cylinder

For flow of a fluid at right angle to a circular cylinder, the average heat-transfer coefficient—averaged around the periphery of the cylinder—is given by (Eckert and Drake, 1959)

$$Nu_{av} = 0.43 + 0.532\, Re^{0.5}\, Pr^{0.31}$$
$$1 \leq Re \leq 4000$$

where Nu_{av} and Re are computed using the cylinder diameter as the characteristic length, and the fluid properties are evaluated at the average conditions of the stagnant fluid film surrounding the solid (average temperature and average concentration). Estimate the rate of sublimation of a cylinder of UF_6 (molecular weight = 352), 1.0 cm diameter and 10 cm long exposed to an airstream that flows normal to the cylinder axis at a velocity of 1.0 m/s. The surface temperature of the solid is 303 K, at which temperature the vapor pressure of UF_6 is 27 kPa (Perry and Chilton, 1973). The bulk air is at 1 atm and 325 K.

Solution

Replacing Nu_{av} with Sh_{av} and Pr with Sc in the given heat-transfer correlation, the analogous expression for the mass-transfer coefficient is

$$Sh_{av} = 0.43 + 0.532\, Re^{0.5}\, Sc^{0.31}$$
$$1 \leq Re \leq 4000$$

Along the mass-transfer path—cylinder surface (point 1) to bulk air (point 2) in Figure 2.3—the average temperature is T_{av} = (303 + 325)/2 = 314 K. At point 1, the gas is saturated with UF_6 vapor, while at point 2 the gas is virtually free of UF_6. Then, the average partial pressure of UF_6 is (27 + 0)/2 = 13.5 kPa corresponding to y_A = 13.5/101.3 = 0.133 mol fraction UF_6.

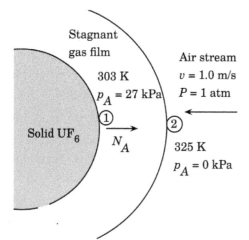

Figure 2.3 Schematic diagram for Example 2.6.

2.5 Mass- and Heat-Transfer Analogies

Use the ideal gas law to estimate the average density of the gas film. The average molecular weight is $M_{av} = 352\,(0.133) + 29\,(1 - 0.133) = 72$. The average density is

$$\rho = \frac{PM_{av}}{RT_{av}} = \frac{101.3 \times 72}{8.314 \times 314} = 2.8 \text{ kg/m}^3$$

To calculate the average viscosity of the mixture, μ, use the corresponding states method of Lucas (1980) for nonpolar gases:

$$\mu_{rM} = \xi_M \mu = f\left(T_{rM}\right) \tag{2-49}$$

where

$$\xi_M = 0.176 \left[\frac{T_{cM}}{M_{av}^3 P_{cM}^4} \right]^{1/6} \tag{2-50}$$

ξ_M = reduced inverse viscosity of the mixture, $(\mu P)^{-1}$

T_{cM} = critical temperature of the mixture, K

P_{cM} = critical pressure of the mixture, bar

$T_{rM} = T / T_{cM}$

$$f\left(T_{rM}\right) = 0.807 T_{rM}^{0.618} - 0.357 \exp\left(-0.449 T_{rM}\right) \\ + 0.340 \exp\left(-4.058 T_{rM}\right) + 0.018 \tag{2-51}$$

The mixture properties are defined as follows:

$$T_{cM} = \sum_i y_i T_{ci} \tag{2-52}$$

$$P_{cM} = RT_{cM} \frac{\sum_i y_i Z_{ci}}{\sum_i y_i V_{ci}} \tag{2-53}$$

$$M_{av} = \sum_i y_i M_i \tag{2-54}$$

Polar and quantum effects (in H_2, D_2, and He) can be considered multiplying equation (2-51) by factors given by Lucas (1980). The data needed to apply the Lucas method for this example are tabulated next (Reid et al., 1987).

Parameter	O_2	N_2	UF_6
y	$0.21 \times 0.867 = 0.182$	$0.79 \times 0.867 = 0.685$	0.133
T_c, K	154.6	126.2	505.8
P_c, bar	50.4	33.9	46.6
M, g/mol	32.0	28.0	352.0
V_c, cm³/mol	73.4	89.8	250.0
Z_c	0.288	0.290	0.277

Substituting the corresponding parameters in equations (2-52) and (2-53), T_{cM} = 181.9 K, P_{cM} = 40.3 bar, T_{rM} = 314/181.9 = 1.726. From equation (2-50), ξ_M = 4.207 × 10⁻³ $(\mu P)^{-1}$; and from equation (2-51), $f(1.726)$ = 0.962; from equation (2-49), we obtain the mixture viscosity μ = 0.962/4.207 × 10⁻³ = 229 μP = 2.29 × 10⁻⁵ kg/m·s. Calculate the Reynolds number, based—as specified—on the cylinder diameter, d.

$$\mathrm{Re} = \frac{dv\rho}{\mu} = \frac{0.01 \times 1.0 \times 2.8}{2.29 \times 10^{-5}} = 1223$$

Notice that this value of Re is within the range specified for the original correlation.

The next step is to calculate the diffusivity of UF_6 vapors in air at 314 K and 1 atm. The Lennard–Jones parameters for these gases are obtained from Appendix B. The Mathcad routine developed in Example 1.6 gives D_{AB} = 0.0904 cm²/s.

Calculate the Schmidt number:

$$\mathrm{Sc} = \frac{\mu}{\rho D_{AB}} = \frac{2.29 \times 10^{-5}}{2.8 \times 0.904 \times 10^{-5}} = 0.905$$

Substituting in the mass-transfer coefficient correlation gives

$$\mathrm{Sh}_{av} = 0.43 + 0.532 \times (1223)^{0.5} \times (0.905)^{0.31} = 18.4$$

To decide what type of mass-transfer coefficient is needed in this problem, we must explore the details of migration of the two species through the stagnant gas film surrounding the solid surface. Evidently, UF_6 vapors (A) are generated at the interface by sublimation of the solid, and these vapors migrate through the film toward the bulk of the gas phase. On the other hand, there is no source or sink of air at the interface since air does not react with UF_6, nor does the solid adsorb air. Therefore, air does not migrate through the stagnant gas film surrounding the solid surface, and N_B = 0. Notice that there is a net movement in the bulk of the gas phase toward and around the solid, but not so in the stagnant air film! Therefore, this is a case of diffusion of A through stagnant B (Ψ_A = 1). However, the solution is

2.5 Mass- and Heat-Transfer Analogies

not dilute since the average A content along the diffusion path is 0.133 mol fraction. For that reason, we must use the F form of the mass-transfer coefficient, instead of the k-type. From Table 2.1,

$$\text{Sh}_{av} = \frac{F_{av}d}{c D_{AB}} = 18.4$$

From equation (1-7)

$$c = \frac{P}{RT_{av}} = \frac{101.3}{8.314 \times 314} = 0.039 \text{ kmol/m}^3$$

Then,

$$F_{av} = \frac{18.4 \times 0.039 \times 0.904 \times 10^{-5}}{0.01} = 6.52 \times 10^{-4} \frac{\text{kmol}}{\text{m}^2 \cdot \text{s}}$$

The UF_6 flux is from equation (2-2),

$$N_A = 6.52 \times 10^{-4} \ln \left[\frac{1.0 - 0.0}{1.0 - (27/101.3)} \right] = 2.02 \times 10^{-4} \frac{\text{kmol}}{\text{m}^2 \cdot \text{s}}$$

This average flux is based on the total surface area of the cylinder, S, given by

$$S = 2 \frac{\pi d^2}{4} + \pi \times d \times L = \pi d \left[\frac{d}{2} + L \right] = 3.3 \times 10^{-3} \text{ m}^2$$

where L is the cylinder length. Therefore, the mass rate of sublimation of the solid, w_A, is given by

$$w_A = N_A S M_A = 2.35 \times 10^{-4} \text{ kg/s}$$

The calculated rate of sublimation is an instantaneous value. Mass transfer will rapidly reduce the cylinder diameter, so that Re and hence F_{av} will change with time. Furthermore, the surface will not remain in the form of a circular cylinder owing to the variation of the local F about the perimeter, so that the correlation for Nu_{av} and Sh_{av} will no longer apply.

Example 2.7 The Chilton–Colburn Analogy

Chilton and Colburn (1934), based on experimental data, defined the *j-factor for mass transfer*, and established the analogy with heat transfer

$$j_D = \text{St}_D \text{Sc}^{2/3} = \Psi \left(\text{Re} \right) = j_H = \text{St}_H \text{Pr}^{2/3} \tag{2-55}$$

This is an extremely powerful analogy to relate heat- and mass-transfer data when the effects of Pr and Sc are not given explicitly, as will be shown next.

114   **Convective Mass Transfer**

We wish to estimate the rate at which benzene will evaporate from a wetted surface of unusual shape when nitrogen at 1 bar and 300 K blows over the surface at a velocity of 10 m/s. No mass-transfer data are available for this situation, but heat-transfer data indicate that for CO_2 at 300 K and 1 bar, the heat-transfer coefficient h between the gas and the surface is given empirically by

$$h = 20\left(G_y\right)^{0.5}$$

where h is in units of $W/m^2 \cdot K$, and $G_y = \rho v$ is the superficial gas mass velocity, in $kg/m^2 \cdot s$. Estimate the required mass-transfer coefficient.

Solution

The experimental heat-transfer correlation given does not include the effect of changing the Prandtl number. The j_H factor will satisfactorily describe the effect of Prandtl on the heat-transfer coefficient. Combining the given correlation with the definitions of j_H and St_H from Table 2.1,

$$j_H = \frac{h}{C_p \rho v}Pr^{2/3} = \frac{h}{C_p G_y}Pr^{2/3} = \Psi(Re)$$

$$h = \frac{C_p G_y}{Pr^{2/3}}\Psi(Re) = 20\left(G_y\right)^{0.5} \text{ for carbon dioxide}$$

Since $Re = \rho v l / \mu = G_y l / \mu$, where l is a characteristic length, the function $\Psi(Re)$ must be compatible with $20 G_y^{0.5}$. Therefore, let $\Psi(Re) = b Re^n$, where b and n are constants to be evaluated. Then,

$$h = \frac{C_p G_y b}{Pr^{2/3}}\left(\frac{l G_y}{\mu}\right)^n = 20\left(G_y\right)^{0.5} \text{ or}$$

$$\frac{C_p b}{Pr^{2/3}}\left(\frac{l}{\mu}\right)^n G_y^{n+1} = 20\left(G_y\right)^{0.5}$$

Comparing both sides of the equation,

$$n + 1 = 0.5 \qquad n = -0.5$$

$$\frac{C_p b}{Pr^{2/3}}\left(\frac{l}{\mu}\right)^{-0.5} = 20 \qquad \text{or}$$

$$b = 20\left(\frac{l}{\mu}\right)^{0.5}\frac{Pr^{2/3}}{C_p}$$

To evaluate b, we need data on the properties of CO_2 at 300 K and 1 bar. The values are: $\mu = 150$ μP, $Pr = 0.77$, and $C_p = 853$ $J/kg \cdot K$ (Welty et al., 1984). Substituting these values in the previous equation,

$$b = 20 \times \left(\frac{l}{1.5 \times 10^{-5}}\right)^{0.5}\frac{\left(0.77\right)^{2/3}}{853} = 5.086 l^{0.5}$$

2.5 Mass- and Heat-Transfer Analogies

Therefore,

$$j_D = j_H = \Psi\left(\text{Re}\right) = 5.086 l^{0.5}\,\text{Re}^{-0.5}$$

From Table 2.1

$$F = \frac{j_D c\upsilon}{\text{Sc}^{2/3}} = \frac{5.086 l^{0.5} c\upsilon}{\text{Re}^{0.5}\,\text{Sc}^{2/3}} = \frac{5.086\left(\rho\upsilon\mu\right)^{0.5}}{M_{av}\text{Sc}^{2/3}}$$

The physical-property data in the previous equation will correspond to the average film conditions. The inner edge of the film is a saturated mixture of benzene vapors and nitrogen at 300 K and 1 bar, while the outer edge is virtually pure nitrogen at the same temperature and pressure. The vapor pressure of benzene, P_A, at 300 K is from the Antoine equation (Himmelblau, 1989)

$$\ln P_A = 15.9008 - \frac{2788.51}{300 - 52.36}$$
$$P_A = 104\,\text{mm Hg} = 13.9\ \text{kPa}$$

Therefore, the average partial pressure of benzene in the film is 13.9/2 = 6.95 kPa, corresponding to 6.95/100 = 0.07 mol fraction. The data needed to estimate the physical properties are tabulated next (Reid et al., 1987).

Parameter	C_6H_6	N_2
y	0.07	0.93
T_c, K	562.2	126.2
P_c, bar	48.9	33.9
M, g/mol	78.1	28.0
V_c, cm^3/mol	259.0	89.8
σ, Å	5.349	3.798
ε/κ, K	412.3	71.4
Z_c	0.271	0.290

The estimated average properties of the film at 300 K and 1 bar are

M_{av} = 31.4 kg/kmol

ρ = 1.26 kg/m^3 (ideal gas law)

μ = 161 mP (method of Lucas)

D_{AB} = 0.0986 cm^2/s (Mathcad routine of Example 1.4)

Sc = 1.3

Substituting in the equation for F derived above yields

$$F = \frac{5.086\left(1.26 \times 10 \times 1.61 \times 10^{-5}\right)^{0.5}}{31.4 \times 1.3^{2/3}} = 1.94 \times 10^{-3}\ \text{kmol/m}^2 \cdot \text{s}$$

116   **Convective Mass Transfer**

2.6 CONVECTIVE MASS-TRANSFER CORRELATIONS

Your objectives in studying this section are to be able to:

1. Estimate convective mass-transfer coefficients for the following situations: (a) flow parallel to a flat surface, (b) flow past a single sphere, (c) flow normal to a single cylinder, (d) turbulent flow in circular pipes, (e) flow through packed and fluidized beds, and (f) flow through the shell side of a hollow-fiber membrane module.
2. Use the corresponding coefficients to solve typical mass-transfer problems.

Thus far, we have considered mass-transfer correlations developed from analogies with heat transfer. In this section, we present a few of the correlations developed directly from experimental mass-transfer data in the literature. Others more appropriate to particular types of mass-transfer equipment will be introduced as needed.

Experimental data are usually obtained by blowing gases over various shapes wet with evaporating liquids or causing liquids to flow past solids which dissolve. Average rather than local, mass-transfer coefficients are usually obtained. In most cases, the data are reported in terms of the k-type coefficients applicable to the binary systems used with $N_B = 0$, without details concerning the actual concentrations of solute during the experiments. Fortunately, the experimental concentrations are usually fairly low, so that, if necessary, conversion of the data to the corresponding F is possible by taking $p_{B,M}/P$, $x_{B,M}$, etc., equal to unity.

2.6.1 Mass-Transfer Coefficients for Flat Plates

Several investigators have measured the evaporation from a free liquid surface—or the sublimation from a flat solid surface—of length L into a controlled air stream under both laminar and turbulent conditions. These data are correlated by (Welty et al., 1984)

$$\mathrm{Sh}_L = 0.664\,\mathrm{Re}_L^{0.5}\,\mathrm{Sc}^{1/3} \quad \text{(laminar)} \quad \mathrm{Re}_L < 3\times10^5 \qquad (2\text{-}56)$$

$$\mathrm{Sh}_L = 0.036\,\mathrm{Re}_L^{0.8}\,\mathrm{Sc}^{1/3} \quad \text{(turbulent)} \quad \mathrm{Re}_L \geq 3\times10^5 \qquad (2\text{-}57)$$

where the characteristic length in Re_L and Sh_L is L. These equations may be used if the Schmidt number is in the range of $0.6 < \mathrm{Sc} < 2500$. Their use is illustrated in Example 2.8. Both equations (2-56) and (2-57) give the average mass-transfer coefficient along the surface, and can be expressed in terms of the j_D-factor

$$j_D = 0.664\,\mathrm{Re}_L^{-0.5} \quad \text{(laminar)}\,\mathrm{Re}_L < 3\times10^5 \qquad (2\text{-}58)$$

$$j_D = 0.036\,\mathrm{Re}_L^{-0.2} \quad \text{(turbulent)}\,\mathrm{Re}_L \geq 3\times10^5 \qquad (2\text{-}59)$$

2.6 Convective Mass-Transfer Correlations 117

Example 2.8 Benzene Evaporation along a Vertical Flat Plate

Liquid benzene, C_6H_6, flows in a thin film down the outside surface of a vertical plate, 1.5 m wide and 3 m long. The liquid temperature is 300 K. Benzene-free nitrogen at 300 K and 1 bar pressure flows across the width of the plate parallel to the surface at a speed of 5 m/s. Calculate the rate at which the liquid should be supplied at the top of the plate so that evaporation will just prevent it from reaching the bottom of the plate. The density of liquid benzene at 300 K is 0.88 g/cm^3 (Himmelblau, 1989).

Solution
The film conditions, and average properties, are identical to those in Example 2.7, only the geometry is different. To determine whether the flow is laminar or turbulent, calculate Re_L. Notice that, since the flow is across the width of the plate, L refers to the width, 1.5 m:

$$\text{Re}_L = \frac{\rho v L}{\mu} = \frac{1.2 \times 5 \times 1.5}{1.61 \times 10^{-5}} = 5.59 \times 10^5 \text{ (turbulent)}$$

The Schmidt number calculated in Example 2.7, Sc = 1.3, is within the limits of applicability of equations (2-56) and (2-57). Substituting in (2-57) gives us

$$\text{Sh}_L = 0.036 \times \left(5.59 \times 10^5\right)^{0.8} \times \left(1.3\right)^{1/3} = 1557$$

Nitrogen (component B) does not react with benzene (component A), neither dissolves in the liquid; therefore, $N_B = 0$ and $\Psi_A = 1$. Since, at least at the inner edge of the film, the benzene concentration is relatively high (0.139 mole fraction), the F-form of the mass-transfer coefficient should be used:

$$F = \frac{\text{Sh}_L c \text{D}_{AB}}{L} = \frac{\text{Sh}_L \rho \text{D}_{AB}}{M_{av} L} = \frac{1557 \times 1.26 \times 0.986 \times 10^{-5}}{31.4 \times 1.5}$$
$$= 4.11 \times 10^{-4} \text{ kmol/m}^2 \cdot \text{s}$$

$$N_A = F \ln\left[\frac{1 - y_{A2}}{1 - y_{A1}}\right] = 4.11 \times 10^{-4} \ln\left[\frac{1.0 - 0.0}{1.0 - 0.139}\right]$$
$$= 6.15 \times 10^{-5} \text{ kmol/m}^2 \cdot \text{s}$$

Calculate the total mass rate of evaporation over the surface of the plate,

$$w_A = N_A S M_A = 6.15 \times 10^{-5} \times 3 \times 1.5 \times 78.1 \times 60 \times 1000$$
$$= 1296 \text{ g/min}$$

Therefore, liquid benzene should be supplied at the top of the plate at the rate of 1296/0.88 = 1470 mL/min, or about 1.5 L/min, so that evaporation will just prevent it from reaching the bottom of the plate.

118   **Convective Mass Transfer**

2.6.2 Mass-Transfer Coefficients for a Single Sphere

Investigators have studied the mass transfer from single spheres and have correlated the Sherwood number by direct addition of terms representing transfer by purely molecular diffusion and transfer by forced convection in the form

$$Sh = Sh_0 + C\,Re^m\,Sc^{1/3} \qquad (2\text{-}60)$$

where C and m are correlating constants. For extremely low Reynolds number, when there are no natural convection effects, the Sherwood number approaches a theoretical value of 2.0 (Bird et al., 1960). Accordingly, the generalized correlation becomes

$$Sh = 2 + C\,Re^m\,Sc^{1/3} \qquad (2\text{-}61)$$

For transfer into liquid streams, the equation of Brian and Hales (1969),

$$Sh = \left(4 + 1.21Pe_D^{2/3}\right)^{0.5} \qquad (2\text{-}62)$$

correlates data that are obtained for $Pe_D < 10,000$. For $Pe_D > 10,000$, Levich (1962) recommended

$$Sh = 1.01Pe_D^{1/3} \qquad (2\text{-}63)$$

The equation by Froessling (1939) and Evnochides and Thodos (1959)

$$Sh = 2.0 + 0.522Re^{1/2}\,Sc^{1/3} \qquad (2\text{-}64)$$

correlates the data for transfer into gases at Reynolds numbers ranging from 2 to 12,000, and Schmidt numbers ranging from 0.6 to 2.7.

> Equations (2-62), (2-63), and (2-64) can be used only when the effects of natural convection are negligible, that is, when
>
> $$Re \geq 0.4Gr_D^{1/2}Sc^{-1/6} \qquad (2\text{-}65)$$

The following equations by Steinberger and Treybal (1960) are recommended when the transfer occurs in the presence of natural convection:

$$Sh = Sh_{nc} + 0.347\left(Re\,Sc^{1/2}\right)^{0.62}$$
$$1 \leq Re \leq 30,000 \qquad (2\text{-}66)$$
$$0.6 \leq Sc \leq 3200$$

where

$$Sh_{nc} = 2.0 + 0.569\left(Gr_D Sc\right)^{0.25} \qquad Gr_D Sc < 10^8$$
$$= 2.0 + 0.0254\left(Gr_D Sc\right)^{1/3} Sc^{0.244} \qquad Gr_D Sc \geq 10^8 \qquad (2\text{-}67)$$

2.6 Convective Mass-Transfer Correlations   119

Example 2.9 Evaporation of a Drop of Water Falling in Air

Estimate the distance a spherical drop of water, originally 1.0 mm in diameter, must fall in quiet, dry air at 323 K and 1 atm to reduce its volume by evaporation by 50%. Assume that the velocity of the drop is its terminal velocity evaluated at its mean diameter during the process, and that the water temperature remains at 293 K.

Solution
The arithmetic mean diameter during the process is evaluated by

$$d_p = \frac{d_{p1} + d_{p2}}{2} = \frac{d_{p1} + (0.5)^{1/3} d_{p1}}{2}$$
$$= 0.897 d_{p1} = 0.897 \text{ mm}$$

The density of the drop of water at 293 K is ρ_p = 995 kg/m^3, the density of air at 323 K and 1 atm is ρ = 1.094 kg/m^3, its viscosity is μ = 195 mP (Holman, 1990). By considering a force balance on a spherical particle falling in a fluid medium, we can show that the terminal velocity of the particle is

$$v_t = \left[\frac{4 d_p (\rho_p - \rho) g}{3 C_D \rho} \right] = \left[\frac{4 d_p \rho_{pr} g}{3 C_D \rho} \right]^{0.5} \tag{2-68}$$

where $\rho_{pr} = (\rho_p - \rho)$, g is the acceleration of gravity, and C_D is the drag coefficient. The drag coefficient is a complicated function of the Reynolds number of the particle, which is proportional to the terminal velocity. To avoid having to resort to a trial-and-error calculation to estimate the terminal velocity, the following computational scheme has been proposed (Benítez, 1993):

(a) Define a new dimensionless number, the *Galileo number*, Ga

$$\text{Ga} = \text{Re}^2 C_D = C_D v_t^2 \left(\frac{d_p \rho}{\mu} \right)^2 \tag{2-69}$$

Combining equations (2-68) and (2-69), we have

$$\text{Ga} = \frac{4 d_p^3 \rho \rho_{pr} g}{3 \mu^2} \tag{2-70}$$

(b) Establish another useful relation between Re and C_D:

$$\frac{\text{Re}}{C_D} = \frac{\text{Re}^3}{\text{Ga}} = \frac{3 \rho^2 v_t^3}{4 g \rho_{pr} \mu} \tag{2-71}$$

120   **Convective Mass Transfer**

(c) The following correlation is used to relate Re/C_D to Ga (Benítez, 1993):

$$\ln\left(\frac{Re}{C_D}\right)^{1/3} = -3.194 + 2.153\ln(Ga)^{1/3} - 0.238\left[\ln(Ga)^{1/3}\right]^2$$
$$+0.01068\left[\ln(Ga)^{1/3}\right]^3 \tag{2-72}$$

(d) To calculate v_t for a particle of a given diameter, first calculate Ga from equation (2-70), calculate Re/C_D from equation (2-72), and then v_t from equation (2-71).

Applying the computational scheme to calculate the terminal velocity of the falling drop of water,

$$Ga^{1/3} = 8.97 \times 10^{-4}\left[\frac{4 \times 1.094 \times 994 \times 9.8}{3 \times (1.95 \times 10^{-5})^2}\right]^{1/3} = 30.837$$

$$\ln(Ga^{1/3}) = 3.429$$

From equations (2-72) and (2-71)

$$\left(\frac{Re}{C_D}\right)^{1/3} = 6.178 = v_t\left[\frac{3\rho^2}{4g\rho_{pr}\mu}\right]$$

$$v_t = 3.56 \text{ m/s}$$

The vapor pressure of water at 293 K is 2.34 kPa (Smith et al., 1996). Therefore, the water concentration at the inner edge of the gas film is 2.34/101.3 = 0.0231 mol fraction. The average molecular weight of the mixture is 28.75 kg/kmol, and the density is 1.195 kg/m^3. The outer edge of the film is dry air at 323 K and 1 atm, with a density of 1.094 kg/m^3. Then, $\Delta\rho$ = 0.101 kg/m^3. To determine whether natural convection effects are important, calculate Gr_D. The density and viscosity in the Grashof number are the properties at the average film conditions. Since the water vapor content of the film is so low, we may assume that the average film conditions are basically those of dry air at 1 atm and (293 + 323)/2 = 308 K. Under those conditions, ρ = 1.14 kg/m^3 and μ = 192 µP (Holman, 1990). Then,

$$Gr_D = \frac{gd_p^3\rho\Delta\rho}{\mu^2} = \frac{9.8 \times (8.97 \times 10^{-4})^3 \times 1.14 \times 0.101}{(1.92 \times 10^{-5})^2} = 2.232$$

Estimate the diffusivity of water in air at 308 K and 1 atm: D_{AB} = 0.242 cm^2/s. The Schmidt number is then

$$Sc = \frac{\mu}{\rho D_{AB}} = \frac{1.92 \times 10^{-5}}{1.14 \times 2.42 \times 10^{-5}} = 0.692$$

The Reynolds number is

$$Re = \frac{\rho v_t d}{\mu} = \frac{1.14 \times 3.56 \times 8.97 \times 10^{-4}}{1.92 \times 10^{-5}} = 191$$

2.6 Convective Mass-Transfer Correlations

121

In equation (2-65)

$$0.4 \mathrm{Gr}_D^{0.5} \mathrm{Sc}^{-1/6} = 0.4 \times (2.232)^{0.5} \times (0.692)^{-1/6} = 0.635$$

This is much smaller than the Reynolds number, indicating that natural convection effects are negligible. The Froessling equation (2-64) can be used to evaluate the mass-transfer coefficient

$$\mathrm{Sh} = 2.0 + 0.552 \times (191)^{0.5} \times (0.692)^{1/3} = 8.748$$

Given that the solubility of air (component B) in water is negligible, $N_B = 0$. We have already shown that the gas phase is very dilute in water vapor (component A); therefore, it is appropriate to use k-type mass-transfer coefficients. From Table 2.1

$$\mathrm{Sh} = \frac{k_c p_{B,M} d_p}{P D_{AB}} \cong \frac{k_c d_p}{D_{AB}} = 8.748 \quad \text{since } p_{B,M} \cong P$$

$$k_c = \frac{\mathrm{Sh} D_{AB}}{d_p} = \frac{8.748 \times 2.42 \times 10^{-5}}{8.97 \times 10^{-4}} = 0.236 \text{ m/s}$$

The average rate of evaporation is

$$w_A = \pi d_p^2 N_A M_A = \pi d_p^2 k_c \left(c_{A1} - c_{A2} \right) M_A$$

The dry-air concentration, $c_{A2} = 0$. The interface concentration is evaluated from the vapor pressure of water at 293 K:

$$c_{A1} = \frac{P_A}{RT_1} = \frac{2.34}{8.314 \times 293} = 9.61 \times 10^{-4} \text{ kmol/m}^3$$

Substituting the known values into the average rate of evaporation equation,

$$w_A = \pi \times \left(8.97 \times 10^{-4} \right)^2 \times 18 \times 0.236 \times \left(9.61 \times 10^{-4} - 0 \right)$$
$$= 1.031 \times 10^{-8} \text{ kg/s} = 1.031 \times 10^{-5} \text{ g/s}$$

The total amount of water evaporated, m, when the volume is reduced by 50% is given by

$$m = \rho \Delta V = \frac{\rho V_1}{2} = \frac{\rho \pi d_p^3}{12} = \frac{995 \times \pi \times (0.001)^3}{12}$$
$$= 2.61 \times 10^{-7} \text{ kg} = 2.61 \times 10^{-4} \text{ g}$$

The time necessary to reduce the volume by 50% at the average rate of evaporation is

$$t = \frac{m}{w_A} = \frac{2.61 \times 10^{-4} \text{ g}}{1.031 \times 10^{-5} \text{ g/s}} = 25.3 \text{ s}$$

The distance of fall is equal to the product of time and terminal velocity (assuming that the terminal velocity is achieved instantaneously): $t \times v = 25.3 \times 3.56 = 90.0$ m.

122 Convective Mass Transfer

2.6.3 Mass-Transfer Coefficients for Single Cylinders

Several investigators have studied the sublimation from a solid cylinder into air flowing normal to its axis. Additional results on the dissolution of solid cylinders into a turbulent water stream have been reported. Bedingfield and Drew (1950) correlated the available data, evaluating properties at the average film conditions, by

$$\frac{k_G P \mathrm{Sc}^{0.56}}{G_M} = 0.281\,\mathrm{Re}^{-0.4}$$
$$400 \leq \mathrm{Re} \leq 25{,}000$$
$$0.6 < \mathrm{Sc} < 2.6$$

$$(2\text{-}73)$$

where Re is the Reynolds number based on the diameter of the cylinder, $G_M = vc$ is the molar velocity of the gas, and P is the total pressure. The use of this correlation is illustrated in Example 2.10.

Example 2.10 Mass Transfer for Single Cylinder

Repeat Example 2.6 using equation (2-73) if it applies.

Solution

The values of the dimensionless parameters calculated in Example 2.6 (Re = 1223 and Sc = 0.905) are within the limits of applicability of equation (2-73). The molar density calculated in that example was $c = 0.039$ kmol/m^3, and the gas velocity was specified as $v = 1.0$ m/s. Therefore, $G_M = vc = (1.0)(0.039) = 0.039$ kmol/m^2·s. Substituting in (2-73) we obtain

$$k_G P = \frac{0.281\,\mathrm{Re}^{-0.4}\,G_M}{\mathrm{Sc}^{0.56}} = \frac{0.281 \times 0.039 \times (1223)^{-0.4}}{(0.905)^{0.56}}$$
$$= 6.75 \times 10^{-4}\ \mathrm{kmol/m^2 \cdot s}$$

From equation (2-9), $k_G P = k_y$. However, in this problem k-type coefficients cannot be used because we are not dealing with dilute solutions. Therefore, we must try to relate the results obtained so far with the F coefficient. We know from equation (2-6) that $F = k_y y_{B,M}$. The fact that the correlation in equation (2-73) was developed in terms of k_G is evidence that the experimental data used were obtained under very dilute conditions for which $y_{B,M} \approx 1.0$, and $F \approx k_y$. Then, we may conclude that $F = 6.75 \times 10^{-4}$ kmol/m^2·s, which is only about 3% higher than the value estimated in Example 2.6 ($F = 6.52 \times 10^{-4}$ kmol/m^2·s) using an analogy between heat and mass transfer.

2.6.4 Turbulent Flow in Circular Pipes

Mass transfer from the inner wall of a tube to a moving fluid has been studied extensively, and most experimental data come from wetted-wall towers. In Figure 2.4, a volatile pure liquid is allowed to flow down the inside surface of a circular pipe while

2.6 Convective Mass-Transfer Correlations

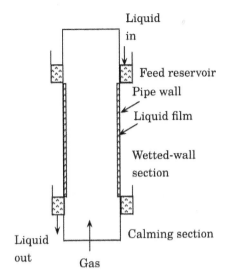

Figure 2.4 Wetted-wall tower.

a gas is blown upward or downward through the central core. Measurement of the rate of evaporation of the liquid into the gas stream over the known surface permits calculation of the mass-transfer coefficient for the gas phase. Use of different gases and liquids provides variations of Sc.

Gilliland and Sherwood (1934) studied the vaporization of nine different liquids into air. Their correlation is

$$\text{Sh} = 0.023\,\text{Re}^{0.83}\,\text{Sc}^{0.44}$$
$$2000 < \text{Re} < 35{,}000 \tag{2-74}$$
$$0.6 < \text{Sc} < 2.5$$

where the characteristic length in Sh and Re is the tube diameter, *and the physical properties of the gas are evaluated at the bulk conditions of the flowing gas stream.*

In a subsequent study, Linton and Sherwood (1950) extended the range of Schmidt number when they investigated the rate of dissolution of benzoic acid, cinnamic acid, and β-naphthol. The combined results were correlated by

$$\text{Sh} = 0.023\,\text{Re}^{0.83}\,\text{Sc}^{1/3}$$
$$4000 < \text{Re} < 70{,}000 \tag{2-75}$$
$$0.6 < \text{Sc} < 3000$$

Notice that equation (2-75) is the best correlation for all the data, including both gases and liquids. However, the data for gases only are best correlated by equation (2-74).

Example 2.11 Simultaneous Heat and Mass Transfer in Pipe

Mass transfer may occur simultaneously with the transfer of heat, either because of an externally imposed temperature difference or because of the absorption or evolution of heat which generally occurs when a substance is transferred from one phase to another. In such cases, within one phase, the heat transferred is the result not only of the conduction or convection by virtue of the temperature difference which would happen in the absence of mass transfer, but also includes the sensible heat carried by the diffusing matter.

Consider a fluid consisting of a mixture of substances A and B at a bulk temperature T_1 that flows past a second phase under conditions causing mass transfer from one phase to the other through an interface at a temperature T_i. The total sensible heat flux to the interface q_s is given by (Treybal, 1980)

$$q_s = \frac{N_A M_A C_{p,A} + N_B M_B C_{p,B}}{1 - \exp\left[-\dfrac{N_A M_A C_{p,A} + N_B M_B C_{p,B}}{h}\right]}(T_1 - T_i) \qquad (2\text{-}76)$$

The total heat release at the interface q_t will include additionally the thermal effect produced when the transferred mass passes through the interface. This may be a latent heat of vaporization, a heat of solution, or both, depending upon the circumstances.

$$q_t = q_s + \lambda_A N_A + \lambda_B N_B \qquad (2\text{-}77)$$

In some cases, where the mass transfer in the fluid is in the direction opposite to the sensible-heat transfer, it is possible that the diffusing mass may carry energy into the fluid as fast as it is released, in which case $q_t = 0$.

To illustrate the concept of simultaneous mass and heat transfer, consider the following air humidification problem. Water flows down the inside wall of a 25.4-mm diameter wetted-wall tower of the design of Figure 2.4, while air flows upward through the core. At a point in the tower humid air flows at a mass velocity of 5.0 kg/m²-s. The temperature of the gaseous mixture is 308 K, the pressure is 1 atm, and the partial pressure of water vapor in the mixture is 1.95 kPa. Assuming that the process is virtually adiabatic, estimate the temperature of the liquid water at that point, and the rate of water evaporation.

Solution

In this problem, water (A) evaporates into the air (B) stream. Since there is no external source of energy, the latent heat of vaporization of water must be supplied by sensible heat flowing from the gas stream to the liquid water. Therefore, the liquid water temperature (T_i) must be lower than the air bulk temperature (T_1), and $q_t = 0$. From the point of view of mass transfer in the gas phase, $N_B = 0$, the solution is dilute, and the use of k-type coefficients is justified. Therefore, the rates of mass and heat transfer can be written in terms of the unknown liquid water temperature as

$$N_A(T_i) = k_G\left[p_{A,1} - P_A(T_i)\right] \qquad (2\text{-}78)$$

$$q_t(T_i) = \frac{N_A(T_i) M_A C_{p,A}}{1 - \exp\left[-\dfrac{N_A(T_i) M_A C_{p,A}}{h}\right]}(T_1 - T_i) + N_A(T_i)\lambda_A(T_i) = 0 \qquad (2\text{-}79)$$

2.6 Convective Mass-Transfer Correlations 125

where

$p_{A,1}$ = partial pressure of water vapor in the bulk gas = 1.95 kPa
$P_A(T_i)$ = vapor pressure of water at T_i
T_1 = air bulk temperature = 308 K

To estimate the coefficients k_G and h, the properties of the bulk gas phase are needed. Since the solution is so dilute, assume that the properties are those of dry air at 308 K and 1 atm: μ = 192 mP, ρ = 1.14 kg/m^3, D_{AB} = 0.242 cm^2/s, Sc = 0.696, C_p = 1.007 kJ/kg·K, k = 0.027 W/m·K, Pr = 0.70. Calculate the Reynolds number

$$\mathrm{Re} = \frac{DG_y}{\mu} = \frac{0.0254 \times 5.0}{1.92 \times 10^{-5}} = 6615$$

From equation (2-74), Sh = 29.1; from Table 2.1, k_G = 0.0108 mol/(m^2·s·kPa). To estimate the heat-transfer coefficient, use the *Dittus–Boelter equation* for cooling (Incropera et al., 2007)

$$\mathrm{Nu} = 0.023\,\mathrm{Re}^{0.8}\,\mathrm{Pr}^{0.3} \tag{2-80}$$

From equation (2-80), Nu = 23.53; from Table 2.1, h = 25.01 W/(m^2·K).

To solve equations (2-78) and (2-79), expressions for the vapor pressure and latent heat of vaporization of water as a function of temperature are needed. Use the Antoine equation for the vapor pressure (Smith et al., 1996)

$$P_A\left(T_i\right) = \exp\left[16.3872 - \frac{3885.7}{T_i - 42.98}\right]\ \mathrm{kPa} \tag{2-81}$$

Use the method proposed by Watson (Smith et al., 1996) for the latent heat of vaporization knowing that, for water at T_n = 373.15 K, λ_A = 40.63 kJ/mol; T_c = 647.1 K:

$$\lambda_A(T_i) = \lambda_A(T_n)\left[\frac{1 - \left(T_i/T_c\right)}{1 - \left(T_n/T_c\right)}\right]^{0.38} \tag{2-82}$$

Assume that, in the temperature range $T_i \le T \le T_1$, the heat capacity of water vapor is constant at $C_{p,A}$ = 1.88 kJ/kg·K. Equations (2-78), (2-79), (2-81), and (2-82) are easily solved using the "solve block" feature of Mathcad. The results are as follows: T_i = 295 K and N_A = −0.007 mol/m^2·s (the negative sign means that mass and heat are transferred in opposite directions). The corresponding rate of water evaporation is 7.9 g/min·m^2. Appendix 2.2 presents the solution using Python.

Example 2.12 Air Humidification in Wetted-Wall Column

Water flows down the inside wall of a wetted-wall tower of the design of Figure 2.4, while air flows upward through the core. The ID of the tower is 25.4 mm, and the length of the wetted section is 1.5 m. Dry air enters the wetted section at a mass velocity of 5.0 kg/m^2·s, a temperature of 308 K, and a pressure of 1 atm.

The water enters at 295 K. Estimate the partial pressure of water in the air leaving the tower. Assume that the temperature of the air and of the water remain constant.

Solution
In this problem, dry air enters the bottom of the tower. As it flows upward, water vapor is transferred from the gas–liquid interface toward the bulk of the gas phase due to a partial pressure driving force, $(p_{A1} - p_{A2})$, which changes continuously along the wetted section. The water vapor partial pressure at the interface remains constant at the vapor pressure of liquid water at 295 K, which is $p_{A1} = P_A = 2.64$ kPa. However, the water vapor partial pressure at the bulk of the gas phase increases from $p_{A2} = p_{Ain} = 0$, for the dry inlet air, to $p_{A2} = p_{Aout}$ for the air leaving the tower. Not only does the driving force change, but also the total gas flow rate increases as water evaporates along the wetted section. Since the mass-transfer coefficient is a function of gas flow rate, it also changes along the column.

For the case of the system air-water at near-ambient conditions, we are dealing with very dilute solutions of water vapor in air. For this reason, it is justified to assume that the total gas flow rate will remain relatively constant at its inlet value, and that the gas phase will behave as if it were dry air. If we assume that the temperature and pressure will remain constant, it follows that the dimensionless numbers Re and Sc will not change along the column, justifying the assumption of a constant value for the mass-transfer coefficient. Dilute gas solutions, and $N_B = 0$, also justify the use of a k-type coefficient coupled to a linear concentration driving force. These assumptions simplify considerably the analysis of the problem.

To consider the fact that the driving force, hence, the mass-transfer rate, changes with position along the column, consider a segment of the wetted section of thickness Δz, as shown in Figure 2.5. A water (component A) balance, assuming steady-state and constant molar mass gas velocity (G_M), is

$$\frac{\pi D^2}{4}\left[G_M y_A\right]_{z+\Delta z} - \frac{\pi D^2}{4}\left[G_M y_A\right]_z = N_A \Delta S = a \frac{\pi D^2}{4} \Delta z N_A \quad (2\text{-}83)$$

where $a = \Delta S / \Delta V$ is the mass-transfer surface area per unit volume of column. Divide both sides by $\pi D^2 \Delta z / 4$, rearrange, and take the limit as Δz tends to zero:

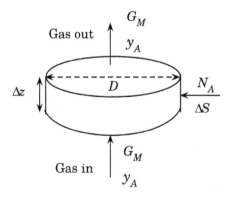

Figure 2.5 Differential section for Example 2.12.

2.6 Convective Mass-Transfer Correlations

$$G_M \frac{dy_A}{dz} = aN_A = k_G a \left(P_A - p_A \right) \tag{2-84}$$

where P_A is the vapor pressure at the water temperature. But $y_A = p_A/P$, then

$$\frac{dp_A}{P_A - p_A} = \frac{k_G aP}{G_M} dz \tag{2-85}$$

Integrating equation (2-85) within the limits $z = 0$, $p_A = 0$, and $z = Z$, $p_A = p_{Aout}$ gives

$$\ln \left[\frac{P_A}{P_A - p_{Aout}} \right] = \frac{k_G aPZ}{G_M} \tag{2-86}$$

where Z is the total height of the wetted section. Rearranging equation (2-86) yields

$$p_{Aout} = P_A \left[1 - \exp \left(-\frac{k_G aPZ}{G_M} \right) \right] \tag{2-87}$$

From Table 2.1, for dilute solutions, $\text{Sh} \approx k_G RTD / D_{AB}$. Then,

$$\frac{k_G aPZ}{G_M} = \frac{\text{Sh}acZD_{AB}}{G_M D} \quad \text{where } c = \frac{P}{RT}$$

For the wetted-wall geometric arrangement,

$$a = \frac{\text{surface area}}{\text{volume}} = \frac{\pi DZ}{\pi D^2 Z / 4} = \frac{4}{D} \tag{2-88}$$

Substituting in equation (2-87)

$$p_{Aout} = P_A \left\{ 1 - \exp \left[-\frac{4\text{Sh}cZD_{AB}}{G_M D^2} \right] \right\} \tag{2-89}$$

Equation (2-89) is the relation needed to estimate the water vapor concentration in the air leaving the tower. Assuming that the gas phase is basically dry air, $G_M = G_y/M_{av} = 5/29 = 0.172$ kmol/m²·s. The properties of dry air at 308 K and 1 atm are $\rho = 1.14$ kg/m³ and $\mu = 192$ µP, $D_{AB} = 0.242$ cm²/s, $Sc = 0.696$ (from Example 2.9). Calculate $Re = G_y D/\mu = (5)(0.0254)/1.92 \times 10^{-5} = 6615$. From equation (2-74), $Sh = 29.07$. Calculate $c = P/RT = (101.3)/[(8.314)(308)] = 0.040$ kmol/m³. Substituting in equation (2-89),

$$p_{Aout} = 2.64 \left\{ 1 - \exp \left[-\frac{4 \times 29.07 \times 1.5 \times 0.04 \times 0.242}{0.172 \times (2.54)^2} \right] \right\}$$

$$= 2.05 \text{ kPa}$$

128   **Convective Mass Transfer**

This is 77.7% of the maximum possible partial pressure of water in the air, which is $p_{Amax} = P_A = 2.64$ kPa.

2.6.5 Mass Transfer in Packed and Fluidized Beds

Packed and fluidized beds are commonly used in industrial mass-transfer operations, including adsorption, ion exchange, chromatography, and gaseous reactions that are catalyzed by solid surfaces. They offer a dramatic increase in the surface area available for heat and mass transfer for a given volume, compared to an empty tube. Numerous investigations have been conducted for measuring mass-transfer coefficients in packed beds and correlating the results. Sherwood et al. (1975) suggested the following correlation for gases:

$$j_D = 1.17\,\mathrm{Re}^{-0.415}$$
$$10 < \mathrm{Re} < 2500 \tag{2-90}$$

where $\mathrm{Re} = d_p G_y/\mu$, where G_y is the gas mass velocity based on the total cross-sectional area of the tower, and d_p is the diameter of a sphere with the same surface area per unit volume as the particle.

Mass transfer in both gas and liquid fixed and fluidized beds of spheres has been correlated by Gupta and Thodos (1962) with the equation

$$\varepsilon j_D = 0.010 + \frac{0.863}{\mathrm{Re}^{0.58} - 0.483} \tag{2-91}$$

for Re between 1 and 2100, where ε is the void fraction of the bed.

For packed and fluidized beds, the area for mass transfer is usually expressed in terms of a, defined in this case as the area for mass transfer per unit volume of packed bed. It is easy to demonstrate that

$$a = \frac{6\left(1 - \varepsilon\right)}{d_p} \tag{2-92}$$

According to equation (2-92), the smaller the packing size, the higher the area available for mass transfer per unit of packed volume. However, pressure drop through the packed bed becomes a limiting factor as the size of the packing material diminishes.

Example 2.13 Air Humidification in a Packed Bed

The column of Example 2.12 is packed with spherical glass beads, with diameter of 3.5 mm. The water is now supplied as a fine mist that flows down through the packed bed at a flow rate just enough to replace the water lost by vaporization, and to keep the surface of the packing wet. Under these conditions, it may be assumed that equation (2-90) applies.

2.6 Convective Mass-Transfer Correlations

(a) Estimate the depth of packing required if the water partial pressure in the air leaving the bed is to be 99% of the water vapor pressure P_A.

(e) Estimate the gas pressure drop through the bed.

Solution
(a) Begin with the integrated mass-balance of equation (2-87) and specialize it for the given packed-bed geometry using

$$a = \frac{6(1-\varepsilon)}{d_p} \quad k_G = \frac{ShD_{AB}}{RTd_p}$$

Substituting in equation (2-87) we have

$$p_{Aout} = P_A \left\{ 1 - \exp\left[-\frac{6(1\text{-}\varepsilon)ShcZD_{AB}}{G_M d_p^2} \right] \right\} \tag{2-93}$$

Calculate the Reynolds number, based on the particle size:

$$Re = \frac{5 \times 3.5 \times 10^{-3}}{1.92 \times 10^{-5}} = 911.5$$

Substituting in equation (2-90), j_D = 0.069. From Example 2.12, Sc = 0.696. Then, St_D = 0.069/0.696$^{2/3}$ = 0.088; Sh = St_D × Re × Sc = 55.8. To estimate the bed porosity, use the following correlation developed for uniform spherical particles randomly packed in a cylindrical container of diameter D (Nandakumar et al., 1999):

$$\varepsilon = 0.406 + 0.571 \frac{d_p}{D} \quad \text{for } \frac{d_p}{D} < 0.14 \tag{2-94}$$

Substituting yields ε = 0.406 + 0.571(3.5/25.4) = 0.485. To determine the bed depth required to achieve 99% of the water vapor pressure P_A at the exit, substitute p_{Aout} = 0.99P_A in equation (2-93) and solve for Z:

$$Z = \frac{-0.172 \times (0.35)^2 \times \ln(1.00 - 0.99)}{6 \times (1.0 - 0.485) \times 55.8 \times 0.04 \times 0.242} = 0.058 \text{ m} = 5.8 \text{ cm}$$

Although the packed bed is 26 times smaller than the wetted-wall column, the water vapor partial pressure in the air leaving the bed is virtually the maximum achievable, while in the air leaving the wetted-wall column it was only 77.7% of the maximum. This explains why the wetted-wall arrangement is used only in experimental setups, and never for practical industrial applications.

130   **Convective Mass Transfer**

(b) To estimate the gas-pressure drop through the bed, assume that—for the given conditions of extremely low liquid flow—it is a problem of single-phase flow, reasonably well correlated by (Ergun, 1952)

$$\frac{\Delta P}{Z}\frac{\varepsilon^3 d_p \rho}{(1-\varepsilon)G_y^2} = \frac{150(1-\varepsilon)}{\text{Re}} + 1.75 \tag{2-95}$$

Substituting in equation (2-95), we obtain

$$\Delta P = \left[\frac{150\times(1-0.485)}{911.5} + 1.75\right] \times \frac{(1-0.485)\times 5^2 \times 0.058}{0.485^3 \times 0.0035 \times 1.14}$$

$$= 3010 \text{ Pa}$$

This is an extremely high pressure drop, even though the packed bed is very shallow! For comparison purposes, let us estimate the pressure drop for the wetted-wall column of Example 2.12. To estimate the pressure drop in turbulent flow through a circular "smooth" pipe of diameter D and length Z (Welty et al., 1984)

$$\Delta P = 2f\frac{Z}{D}\frac{G_y^2}{\rho}$$

$$\text{where } \frac{1}{\sqrt{f}} = 4\log_{10}(\text{Re}\sqrt{f}) - 0.4 \tag{2-96}$$

From Example 2.12, Re = 6615, Z = 1.5 m, D = 0.0254 m, G_y = 5 kg/m^2·s. Equation (2-96) yields f = 0.00864, ΔP = 22.4 Pa. Notice the dramatic difference between this result and the pressure drop estimated for the packed bed!

2.6.6 Mass Transfer in Hollow-Fiber Membrane Modules

Mass transfer in fiber bundles is a problem of great practical importance for membrane separation processes. Such processes commonly utilize a bundle of randomly packed hollow fibers enclosed in a case to contact two process streams. Ports on the case permit one to introduce and remove streams from the space inside the fibers, the *lumen*, and the space outside the fibers, the *shell*. Figure 2.6 illustrates the construction of a typical hollow-fiber membrane module (Bao et al., 1999).

Hollow-fiber membrane modules are the mass-transfer equivalent of shell-and-tube heat exchangers. As fluids flow through the shell and lumen, mass is transferred from one stream to the other across the fiber wall. In contrast to heat exchangers, though, mass transfer may involve a combination of diffusion and convection, depending on the nature of the membrane. These modules are used for a wide range of membrane processes, including gas separation, reverse osmosis, filtration, and dialysis.

The literature contains numerous correlations for shell-side mass-transfer coefficients in hollow-fiber membrane modules (Bao et al., 1999). A representative relationship for liquid flowing through the shell parallel to the fibers is (Costello et al., 1993)

2.6 Convective Mass-Transfer Correlations

$$\text{Sh} = 0.53(1-1.1\phi)\left[\frac{1-\phi}{\phi}\right]^{-0.47} \text{Re}^{0.53}\text{Sc}^{0.33}$$

$$0.3 < \phi < 0.75 \qquad (2\text{-}97)$$

$$20 < \text{Re}\left[\frac{1-\phi}{\phi}\right] < 350$$

where $\text{Sh} = 2Rk_L/\text{D}_{AB}$ is the Sherwood number calculated using the effective mass-transfer coefficient for a module length L; R is the fiber radius; ϕ is the fiber packing fraction; the Reynolds number is defined as $\text{Re} = 2Rv_0\rho/\mu$, where v_0 is the superficial velocity based on an empty shell. Packing fraction refers to the fraction of the cross-sectional area that is occupied by the fibers.

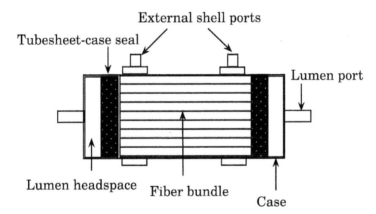

Figure 2.6 Construction of a typical hollow-fiber membrane module.

Flow through the lumen usually corresponds to laminar flow inside a circular pipe. Theoretical solutions have been developed for this situation, with two different boundary conditions: (a) constant wall concentration and (b) constant wall flux. Constant wall concentration is much more common than constant flux. The corresponding correlation is (Cussler, 1997)

$$\text{Sh} = 1.62\text{Pe}_D^{1/3}\left(\frac{d_i}{L}\right)^{1/3} = 1.62\text{Gz}^{1/3} \qquad (2\text{-}98)$$

$$\text{Re} < 2100$$

where $\text{Sh} = d_i\, k_L/\text{D}_{AB}$ is the Sherwood number calculated using the effective mass-transfer coefficient for a module length L; d_i is the fiber inside diameter; the Reynolds number is defined as $\text{Re} = d_i v\rho/\mu$, where v is the average fluid velocity; the *Graetz number* is another dimensionless quantity defined as $\text{Gz} = \text{Pe}_D(d_i/L)$.

132   **Convective Mass Transfer**

Example 2.14 Design of a Hollow-Fiber Boiler Feed Water (BFW) Deaerator

Boiler feed water (BFW) must be deaerated to avoid corrosion problems in the boilers. Hollow fibers made of microporous polypropylene can be used for extremely fast removal of dissolved oxygen from water, therefore making a compact BFW deaerator possible (Yang and Cussler, 1986). For a given boiler, 40,000 kg/h of BFW is needed. Design a hollow-fiber membrane module for that purpose, if the unit can remove 99% of the dissolved oxygen present in natural waters at 298 K.

Solution

To design the deaerator, we must first decide on the geometry for the hollow-fiber module. We will use commercially available microporous polypropylene hollow fibers in a module like that shown in Figure 2.4. Water at 298 K will flow through the shell side, parallel to the fibers, at a superficial velocity of 10 cm/s. Pure nitrogen at 298 K and 1 atm at the rate of 40 L/min will be used as a sweep gas in countercurrent flow through the lumen. The outside diameter of the available fibers is 290 μm; the packing factor is 40%; and the surface area per unit volume is $a = 46.84$ cm^{-1} (Prasad and Sirkar, 1988).

First, calculate the volumetric water flow rate, Q_L, if the density of liquid water at 298 K is approximately 1000 kg/m^3:

$$Q_L = \frac{40,000}{3600 \times 1000} = 0.0111 \text{ m}^3 / \text{s}$$

Calculate the shell diameter, D, for the calculated water flow rate and the specified superficial velocity, $v_0 = 10$ cm/s.

$$D = \left[\frac{4Q_L}{\pi v_0}\right]^{0.5} = 0.376 \text{ m}$$

Estimate the properties of dilute mixtures of oxygen in water at 298 K: $\rho = 1000$ kg/m^3, $\mu = 0.9$ cP. The diffusivity is from equation (1-53), $D_{AB} = 1.93 \times 10^{-5}$ cm^2/s. Calculate the dimensionless numbers, Sc = 467 and Re = $dv_0\rho/\mu$ = 32.2. Substituting in equation (2-97),

$$\text{Sh} = 0.53 \times \left(1.0 - 1.1 \times 0.4\right) \times \left[\frac{1.0 - 0.4}{0.4}\right]^{-0.47} \times 32.2^{0.53} \times 467^{0.33}$$

$$= 11.74$$

Calculate the mass-transfer coefficient on the shell side, $k_L = 11.74 D_{AB}/d = 0.0078$ cm/s.

The design equation is based on the observation by Yang and Cussler (1986) that, for this situation, all the resistance to mass transfer resides on the shell side. Then—as will be shown in Chapter 9—the total volume of the module, V_T, for countercurrent flow is given by

2.7 Multicomponent Mass-Transfer Coefficients

$$V_t = \frac{L}{k_L a c (1-A)} \ln\left[\frac{x_{in}}{x_{out}}(1-A)+A\right]$$ (2-99)

where

$$A = \frac{L}{mV}$$ (2-100)

where L and V are the molar flow rates of liquid and gas, respectively, and m is the equilibrium ratio of gas concentration to liquid concentration. From the specified BFW flow rate,

$$L = 40,000\,\frac{\text{kg}}{\text{h}} \times \frac{1\,\text{h}}{3600\,\text{s}} \times \frac{1\,\text{kmol}}{18\,\text{kg}} = 0.617\,\text{kmol/s}$$

From the ideal gas law,

$$V = 40\,\frac{\text{L}}{\text{min}} \times \frac{1\,\text{min}}{60\,\text{s}} \times \frac{1\,\text{m}^3}{1000\,\text{L}} \times \frac{101.3\,\text{kPa}}{298\,\text{K}} \times \frac{\text{kmol}\cdot\text{K}}{8.314\,\text{kPa}\cdot\text{m}^3}$$
$$= 2.726\times10^{-5}\,\text{kmol/s}$$

From the solubility of oxygen in water at 298 K, $m = 4.5 \times 10^4$ (Davis and Cornwell, 1998). Then, $A = 0.617/(4.5 \times 10^4 \times 2.726 \times 10^{-5}) = 0.503$. For 99% removal of the dissolved oxygen, $x_{in}/x_{out} = 100$. Substituting in equation (2-99),

$$V_T = \frac{0.617 \times \ln\left[100\times(1.0-0.503)+0.503\right]}{0.0078\times46.84\times55.5\times(1.0-0.503)} = 0.24\,\text{m}^3$$

The module length, Z, is

$$Z = \frac{4V_T}{\pi D^2} = \frac{4\times0.24}{\pi\times0.376^2} = 2.16\,\text{m}$$

2.7 MULTICOMPONENT MASS-TRANSFER COEFFICIENTS

Your objectives in studying this section are to be able to:

1. Estimate convective mass-transfer coefficients for multicomponent problems based on binary correlations.
2. Write and solve the MS equations for multicomponent mixtures of ideal gases in terms of the corresponding binary mass-transfer coefficients.

Most published empirical correlations to estimate mass-transfer coefficients have concentrated on binary systems and there are no correlations available for the multicomponent case. The need to estimate multicomponent mass-transfer

134   **Convective Mass Transfer**

coefficients is very real, however. Various approximate methods have been proposed to estimate multicomponent mass-transfer coefficients based on binary correlations.

The simplest approach is to calculate binary mass-transfer coefficients $F_{i,j}$, from the corresponding empirical correlation, substituting the MS diffusivity $D_{i,j}$, for the Fick diffusivity in the Sc and Sh numbers, as illustrated in Example 2.15. The MS equations are, then, written in terms of the binary mass-transfer coefficients. For ideal gas multicomponent mixtures, and one-dimensional fluxes, they become

$$\frac{dy_i}{d\eta} = \sum_{j=1}^{n} \frac{y_i N_j - y_j N_i}{F_{i,j}} \qquad i = 1, 2, \cdots, n-1 \qquad (2\text{-}101)$$

where $\eta = z/\delta$ is the dimensionless distance along the gas film.

Example 2.15 Ternary Distillation in a Wetted-Wall Column.

Dribicka and Sandall (1979) distilled ternary mixtures of benzene (1), toluene (2), and ethylbenzene (3) in a wetted-wall column of 2.21-cm inside diameter. Samples of the vapor and liquid phases were taken from various points along the column. During one of their experiments the bulk of the vapor phase at a height of 300 mm from the bottom of the column had the following molar composition: 74.71% benzene, 20.72% toluene, and 4.57% ethylbenzene. From the experimental conditions, Taylor and Krishna (1993) estimated that the corresponding vapor composition at the interface was 89.06% benzene, 9.95% toluene, and 0.99% ethylbenzene. Estimate the individual molar fluxes, assuming that the total molar flux is zero. The following data apply:

> Viscosity of the vapor, $\mu = 8.82 \times 10^{-6}$ kg/m·s
> Density of the vapor, $\rho = 2.81$ kg/m^3
> Total molar density, $c = 34.14$ mol/m^3
> $D_{12} = 2.228$ mm^2/s; $D_{13} = 2.065$ mm^2/s; $D_{23} = 1.832$ mm^2/s
> $G_y = 8.68$ kg/m^2·s

Solution

The Sherwood–Gilliland correlation in the form of equation (2-74) may be used to estimate the binary mass-transfer coefficients in the vapor phase. Calculate the Reynolds number:

$$\text{Re} = \frac{G_y D}{\mu} = 21,750$$

Compute the Schmidt number for each binary pair using the MS diffusivities:

$$\text{Sc}_{i,j} = \frac{\mu}{\rho D_{i,j}}$$

$$\text{Sc}_{1,2} = 1.409 \qquad \text{Sc}_{1,3} = 1.520 \qquad \text{Sc}_{2,3} = 1.713$$

Problems 135

The corresponding Sherwood numbers are from equation (2-74)

$$\text{Sh}_{i,j} = 0.023\,\text{Re}^{0.83}\,\text{Sc}_{i,j}^{0.44}$$

$$\text{Sh}_{1,2} = 106.5 \qquad \text{Sh}_{1,3} = 110.1 \qquad \text{Sh}_{2,3} = 116.1$$

Calculate the binary mass-transfer coefficients from

$$F_{i,j} = \frac{\text{Sh}_{i,j}\,cD_{i,j}}{D}$$

$$F_{1,2} = 0.367 \text{ mol/m}^2 \cdot \text{s}$$
$$F_{1,3} = 0.351 \text{ mol/m}^2 \cdot \text{s}$$
$$F_{2,3} = 0.328 \text{ mol/m}^2 \cdot \text{s}$$

We have now all the information needed to modify the Mathcad program of Example 1.17 to estimate the individual molar fluxes. In this case, there is an additional unknown value, namely N_3. The additional relation is the statement that the total flux is zero:

$$N_1 + N_2 + N_3 = 0$$

The program yields the following results:

$$N_1 = -0.0527 \text{ mol/m}^2 \cdot \text{s}$$
$$N_2 = 0.0395 \text{ mol/m}^2 \cdot \text{s}$$
$$N_3 = 0.0132 \text{ mol/m}^2 \cdot \text{s}$$

PROBLEMS

The problems at the end of each chapter have been grouped into four classes (designated by a superscript after the problem number).

Class a: Illustrates direct numerical application of the formulas in the text.
Class b: Requires elementary analysis of physical situations, based on the subject material in the chapter.
Class c: Requires somewhat more mature analysis.
Class d: Requires computer solution.

2.1[a]. Mass-transfer coefficients in a gas absorber

A gas absorber is used to remove benzene (C_6H_6) vapors from air by scrubbing the gas mixture with a nonvolatile oil at 300 K and 1 atm. At a certain point

136   **Convective Mass Transfer**

in the absorber, the benzene mol fraction in the bulk of the gas phase is 0.02, while the corresponding interfacial benzene gas-phase concentration is 0.0158. The benzene flux at that point is measured as 0.80 g/m^2·s.

(a) Calculate the mass-transfer coefficient in the gas phase at that point in the equipment, expressing the driving force in terms of mole fractions.

Answer: k_y = 2.442 mol/m^2·s

(b) Calculate the mass-transfer coefficient in the gas phase at that point in the equipment, expressing the driving force in terms of molar concentrations, kmol/m^3.

Answer: k_c = 0.060 m/s

(c) At the same place in the equipment, the benzene mol fraction in the bulk of the liquid phase is 0.125, while the corresponding interfacial benzene liquid-phase concentration is 0.158. Calculate the mass-transfer coefficient in the liquid phase, expressing the driving force in terms of mol fractions.

Answer: F_L = 0.267 mol/m^2·s

2.2^a. Mass-transfer coefficients from naphthalene sublimation data

In a laboratory experiment, air at 347 K and 1 atm is blown at high speed around a single naphthalene ($C_{10}H_8$) sphere, which sublimates partially. When the experiment begins, the diameter of the sphere is 2.0 cm. At the end of the experiment, 14.32 min later, the diameter of the sphere is 1.85 cm.

(a) Estimate the mass-transfer coefficient, based on the average surface area of the particle, expressing the driving force in terms of partial pressures. The density of solid naphthalene is 1.145 g/cm^3 and its vapor pressure at 347 K is 670 Pa (Perry and Chilton, 1973).

Answer: k_G = 0.012 mol/m^2·s·kPa

(b) Calculate the mass-transfer coefficient, for the driving force in terms of molar concentration.

Answer: k_c = 0.034 m/s

2.3^a. Mass-transfer coefficients from acetone evaporation data

In a laboratory experiment, air at 300 K and 1 atm is blown at high speed parallel to the surface of a rectangular shallow pan that contains liquid acetone (C_3H_6O), which evaporates partially. The pan is 1 m long and 50 cm wide. It is connected to a reservoir containing liquid acetone which automatically replaces the acetone evaporated, maintaining a constant liquid level in the pan. During an experimental run, it was observed that 2.0 L of acetone evaporated in 5 min. Estimate the mass-transfer coefficient. The density of liquid acetone at 300 K is 0.79 g/cm^3; its vapor pressure is 27 kPa (Perry and Chilton, 1973).

Answer: F_G = 0.586 mol/m^2·s

Problems

2.4[b]. Mass-transfer coefficients from wetted-wall experimental data

A wetted-wall experimental setup consists of a glass pipe, 50 mm in diameter and 1.0 m long. Water at 300 K flows down the inner wall. Dry air enters the bottom of the pipe at the rate of 1.04 m³/min, measured at 308 K and 1 atm. It leaves the wetted section at 308 K and with a relative humidity of 21.5%. With the help of equation (2-87), estimate the average mass-transfer coefficient, with the driving force in terms of molar fractions.

Answer: k_y = 1.809 mol/m²·s

2.5[b]. Dimensional analysis: aeration

Aeration is a common industrial process and yet one in which there is often serious disagreement about correlations (Cussler, 1997). This is especially true for deep-bed fermentors and for sewage treatment, where the rising bubbles can be the chief means of stirring. We expect that the rate of oxygen mass transfer through the bed can be calculated in terms of a mass-transfer coefficient, k_c, which will depend on the average bubble velocity v, the solution's density ρ and viscosity μ, the bubble diameter d, the depth of the bed z, and the diffusivity D_{AB}. Use the Buckingham method to determine the dimensionless groups to be formed from the variables significant to this problem. Choose D_{AB}, d, and ρ as core variables.

2.6[b]. Dimensional analysis: the artificial kidney

In the artificial kidney, blood flowing inside a tubular membrane is dialyzed against well-stirred saline solution. Toxins in the blood diffuse across the membrane into the saline solution, thus purifying the blood. This dialysis is often slow; it can take more than 40 hours per week (Cussler, 1997). Thus, increasing the mass transfer in this system would greatly improve its clinical value. It has been found that in modern designs of artificial kidneys (thin membranes and vigorous stirring of the saline solution) most of the resistance to toxin removal resides inside the membrane, in the blood. The mass-transfer coefficient, k_c, for a particular toxin varies with v, ρ, and μ (velocity, density, and viscosity of the blood), the diffusivity, D_{AB}, the diameter d, and the length L, of the tube. Use the Buckingham method to determine the dimensionless groups to be formed from the variables significant to this problem. Choose D_{AB}, d, and ρ as core variables.

2.7[c]. Mass transfer in an annular space

(a) In studying rates of diffusion of naphthalene into air, an investigator replaced a 30.5-cm section of the inner pipe of an annulus with a naphthalene rod. The annulus was composed of a 51-mm-OD brass inner pipe that is surrounded by a 76-mm-ID brass pipe. While operating at a mass velocity within the annulus of 12.2 kg of air/m²·s at 273 K and 1 atm, the investigator determined that the partial pressure of naphthalene in the exiting gas stream was 0.041 Pa. Under the conditions of the investigation, the Schmidt number of the gas was 2.57, the viscosity

138   **Convective Mass Transfer**

was 175 μP, and the vapor pressure of naphthalene was 1.03 Pa. Estimate the mass-transfer coefficient from the inner wall for this set of conditions. Assume that equation (2-87) applies.

Answer: $k_y = 0.872$ mol/m^2·s

(b) Monrad and Pelton (1942) presented the following correlation for the heat-transfer coefficient in an annular space:

$$\frac{h}{C_p G_y} = 0.023 \left[\frac{d_o}{d_i}\right]^{0.5} \left[\frac{d_e G_y}{\mu}\right]^{-0.2} \left[\frac{C_p \mu}{k}\right]^{-2/3} \tag{2-102}$$

where d_o and d_i are the outside and inside diameters of the annulus, and d_e is the equivalent diameter defined as

$$d_e = 4 \times \frac{\text{cross-sectional area of flow}}{\text{wetted perimeter}} \tag{2-103}$$

Write down the analogous expression for mass transfer and use it to estimate the mass-transfer coefficient for the conditions of part (a). Compare the two results.

Answer: 2.4% difference

2.8^c. The Chilton–Colburn analogy: flow across tube banks

Winding and Cheney (1948) passed air at 310 K and 1 atm through a bank of rods of naphthalene. The rods were in a staggered arrangement, with the air flowing at right angles to the axes of the rods. The bank consisted of 10 rows containing alternately five and four 38-mm-OD tubes (d = 38 mm) spaced on 57-mm centers, with the rows 76 mm apart. The mass-transfer coefficient was determined by measuring the rate of sublimation of the naphthalene. The data could be correlated by

$$k_G = 3.86 \times 10^{-9} G_y^{0.56}$$
$$5 < G_y < 40 \text{ kg/m}^2 \cdot \text{s} \tag{2-104}$$

where G_y is the maximum mass velocity through the tube bank, in kg/m^2·s, and k_G is in kmol/m^2·s·Pa.

(a) Rewrite equation (2-104) in terms of the Colburn j_D factor. The diffusivity of naphthalene in air at 310 K and 1 atm is 0.074 cm^2/s.

Answer: $j_D = 0.551 \text{Re}^{-0.44}$ where $\text{Re} = G_y d / \mu$

(b) Estimate the mass-transfer coefficient for evaporation of n-propyl alcohol into carbon dioxide over the outer surface of the tubes, for the same geometrical arrangement, when the carbon dioxide flows at a maximum velocity of 10 m/s at 300 K and 1 atm. The vapor pressure of n-propyl alcohol at 300 K is 2.7 kPa. The diffusivity of n-propyl alcohol in CO$_2$ at 300 K and 1 atm is 0.076 cm^2/s.

Answer: $k_G = 0.019$ mol/m^2·s·kPa

Problems 139

(c) Zhukauskas (1972) proposed the following correlation for the heat-transfer coefficient in a staggered tube bank arrangement like that studied by Winding and Cheney:

$$Nu = 0.453\,Re^{0.568}\,Pr^{0.36}$$
$$10 < Re < 10^6 \qquad\qquad (2\text{-}105)$$
$$0.7 < Pr < 500$$

Use the mass-transfer expression analogous to equation (2-105) to estimate the mass-transfer coefficient of part (b). Compare the results.

Answer: $k_G = 0.017$ mol/m^2·s·kPa

2.9^b. Mass transfer from a flat plate

A 1-m square thin plate of solid naphthalene is oriented parallel to a stream of air flowing at 20 m/s. The air is at 330 K and 101.3 kPa. The naphthalene remains at 290 K; at this temperature, the vapor pressure of naphthalene is 26 Pa. Estimate the moles of naphthalene lost from the plate per hour if the end effects can be ignored.

Answer: 1.86 mol/h

2.10^b. Mass transfer from a flat plate

A thin plate of solid salt, NaCl, measuring 15 by 15 cm, is to be dragged through seawater at a velocity of 0.6 m/s. The 291 K seawater has a salt concentration of 0.0309 g/cm^3 and a density of 1.022 g/cm^3. Estimate the rate at which the salt goes into solution if the edge effects can be ignored. Assume the kinematic viscosity at the average liquid film conditions is 1.02×10^{-6} m^2/s, and the diffusivity is 1.25×10^{-9} m^2/s. The solubility of NaCl in water at 291 K is 0.35 g/cm^3, and the density of the saturated solution is 1.22 g/cm^3 (Perry and Chilton, 1973).

Answer: 0.86 kg/h

2.11^b. Mass transfer from a flat liquid surface

During the experiment described in Problem 2.3, the air velocity was measured at 6 m/s, parallel to the longest side of the pan. Estimate the mass-transfer coefficient predicted by equation (2-56) or (2-57) and compare it to the value measured experimentally. Notice that, due to the high volatility of acetone, the average acetone concentration in the gas film is relatively high. Therefore, properties such as density and viscosity should be estimated carefully. The following data for acetone might be needed: $T_c = 508.1$ K, $P_c = 47.0$ bar, $M = 58$, $V_c = 209$ cm^3/mol, $Z_c = 0.232$ (Reid et al., 1987).

Answer: $F_G = 0.617$ mol/m^2·s

140   **Convective Mass Transfer**

2.12[b]. Evaporation of a drop of water falling in air

Repeat Example 2.9 for a drop of water which is originally 2 mm in diameter.

Answer: 390 m

2.13[b]. Dissolution of a solid sphere into a flowing liquid stream

Estimate the mass-transfer coefficient for the dissolution of sodium chloride from a cast sphere, 1.5 cm in diameter, if placed in a flowing water stream. The velocity of the 291 K water stream is 1.0 m/s. Assume that the kinematic viscosity at the average liquid film conditions is 1.02×10^{-6} m^2/s, and the mass diffusivity is 1.25×10^{-9} m^2/s. The solubility of NaCl in water at 291 K is 0.35 g/cm^3, and the density of the saturated solution is 1.22 g/cm^3 (Perry and Chilton, 1973).

Answer: F_L = 1.05 mol/m^2·s

2.14[b]. Sublimation of a solid sphere into a gas stream

During the experiment described in Problem 2.2, the air velocity was measured at 10 m/s. Estimate the mass-transfer coefficient predicted by equation (2-64) and compare it to the value measured experimentally. The following data for naphthalene might be needed: T_b = 491.1 K, V_c = 413 cm^3/mol.

Answer: k_G = 0.011 mol/m^2·s·kPa

2.15[b]. Dissolution of a solid sphere into a flowing liquid stream

The crystal of Problem 1.25 is a sphere 2 cm in diameter. It is falling at terminal velocity under the influence of gravity into a big tank of water at 288 K. The density of the crystal is 1464 kg/m^3 (Perry and Chilton, 1973).

(a) Estimate the crystal's terminal velocity.

Answer: 0.56 m/s

(b) Estimate the rate at which the crystal dissolves and compare it to the answer obtained in Problem 1.25.

2.16[c]. Mass transfer inside a circular pipe

Water flows through a thin tube, the walls of which are lightly coated with benzoic acid ($C_7H_6O_2$). The water flows slowly, at 298 K and 0.1 cm/s. The pipe is 1 cm in diameter. Under these conditions, equation (2-98) applies.

(a) Show that a material balance on a length of pipe Z leads to

$$c_{Aout} = c_A^* \left[1 - \exp\left(\frac{-k_L a Z}{v} \right) \right] \qquad (2\text{-}106)$$

where v is the average fluid velocity, and c_A* is the equilibrium solubility concentration.

Problems   141

(b) What is the average concentration of benzoic acid in the water after a length of pipe of 2 m. The solubility of benzoic acid in water at 298 K is 0.003 g/cm^3, and the mass diffusivity is 1.0×10^{-5} cm^2/s (Cussler, 1997).

Answer: 1.14 mg/cm^3

2.17[b]. Mass transfer in a wetted-wall tower

Water flows down the inside wall of a 25-mm-ID wetted-wall tower of the design of Figure 2.4, while air flows upward through the core. Dry air enters at the rate of 7 kg/m^2·s. Assume the air is everywhere at its average conditions of 309 K and 1 atm, the water at 294 K, and the mass-transfer coefficient constant. Compute the average partial pressure of water in the air leaving a tower 1-m-long.

Answer: 1.516 kPa

2.18[c]. Mass transfer in an annular space

In studying the sublimation of naphthalene into an airstream, an investigator constructed a 3-m-long annular duct. The inner pipe was made from a 25-mm-OD, solid naphthalene rod; this was surrounded by a 50-mm-ID naphthalene pipe. Air at 289 K and 1 atm flowed through the annular space at an average velocity of 15 m/s. Estimate the partial pressure of naphthalene in the airstream exiting from the tube. At 289 K, naphthalene has a vapor pressure of 5.2 Pa, and a diffusivity in air of 0.06 cm^2/s. Use the results of Problem 2.7 to estimate the mass-transfer coefficient for the inner surface and equation (2-74), using the equivalent diameter defined in Problem 2.7, to estimate the coefficient from the outer surface.

Answer: 3.59 Pa

2.19[c]. Benzene evaporation on the outside surface of a single cylinder

Benzene is evaporating at 20 kg/h over the surface of a porous 10-cm-diameter cylinder. Dry air at 325 K and 1 atm flows at right angle to the axis of the cylinder at a velocity of 2 m/s. The liquid is at a temperature of 315 K where it exerts a vapor pressure of 26.7 kPa. Estimate the length of the cylinder. For benzene, $T_c = 562.2$ K, $P_c = 48.9$ bar, $M = 78$, $V_c = 259$ cm^3/mol, and $Z_c = 0.271$ (Reid et al., 1987).

Answer: 1.74 m

2.20[b]. Mass transfer in a packed bed

Wilke and Hougan (1945) reported the mass transfer in beds of granular solids. Air was blown through a bed of porous celite pellets wetted with water, and by evaporating this water under adiabatic conditions, they reported gas-film coefficients for packed beds. In one run, the following data were reported:

142   **Convective Mass Transfer**

Effective particle diameter	5.71 mm
Gas stream mass velocity	0.816 kg/m^2·s
Temperature at the surface	311 K
Pressure	97.7 kPa
k_G	4.415 × 10^{-3} kmol/m^2·s·atm

With the assumption that the properties of the gas mixture are the same as those of air, calculate the gas-film mass-transfer coefficient using equation (2-90) and compare the result with the value reported by Wilke and Hougan.

Answer: 2.1% difference

2.21[b]. Mass transfer and pressure drop in a packed bed

Air at 373 K and 2 atm is passed through a bed 10 cm in diameter composed of iodine spheres 0.7 cm in diameter. The air flows at a rate of 2 m/s, based on the empty cross section of the bed. The porosity of the bed is 40%.

(a) How much iodine will evaporate from a bed 0.1 m long? The vapor pressure of iodine at 373 K is 6 kPa.

Answer: 0.41 kg/min

b) Estimate the pressure drop through the bed.

2.22[b]. Volumetric mass-transfer coefficients in industrial towers

The interfacial surface area per unit volume, a, in many types of packing materials used in industrial towers is virtually impossible to measure. Both a and the mass-transfer coefficient depend on the physical geometry of the equipment and on the flow rates of the two contacting, immiscible streams. Accordingly, they are normally correlated together as the *volumetric mass-transfer coefficient, $k_L a$*.

Empirical equations for the volumetric coefficients must be obtained experimentally for each type of mass-transfer operation. Sherwood and Holloway (1940) obtained the following correlation for the liquid-film mass-transfer coefficient in packed absorption towers:

$$\frac{k_L a}{D_{AB}} = \alpha \left[\frac{G_x}{\mu}\right]^{1-n} \left[\frac{\mu}{\rho D_{AB}}\right]^{0.5} \tag{2-107}$$

The values of α and n to be used in equation (2-107) for various industrial packings are listed in the following table, *when SI units are used exclusively*.

(a) Consider the absorption of sulfur dioxide with water at 294 K in a tower packed with 25-mm Raschig rings. If the liquid mass velocity is G_x = 2.04 kg/m^2·s, estimate the liquid-film mass-transfer coefficient. The diffusivity of SO$_2$ in water is 1.7 × 10^{-9} m^2/s at 294 K.

Answer: 6.7 × 10^{-3} s^{-1}

Packing	α	n
50-mm Raschig ring	341	0.22
38-mm ring	384	0.22
25-mm ring	426	0.22
13-mm ring	1391	0.35
9.5-mm ring	3116	0.46
38-mm Berl saddle	731	0.28
25-mm saddle	777	0.28
13-mm saddle	686	0.28
76-mm spiral tiles	502	0.28

(b) Whitney and Vivian (1949) measured rates of absorption of SO_2 in water and found the following expression for 25-mm Raschig rings at 294 K

$$k_x a = 0.152 G_x^{0.82} \qquad (2\text{-}108)$$

where $k_x a$ is in kmol/m^3·s, and G_x is in kg/m^2·s. For the conditions described in part (a), estimate the liquid-film mass-transfer coefficient using equation (2-108). Compare the results.

2.23[b]. Mass transfer in fluidized beds

Cavatorta et al. (1999) studied the electrochemical reduction of ferricyanide ions, $\{Fe(CN)_6\}^{-3}$, to ferrocyanide, $\{Fe(CN)_6\}^{-4}$, in aqueous alkaline solutions. They studied different arrangements of packed columns, including fluidized beds. The fluidized bed experiments were performed in a 5-cm-ID circular column, 75 cm high. The bed was packed with 0.534-mm spherical glass beads, with a particle density of 2.612 g/cm^3. The properties of the aqueous solutions were density = 1083 kg/m^3, viscosity = 1.30 cP, diffusivity = 5.90×10^{-10} m^2/s. They found that the porosity of the fluidized bed, ε, could be correlated with the superficial liquid velocity based on the empty tube, v_s, through

$$v_s = 8.88\varepsilon^{3.17}$$
$$0.35 < \varepsilon < 1.0$$

where v_s is in cm/s.

(a) Using equation (2-91), estimate the mass-transfer coefficient, k_L, if the porosity of the bed is 60%.

Answer: $k_L = 5.78 \times 10^{-5}$ m/s

(b) Cavatorta et al. (1999) proposed the following correlation to estimate the mass-transfer coefficient for their fluidized bed experimental runs:

144   **Convective Mass Transfer**

$$\varepsilon j_D = 0.333 \left[\frac{Re}{\varepsilon(1-\varepsilon)} \right]^{-0.364}$$

$$j_D = \frac{k_L}{v_s} Sc^{2/3}$$

where Re is based on the empty tube velocity. Using this correlation, estimate the mass-transfer coefficient, k_L, if the porosity of the bed is 60%. Compare your result to that of part (a).

2.24[c, d]. Simultaneous heat and mass transfer: film condensation

An air–water vapor mixture flows upward through a vertical copper tube 25.4-mm OD, 1.65-mm wall thickness, which is surrounded by flowing cold water. As a result, water vapor condenses and flows as a liquid film down the inside of the tube. At one point in the condenser, the average velocity of the gas is 10 m/s, its bulk-average temperature is 339 K, the pressure is 1 atm, and the bulk-average partial pressure of water vapor is 0.24 atm. The film of condensed liquid is such that its heat-transfer coefficient is 6.0 kW/m²·K. The cooling water is at a bulk-average temperature of 297 K and has a heat-transfer coefficient of 0.57 kW/m²·K. Calculate the local rate of condensation of water from the airstream. For the gas mixture, C_p = 1.145 kJ/kg·K, Sc = 0.6, Pr = 0.7, and μ = 175 μP. For the water vapor, C_p = 1.88 kJ/kg·K. The thermal conductivity of copper is 0.381 kW/m·K.

Answer: 21.3 kg/m²·h

2.25[c]. Simultaneous heat and mass transfer: transpiration cooling

Equation (2-76) can also be used for calculations of *transpiration cooling*, a method of cooling porous surfaces which are exposed to extremely hot gases by forcing a cold gas or evaporating liquid through the surface into the gas stream. Consider the following application of transpiration cooling. Air at 900 K and 1 atm flows past a flat, porous surface. Saturated steam at 125 kPa flows through the surface into the airstream to keep the surface temperature at 523 K. The air velocity is such that the heat-transfer coefficient would be 1.1 kW/m²·s if no steam were used. From 523 K to 900 K, $C_{p,A}$ = 2.09 kJ/kg·K.

(a) What rate of steam flow is required?

Answer: 0.69 kg/m²·s

(b) Repeat if, instead of steam, liquid water at 298 K is forced through the porous surface.

2.26[b]. Mass transfer in a hollow-fiber boiler feedwater deaerator

Consider the hollow-fiber BFW deaerator described in Example 2.14. If the water flow rate increases to 60,000 kg/h while everything else remains constant, calculate the fraction of the entering dissolved oxygen that can be removed.

Answer: 94.2%

Problems

2.27[b]. Mass transfer in a hollow-fiber boiler feedwater deaerator

(a) Consider the hollow-fiber BFW deaerator described in Example 2.14. If only oxygen diffuses across the membrane, calculate the gas volume flow rate and composition at the lumen outlet. The water enters the shell side at 298 K saturated with atmospheric oxygen, which means a dissolved oxygen concentration of 8.38 mg/L.

Answer: 44.2 L/min, 9.6% O_2

(b) Calculate the mass-transfer coefficient at the average conditions inside the lumen. Neglect the thickness of the fiber walls when estimating the gas velocity inside the lumen.

Answer: k_c = 0.37 cm/s

2.28[b]. Intravascular resistance to oxygen diffusion in Krogh model

Because of intravascular resistance to oxygen diffusion, the oxygen partial pressure at the blood–tissue interface may be less than the average oxygen partial pressure in the blood. This effect can be represented approximately by the expression (McGuire and Secomb, 2001):

$$q_L = k_{L,G}\left[p_{A,b} - p_A(R_c)\right] \qquad (2\text{-}109)$$

where q_L is the rate of oxygen diffusion from the capillary into the surrounding tissue per unit length of capillary, and $k_{L,G}$ is an intravascular mass-transfer coefficient. For this case, a Sherwood number has been defined as

$$\mathrm{Sh} = \frac{k_{L,G}}{\pi K_{pl}} \qquad (2\text{-}110)$$

where K_{pl} is the Krogh oxygen diffusion constant in plasma = 8.3×10^{-10} cm^3 of O_2/cm·s·torr. The value of Sh depends on the vessel diameter. For example, for a capillary with a 5-μm diameter, Sh = 2.5 (McGuire and Secomb, 2001).

(a) Show that, for the Krogh cylinder model described in Example 1.28,

$$p_A\left(R_c\right) = p_{A,b} - \frac{M_0}{\mathrm{Sh}K_{pl}}\left(R_t^2 - R_c^2\right)$$

(b) For the conditions described in Example 1.28, at the capillary entrance, calculate the oxygen partial pressure at the capillary wall.

Answer: 78.5 torr

2.29[b, d]. Ternary condensation inside a vertical tube

Isopropanol (1) and water (2) are condensing in the presence of nitrogen (3) inside a vertical tube of 2.54 cm inside diameter. At the vapor inlet, the bulk composition of the gas phase is 11.23 mol% isopropanol, 42.46% water, and

146   **Convective Mass Transfer**

46.31% nitrogen. The interfacial vapor composition is 14.57 mol% isopropanol, 16.40% water, and 69.03% nitrogen. Estimate the molar rates of condensation of isopropanol and water at that point. The following data apply:

> viscosity of the vapor, $\mu = 1.602 \times 10^{-5}$ kg/m·s
> density of the vapor, $\rho = 0.882$ kg/m^3
> $D_{12} = 16.00$ mm^2/s; $D_{13} = 14.44$ mm^2/s; $D_{23} = 38.73$ mm^2/s
> vapor-phase Reynolds number, Re = 9574

Answer: $N_2 = 0.575$ mol/m^2·s

2.30[c, d]. Simultaneous heat and mass transfer: surface evaporation

A hollow, porous cylinder 25 mm in OD, 15 mm in ID, is fed internally with liquid diethyl ether at 293 K. The ether flows radially outward and evaporates on the outer surface. Pure nitrogen at 373 K and 1 atm flows perpendicular to the cylinder axis at 3 m/s carrying away the evaporated ether. The ether flow is to be that which will just keep the outer surface of the cylinder wet with liquid. Since mass-transfer rates vary about the periphery, the cylinder will be rotated slowly to keep the surface uniformly wet. Calculate the surface temperature and the rate of ether flow, in kg/h per meter of cylinder length.

Answer: 3.47 kg/h·m

2.31[d]. Design of a packed-bed air humidifier

Design a packed-bed air humidifier to process 9.12 kg/h of dry air. Assume that conditions are like those of Example 2.13. The packing will consist of spherical glass beads with a particle-to-bed diameter ratio of 1/20. The partial pressure of water in the outlet air must be 99.9% of the maximum and the total gas pressure drop is not to exceed 500 Pa. Assuming a cylindrical bed, determine the diameter and depth of the bed, the gas mass velocity, and the diameter of the glass beads.

2.32[d]. Influence of mass transfer on the dynamics of a rising and sinking droplet in water

The occurrence of mass transfer to and from droplets driven by buoyancy can be found in applications such as liquid–liquid extraction, and during environmental accidents such as deep water oil spills. (During the "Deepwater Horizon" accident in 2010 about 4.9 million barrels of oil was released into the Gulf of Mexico from a point of release located at a depth of around 1500 meters.) The oil droplets entering the oceanic environment during an accidental release are a mixture of many hydrocarbons. They can be treated as a pseudobinary mixture; the first component includes all the lighter hydrocarbons which are soluble in the surrounding water, and the second component represents all the heavier hydrocarbons practically insoluble in water.

Problems 147

When the droplets are released with an initial mixture density lower than that of the surrounding water, they rise due to buoyancy. However, as the droplet rises, the mass transfer of the lighter soluble components into the surrounding water causes the droplet density to increase gradually. As the density of the droplet approaches that of the water, the buoyancy effect disappears, and the droplet reaches a stationary stage. Further loss of the lighter components makes the density of the droplet higher than that of the water and the droplet starts to sink until it reaches the bottom of the ocean.

Rao et al. (2015) investigated this phenomenon experimentally as well as numerically. Their droplets were mixtures of acetonitrile (light, soluble component) and chlorobenzene (heavy, insoluble component). The droplets were released near the bottom of a glass tank filled with stagnant water. The motion of the droplets was captured using a high-speed camera. They measured the rate at which aceto-nitrile was transferred from the droplet (the dispersed phase) to the water (the continuous phase). They found that during both the ascent and descent stages, mass transfer is dominated by diffusion and forced convection. Their experimental data on the overall mass-transfer coefficient were well correlated by

$$\mathrm{Sh} = 2 + 0.573\,\mathrm{Re}^{0.31}\,\mathrm{Sc}^{0.44}$$
$$20 < \mathrm{Re} < 200 \qquad \mathrm{Sc} < 800$$

$$(2\text{-}111)$$

Near the stationary stage, diffusion, forced convection, and natural convection are all important, and their correlation was

$$\mathrm{Sh} = 2 + 0.213\,\mathrm{Re}^{0.45}\,\mathrm{Sc}^{0.31} + 0.4\mathrm{Gr}_D^{0.28}\mathrm{Sc}^{0.31}$$
$$\mathrm{Re} < 20 \qquad \mathrm{Sc} < 800$$

$$(2\text{-}112)$$

The fluid properties were evaluated at the bulk conditions of the continuous phase, virtually pure water at 298 K.

During Experiment #4 of Rao et al. (2015), 3.6 s after being released, a droplet (27.5 wt% acetonitrile, 72.5% chlorobenzene) had a diameter of 4.8 mm and was rising in the water tank at a velocity of 0.015 m/s. Estimate the value of the overall mass-transfer coefficient at that instant.

2.33[b]. Volumetric mass-transfer coefficients in high-viscosity biochemical reactors

The volumetric mass-transfer coefficient, $k_L a$, is one of the most important transport characteristics used in the design of mechanically agitated gas–liquid contactors. These apparatuses are frequently used in food and biochemical indus-tries as fermenters and as hydrogenation and chlorination reactors. In these appli-cations liquids often exhibit extremely high viscosities and there is a lack of reliable data for predicting accurate values of $k_L a$. Recently, Labik et al. (2017) performed over 1000 experiments covering a wide range of process conditions under which

148   **Convective Mass Transfer**

industrial fermenters handling high-viscosity liquids operate. They proposed the
following dimensional correlation for gassed agitated biochemical reactors (with
liquid viscosities close to 10 cP), independent of the impeller type:

$$k_L a = 0.295 (\Omega \times D)^{2.083} v_s^{0.461} \text{Po}^{0.737} \qquad (2\text{-}113)$$

Here

$k_L a$ = volumetric mass-transfer coefficient (s^{-1})
Ω = impeller frequency (s^{-1})
D = impeller diameter (m)
v_s = gas superficial velocity (based on total cross-section area of vessel, m/s)
Po = impeller power number [$P/(\rho\, \Omega^3 D^5)$, dimensionless]
P = power dissipated by the impeller (W)
ρ = liquid density (kg/m^3)

Consider a cylindrical gassed stirred pilot-plant fermenter handling a very viscous
batch. The vessel dimensions are 0.6-m diameter and 1.8-m height. The fermenter
contents are agitated by impellers 0.2 m in diameter. The rate of rotation of the
impellers is 360 rpm, and the specific power dissipated by them is 1.0 kW/m^3. The
rate of aeration is 72.0 L/min. The temperature is kept at 293 K. At this temperature, the density of the liquid is 1004 kg/m^3 and its density is 10.2 cP.

(a) Calculate the gas superficial velocity.

Answer: 4.24 mm/s

(b) Calculate the power number for the impellers.

Answer: 7.2

(c) Estimate the liquid volumetric mass-transfer coefficient.

Answer: $k_L a$ = 0.149 s^{-1}

2.34[b]. Estimation of mass-transfer coefficients at a smooth air–water interface

Air and water are the two most important fluids on Earth. The physical
processes at the surfaces of oceans, estuaries, rivers, and lakes relevant to the
exchange of chemicals between the two phases are extremely complicated and variable. For relatively calm conditions, the air-surface water interface is smooth, and
a mixture of theoretical concepts and empirical knowledge allows reasonable prediction of the rates of interface mass transfer.

References 149

Traditionally, water vapor is used as the test substance for determining the air phase mass-transfer coefficient, k_{cia}, for any component i. The corresponding relation is (Johnson, 2010)

$$k_{cia} = k_{cwa} \left(\frac{Sc_{ia}}{Sc_{wa}} \right)^{-2/3} \tag{2-114}$$

Here

k_{cwa} air phase mass-transfer coefficient for water vapor in air (cm/s)
k_{cia} air phase mass-transfer coefficient for component i in air (cm/s)
Sc_{wa} Schmidt number for water vapor in air
Sc_{ia} Schmidt number for component i in air

Based on the theory of wind friction in a boundary layer, a model called the Coupled Ocean-Atmospheric Response Experiment (COARE) is used to estimate k_{cwa} as follows (Johnson, 2010):

$$k_{cwa} = 0.1 + \frac{\alpha(u_{10}) \times u_{10}}{13.3 Sc_{wa}^{0.5} + \dfrac{100}{\alpha(u_{10})} - 5 + 1.25 \ln(Sc_{wa})} \tag{2-115}$$

$$\alpha(u_{10}) = (6.1 + 0.63 u_{10})^{0.5}$$

Here: k_{cwa} air phase mass-transfer coefficient for water vapor in air (cm/s)
u_{10} wind speed measured 10 m above the water surface (m/s)

Estimate the air phase mass-transfer coefficient for methanol when the atmospheric conditions are 298 K, 1 atm, and measured wind speed of 15 m/s at an elevation of 10 m. Under these conditions, the kinematic viscosity of air is $\nu = 0.155$ cm^2/s. Use the mass diffusivities values listed in Appendix A.

Answer: 1.5 cm/s

REFERENCES

Bao, L., B. Liu, and G. G. Lipscomb, *AIChE J.*, **45**, 2346 (1999).
Bedingfield, C. H., and T. B. Drew, *Ind. Eng. Chem.*, **42**, 1164 (1950).
Benítez, J., *Process Engineering and Design for Air Pollution Control*, Prentice Hall, Englewood Cliffs, NJ (1993).
Bird, R. B., W. E. Stewart, and E. N. Lightfoot, *Transport Phenomena*, Wiley, New York, NY (1960).
Blasius, H., *Z. Math. U. Phys. Sci.*, **56,** 15656 (1908).
Brian, P. L. T., and H. B. Hales, *AIChE J.*, **15**, 419 (1969).
Buckingham, E., *Phys. Rev.*, **2**, 345 (1914).

Cavatorta, O. N., et al., *AIChE J.*, **45**, 938 (1999)

Chilton, T. H., and A. P. Colburn, *Ind. Eng. Chem.*, **26**, 1183 (1934).

Costello, M. J., et al., *J. Memb. Sci.*, **80**, 1 (1993).

Cussler, E. L., *Diffusion*, 2nd ed., Cambridge University Press, New York, NY (1997).

Davis, M. L., and D. A. Cornwell, *Introduction to Environmental Engineering*, 3rd ed., McGraw-Hill, New York, NY (1998).

Dribicka, M. M., and O. C. Sandall, *Chem. Eng. Sci.*, **34**, 733 (1979).

Eckert, E. R. G., and R. M. Drake, *Heat and Mass Transfer*, 2nd ed., McGraw-Hill, NY (1959).

Ergun, S., *Chem. Eng. Prog.*, **48**, 89 (1952).

Evnochides, S., and G. Thodos, *AIChE J.*, **5**, 178 (1959).

Froessling, N., *Gerlands Beitr. Geophys.*, **52**, 170 (1939).

Gilliland, E. R., and T. K. Sherwood, *Ind. Eng. Chem.*, **26**, 516 (1934).

Gupta, A. S., and G. Thodos, *AIChE J*, **8**, 608 (1962).

Himmelblau, D. M., *Basic Principles and Calculations in Chemical Engineering*, 5th ed., Prentice Hall, Englewood Cliffs, NJ (1989).

Holman, J. P., *Heat Transfer*, 7th ed., McGraw-Hill, New York, NY (1990).

Howarth, L., *Proc. Roy. Soc. London*, **A164**, 547 (1938).

Incropera, F. P., et al., *Introduction to Heat Transfer*, 5th ed., John Wiley and Sons, Hoboken, NJ (2007).

Johnson, M. T., *Ocean Sci.*, **6**(4), 913 (2010).

Labik, L., et al., *Chemical Engineering Science*, **170**, 451 (2017).

Levich, V. G., *Physicochemical Hydrodynamics*, Prentice Hall, Englewood Cliffs, NJ (1962).

Linton, W. H., and T. K. Sherwood, *Chem. Eng. Prog.*, **46**, 258 (1950).

Lucas, K., *Phase Equilibria and Fluid Properties in the Chemical Industry*, p. 573, Dechema, Frankfurt (1980).

McGuire, B. J., and T. W. Secomb, *J. Appl. Physiol.*, **91**, 2255 (2001).

Monrad, C. C., and J. F. Pelton, *Trans. AIChE*, **38**, 593 (1942).

Nandakumar, K., Y. Shu, and K. T. Chuang, *AIChE J.*, **45**, 2286 (1999).

Perry, R. H., and C. H. Chilton (eds.), *Chemical Engineers' Handbook*, 5th ed., McGraw-Hill, New York, NY (1973).

Prasad, R., and K. K. Sirkar, *AIChE J.*, **34**, 177 (1988).

Rao, A., et al., *AIChE J.*, **61**, 342 (2015).

Reid, R. C., J. M. Prausnitz, and B. E. Poling, *The Properties of Gases and Liquids*, 4th ed., McGraw-Hill, Boston, MA (1987).

Sherwood, T. K., and F. A. Holloway, *Trans. AIChE*, **36**, 21, 39 (1940).

Sherwood, T. K., R. L. Pigford, and C. R. Wilke, *Mass Transfer*, McGraw-Hill, New York, NY (1975).

Smith, J. M., H. C. Van Ness, and M. M. Abbott, *Introduction to Chemical Engineering Thermodynamics*, 5th ed., McGraw-Hill, New York, NY (1996).

Steinberger, R. L., and R. E. Treybal, *AIChE J.*, **6**, 227 (1960).

Taylor, R., and R. Krishna, *Multicomponent Mass Transfer*, Wiley, New York, NY (1993).

Treybal, R. E., *Mass-Transfer Operations*, 3rd ed., McGraw-Hill, New York, NY (1980).

References 151

Welty, J. R., C. E. Wicks, and R. E. Wilson, *Fundamentals of Momentum, Heat, and Mass Transfer*, 3rd ed., Wiley, New York, NY (1984).

Whitney, R. P., and J. E. Vivian, *Chem. Eng. Progr.*, **45**, 323 (1949).

Wilke, C. R., and O. A. Hougan, *Trans. AIChE*, **41**, 445 (1945).

Winding, C. C., and A. J. Cheney, *Ind. Eng. Chem.*, **40**, 1087 (1948).

Yang, M. C., and E. L. Cussler, *AIChE J.*, **32**, 1910 (1986).

Zhukauskas, A., *Adv. Heat Transfer*, **8**, 93 (1972).

APPENDIX 2.1

Python routine to solve Blasius problem.

```
Solution of the Blasius Problem by the Shooting Method
import numpy as np
import scipy as sp
from scipy.optimize import fsolve
import matplotlib.pyplot as plt
from scipy.integrate import odeint
def model(z,t):
    return [z[1], z[2], - z[2]*z[0]/2]
def r(p):
    a = p[0]
    b = p[1]
    time = np.arange(0,1.1*a,0.1*a)
    z = odeint(model, [0, 0, b], time)
    return z[10,1] - 0.99, z[10,2] - 0.016
p = np.zeros(2,dtype=float)
p[0] = 1
p[1] = 1
p = fsolve(r,p)
print('The dimensionless boundary layer thickness = %5.1f' %p[0])
print('The dimensionless surface velocity gradient = %5.3f' %p[1])
a = p[0]; b = p[1]
time = np.arange(0,1.1*a,0.1*a)
z = odeint(model, [0, 0, b], time)
plt.plot(time, z[:,1], '-k')
plt.grid(True)
plt.xlabel('Dimensionless Position')
plt. ylabel('Dimensionless Velocity')
plt.show()
```

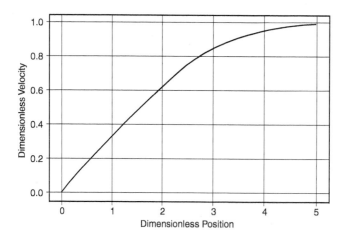

APPENDIX 2.2

Python routine to solve Example 2.11.

```
Solving Example 2-11

import numpy as np
from scipy.optimize import fsolve
# Enter data:
pA1 = 1.95; kG = 0.0108; h = 25.01 # Units of kPa, mol/m^2-s-kPa, W/m^2-K
MA = 18; CpA = 1.88; Tn = 373.15 # Units of g/mol, J/g-K, K
lambn = 40630; Tc = 647.1; T1 = 308   #Units of J/mol, K

def PA(Ti):
    return np.exp(16.3872-3885.7/(Ti-42.98))
def NA(Ti):
    return kG*(pA1-PA(Ti))
def lambd(Ti):
    return lambn*((1-(Ti/Tc))/(1-Tn/Tc))**0.38
def qt(Ti):
    qt1 = NA(Ti)*MA*CpA*(T1-Ti)/(1-np.exp(-NA(Ti)*MA*CpA/h))
    qt2 = NA(Ti)*lambd(Ti)
    return qt1+qt2
Ti = fsolve(qt, 330)
wi = abs(NA(Ti)*MA*60)   #Rate of water evaporation, in g/m^2-min
print('The interface temperature, in K = %5.2f' %Ti)
print('The water molar flux is, in mol/m^2-s = %6.5f' %NA(Ti))
print('The rate of water evaporation is, in g/m^2-min = %5.2f' %wi)
```

Program results:

The interface temperature, in K = 294.93

The water molar flux is, in mol/$m^2 \cdot$ s = -0.00728

The rate of water evaporation is, in g/$m^2 \cdot$ s = 7.86

3

Interphase
Mass Transfer

3.1 INTRODUCTION

Thus far, we have considered only the diffusion of substances within a single phase. In most of the mass-transfer operations, however, two insoluble phases are brought into contact in order to permit transfer of constituent substances between them. Therefore we are now concerned with the simultaneous application of the diffusional mechanism for each phase to the combined system. We have seen that the rate of diffusion within each phase is dependent upon the concentration gradient existing within it. At the same time, the concentration gradients of the two-phase system are indicative of the departure from equilibrium which exists between the phases. Should equilibrium be established, the concentration gradients and hence the rate of diffusion will fall to zero. It is necessary, therefore, to consider both the diffusional phenomena and the equilibria in order to describe the various situations fully.

3.2 EQUILIBRIUM CONSIDERATIONS IN CHEMICAL AND BIOCHEMICAL SYSTEMS

Your objectives in studying this section are to be able to:

1. Understand the concept of equilibrium between two insoluble phases, and its importance from the point of view of mass transfer.

Principles and Applications of Mass Transfer: The Design of Separation Processes for Chemical and Biochemical Engineering, Fourth Edition. Jaime Benitez.
© 2023 John Wiley & Sons, Inc. Published 2023 by John Wiley & Sons, Inc.

156   **Interphase Mass Transfer**

2. Represent interphase equilibrium data in the form of an equilibrium distribution curve.
3. Review the concepts of Raoult's law, modified Raoult's law, and Henry's law and use them in phase equilibria calculations.

3.2.1 Chemical Phase Equilibria

It is convenient first to consider the equilibrium characteristics of a particular operation and then to generalize the result for others. As an example, consider the gas-absorption operation which occurs when ammonia is dissolved from an ammonia–air mixture by liquid water. Suppose that a fixed amount of liquid water is placed in a closed container together with a gaseous mixture of ammonia and air, the whole arranged so that the system can be maintained at constant temperature and pressure. Since ammonia is very soluble in water, some ammonia molecules will instantly transfer from the gas into the liquid, crossing the interphase. A portion of the ammonia molecules escapes back into the gas, at a rate proportional to their concentration in the liquid. As more ammonia moves into the liquid, with the consequent increase in concentration within the liquid, the rate at which ammonia returns to the gas increases until eventually the rate at which it enters the liquid exactly matches that at which it leaves. At the same time, through the mechanism of diffusion, the concentrations throughout each phase become uniform. A dynamic equilibrium develops, and while ammonia molecules continue to transfer back and forth from one phase to the other, the net transfer falls to zero. The concentrations within each phase no longer change. To the observer who cannot see the individual molecules, the diffusion has apparently stopped.

If we now inject additional ammonia into the container, a new set of equilibrium concentrations will be eventually established, with higher concentrations in each phase than were initially obtained. In this manner, we can eventually obtain the complete relationship between the equilibrium concentrations in both phases. If the ammonia is designated as substance A, the equilibrium mol fractions in the gas (y_A) and liquid (x_A) give rise to an *equilibrium-distribution curve* as shown in Figure 3.1. This curve results irrespective of the amounts of air and water that we start with and is influenced only by the temperature and pressure of the system. It is important to note that at equilibrium the concentrations in the two phases are not equal; instead, the chemical potential of the ammonia is the same in both phases. It is this equality of chemical potentials which causes the net transfer of ammonia to stop. The curve of Figure 3.1 does not, of course, show all the equilibrium concentrations existing within the system. For example, water will partially vaporize into the gas phase, the components of the air will also dissolve to a small extent into the liquid, and equilibrium concentrations for these substances will also be established. For the moment, we need not consider these equilibria since they are of minor importance.

Equations relating the equilibrium concentrations in the two phases have been developed and are presented in textbooks on thermodynamics. In cases involving ideal gas and liquid phases, the fairly simple, yet useful relation known as *Raoult's law* applies:

$$y_A P = x_A P_A \qquad (3-1)$$

where P_A is the vapor pressure of pure A at the equilibrium temperature, and P is the equilibrium pressure. If the liquid phase does not behave ideally, a *modified form of Raoult's law* is

3.2 Equilibrium Considerations in Chemical and Biochemical Systems

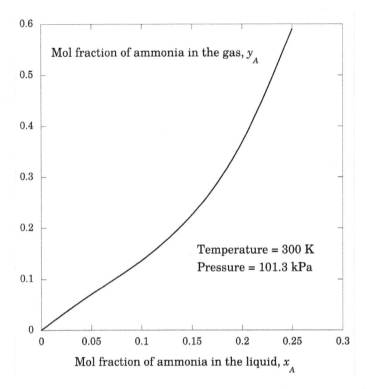

Figure 3.1 Equilibrium distribution of ammonia between air and water at constant temperature and pressure. Data from Perry and Chilton (1973).

$$y_A = x_A \gamma_A P_A \tag{3-2}$$

where γ_A is the *activity coefficient* of species A in solution. Another equilibrium relation which is found to be true for dilute solutions is *Henry's law*, expressed by

$$p_A = y_A P = H x_A \tag{3-3}$$

where p_A is the equilibrium partial pressure of component A in the vapor phase, H is the Henry's law constant. An equation similar to Henry's law relation describes the distribution of a solute between two immiscible liquids. This equation, the "distribution-law" equation, is

$$c_{Aliquid1} = K_{DA} c_{Aliquid2} \tag{3-4}$$

where K_{DA} is the distribution or partition coefficient of component A.

A detailed discussion of the characteristic shapes of the equilibrium curves for the various situations and the influence of conditions such as temperature and pressure must be left for the studies of the individual unit operations. Nevertheless, the following principles are common to all systems involving the distribution of a component between two phases:

158 Interphase Mass Transfer

1. At a fixed set of conditions, such as temperature and pressure, Gibbs' phase rule stipulates that a set of equilibrium relations exists which may be shown in the form of an equilibrium-distribution curve.

2. When a system is in equilibrium, there is no net mass transfer between the phases.

3. When a system is not in equilibrium, diffusion of the components between the phases will occur in such a manner as to cause the system composition to shift toward equilibrium. If sufficient time is allowed, the system will eventually reach equilibrium.

Examples (3.1), (3.2), and (3.3) illustrate the application of equilibrium relations for determining equilibrium concentrations.

Example 3.1 Application of Raoult's Law to a Binary System

Raoult's law may be used to determine phase compositions for the binary system of benzene and toluene, at low temperatures and pressures. Determine the composition of the vapor in equilibrium with a liquid containing 0.4 mol fraction of benzene at 300 K and calculate the total equilibrium pressure. Estimate the vapor pressure of benzene and toluene at 300 K from the Antoine equation,

$$\ln P_i = A_i - \frac{B_i}{C_i + T} \tag{3-5}$$

where P_i is the vapor pressure of component i, in mm Hg, and T is the temperature in K. The Antoine constants for benzene and toluene are given in the following table (Himmelblau, 1989):

Component	A	B	C
Benzene	15.9008	2788.51	−52.36
Toluene	16.0137	3096.52	−53.67

Solution

Benzene is more volatile than toluene and will be designated component A. From equation (3-5), at 300 K, the vapor pressures of benzene and toluene are $P_A = 103.5$ mm Hg = 13.8 kPa and $P_B = 31.3$ mm Hg = 4.17 kPa. The partial pressures are $p_A = x_A P_A = (0.4)(13.8) = 5.52$ kPa; $p_B = x_B P_B = (0.6)(4.17) = 2.50$ kPa. The total pressure is $P = p_A + p_B = 8.02$ kPa. The molar fraction of benzene in the gas phase is $y_A = p_A/P = 5.52/8.02 = 0.688$.

Example 3.2 Henry's Law: Saturation of Water with Oxygen

The Henry's law constant for oxygen dissolved in water at 298 K is 4.5×10^4 atm/mol fraction. Determine the saturation concentration of oxygen in water exposed to dry air at 298 K and 1 atm.

3.2 Equilibrium Considerations in Chemical and Biochemical Systems 159

Solution

Dry air contains 21% oxygen, then $p_A = y_A P = 0.21$ atm. The equilibrium liquid concentration is from Henry's law:

$$x_A = \frac{p_A}{H} = \frac{0.21}{4.5 \times 10^4} = 4.67 \times 10^{-6}$$

Basis: 1 L of saturated solution

For 1 L of very dilute solution of oxygen in water, the total moles of solution, n_T, will be approximately equal to the moles of water. The density is around 1 kg/L, then $n_T = 1000/18 = 55.6$ mol. The moles of oxygen in 1 L of solution are $n_O = 55.6 \times 4.67 \times 10^{-6} = 2.6 \times 10^{-4}$ mol. Then, the saturation concentration is $(2.6 \times 10^{-4}$ mol/L) $(32$ g/mol)$(1000$ mg/g) $= 8.32$ mg/L.

Example 3.3 Material Balances Combined with Equilibrium Relations: Algebraic Solution

Ten kilograms of dry gaseous ammonia, NH_3, and 15 m^3 of dry air measured at 300 K and 1 atm are mixed together and then brought into contact with 45 kg of water at 300 K in a piston/cylinder device. After a long period of time, the system reaches equilibrium. Assuming that the temperature and pressure remain constant, calculate the equilibrium concentrations of ammonia in the liquid and gas phases. Assume that the amount of water that evaporates and the amount of air that dissolves in the water are negligible. At 300 K and 1 atm, the equilibrium solubility of ammonia in air can be described by equation (3-2) with $P_A = 10.51$ atm, and the activity coefficient of ammonia given by (for $x_A < 0.3$)

$$\gamma_A = 3.5858x_A^2 - 0.622x_A + 0.156 \tag{3-6}$$

Solution

Convert the given quantities of ammonia, air, and water to moles. From the ideal gas law, $n_{air} = 15 \times 101.3/(8.314 \times 300) = 0.609$ kmol; $n_{water} = 45/18 = 2.5$ kmol. The total amount of ammonia in the system is $n_A = 10/17 = 0.588$ kmol; in equilibrium, part of it will be in the liquid phase, and the rest in the gas phase. Define L_A as the number of kmol of ammonia in the liquid phase when equilibrium is achieved. Then, at equilibrium, $0.588 - L_A$ kmol of ammonia will remain in the gas phase. Assuming that all the air will remain in the gas phase, and all the water will remain in the liquid phase, the equilibrium concentrations can be written in terms of L_A:

$$x_A = \frac{L_A}{2.5 + L_A} \qquad y_A = \frac{0.588 - L_A}{0.609 + 0.588 - L_A} = \frac{0.588 - L_A}{1.197 - L_A} \tag{3-7}$$

The equilibrium relation for 300 K and 1 atm is

$$y_A = 10.51\gamma_A x_A \tag{3-8}$$

Equations (3-6), (3-7), and (3-8) must be solved simultaneously for L_A, x_A, γ_A, and y_A.

This system of simultaneous nonlinear algebraic equations is easily solved using the "solve block" capabilities of Mathcad, as shown in Figure 3.2. Initial guesses of the values of the four variables must be supplied. However, the computational algorithm used by Mathcad is very robust and convergence to the true solution is virtually independent of the initial guesses. Figure 3.2 shows that, at equilibrium, the ammonia content of the liquid phase will be 14.4% by mol, and that of the gas phase will be 21.4% by mol. The amount of ammonia absorbed by the water will be 0.422 kmol, which is 71.8% of all the ammonia present in the system. Appendix 3.1 shows the Python solution of this example.

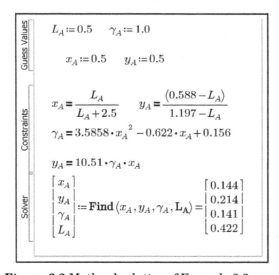

Figure 3.2 Mathcad solution of Example 3.3.

3.2.2 Biochemical Equilibrium Concepts (Seader et al., 2011)

Your objectives in studying this section are to be able to:

1. For a weak acid or base (including amino acids) define the concept of acid-ionization constant.
2. Calculate the isoelectric point of amino acids.
3. Define electrophoresis.
4. Calculate the effect of ionic strength of a solution on the acid-ionization constant.
5. Calculate the effect of solution pH on the water solubility of weak acids (including amino acids).

Most chemical separations involve creating or adding a second phase, usually a vapor. This is seldom possible with bioproducts separations since most products of biological origin are unstable in the vapor phase. Equilibrium thermodynamics

3.2 Equilibrium Considerations in Chemical and Biochemical Systems   161

of bioseparations therefore emphasizes those factors that influence biological activity. Some of these factors are: (1) pH buffering, (2) ionic strength, and (3) solubility.

Controlling pH by adding a suitable buffer to absorb or release protons produced or consumed in biochemical reactions is important to maintain activity of biological products. Suitability is determined by several buffer attributes, including its *acid-ionization* constant. A proton transfer can occur only if an acid (HA) reacts with a base (B) since isolated protons are very unstable species:

$$\begin{aligned} HA &\rightleftharpoons A^- + H^+ \\ \underline{H^+ + B} &\underline{\rightleftharpoons BH^+} \\ HA + B &\rightleftharpoons BH^+ + A^- \end{aligned} \qquad (3\text{-}9)$$

Note that A^- is called the *conjugate* base of HA and BH^+ is called the conjugate acid of B. The acid ionization constant for a weak acid, K_a, is defined as

$$K_a = \frac{\left[H^+\right]\left[A^-\right]}{\left[HA\right]} \qquad (3\text{-}10)$$

Here, the concentrations are expressed in terms of mol/L of solution (*molarity*, M). Similarly, the acid ionization constant of the conjugate acid of base B is defined as

$$K_a = \frac{\left[H^+\right]\left[B\right]}{\left[HB^+\right]} \qquad (3\text{-}11)$$

Adding a small volume of simple, dilute, weak acid (such as acetic acid, ≤ 1.0 M) or a weak base (such as Tris, ≤ 1.0 M) to a well-stirred protein solution, for example, allows its pH to be adjusted between 5 and 8 in the presence of buffering salts with minimal risk of inactivation.

Since values of K_a are usually extremely small, they are conveniently expressed in logarithmic form as

$$pK_a = -\log(K_a) \qquad (3\text{-}12)$$

For acetic acid, for example, $pK_a = 4.75$, while for the conjugate acid of Tris, $pK_a = 8.07$, both at $25\,°C$. Substituting (3-10) or (3-11) into (3-12) and using the definition of pH,

$$pH = -\log\left[H^+\right] \qquad (3\text{-}13)$$

yields, upon rearrangement

$$pH = pK_a + \log\left|\frac{\text{basic form}}{\text{acid form}}\right| \qquad (3\text{-}14)$$

which gives the pH of a solution containing both forms of a buffer. Equation (3-14) also shows that the pK_a of an acid or base is the pH at which it is half-dissociated.

162 Interphase Mass Transfer

Amino acids each contain a negatively charged carboxylate group ($-COO^-$) that ionizes at pH between 2 and 3 and a positively charged amino group ($-NH_3^+$) that ionizes at pH between 8 and 10. In some cases another ionizable group appears on a side chain. The *isoelectric point* (pI) is the pH of a solution at which an amino acid carries zero net electric charge. An amino acid carrying no net charge is called a *zwitterion* (from the German word *zwitter* meaning "hybrid"). In an aqueous solution having a pH equal to the isoelectric point, the amino acids show minimum solubility. At a pH lower than the isoelectric point, amino acids have a positive charge while at a pH higher than the isoelectric point they have a negative charge. Zwitterions buffer solutions at high pH by releasing H^+, as well as at low pH by accepting H^+.

For amino acids having a neutral side chain, the pI value is calculated as the mean of the pK_a of the carboxylate and the pK_a of the ammonium group (see Problem 3.6). For amino acids having a charged or ionizable side chain, the pI value is calculated as the mean of the pK_as of similarly ionizable groups, as illustrated in Example (3.4).

Example 3.4 Isoelectric Points of Amino Acids at 25 °C

Amino acids are the basic building blocks of proteins, and they serve as the nitrogenous backbones for compounds like neurotransmitters and hormones. Amino acids that can be synthesized by the human body are called *nonessential*. *Essential* amino acids, on the other hand, cannot be synthesized by the body and must be supplied by the diet. Calculate the isoelectric point of the following amino acids at 25 °C.

(a) Alanine is a nonessential amino acid that occurs at high levels in its free state in plasma. It is involved in sugar and acid metabolism, increases immunity, and provides energy for muscle tissue, brain, and the central nervous system. It carries a neutral methyl ($-CH_3$) side chain. At 25 °C, the pK_a for its carboxylic acid group is 2.3 and for its ammonium group is 9.9.

(b) Aspartic acid is another nonessential amino acid that is used by the body in the biosynthesis of proteins. It carries an acidic ($-CH_2COOH$) side chain. At 25 °C, the pK_a for its carboxylic acid group is 2.0, for its ammonium group is 10.0, and for its side chain is 3.9.

(c) Lysine is an essential amino acid used in the human body in the synthesis of proteins, in the crosslinking of collagen polypeptides, uptake of essential mineral nutrients, and in the production of carnitine (key in fatty acid metabolism). It carries the basic side chain lysyl [$(CH_2)_4NH_2$]. At 25 °C K, the pK_a for its carboxylic acid group is 2.2, for its ammonium group is 9.2, and for its side chain is 10.8.

Solution

(a) Since alanine has a neutral side chain, the isoelectric point will be calculated as the mean of the pK_a of the carboxylate and the pK_a of the ammonium group: pI = (2.3 + 9.9)/2 = 6.1.

(b) Aspartic acid have an acidic side chain. To calculate the pI, the pK_a of the similarly ionizable groups will be considered which in this case are the carboxylic acid group (pK_a = 2.0) and the acidic side chain (pK_a = 3.9): pI = (2.0 + 3.9)/2 = 2.95.

3.2 Equilibrium Considerations in Chemical and Biochemical Systems 163

(c) Lysine has a basic side chain. To calculate the pI, the pK_a of the similarly ionizable groups will be considered which in this case are the ammonium group ($pK_a = 9.20$) and the basic side chain ($pK_a = 10.8$): pI = (9.2 + 10.8)/2 = 10.0.

Electrophoresis refers to the separation of molecules based on their net charge under the influence of an electric field. At a given pH, an amino acid will exist in either neutral, positively, or negatively charged forms. When a mixture of different amino acids is placed on a fluid or paper and a constant electric field is applied, the amino acids having a net negative charge will move toward the anode (positive electrode) while the positively charged amino acids will move toward the cathode (negative electrode). Thus, electrophoresis can separate amino acids based on their net charge at a particular pH.

Equation (3-14) shows that the pH of a solution changes less per proton absorbed or released as the ratio [basic form]/[acid form] approaches 1.0; for this reason it is preferable to use a buffer whose pK_a is ±0.5 unit of the desired pH. Values of pK_a change more with buffer concentrations in solutions of multivalent buffers like phosphate or citrate than is simple monovalent buffers like acetate or Tris. The amount of change can be calculated by the simplified Debye-Hückel equation

$$pK_a = pK_a^0 + \frac{0.5n\sqrt{I}}{1+1.6\sqrt{I}} \quad \text{(at 298 K)} \tag{3-15}$$

where I represents the ionic strength of the solution given by

$$I = \frac{1}{2}\sum_i c_i z_i^2 \tag{3-16}$$

where c_i is the molar concentration (mol/L) of ionic species i, which has charge z_i; pK_a^o is a value of pK_a extrapolated to zero ionic strength; and $n = 2z - 1$ for a given valence z on the buffer acid form. Example 3.5 illustrates these calculations.

Example 3.5 Phosphate-Buffered Saline (PBS)

Phosphate-buffered saline (PBS) is a buffered solution commonly used in biological research. It is a water-based salt solution containing disodium hydrogen phosphate ($Na_2HPO_4.2H_2O$), sodium chloride and, in some formulations, potassium chloride and potassium dihydrogen phosphate (KH_2PO_4). The osmolarity and ion concentrations of the solution match those of the human body (isotonic). PBS has shown to be an acceptable alternative to viral transport medium regarding transport and storage of RNA viruses such as SARS-CoV-2 (Perchetti et al., 2020).

There are many different ways to prepare PBS solutions. One common formulation is to prepare 10-liter stock of 10 × PBS by dissolving 800 g of NaCl, 20 g of KCl, 144 g of $Na_2HPO_4.2H_2O$, and 24 g of KH_2PO_4 in 8 L of distilled water, and topping up to 10 L Estimate the pH of this solution in its concentrated form at 298 K. For monobasic sodium phosphate (NaH_2PO_4), $z = -1$ and $pK_a^o = 7.2$ (Seader et al., 2011).

Solution

The first step is to calculate the ionic strength of the solution. The two chemical species that contribute to the ionic strength are NaCl [$c = 800/(58.44 \times 10) = 1.369$ mol/L] and KCl [$c = 20/(74.55 \times 10) = 0.027$ mol/L]. Since both the positive and negative ions contribute to the ionic strength, substituting in equation (3-16) we obtain

$$I = \frac{1}{2}\left[1.369 \times (+1)^2 + 1.369 \times (-1)^2 + 0.027 \times (+1)^2 + 0.027 \times (-1)^2\right]$$
$$= 1.396 \text{ mol/L}$$

For the acid form of the buffer (NaH_2PO_4), $n = 2 \times (-1) - 1 = -3$. Substituting in equation (3-15)

$$pK_a = 7.2 + \frac{0.5 \times (-3) \times \sqrt{1.396}}{1 + 1.6 \times \sqrt{1.396}} = 6.59$$

The pH of the solution is from equation (3-14) given that the molecular weight of the basic form ($Na_2HPO_4.2H_2O$) is 178.1 g/mol and of the acid form (KH_2PO_4) is 136.1 g/mol:

$$pH = 6.59 + \log\left[\frac{\dfrac{144}{178.1 \times 10}}{\dfrac{24}{136.1 \times 10}}\right] = 7.25$$

For comparison purposes, after dilution and adjusting to a physiologic temperature of 310 K, the pH of the resultant 1 × PBS is 7.48, very close to that of arterial blood plasma, pH = 7.4.

Water solubility of ionized species is very high compared to uncharged species. An uncharged or undissociated weak acid, HA, exhibits a partial solubility, S_0. In a solution saturated with undissolved HA, the partial solubility is $S_0 = M_{HA}$ where M_i represents molality (moles of solute per kg of solvent) of species i. The total concentration of both undissociated and ionized acid, S_T, is given by

$$S_T = M_{HA} + M_{A^-} \tag{3-17}$$

Combining equations (3-10), (3-17), and the definitions of pH and pK_a we obtain

$$S_T(pH) = S_0\left[1 + 10^{-(pK_{HA} - pH)}\right] \tag{3-18}$$

Relative to a solubility value reported for pure water, $S_T^{pH=7}$, the effect of pH on total solubility is given by

$$S_T(pH) = S_T^{pH=7}\frac{\left[1 + 10^{-(pK_{HA} - pH)}\right]}{\left[1 + 10^{-(pK_{HA} - 7)}\right]} \tag{3-19}$$

3.2 Equilibrium Considerations in Chemical and Biochemical Systems 165

It can be shown that the total solubility of a zwitterionic amino acid relative to water solubility is given by

$$S_T\left(\text{pH}\right) = S_T^{\text{pH}=7} \frac{\left[1 + 10^{-\left(\text{p}K_a^a - \text{pH}\right)} + 10^{-\left(\text{pH} - \text{p}K_a^c\right)}\right]}{\left[1 + 10^{-\left(\text{p}K_a^a - 7\right)} + 10^{-\left(7 - \text{p}K_a^c\right)}\right]} \tag{3-20}$$

The superscripts c and a in the $\text{p}K_a$ values refer to the carboxylic acid and amino base, respectively. Examples 3.6 and 3.7 illustrate these calculations.

Example 3.6 Effect of pH on Solubility of Caprylic Acid

Caprylic acid (from the Latin word *capra*, meaning "goat") is a naturally occurring saturated fatty acid with the formula $C_8H_{16}O_2$, present in goat milk and coconuts. It plays an important role in the human body's regulation of energy input and output, a function which is performed by the hormone ghrelin. Caprylic acid participates in the process by linking at a specific site on ghrelin molecules.

The water solubility of caprylic acid is 0.68 g/kg at 293 K with a value of $\text{p}K_a$ = 4.89. Calculate the solubility of caprylic acid at pH = 2, pH = $\text{p}K_a$, and pH = 8. Compare your results to the solubility in water (pH = 7).

Solution
Substituting in equation (3-19), we obtain S_T(pH = 2) = 0.0052 g/kg; S_T(pH = 4.89) = 0.01 g/kg; and S_T(pH = 8) = 6.74 g/kg. When compared to the solubility in water (0.68 g/kg), it is evident that the pH of the solution has a dramatic effect on the total solubility of this fatty acid.

Example 3.7 Effect of pH on Solubility of the Amino Acid Valine

Due to a mutation, in patients with sickle-cell anemia the hydrophobic amino acid valine ($C_5H_{11}NO_2$) substitutes for hydrophilic glutamic acid in hemoglobin. This results in deformation of the red blood cells into a sickle-like shape making them relatively inflexible and unable to traverse the capillary beds.

The water solubility of valine at 298 K is 88.5 g/kg and its ionization constants are $\text{p}K_a^c = 2.3$, $\text{p}K_a^a = 9.6$. Calculate the solubility of valine at its isoelectric point, pI, at pH = 2, and at pH = 10.

Solution
The first step is to calculate $\text{p}I = (\text{p}K_a^c + \text{p}K_a^a)/2 = 5.95$. Then, use equation (3-20) to calculate the valine total solubility at the specified pH values: S_T(pH = 2) = 264.4 g/kg; S_T(pH = 5.95) = 88.3 g/kg; and S_T(pH = 10) = 310 g/kg. The minimum water solubility happens at the isoelectric point. Reduced water solubility at the pI of a biomolecule may be used for extraction into a less-polar phase as will be shown later.

---

166   Interphase Mass Transfer

3.3 DIFFUSION BETWEEN PHASES

Your objectives in studying this section are to be able to:

1. Calculate interfacial mass-transfer rates in terms of the local mass-transfer coefficients for each phase.
2. Define and use, where appropriate, overall mass-transfer coefficients.

Having established that departure from equilibrium provides the driving force for diffusion, we can now study the rates of diffusion in terms of the driving forces. Many of the mass-transfer operations are carried out in steady-flow fashion, with continuous and invariant flow of the contacted phases and under circumstances such that concentrations at any position in the equipment used do not change with time. It will be convenient at this point to use one of these operations as an example with which to establish the principles, and to generalize with respect to other operations later.

For this purpose, let us consider the absorption of a soluble gas such as ammonia (substance A) from a mixture with air, by liquid water, in a wetted-wall tower. The ammonia–air mixture may enter at the bottom of the tower and flow upward, while the water flows downward around the inside of the pipe. The ammonia concentration in the gas mixture diminishes as it flows upward, while the water absorbs the ammonia as it flows downward and leaves at the bottom as an aqueous ammonia solution. Uder steady-state conditions, the concentrations at any point in the apparatus do not change with passage of time.

3.3.1 Two-Resistance Theory

Consider the situation at a particular level along the tower. Since the solute is diffusing from the gas phase into the liquid, there must be a concentration gradient in the direction of mass transfer within each phase. This can be shown graphically in terms of the distance through the phases, as in Figure 3.3, where a section through the two phases in contact is shown. The concentration of A in the main body of the gas is $y_{A,G}$ mol fraction and it falls to $y_{A,i}$ at the interface. In the liquid, the concentration falls from $x_{A,i}$ at the interface to $x_{A,L}$ in the bulk liquid. The bulk concentrations $y_{A,G}$ and $x_{A,L}$ are clearly not equilibrium values, since otherwise diffusion of the solute would not occur. At the same time, these bulk concentrations cannot be used directly with a mass-transfer coefficient to describe the rate of interphase mass transfer since the two concentrations are differently related to the chemical potential, which is the real "driving force" of mass transfer.

To get around this problem, Lewis and Whitman (1924) assumed that the only diffusional resistances are those residing in the fluids themselves. There is then no resistance to solute transfer across the interface separating the phases, and as a result the concentrations $y_{A,i}$ and $x_{A,i}$ are equilibrium values given by the system's equilibrium-distribution curve. This concept has been called the *two-resistance theory*.

3.3 Diffusion Between Phases

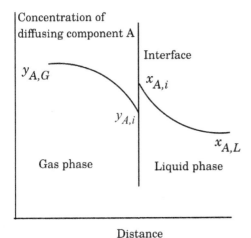

Figure 3.3 The two-resistance concept.

The reliability of this theory has been the subject of a great amount of study. A careful review of the results indicates that departure from concentration equilibrium at the interface must be a rarity (Treybal, 1980). Consequently, in most situations the interfacial concentrations in Figure 3.3 are those corresponding to a point on the equilibrium-distribution curve. Notice that the concentration rise at the interface, from $y_{A,i}$ to $x_{A,i}$, is not a barrier to diffusion in the direction gas to liquid. They are equilibrium concentrations and hence correspond to equal chemical potentials of substance A in both phases at the interface.

The various concentrations can also be shown graphically, as in Figure 3.4, whose coordinates are those of the equilibrium-distribution curve. Point P represents the two bulk-phase concentrations and point M those at the interface. For steady-state mass transfer, the rate at which A reaches the interface from the gas must be equal to the rate at which it diffuses to the bulk liquid, so that no accumulation or depletion of A at the interface occurs. We can, therefore, write the flux of A in terms of the mass-transfer coefficients for each phase and the concentration changes appropriate to each. Let us assume at this point that we are dealing with dilute solutions in both the liquid and gas phases. Furthermore, in both phases, substance A (ammonia) diffuses through nondiffusing B (air in the gas phase, and water in the liquid phase). Therefore, the development that follows will be done in terms of the k-type coefficients. The general results, in terms of the F-type coefficients, will be presented later. Thus, when k_x and k_y are the locally applicable coefficients,

$$N_A = k_y \left(y_{A,G} - y_{A,i} \right) = k_x \left(x_{A,i} - x_{A,L} \right) \quad (3\text{-}21)$$

Rearrangement as

$$\frac{y_{A,G} - y_{A,i}}{x_{A,L} - x_{A,i}} = -\frac{k_x}{k_y} \quad (3\text{-}22)$$

provides the slope of the line *PM*. If the mass-transfer coefficients are known, the interfacial concentrations and hence the flux N_A can be determined, either graphically by plotting the line *PM* on the equilibrium-distribution diagram or analytically by solving equation (3-22) simultaneously with an algebraic expression for the equilibrium-distribution curve,

$$y_{A,i} = f(x_{A,i}) \tag{3-23}$$

3.3.2 Overall Mass-Transfer Coefficients

In experimental determination of the rate of mass transfer, it is usually possible to determine the solute concentrations in the bulk of the fluids by sampling and analyzing. Successful sampling of the fluids at the interface, however, is ordinarily quite difficult since the greatest part of the concentration gradients takes place over extremely small distances. Any ordinary sampling device will be so large in comparison that it would be practically impossible to get close enough to the interface. Sampling and analyzing, therefore, will yield $y_{A,G}$ and $x_{A,L}$ but not $y_{A,i}$ and $x_{A,i}$. Under these circumstances, only an overall effect, in terms of the bulk concentrations, can be determined. The bulk concentrations, however, are not by themselves on the same basis in terms of chemical potential.

Consider the situation shown in Figure 3.5. Since the equilibrium-distribution curve for the system is unique at fixed temperature and pressure, then y_A^*, in equilibrium with $x_{A,L}$, is as good a measure of $x_{A,L}$ as $x_{A,L}$ itself, and moreover it is on the same thermodynamic basis as $y_{A,G}$. The entire two-phase mass-transfer effect can then be measured in terms of an overall mass-transfer coefficient, K_y, which includes the resistance to diffusion in both phases in terms of a gas phase molar fraction driving force

$$N_A = K_y(y_{A,G} - y_A^*) \tag{3-24}$$

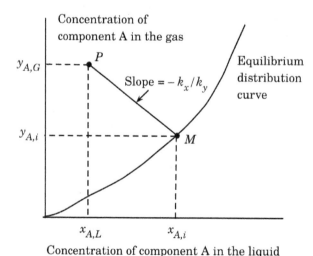

Figure 3.4 Interfacial concentrations as predicted by the two-resistance theory.

3.3 Diffusion Between Phases

In similar fashion, x_A^* is a measure of $y_{A,G}$ and can be used to define another overall mass-transfer coefficient, K_x

$$N_A = K_x\left(x_A^* - x_{A,L}\right) \tag{3-25}$$

A relation between the overall coefficients and the individual phase coefficients can be obtained when the equilibrium relation is linear as expressed by

$$y_{A,i} = m\, x_{A,i} \tag{3-26}$$

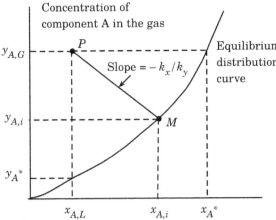

Figure 3.5 Overall concentration differences.

This condition is always encountered at low concentrations, where Henry's law is obeyed; the proportionality constant is then $m = H/P$. Utilizing equation (3-26), we may relate the gas- and liquid-phase concentrations by

$$y_A^* = m x_{A,L} \tag{3-27}$$

$$y_{A,G} = m x_A^* \tag{3-28}$$

Rearranging equation (3-24) we obtain

$$\frac{1}{K_y} = \frac{y_{A,G} - y_A^*}{N_A} \tag{3-29}$$

From the geometry of Figure 3.5

$$y_{A,G} - y_A^* = \left(y_{A,G} - y_{A,i}\right) + \left(y_{A,i} - y_A^*\right) \tag{3-30}$$

Substituting equation (3-30) into equation (3-29) gives us

Interphase Mass Transfer

$$\frac{1}{K_y} = \frac{y_{A,G} - y_A{}^*}{N_A} = \frac{y_{A,G} - y_{A,i}}{N_A} + \frac{y_{A,i} - y_A{}^*}{N_A} \qquad (3\text{-}31)$$

or in terms of m,

$$\frac{1}{K_y} = \frac{y_{A,G} - y_A{}^*}{N_A} = \frac{y_{A,G} - y_{A,i}}{N_A} + \frac{m\left(x_{A,i} - x_{A,L}\right)}{N_A} \qquad (3\text{-}32)$$

The substitution of equation (3-21) into the above relation relates K_y to the individual phase coefficients by

$$\frac{1}{K_y} = \frac{1}{k_y} + \frac{m}{k_x} \qquad (3\text{-}33)$$

A similar expression for K_x may be derived as follows:

$$\frac{1}{K_x} = \frac{x_A{}^* - x_{A,L}}{N_A} = \frac{y_{A,G} - y_{A,i}}{mN_A} + \frac{x_{A,i} - x_{A,L}}{N_A} \qquad (3\text{-}34)$$

or

$$\frac{1}{K_x} = \frac{1}{m k_y} + \frac{1}{k_x} \qquad (3\text{-}35)$$

Equations (3-33) and (3-35) lead to the following relationships between the mass transfer resistances:

$$\frac{\text{resistance in gas phase}}{\text{total resistance in both phases}} = \frac{1/k_y}{1/K_y} \qquad (3\text{-}36)$$

$$\frac{\text{resistance in liquid phase}}{\text{total resistance in both phases}} = \frac{1/k_x}{1/K_x} \qquad (3\text{-}37)$$

Assuming that the numerical values of k_x and k_y are roughly the same, the importance of the solubility of the gas—as indicated by the slope of the equilibrium curve, m—can readily be demonstrated. If m is small (solute A is very soluble in the liquid), the term m/k_x in equation (3-33) becomes minor, the major resistance is represented by $1/k_y$, and it is said that the rate of mass transfer is gas-phase-controlled. In the extreme, this becomes

$$\frac{1}{K_y} \approx \frac{1}{k_y} \qquad (3\text{-}38)$$

Under such circumstances, even fairly large percentage changes in k_x will not significantly affect K_y, and efforts to increase the rate of mass transfer would best be

3.3 Diffusion Between Phases
171

directed toward decreasing the gas-phase resistance. Conversely, when m is very large (solute A relatively insoluble in the liquid), with k_x and k_y nearly equal, the first term on the right of equation (3-35) becomes minor and the major resistance to mass transfer resides within the liquid, which is then said to control the rate. Ultimately, this becomes

$$\frac{1}{K_x} \approx \frac{1}{k_x} \qquad (3\text{-}39)$$

> Remember that the relationships derived between the individual coefficients and the overall coefficients are valid only for a straight equilibrium-distribution line. Under such circumstances, the use of overall coefficients will eliminate the need to calculate the concentrations at the interface. If the distribution line is not straight, the overall mass-transfer coefficients will change with changing concentrations of the two phases, and the local coefficients should be used instead.

Example 3.8 illustrates the concept of mass-transfer resistances in two phases.

Example 3.8 Mass-Transfer Resistances during Absorption of Ammonia by Water

In an experimental study of the absorption of ammonia by water in a wetted-wall column, the value of K_G was found to be 2.75×10^{-6} kmol/m^2·s·kPa. At one point in the column, the composition of the gas and liquid phases was 8.0 and 0.115 mol% NH_3, respectively. The temperature was 300 K, and the total pressure was 1 atm. Eighty-five percent of the total resistance to mass transfer was found to be in the gas phase. At this temperature, ammonia–water solutions follow Henry's law up to 5 mol% ammonia in the liquid, with $m = 1.64$ when the total pressure is 1 atm. Calculate the individual film coefficients and the interfacial concentrations.

Solution
The first step in the solution is to convert the given overall coefficient from K_G to K_y. The relations among the overall coefficients are the same as those among the individual coefficients, presented in Chapter 2. Therefore, $K_y = K_G P = 2.75 \times 10^{-6} \times 101.3 = 2.786 \times 10^{-4}$ kmol/m^2·s. From equation (3-36), for a gas-phase resistance that accounts for 85% of the total resistance,

$$k_y = \frac{K_y}{0.85} = 3.28 \times 10^{-4} \text{ kmol/m}^2 \cdot \text{s}$$

From equation (3-33)

$$\frac{m}{k_x} = \frac{1}{K_y} - \frac{1}{k_y} = \frac{0.15}{K_y}$$

172   **Interphase Mass Transfer**

or

$$k_x = \frac{mK_y}{0.15} = \frac{1.64 \times 2.786 \times 10^{-4}}{0.15} = 3.05 \times 10^{-3} \text{ kmol/m}^2 \cdot \text{s}$$

To estimate the ammonia flux and the interfacial concentrations at this particular point in the column, use equation (3-27) to calculate

$$y_A{}^* = mx_{A,L} = 1.64 \times 1.15 \times 10^{-3} = 1.886 \times 10^{-3}$$

The flux is from equation (3-24)

$$N_A = 2.786 \times 10^{-4} \left(0.080 - 1.886 \times 10^{-3} \right) = 2.18 \times 10^{-5} \text{ kmol/m}^2 \cdot \text{s}$$

Calculate the gas-phase interfacial concentration from equation (3-21)

$$y_{A,i} = y_{A,G} - \frac{N_A}{k_y} = 0.080 - \frac{2.18 \times 10^{-5}}{3.28 \times 10^{-4}} = 0.01362$$

Since the interfacial concentrations lie on the equilibrium line,

$$x_{A,i} = \frac{y_{A,i}}{m} = \frac{0.01362}{1.64} = 8.305 \times 10^{-3}$$

As a double check on the calculations so far, calculate the ammonia flux based on the conditions in the liquid phase,

$$N_A = 3.05 \times 10^{-3} \left(8.305 \times 10^{-3} - 1.15 \times 10^{-3} \right) = 2.18 \times 10^{-5} \text{ kmol/m}^2 \cdot \text{s}$$

The two estimates of the flux agree within the round-off error. Notice that, in this example, it was not necessary to calculate the interfacial concentrations to estimate the ammonia flux since the use of overall coefficients was appropriate. They were calculated just for illustration purposes. Usually, for the dilute conditions inherent to the use of k-type coefficients, the equilibrium relation follows Henry's law, and overall coefficients are justified.

3.3.3 Local Mass-Transfer Coefficients: General Case

When we deal with situations which do not involve either diffusion of only one substance or equimolar counterdiffusion, or if mass-transfer rates are large, F-type coefficients should be used. The general approach is the same, although the resulting expressions are more cumbersome than those developed above. Thus, in a situation like that shown in Figures 3.3–3.5, the mass-transfer flux is

$$N_A = \Psi_{A,G} F_G \ln \left[\frac{\Psi_{A,G} - y_{A,i}}{\Psi_{A,G} - y_{A,G}} \right] = \Psi_{A,L} F_L \ln \left[\frac{\Psi_{A,L} - x_{A,L}}{\Psi_{A,L} - x_{A,i}} \right] \qquad (3\text{-}40)$$

3.3 Diffusion Between Phases

Rearranging equation (3-40) yields

$$\left[\frac{\Psi_{A,G} - y_{A,i}}{\Psi_{A,G} - y_{A,G}}\right] = \left[\frac{\Psi_{A,L} - x_{A,L}}{\Psi_{A,L} - x_{A,i}}\right]^{(F_L \Psi_{A,L}/F_G \Psi_{A,G})} \tag{3-41}$$

The interfacial compositions $y_{A,i}$ and $x_{A,i}$ must satisfy simultaneously both equation (3-41) and the equilibrium-distribution relation. If the equilibrium relation is in the form of an algebraic expression, then the interfacial compositions will be obtained solving simultaneously the equilibrium expression and equation (3-41). If the equilibrium relation is in the form of a distribution diagram, plot equation (3-41)—with $y_{A,i}$ replaced by y_A and $x_{A,i}$ by x_A—on the distribution diagram and determine graphically the intersection of the resulting curve with the distribution curve.

Example 3.9 Absorption of Ammonia by Water: Use of F-Type Mass-Transfer Coefficients

A wetted-wall absorption tower is fed with water as the wall liquid and an ammonia–air mixture as the central-core gas. At a particular point in the tower, the ammonia concentration in the bulk gas is 0.60 mol fraction, that in the bulk liquid is 0.12 mol fraction. The temperature is 300 K, and the pressure is 1 atm. Ignoring the vaporization of water, calculate the local ammonia mass-transfer flux. The rates of flow are such that $F_L = 3.5$ mol/m^2-s, and $F_G = 2.0$ mol/m^2-s. The equilibrium-distribution data for the system at 300 K and 1 atm are those shown graphically in Figure 3.1, and algebraically in Example 3.3.

Solution

In this case, the ammonia concentrations are too high ($y_{A,G} = 0.60$, $x_{A,L} = 0.12$), particularly in the gas phase, to justify the use of k-type coefficients. Therefore, equations (3-40) and (3-41) will be used, together with the equilibrium-distribution data, to calculate the interfacial ammonia compositions and the local flux. In the gas phase, substance A is ammonia and substance B is air. Since the solubility of air in water is so low, we may assume that $N_{B,G} = 0$, and $\Psi_{A,G} = 1.0$. In the liquid phase, substance B is water. Ignoring the vaporization of water (the vapor pressure of water at 300 K is only 3.5 kPa), we may assume that $N_{B,L} = 0$, and $\Psi_{A,L} = 1.0$. Then, equation (3-41) reduces to

$$\frac{1 - y_{A,i}}{1 - y_{A,G}} = \left[\frac{1 - x_{A,L}}{1 - x_{A,i}}\right]^{F_L/F_G} \tag{3-42}$$

Substituting into equation (3-30) the numerical values given, and rearranging:

$$y_{A,i} = 1 - \left(1 - y_{A,G}\right)\left[\frac{1 - x_{A,L}}{1 - x_{A,i}}\right]^{F_L/F_G} = 1 - 0.4\left[\frac{0.88}{1 - x_{A,i}}\right]^{1.75} \tag{3-43}$$

174   **Interphase Mass Transfer**

Equation (3-43) must be combined with the equilibrium data to calculate the interfacial concentrations. Next, we will illustrate both the algebraic and graphical solutions.

(a) Algebraic solution:

The equilibrium relations for this case are equations (3-8) and (3-6), which become

$$y_{A,i} = 10.51 \gamma_A x_{A,i} \tag{A}$$

$$\gamma_A = 3.5858 x_{A,i}^2 - 0.622 x_{A,i} + 0.156 \tag{B}$$

Equations (3-43), (A), and (B) constitute a system of simultaneous nonlinear algebraic equations which is easily solved using the "solve block" capabilities of Mathcad, as shown in Figure 3.6. Initial guesses of the values of the three variables must be supplied. Figure 3.6 shows that the local interfacial concentrations are $x_{A,i} = 0.231$ and $y_{A,i} = 0.494$. Appendix 3.2 shows the Python solution of this example.

(b) Graphical solution:

Equation (3-43) relates the interfacial concentrations to each other. So, it is valid only at that point on the equilibrium-distribution curve which represents the local interfacial compositions. For the purpose of locating that point, rewrite equation (3-43) as a continuous relation between the variables in the equilibrium-distribution diagram, namely x_A and y_A:

Guess Values

$$x_A := 0.2 \qquad \gamma_A := 1.0 \qquad y_A := 0.5$$

Constraints

$$y_A = 1 - 0.4 \cdot \left(\frac{0.88}{1 - x_A} \right)^{1.75}$$

$$\gamma_A = 3.5858 \cdot \left(x_A^2 \right) - 0.622 \cdot \left(x_A \right) + 0.156$$

$$y_A = 10.51 \cdot \gamma_A \cdot x_A \qquad x_A < 0.3$$

Solver

$$\begin{bmatrix} x_A \\ y_A \\ \gamma_A \end{bmatrix} := \text{Find}\,(x_A, y_A, \gamma_A) = \begin{bmatrix} 0.231 \\ 0.494 \\ 0.204 \end{bmatrix}$$

Figure 3.6 Mathcad solution of Example 3.9a.

$$y_A = 1 - 0.4 \left[\frac{0.88}{1 - x_A} \right]^{1.75} \tag{3-44}$$

3.3 Diffusion Between Phases

Equation (3-44) represents a continuous curve on the xy diagram; its intersection with the equilibrium curve will yield the interfacial concentrations. To plot equation (3-44), generate the following table of (x_A, y_A) values:

x_A	0.12	0.15	0.20	0.25	0.30
y_A	0.600	0.575	0.527	0.471	0.403

These values are plotted on the equilibrium-distribution diagram, as shown in Figure 3.7. The resulting curve intersects the equilibrium curve to give the interface compositions $x_{A,i} = 0.231$ and $y_{A,i} = 0.494$. This solution agrees with the solution obtained previously by the algebraic method.

Once the interfacial compositions have been calculated by either method, the local flux is from equation (3-40), which, in this case, becomes

$$N_A = F_G \ln\left[\frac{1-y_{A,i}}{1-y_{A,G}}\right] = 2.0 \ln\left[\frac{1-0.494}{1-0.600}\right]$$

$$= 0.47 \text{ mol/m}^2 \cdot \text{s}$$

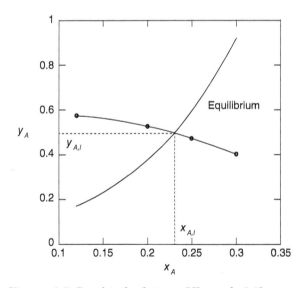

Figure 3.7 Graphical solution of Example 3.9b.

Example 3.10 Distillation of a Mixture of Methanol and Water in a Packed Tower: Use of F-Type Mass-Transfer Coefficients

A mixture of methanol (substance A, the more volatile) and water (substance B) is being distilled in a packed tower at constant pressure of 1 atm. At a point along the tower, the methanol content of the bulk of the gas phase is 36 mol%; that of the bulk of the liquid phase is 20 mol%. The temperature at that point in

176   **Interphase Mass Transfer**

the tower is around 360 K. At that temperature, the molar latent heat of vaporization of methanol is $\lambda_A = 33.3$ MJ/kmol, and that of water is $\lambda_B = 41.3$ MJ/kmol. The and flow conditions at that point are such that $F_G = 0.0017$ kmol/m^2·s, and $F_L = 0.0149$ kmol/m^2·s. Estimate the local flux of methanol from the liquid to the gas phase.

The VLE for this system is adequately described by the modified Raoult's law, equation (3-2), using the Wilson equation to estimate the activity coefficients (Smith et al., 1996)

$$\ln \gamma_A = -\ln\left(x_A + x_B \Delta_{A,B}\right) + x_B \left[\frac{\Delta_{A,B}}{x_A + x_B \Delta_{A,B}} - \frac{\Delta_{B,A}}{x_B + x_A \Delta_{B,A}}\right] \quad (3\text{-}45)$$

$$\ln \gamma_B = -\ln\left(x_B + x_A \Delta_{B,A}\right) - x_A \left[\frac{\Delta_{A,B}}{x_A + x_B \Delta_{A,B}} - \frac{\Delta_{B,A}}{x_B + x_A \Delta_{B,A}}\right] \quad (3\text{-}46)$$

where

$$\Delta_{i,j} = \frac{V_j}{V_i}\exp\left[-\frac{a_{i,j}}{RT}\right] \quad \left(i \neq j\right) \quad (3\text{-}47)$$

where V_j and V_i are the molar volumes at temperature T of pure liquids j and i and $a_{i,j}$ is a constant independent of composition and temperature. Recommended values of the parameters for this system are $V_A = 40.73$ cm^3/mol, $V_B = 18.07$ cm^3/mol, $a_{A,B} = 107.38$ cal/mol, $a_{B,A} = 469.55$ cal/mol (Smith et al., 1996).

Solution
Assume that the distillation operation takes place inside the tower adiabatically, in such a way that the energy required to evaporate methanol is supplied by condensation of water vapor. Therefore, from an energy balance

$$N_B = -\frac{\lambda_A}{\lambda_B}N_A = -\frac{33.3}{41.3}N_A = -0.806 N_A$$

$$\Psi_{A,G} = \Psi_{A,L} = \frac{1}{1 - 0.806} = 5.155$$

Equation (3-41) becomes

$$\frac{5.155 - y_{A,i}}{5.155 - 0.360} = \left|\frac{5.155 - 0.200}{5.155 - x_{A,i}}\right|^{0.0149/0.0017}$$

Rearranging, we obtain

$$y_{A,i} = 5.155 - 4.795 \left[\frac{4.955}{5.155 - x_{A,i}}\right]^{8.765} \quad (3\text{-}48)$$

3.3 Diffusion Between Phases

The modified form of Raoul's law is written for both components:

$$y_{A,i} = \frac{x_{A,i}\gamma_A P_A}{P} \quad y_{B,i} = \frac{x_{B,i}\gamma_B P_B}{P} \tag{3-49}$$

For a binary system, such as this

$$y_{A,i} + y_{B,i} = 1.0 \qquad x_{A,i} + x_{B,i} = 1.0 \tag{3-50}$$

Antoine equations for methanol and water are (Smith et al., 1996):

$$\ln P_A = 16.5938 - \frac{3644.3}{T - 33} \quad \ln P_B = 16.2620 - \frac{3800.0}{T - 47} \tag{3-51}$$

where P_i is the vapor pressure in kPa, and T is in K. Equations (3-48) to (3-50) constitute a system of five equations in five unknowns: the equilibrium temperature, and the four interfacial mol fractions. They are solved, with the help of equations (3-45) to (3-47), and (3-51), using Mathcad as shown in Figure 3.8. Appendix 3.3 shows the Python solution of this example.

The solution is $x_{A,i} = 0.177$, $y_{A,i} = 0.549$, $T = 356.5$ K. The local methanol flux is, from equation (3-40),

$$N_A = 5.155 \times 1.7 \times \ln\left[\frac{5.155 - 0.549}{5.155 - 0.360}\right]$$
$$= -0.352 \text{ mol/m}^2 \cdot \text{s}$$

The negative sign of the flux implies that the methanol flows from the liquid to the gas phase, which is in the opposite direction to that assumed in the derivation of equation (3-40).

An alternative approximate solution to this problem can be obtained through the use of k'-type coefficients. As mentioned in Example 2.3, the McCabe–Thiele method of analysis of distillation columns assumes equimolar counterdiffusion, which would justify the use of k'-type coefficients regardless of the concentration levels. This would be exactly so only if the molar latent heats of vaporization of both components were equal. Although in this case there is a significant difference between the two heats of vaporization, we will show that the approximate solution is fairly close to the rigorous solution obtained above. It is easily shown that in this case, assuming steady-state equimolar counterdiffusion,

$$\frac{y_{A,G} - y_{A,i}}{x_{A,L} - x_{A,i}} = -\frac{k'_x}{k'_y} = -\frac{F_L}{F_G} \tag{3-52}$$

This is the equation of a straight line on the xy diagram with slope = $-F_L/F_G$.

178   **Interphase Mass Transfer**

Parameters:

$$a12 := 107.38 \; \frac{\text{cal}}{\text{mol}} \qquad a21 := 469.55 \; \frac{\text{cal}}{\text{mol}} \qquad V1 := 40.73 \; \frac{\text{cm}^3}{\text{mol}}$$

$$R := 1.987 \; \frac{\text{cal}}{\text{mol} \cdot \text{K}} \qquad P := 101.3 \text{ kPa} \qquad V2 := 18.07 \; \frac{\text{cm}^3}{\text{mol}}$$

Correlations:

$$\Lambda12(T) := \frac{V2}{V1} \cdot \exp\left(\frac{-a12}{R \cdot T}\right) \qquad\qquad \Lambda21(T) := \frac{V1}{V2} \cdot \exp\left(\frac{-a21}{R \cdot T}\right)$$

$$P1(T) := \exp\left(16.5938 - \frac{3644.3 \text{ K}}{T - 33 \text{ K}}\right) \text{ kPa} \qquad P2(T) := \exp\left(16.2620 - \frac{3800.0 \text{ K}}{T - 47 \text{ K}}\right) \text{ kPa}$$

$$\gamma1(T, x1, x2) := \exp\left(-\ln(x1 + x2 \cdot \Lambda12(T)) + x2 \cdot \left[\frac{\Lambda12(T)}{x1 + x2 \cdot \Lambda12(T)} - \frac{\Lambda21(T)}{x2 + x1 \cdot \Lambda21(T)}\right]\right)$$

$$\gamma2(T, x1, x2) := \exp\left(-\ln(x2 + x1 \cdot \Lambda21(T)) - x1 \cdot \left[\frac{\Lambda12(T)}{x1 + x2 \cdot \Lambda12(T)} - \frac{\Lambda21(T)}{x2 + x1 \cdot \Lambda21(T)}\right]\right)$$

Guess Values

$$x1 := 0.1 \qquad x2 := 0.9$$

$$y1 := 0.2 \qquad y2 := 0.8 \qquad T := 360 \text{ K}$$

Constraints

$$x1 + x2 = 1 \qquad y1 + y2 = 1$$

$$y1 = \frac{x1 \cdot \gamma1(T, x1, x2) \cdot P1(T)}{P} \qquad y2 = \frac{x2 \cdot \gamma2(T, x1, x2) \cdot P2(T)}{P}$$

$$y1 = 5.155 - 4.795 \cdot \left[\frac{4.955}{5.155 - x1}\right]^{8.765}$$

Solver

$$\begin{bmatrix} x1 \\ y1 \\ x2 \\ y2 \\ T \end{bmatrix} := \text{Find}(x1, x2, y1, y2, T) = \begin{bmatrix} 0.177 \\ 0.823 \\ 0.549 \\ 0.451 \\ 356.464 \text{ K} \end{bmatrix}$$

Figure 3.8 Mathcad solution of Example 3.10.

Figure 3.9 is the equilibrium-distribution diagram for this system at a pressure of 1 atm. Point P $(y_{A,G}, x_{A,L})$ represents the bulk concentration of both phases, and point M $(y_{A,i}, x_{A,i})$ on the equilibrium curve represents the interfacial concentrations. The straight line joining those two points has a slope of $-F_L / F_G = -8.765$.

To draw the straight line passing through point P $(y_{A,G}, x_{A,L})$ with the given slope of −8.765, choose arbitrarily a value of y_A on the line, for example, $y_A = 0.6$, and calculate the corresponding value of x_A from the expression for the slope:

3.3 Diffusion Between Phases

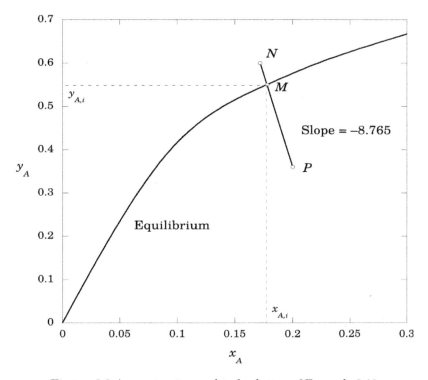

Figure 3.9 Approximate graphical solution of Example 3.10.

$$\frac{0.36 - 0.60}{0.20 - x_A} = -8.765 \Rightarrow x_A = 0.172$$

Therefore, the points N (0.172, 0.600) and P (0.200, 0.360) completely define the straight line that contains the segment PM. The intersection of line PN with the equilibrium curve yields the interfacial concentrations at point M. They are very similar to those obtained by the rigorous method using the F-type coefficients, namely, $x_{A,i}$ = 0.178, $y_{A,i}$ = 0.550. The local methanol flux is

$$N_A = k'_y \left(y_{A,G} - y_{A,i} \right) = F_G \left(y_{A,G} - y_{A,i} \right)$$
$$= 1.7 \times (0.36 - 0.55)$$
$$= -0.323 \text{ mol/m}^2 \cdot \text{s}$$

which is about 8% lower than the flux value obtained through the rigorous solution method. This was to be expected since assuming equimolar counterdiffusion implies that there is no net bulk flow contribution, which underestimates the actual flow in the present situation where $N_A + N_B = 0.194\, N_A$.

3.4 MATERIAL BALANCES

Your objectives in studying this section are to be able to:

1. Understand the importance of material balances in interphase mass-transfer calculations.
2. Define and generate minimum and actual operating lines for batch operations, and for steady-state cocurrent and countercurrent processes.

The concentration-difference driving forces discussed so far are those existing at a given position in the equipment used to contact the immiscible phases. In the case of a steady-state flow process, because of the transfer of solute from one phase to the other, the concentration within each phase changes as it moves through the equipment. Similarly, in the case of a batch process, the concentration in each phase changes with time. These changes produce corresponding variations in the driving forces, and these can be followed with the help of material balances.

3.4.1 Countercurrent Flow

Consider any steady-state mass-transfer operation which involves the countercurrent contact of two immiscible phases as shown schematically in Figure 3.10. The two phases will be identified as phase V and phase L, and for the present consider only the case where a single substance A diffuses from phase V to phase L during their contact.

At the bottom of the mass-transfer device, or plane z_1, the flow rates and concentrations are defined as follows:

V_1: total moles of phase V entering the tower per unit time
L_1: total moles of phase L entering the tower per unit time
y_1: mol fraction of component A in V_1
x_1: mol fraction of component A in L_1

Similarly, at the top of the device, or plane z_2, the total moles of each phase will be V_2 and L_2, and the compositions of each stream will be y_2 and x_2.

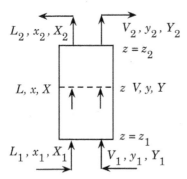

Figure 3.10 Steady-state countercurrent process.

3.4 Material Balances 181

An overall macroscopic mass balance for component A around the steady-state device (from plane $z = z_1$ to plane $z = z_2$), in which there is no chemical production or disappearance of A, requires that

$$\begin{bmatrix} \text{moles of A entering} \\ \text{the device} \end{bmatrix} = \begin{bmatrix} \text{moles of A leaving} \\ \text{the device} \end{bmatrix} \quad (3\text{-}53)$$

or

$$V_1 y_1 + L_2 x_2 = V_2 y_2 + L_1 x_1 \quad (3\text{-}54)$$

A mass balance for component A around plane $z = z_1$ and the arbitrary plane at z stipulates

$$V_1 y_1 + Lx = Vy + L_1 x_1 \quad (3\text{-}55)$$

Simpler relations, and certainly easier equations to use, may be expressed in terms of *solute-free concentration units or mol ratios*. The concentration of each phase will be defined as follows:

Y: moles of A in phase V per mol of A-free V; that is,

$$Y = \frac{y}{1-y} \qquad y = \frac{Y}{1+Y} \quad (3\text{-}56)$$

X: moles of A in phase L per mol of A-free L; that is,

$$X = \frac{x}{1-x} \qquad x = \frac{X}{1+X} \quad (3\text{-}57)$$

When using mol ratios as concentration units, it is convenient to express the flow rates in terms of L_S and V_S. These are defined as

L_S: moles of phase L on a solute-free basis, per unit time
V_S: moles of phase V on a solute-free basis, per unit time.

The overall balance on component A may be written, using the solute-free terms, as

$$V_S Y_1 + L_S X_2 = V_S Y_2 + L_S X_1 \quad (3\text{-}58)$$

Rearranging, we obtain

$$\frac{L_S}{V_S} = \frac{Y_1 - Y_2}{X_1 - X_2} \quad (3\text{-}59)$$

Equation (3-59) is the equation of a straight line on the XY diagram which passes through the points (X_1, Y_1) and (X_2, Y_2) with a slope of L_S/V_S. A mass balance on component A around plane z_1 and the arbitrary plane $z = z$ in solute-free terms is

$$\frac{L_S}{V_S} = \frac{Y_1 - Y}{X_1 - X} \tag{3-60}$$

As before, equation (3-60) is the equation of a straight line which passes through the points (X_1, Y_1) and (X, Y) with a slope of L_S/V_S. It is a general expression relating the bulk compositions of the two phases at any plane in the device. Since it defines operating conditions within the equipment, it is designated the *operating-line equation for countercurrent operation*. Figures 3.11 and 3.12 illustrate the location of the operating line relative to the equilibrium line when the transfer is from phase V to phase L, and from phase L to phase V.

It is important to notice the difference between equations (3-55) and (3-60). Although both equations describe the mass balance for component A, only (3-60) is the equation of a straight line. When written in terms of the solute-free units X and Y, the operating line is straight because the mol-ratio concentrations are based on the constant quantities L_S and V_S. On the other hand, when written in terms of mol-fraction units—x and y—the total moles in a phase—L or V—change as the solute is transferred into or out of the phase. This produces a curved operating line on the xy coordinates.

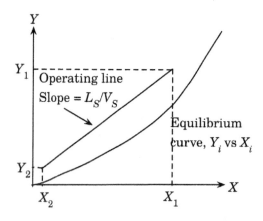

Figure 3.11 Steady-state countercurrent process, transfer from phase V to L.

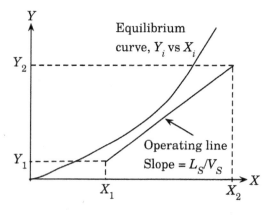

Figure 3.12 Steady-state countercurrent process, transfer from phase L to V.

3.4 Material Balances

In the design of mass-transfer equipment, the flow rate of one phase, and three of the four entering and exiting compositions, must be fixed by the process requirements. The necessary flow rate of the second phase is a design variable. For example, consider the case in which phase V, with a known value of V_S, changes in composition from Y_1 to Y_2 by transferring solute to a second phase which enters the tower with composition X_2. According to equation (3-60), the operating line must pass through point (X_2, Y_2) with a slope L_S/V_S, and must end at the ordinate Y_1. Recall that V_S is fixed, but L_S is subject to choice. If such a quantity L_S is used to give operating line DE on Figure 3.13a, the exiting phase L will have the composition X_1. If less L_S is used, the exit-L phase composition will clearly be greater, as at point F, but, since the operating line will now be closer to the equilibrium curve, the driving forces for diffusion are less and the solute transfer is more difficult. The time of contact between the phases must be greater, and the mass-transfer device must be correspondingly bigger.

The minimum value of L_S which can be used while still achieving the desired amount of solute transfer (a reduction in concentration of the V phase from Y_1 to Y_2) corresponds to the operating line DM, which is tangent to the equilibrium curve at P. Under these conditions, the exiting phase L will have the composition X_1 (max). Any value of L_S smaller than L_S (min) would result in an operating line that would cross and go under the equilibrium curve, a physical impossibility if the solute transfer is to remain from phase V to phase L. At P, the diffusional driving force is zero, the required time of contact for the concentration change desired is infinite, and an infinitely big mass-transfer device results. This then represents a limiting condition, the *minimum L_S/V_S ratio for mass transfer*. The actual L_S/V_S ratio must be higher than the minimum to ensure a finite size of the device; it is usually specified as a multiple of L_S (min)$/V_S$. Therein lies the importance of estimating accurately the ratio L_S (min)$/V_S$.

The equilibrium curve is frequently concave upward, as in Figure 3.13 (b), and $L_S(\text{min})/V_S$ corresponds to the phase L_1 leaving in equilibrium with the entering phase G_1; that is, point M $[X_1(\text{max}), Y_1]$ lies on the equilibrium curve.

Figure 3.14 illustrates the location of the minimum and actual operating lines when solute transfer is from phase L to phase V. In this case, L_S is specified, and V_s is subject to change. The value of V_s to be used must be higher than the minimum.

Example 3.11 Recovery of Benzene Vapors from a Mixture with Air

A waste airstream from a chemical process flows at the rate of 1.0 m^3/s at 300 K and 1 atm, containing 7.4% by volume of benzene vapors. It is desired to recover 85% of the benzene in the gas by a three-step process. First, the gas is scrubbed using a nonvolatile wash oil to absorb the benzene vapors. Then, the wash oil leaving the absorber is stripped of the benzene by contact with steam at 1 atm and 373 K. The mixture of benzene vapor and steam leaving the stripper will then be condensed. Because of the low solubility of benzene in water, two distinct liquid phases will form, and the benzene layer will be recovered by decantation. The aqueous layer will be purified and returned to the process as boiler feedwater. The oil leaving the stripper will be cooled to 300 K and returned to the absorber. Figure 3.15 is a schematic diagram of the process.

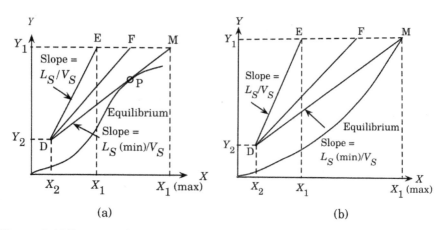

Figure 3.13 Location of minimum and actual operating lines for solute transfer from phase V to phase L.

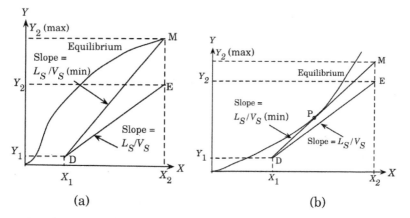

Figure 3.14 Location of minimum and actual operating lines for solute transfer from phase L to phase V.

The wash oil entering the absorber will contain 0.0476 mol fraction of benzene; the pure oil has an average molecular weight of 198. An oil-circulation rate of twice the minimum will be used in the absorber. In the stripper, a steam rate of 1.5 times the minimum will be used.

Compute the oil-circulation rate and the steam rate required for the operation. Wash oil–benzene solutions are ideal. The vapor pressure of benzene at 300 K is 0.136 atm, and 1.77 atm at 373 K.

3.4 Material Balances

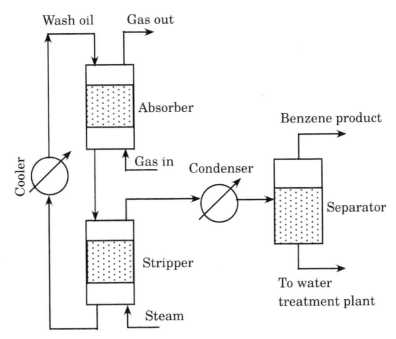

Figure 3.15 Schematic diagram of the benzene-recovery process of Example 3.11.

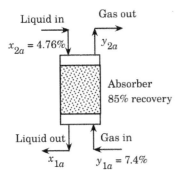

Solution

Consider first the conditions in the absorber. From the flow conditions of the entering gas and the given tower diameter, using the ideal gas, calculate the molar gas velocity entering at the bottom of the tower. (In all calculations in this example, we will use a subscript "a" to indicate flows and concentrations in the absorber, and a subscript "s" to refer to conditions in the stripper.) The entering-gas molar velocity is from the ideal gas law

$$V_{1a} = \frac{101.3 \times 1.0}{8.314 \times 300} = 0.041 \text{ kmol/s}$$

Calculate the inert-gas molar velocity

$$V_{Sa} = V_{1a}(1 - y_{1a}) = 0.041 \times (1 - 0.074) = 0.038 \text{ kmol/s}$$

Convert the entering-gas mol fraction to a mol ratio using equation (3-56)

$$Y_{1a} = \frac{y_{1a}}{1 - y_{1a}} = \frac{0.074}{1 - 0.074} = 0.08 \text{ mol of benzene/mol of dry gas}$$

186 Interphase Mass Transfer

Convert the entering-liquid mol fraction to a mol ratio using equation (3-57)

$$X_{2a} = \frac{x_{2a}}{1-x_{2a}} = \frac{0.0476}{1-0.0476} = 0.05 \text{ mol of benzene/mol of oil}$$

Since the absorber will recover 85% of the benzene in the entering gas, the concentration of the gas leaving it will be

$$Y_{2a} = (1-0.85) \times 0.08 = 0.012 \text{ mol of benzene/mol of dry gas}$$

To determine the minimum amount of wash oil required, we need the equilibrium distribution curve for this system, expressed in terms of molar ratios. The benzene–wash oil solutions are ideal, and the pressure is low, therefore Raoult's law applies. From equations (3-1), (3-56), and (3-57)

$$y_{ia} = 0.136 x_{ia}$$
$$\frac{Y_{ia}}{1+Y_{ia}} = 0.136 \frac{X_{ia}}{1+X_{ia}}$$

The following table of equilibrium values, for the conditions prevailing in the absorber, is generated from the equation above:

X_{ia}, kmol benzene/kmol oil	Y_{ia}, kmol benzene/kmol dry gas
0.00	0.000
0.10	0.013
0.20	0.023
0.30	0.032
0.40	0.040
0.60	0.054
0.80	0.064
1.00	0.073
1.20	0.080
1.40	0.086

Figure 3.16 is the resulting equilibrium distribution curve. Notice that, although Raoult's law is always represented as a straight line on the xy diagram, it is not necessarily so on the XY diagram. We are now ready to locate the minimum operating line. Locate point $D(X_{2a}, Y_{2a})$ on the XY diagram. This is one end of the operating line, and its location is fixed, independent of the amount of liquid used. Starting with any arbitrary operating line, such as DE, rotate it toward the equilibrium curve using D as a pivot point until the operating line touches the equilibrium curve for the first time. In this case, because of the shape of the equilibrium curve, the first contact occurs at point P where the operating line is tangent to the curve. Extending this line all the way to $Y_a = Y_{1a}$, we generate operating line DM, corresponding to the minimum oil rate. From the diagram, we read that $X_{1a}(\max) = 0.91$ kmol benzene/kmol oil. Then,

3.4 Material Balances

$$\frac{L_{Sa}(\min)}{V_{Sa}} = \frac{Y_{1a} - Y_{2a}}{X_{1a}(\max) - X_{2a}} = \frac{0.080 - 0.012}{0.91 - 0.05} = 0.079 \text{ mol of oil/mol of dry gas}$$

$$L_{Sa}(\min) = 0.079 \times 0.038 \times 1000 = 3.0 \text{ mol oil/s}$$

For an actual oil flow rate which is twice the minimum, $L_{Sa} = 6.0$ mol/s. The molecular weight of the oil is 198. The mass rate of flow of the oil, w_{Sa}, is

$$w_{Sa} = L_{Sa} M_{Sa} = 0.006 \times 198 = 1.19 \text{ kg oil/s}$$

The next step is to calculate the actual concentration of the liquid phase leaving the absorber. Rearranging equation (3-59) yields

$$X_{1a} = X_{2a} + \frac{V_{Sa}(Y_{1a} - Y_{2a})}{L_{Sa}} = 0.05 + \frac{0.038(0.080 - 0.012)}{0.006}$$

$$= 0.48 \text{ mol benzene/mol oil}$$

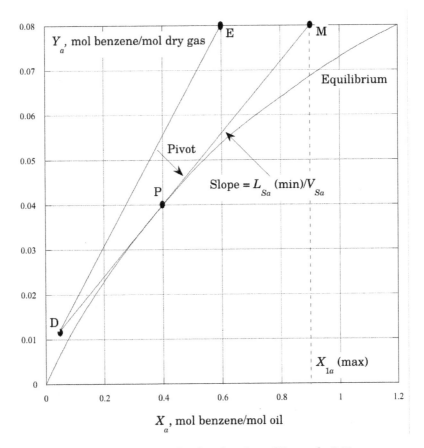

Figure 3.16 *XY* diagram for the absorber of Example 3.11.

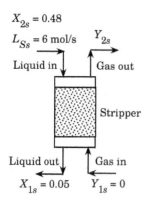

We turn our attention now to conditions in the stripper. Figure 3.15 shows that the wash oil cycles continuously from the absorber to the stripper, and through the cooler back to the absorber. Therefore, $L_{Ss} = L_{Sa} = 0.006$ kmol/s. The concentration of the liquid entering the stripper is the same as that of the liquid leaving the absorber ($X_{2s} = X_{1a} = 0.48$ kmol benzene/kmol oil), and the concentration of the liquid leaving the stripper is the same as that of the liquid entering the absorber ($X_{1s} = X_{2a} = 0.05$ kmol benzene/kmol oil). The gaseous phase entering the stripper is pure steam; therefore, $Y_{1s} = 0$. To determine the minimum amount of steam needed, we need to generate the equilibrium distribution curve for the system at 373 K. Applying Raoult's law at this temperature

$$y_{is} = 1.770 x_{is}$$

$$\frac{Y_{is}}{1+Y_{is}} = 1.770 \frac{X_{is}}{1+X_{is}}$$

The following table of equilibrium values, for the conditions prevailing in the stripper, is generated from the equation above:

X_{is}, kmol benzene/kmol oil	Y_{is}, kmol benzene/kmol steam
0.00	0.000
0.05	0.092
0.10	0.192
0.15	0.300
0.20	0.418
0.25	0.548
0.30	0.691
0.35	0.848
0.40	1.023
0.45	1.219
0.50	1.439

Figure 3.17 is the resulting equilibrium distribution curve. We are now ready to locate the minimum operating line. Locate point D (X_{1s}, Y_{1s}) on the XY diagram. This is one end of the operating line, and its location is fixed, independent of the amount of steam used. Starting with any arbitrary operating line, such as DE, rotate it toward the equilibrium curve using D as a pivot point until the operating line touches the equilibrium curve for the first time. In this case, because of the shape of the equilibrium curve, the first contact occurs at point P where the operating line is tangent to the curve. Extending this line all the way to $X_s = X_{2s}$, we generate operating line DM, corresponding to the minimum oil rate. From the diagram, we read that Y_{2s}(max) = 1.175 kmol benzene/kmol steam. Then,

3.4 Material Balances

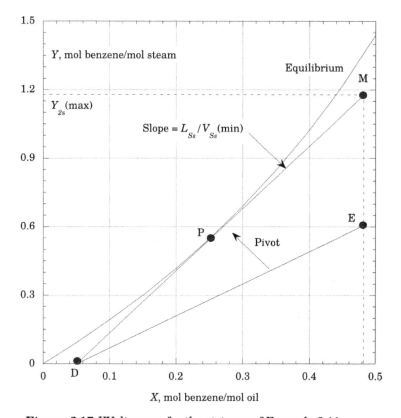

Figure 3.17 XY diagram for the stripper of Example 3.11.

$$\frac{L_{Ss}}{V_{Ss}(\min)} = \frac{Y_{2s}(\max) - Y_{1s}}{X_{2s} - X_{1s}} = \frac{1.175 - 0.0}{0.48 - 0.05} = 2.733 \text{ mol oil/mol steam}$$

$$V_{Ss}(\min) = \frac{0.006}{2.733} \times 1000 = 2.2 \text{ mol/s}$$

$$V_{Ss} = 1.5 V_{Ss}(\min) = 3.3 \text{ mol/s}$$

The mass flow rate of steam used in the stripper, w_{Ss}, is

$$w_{Ss} = V_{Ss} M_{Ss} = 0.0033 \times 18 = 0.059 \text{ kg steam/s}$$

The next step is to calculate the actual concentration of the gas phase leaving the stripper. Rearranging equation (3-59) yields

$$Y_{2s} = Y_{1s} + \frac{L_{Ss}}{V_{Ss}}(X_{2s} - X_{1s}) = 0.0 + \frac{0.0060}{0.0033}(0.48 - 0.05)$$
$$= 0.784 \text{ mol benzene/mol steam}$$

Masses, mass fractions, and mass ratios can be substituted consistently for moles, mol fractions, and mol ratios in the mass balances of equations (3-53) to (3-60). The following example illustrates this concept.

Example 3.12 Adsorption of Nitrogen Dioxide on Silica Gel

Nitrogen dioxide, NO_2, produced by a thermal process for fixation of nitrogen, is to be removed from a dilute mixture with air by adsorption on silica gel in a continuous countercurrent adsorber. The mass flow rate of the gas entering the adsorber is 0.50 kg/s; it contains 1.5% NO_2 by volume, and 85% of the NO_2 is to be removed. Operation is to be isothermal at 298 K and 1 atm. The entering gel will be free of NO_2. If twice the minimum gel rate is to be used, calculate the gel mass flow rate and the composition of the gel leaving the process. The equilibrium adsorption data for this system at this temperature are from Treybal (1980).

Solution

The material balances in this problem are easier to establish in terms of mass flow rates and mass ratios. Define:

$$L_S' = \text{mass flow rate of the "dry" gel, kg/s}$$
$$V_S' = \text{mass flow rate of the "dry" air, kg/s}$$
$$X' = \text{mass ratio in the gel, kg } NO_2/\text{kg gel}$$
$$Y' = \text{mass ratio in the gas, kg } NO_2/\text{kg air}$$

The mass ratio in the entering gas is

$$Y_1' = \frac{y_1}{1-y_1} \times \frac{M_{NO_2}}{M_{air}} = \frac{0.015}{1-0.015} \times \frac{46}{29} = 0.0242 \text{ kg } NO_2 / \text{kg air}$$

For 85% removal of the NO_2, $Y_2' = 0.15 \times 0.0242 = 0.0036$ kg NO_2/kg air. Since the entering gel is free of NO_2, $X_2' = 0$. The original equilibrium data are given in terms the partial pressure of NO_2 (p_A in mm Hg) in equilibrium with the adsorbed solid concentration (m in kg NO_2/100 kg gel). They are converted to mass ratios as follows:

$$Y_i' = \frac{p_A}{760 - p_A} \times \frac{46}{29} \qquad X_i' = \frac{m}{100}$$

to yield the table that follows.

Partial pressure NO_2[a]p_A, mm Hg	Solid concentration[a]m, kg NO_2/100 kg gel	$X_i' \times 100$kg NO_2/kg gel	$Y_i' \times 100$kg NO_2/kg air
0	0	0.00	0.00
2	0.40	0.40	0.42
4	0.90	0.90	0.83
6	1.65	1.65	1.26
8	2.60	2.60	1.69
10	3.65	3.65	2.11
12	4.85	4.85	2.54

[a]Source: Adapted from Treybal (1980).

3.4 Material Balances

Figure 3.18 shows the corresponding equilibrium-distribution curve, together with the minimum operating line DM. From Figure 3.18, X_1' (max) = 0.0375 kg NO_2/kg gel. Then,

$$L_S'(\min) = \frac{Y_1' - Y_2'}{X_1'(\max) - X_2'} \times V_S' = \frac{0.0242 - 0.0036}{0.0375 - 0.0} \times V_S' = 0.550 V_S'$$

To calculate the mass velocity of the inert gas, air in this case, $V_S' = V_1' \times \omega_{B1}$. It is easy to show that

$$\omega_B = \frac{1}{1+Y'}$$

Then, $V_S' = 0.5 \times (1 + 0.0242)^{-1} = 0.488$ kg air/s; $L_S'(\min) = 0.55 \times 0.488 = 0.268$ kg gel/s. The actual mass velocity of the gel to be used is $L_S' = 2L_S'(\min) = 0.536$ kg gel/s. The composition of the gel leaving the adsorber is

$$X_1' = X_2' + \frac{V_S'}{L_S'}(Y_1' - Y_2') = 0.0 + \frac{0.488}{0.536}(0.0242 - 0.0036)$$
$$= 0.0188 \text{ kg } NO_2 \text{ / kg gel}$$

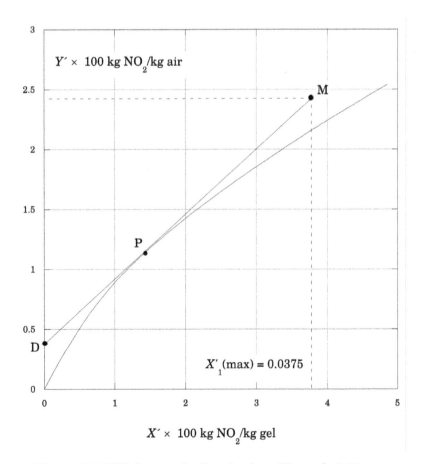

Figure 3.18 $X'Y'$ diagram for the adsorber of Example 3.12.

192   **Interphase Mass Transfer**

Example 3.13 Minimum Flow Rates Determination: Henry's Law

Consider an absorber or stripper in which the equilibrium is described by Henry's law: $y_i = mx_i$. It is easy to show that the equilibrium-distribution curve in terms of mol ratios is given by

$$Y_i = \frac{mX_i}{1+(1-m)X_i} \tag{3-61}$$

The general behavior of the XY distribution curve can be ascertained, for different values of m, by means of the first and second derivatives of equation (3-49):

$$\frac{dY_i}{dX_i} = \frac{m}{\left[1+(1-m)X_i\right]^2} \qquad \frac{d^2Y_i}{dX_i^2} = -\frac{2m(1-m)}{\left[1+(1-m)X_i\right]^3} \tag{3-62}$$

The sign of the second derivative will depend on the value of m. If $m < 1.0$, the second derivative will be negative for all values of Xi and the equilibrium-distribution curve will always be concave downward; therefore, the operating line that gives the minimum liquid rate for an absorber will be tangent to the equilibrium curve at some point $P(XP, YP)$ in the interval $Y_{out} < YP < Yin$. If $m > 1.0$, the second derivative will be positive for all values of X_i and the equilibrium-distribution curve will always be concave upward; therefore, the operating line that gives the minimum vapor rate for a stripper will be tangent to the equilibrium curve at some point $P(X_P, Y_P)$ in the interval $X_{out} < XP < Xin$.

    Consider an absorber with known values of m (<1.0), V_S, X_{in}, Y_{in}, and Y_{out}. The coordinates of the tangent point and the minimum liquid flow rate, LS(min), are determined by simultaneously solving three equations:

$$Y_P = \frac{mX_P}{1+(1-m)X_P} \tag{3-63}$$

$$\frac{L_S(\min)}{V_S} = \frac{Y_P - Y_{out}}{X_P - X_{in}} \tag{3-64}$$

$$\frac{L_S(\min)}{V_S} = \frac{m}{\left[1+(1-m)X_P\right]^2} \tag{3-65}$$

    Equations (3-63) and (3-64) arise from the fact that the tangent point P is located on both the equilibrium curve and the operating line. Equation (3-65) states that, at the tangent point P, the slope of the equilibrium line equals the slope of the operating line.

    Consider now a stripper with known values of m (>1.0), L_S, X_{in}, Y_{in}, and X_{out}. The coordinates of the tangent point and the minimum gas flow rate, V_S(min), are determined by simultaneously solving equation (3-63) and

3.4 Material Balances

$$\frac{L_S}{V_S(\min)} = \frac{Y_P - Y_{in}}{X_P - X_{out}} \qquad (3\text{-}66)$$

$$\frac{L_S}{V_S(\min)} = \frac{m}{\left[1+(1-m)X_P\right]^2} \qquad (3\text{-}67)$$

Figure 3.19 shows a Mathcad program to implement the solution of equations (3-63), (3-64), and (3-65) for an absorber, and equations (3-63), (3-66), and (3-67) for a stripper. For comparison purposes, the program uses data from Example 3.11, where the minimum flow rates were determined by a graphical method. The results

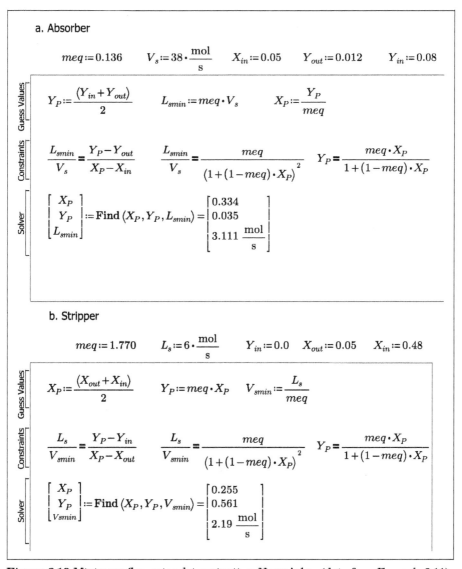

Figure 3.19 Minimum flow rates determination: Henry's law (data from Example 3.11).

shown in Figure 3.19 [L_S (min) = 3.11 mol/s, V_S (min) = 2.19 mol/s] agree with those obtained graphically in Example 3.11. Appendix 3.4 shows the corresponding Python solution.

3.4.2 Cocurrent Flow

For steady-state mass-transfer operations involving cocurrent contact of two insoluble phases, as shown in Figure 3.20, the overall mass balance for component A is

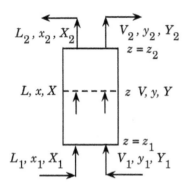

Figure 3.20 Steady-state cocurrent process.

$$V_S Y_1 + L_S X_1 = V_S Y_2 + L_S X_2 = V_S Y + L_S X \qquad (3\text{-}68)$$

Rearranging equation (3-68),

$$\frac{Y_1 - Y_2}{X_1 - X_2} = -\frac{L_S}{V_S} = \frac{Y_1 - Y}{X_1 - X} \qquad (3\text{-}69)$$

This is the equation of a straight line on the XY diagram through the points (X_1, Y_1) and (X_2, Y_2) with slope $-L_S/V_S$, the operating line for cocurrent operation. Figures 3.21 and 3.22 illustrate the location of the operating line relative to the equilibrium line when the transfer is from phase V to phase L, and from phase L to phase V.

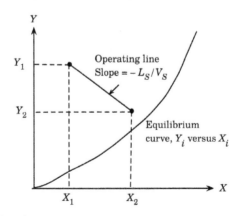

Figure 3.21 Steady-state cocurrent process, transfer from phase V to L.

3.4 Material Balances

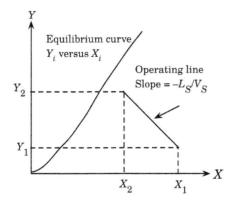

Figure 3.22 Steady-state cocurrent process, transfer from phase L to V.

Example 3.14 Cocurrent Adsorption of NO$_2$ on Silica Gel

Repeat Example 3.12, but for cocurrent contact between the gas mixture and the silica gel.

Solution

Figure 3.23 is the $X'Y'$ diagram for cocurrent operation of the adsorber. In this case, as before, $Y_1' = 0.0242$ kg NO$_2$/kg air, $Y_2' = 0.0036$ kg NO$_2$/kg air; however, now $X_1' = 0.0$. The minimum operating line corresponds to the gel leaving the adsorber in equilibrium with the gas. It goes from point $D(X_1', Y_1')$ to point $M[X_2(max)', Y_2']$ on the equilibrium curve with a slope of $-L_S'(\min)/V_S'$. From Figure 3.23, $X_2(max)' = 0.0034$ kg NO$_2$/kg gel. From equation (3-69),

$$-\frac{L_S'(\min)}{V_S'} = \frac{0.0242 - 0.0036}{0.0 - 0.0034} = -6.09 \text{ kg gel/kg air}$$

Therefore, $L_S'(\min) = 6.09 \times 0.488 = 2.957$ kg/s and $L_S' = 2.0 \times 2.957 = 5.920$ kg/s. Notice that the mass velocity of the silica gel required for cocurrent operation is about 11 times that required for countercurrent operation. This example dramatically illustrates the fact that the driving force for mass transfer is used much more efficiently in countercurrent than in cocurrent operation.

3.4.3 Batch Processes

It is characteristic of batch processes that while there is no flow into and out of the equipment used, the concentrations within each phase change with time. When initially brought into contact, the phases will not be in equilibrium, but they will approach equilibrium with the passage of time. The material balance equation developed for cocurrent steady-state operation also shows the relation between the concentrations X and Y which coexist at any time after the start of operation. If the phases are kept in contact long enough to reach equilibrium, the end of the operating line lies on the equilibrium curve.

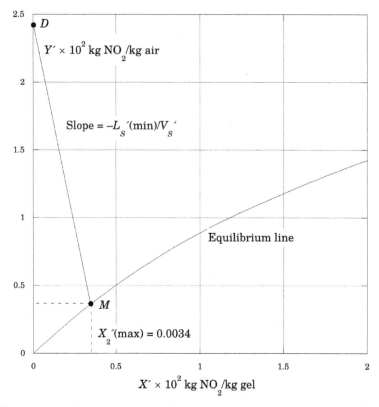

Figure 3.23 $X'Y'$ diagram for the cocurrent adsorber of Example 3.14.

3.5 EQUILIBRIUM-STAGE OPERATIONS

Your objectives in studying this section are to be able to:

1. Define the concepts: stage, ideal stage, and cascade.
2. Calculate the number of ideal stages in a cascade required for a given separation and ratio of flow rates.

One class of mass-transfer devices consists of assemblies of individual units, or stages, interconnected so that the materials being processed pass through each stage in turn. The two streams move countercurrently through the assembly; in each stage they are brought into contact, mixed, and then separated. such multistage systems are called *cascades*. For mass transfer to take place, the streams entering each stage must not be in equilibrium with each other, for it is precisely the departure from equilibrium conditions that provides the driving force for mass transfer. The leaving streams are usually not in equilibrium either but are much closer to being so than are the entering streams. If, as may actually happen in practice, the mass transfer in a given stage is so effective that the leaving streams are

3.5 Equilibrium-Stage Operations

in fact in equilibrium, the stage is, by definition, an ideal stage. Obviously, the cocurrent process and apparatus of Figure 3.20 is that of a single stage, and if the stage were ideal, the effluent compositions would be in equilibrium with each other. A batch mass-transfer operation is also a single stage.

Two or more stages connected so that flow is cocurrent between stages will, of course, never be equivalent to more than one equilibrium stage, although the overall stage efficiency can thereby be increased. For effects greater than one equilibrium stage, the stages in the cascade must be connected for countercurrent flow. This is the most efficient arrangement, requiring the fewest stages for a given change of composition and ratio of flow rates, and is always used in multistage equipment (Foust et al., 1980). Figure 3.24 shows a countercurrent cascade of N equilibrium stages.

In Figure 3.24, the flow rates and compositions are numbered corresponding to the effluent from a stage, so that, for example X_2 is the concentration in the L phase leaving stage 2, etc. Since the stages are ideal, the effluents from each stage are in equilibrium with each other (Y_2 in equilibrium with X_2, etc.). The cascade as a whole has the characteristics of the countercurrent process of Figure 3.12, with an operating line on the XY diagram that goes through the points (X_0, Y_1) and (X_N, Y_{N+1}) with slope L_S/V_S. The number of ideal stages in a cascade can be determined by either graphical or analytical methods. The simplest graphical method is based on the use of an operating line in conjunction with the equilibrium line. The graphical relations are shown in Figure 3.25 for transfer of a solute from phase L to phase V.

The operating line ST may be plotted either by knowing all four of the compositions at both ends of the cascade, or by knowing three compositions and the slope (L_S/V_S) of the operating line. The concentration of the V phase leaving stage 1 is Y_1. Since the stage is ideal, X_1, the concentration of the L phase leaving this stage, is such that the point (X_1, Y_1) must lie on the equilibrium curve. This fact fixes point Q, found by moving horizontally from point T to the equilibrium curve. The operating line is now used to determine Y_2. The operating line passes through all points having coordinates of the type (X_n, Y_{n+1}), and since X_1 is known, Y_2 is found by moving vertically from point Q to the operating line at point B, the coordinates of which are (X_1, Y_2). The step, or triangle, defined by the points T, Q, and B represents one ideal stage, the first one in this cascade. The second stage is located graphically on the diagram by repeating the same construction, passing horizontally to the equilibrium curve at point N, having coordinates (X_2, Y_2), and vertically to the operating line again at point C, having coordinates (X_2, Y_3). This stepwise construction is repeated until point S (X_N, Y_{N+1}) is reached.

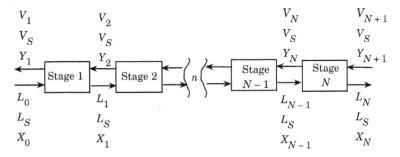

Figure 3.24 Countercurrent cascade of ideal stages.

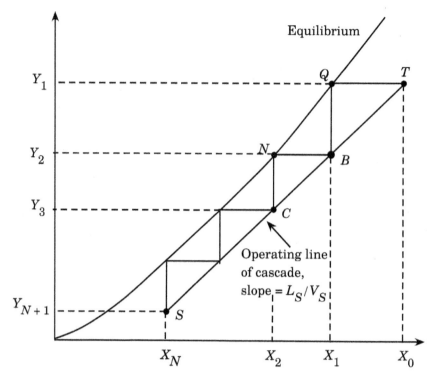

Figure 3.25 Countercurrent cascade of ideal stages, transfer from phase L to V.

We can therefore determine the number of ideal stages required for a countercurrent process by drawing the stairlike construction $TQBNC \ldots S$ and counting the number of steps. If the equilibrium curve and cascade operating line should touch anywhere, the stages become *pinched* and an infinite number are required to bring about the desired composition change. If the solute transfer is from V to L, the entire construction falls above the equilibrium curve of Figure 3.25. Remember that, regardless of the direction of the solute transfer, always in this graphical construction stage 1 refers to that stage where phase L enters the cascade. Therefore, for solute transfer from L to V (such as in a stripper), stage 1 will be located at the high-concentration end of the cascade operating line. On the other hand, for solute transfer from V to L (such as in an absorber), stage 1 will be located at the low-concentration end of the cascade operating line. These concepts are illustrated in Example 3.15.

Example 3.15 Benzene Recovery System: Number of Ideal Stages

Estimate the number of ideal stages in the absorber and stripper of the benzene recovery system of Example 3.11.

Solution
Absorber: Using the nomenclature introduced for multistage cascades and the results of Example 3.11:

3.5 Equilibrium-Stage Operations

$X_0 = X_{2a} = 0.050$ mol benzene/mol oil
$Y_1 = Y_{2a} = 0.012$ mol benzene/mol dry gas
$X_N = X_{1a} = 0.480$ mol benzene/mol oil
$Y_{N+1} = Y_{1a} = 0.080$ mol benzene/mol dry gas

Figure 3.26 shows the operating line for the cascade, DE, located using the four compositions listed above. Since A is transferred in this case from phase V to L, the stairlike construction to determine the number of ideal stages begins at point D moving horizontally toward the equilibrium curve, and so on. Figure 3.26 shows that almost 4 equilibrium stages are required.

Stripper: Using the nomenclature introduced for multistage cascades and the results of Example 3.11:

$X_0 = X_{2s} = 0.480$ mol benzene/mol oil
$Y_1 = Y_{2s} = 0.784$ mol benzene/mol steam
$X_N = X_{1s} = 0.050$ mol benzene/mol oil
$Y_{N+1} = Y_{1s} = 0.0$ mol benzene/mol steam

Figure 3.27 shows the operating line for the cascade, DE, located using the four compositions listed above. Since A is transferred in this case from phase L to V, the stairlike construction to determine the number of ideal stages begins at point E moving horizontally toward the equilibrium curve, and so on. Figure 3.27 shows that almost 6 equilibrium stages are required. A more precise estimate is obtained from

$$N = 5 + \frac{0.070 - 0.050}{0.070 - 0.028} = 5.5 \text{ stages}$$

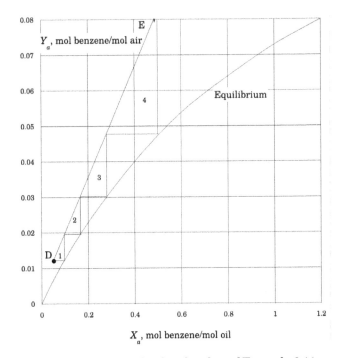

Figure 3.26 Ideal stages for the absorber of Example 3.11.

If the equilibrium distribution curve can be expressed in terms of an algebraic equation, the number of ideal stages in a given cascade can be determined analytically. In that case, the stair like construction (moving from the operating line to the equilibrium curve, and back to the operating line) can be implemented using a simple Mathcad program, as illustrated in the following example.

Example 3.16 Analytical Determination of Number of Ideal Stages

Analytically estimate the number of ideal stages in the absorber and stripper of the benzene recovery system of Example 3.11. Raoult's law applies for this system.

Solution
Absorber: Figure 3.28 shows a Mathcad program for the analytical determination of the number of ideal stages in an absorber using data from Example 3.11.
Stripper: Figure 3.29 shows a Mathcad program for the analytical determination of the number of ideal stages in a stripper using data from Example 3.11. Appendix 3.5 presents the Python solution to both parts (absorber and stripper) of this example.

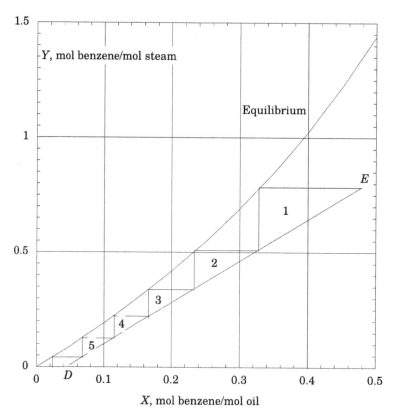

Figure 3.27 Ideal stages for the stripper of Example 3.11.

3.5 Equilibrium-Stage Operations 201

For the special case where the equilibrium curve is a straight line through the origin ($Y_i = mX_i$), an analytical solution was developed by Kremser (1930) which is most useful. Define the *absorption factor A* as

$$A = \frac{L_S}{mV_S} \qquad (3\text{-}70)$$

Notice from equation (3-70) that the absorption factor is really the ratio of the slope of the operating line to that of the equilibrium relation. The Kremser equations are as follows:

For transfer from L to V (stripping of L)

$A = 1$:

$$N = \frac{X_0 - X_N}{X_N - Y_{N+1}/m} \qquad (3\text{-}71)$$

$A \neq 1$:

$$N = \frac{\ln\left[\dfrac{X_0 - Y_{N+1}/m}{X_N - Y_{N+1}/m}\left(1 - A\right) + A\right]}{\ln\left(1/A\right)} \qquad (3\text{-}72)$$

For transfer from V to L (absorption into L)

$A = 1$:

$$N = \frac{Y_{N+1} - Y_1}{Y_1 - mX_0} \qquad (3\text{-}73)$$

$A \neq 1$:

$$N = \frac{\ln\left[\dfrac{Y_{N+1} - mX_0}{Y_1 - mX_0}\left(1 - \dfrac{1}{A}\right) + \dfrac{1}{A}\right]}{\ln\left(A\right)} \qquad (3\text{-}74)$$

The use of these equations is illustrated in Example 3.17.

Example 3.17 Kremser Equations: Overall Cascade Efficiency

An absorber consisting of a cascade of 10 real stages is used to remove 99% of solute A from a gas stream using a pure liquid ($X_0 = 0$). The overall efficiency of the cascade (defined as the ratio of ideal stages to real stages) is 62.4%. The equilibrium expression is $Y_i = 1.2X_i$. The gas enters the cascade at 100 mol/s, with a mol fraction of A of 10%. Calculate the flow rate of liquid required for the specified fractional removal using this absorber.

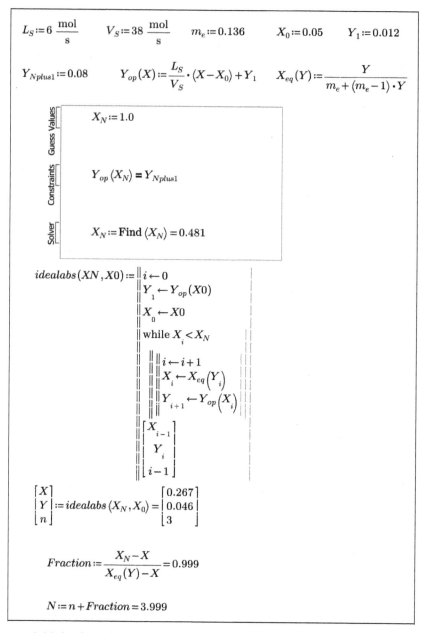

Figure 3.28 Analytical determination of ideal stages for absorber. Data from Example 3.11.

Solution
From the definition of the overall efficiency of the cascade, the number of ideal stages is $N = 0.624 \times 10 = 6.24$ ideal stages. For $X_0 = 0$ and 99% removal ($Y_{N+1}/Y_1 = 100$), equation (3-74) becomes

3.5 Equilibrium-Stage Operations

$$X_0 := 0.48 \qquad X_N := 0.05 \qquad Y_{Nplus1} := 0.0 \qquad L_S := 6\,\frac{\text{mol}}{\text{s}}$$

$$V_S := 3.3\,\frac{\text{mol}}{\text{s}} \qquad m_e := 1.77 \qquad Xeq(Y) := \frac{Y}{m_e + (m_e - 1) \cdot Y}$$

$$Yop(X) := Y_{Nplus1} + \frac{L_S}{V_S} \cdot (X - X_N) \qquad\qquad Y_1 := Yop(X_0) = 0.782$$

$$idealstrip(X_0, X_N) := \begin{Vmatrix} i \leftarrow 0 \\ X_0 \leftarrow X_0 \\ Y_1 \leftarrow Yop(X_0) \\ \text{while } X_i > X_N \\ \quad \begin{Vmatrix} i \leftarrow i + 1 \\ X_i \leftarrow Xeq(Y_i) \\ Y_{i+1} \leftarrow Yop(X_i) \end{Vmatrix} \\ \begin{bmatrix} X_{i-1} \\ Y_i \\ i-1 \end{bmatrix} \end{Vmatrix}$$

$$\begin{bmatrix} X \\ Y \\ n \end{bmatrix} := idealstrip(X_0, X_N) = \begin{bmatrix} 0.058 \\ 0.015 \\ 5 \end{bmatrix}$$

$$Fraction := \frac{X - X_N}{X - Xeq(Y)} = 0.16$$

$$N := n + Fraction = 5.16 \qquad\qquad N = 5.16$$

Figure 3.29 Analytical determination of ideal stages for stripper. Data from Example 3.11.

$$6.24 = \frac{\ln\left[100\left(1 - \dfrac{1}{A}\right) + \dfrac{1}{A}\right]}{\ln(A)}$$

Solving by trial-and-error, $A = 1.85$. From the equilibrium expression, $m = 1.2$. From the given inlet gas flow rate and composition, $V_S = 100 \times (1 - 0.1) = 90$ mol/s. Substituting in equation (3-70), $L_S = 90 \times 1.85 \times 1.2 = 200$ mol/s.

204   Interphase Mass Transfer

PROBLEMS

The problems at the end of each chapter have been grouped into four classes (designated by a superscript after the problem number).

Class a: Illustrates direct numerical application of the formulas in the text.
Class b: Requires elementary analysis of physical situations, based on the subject material in the chapter.
Class c: Requires somewhat more mature analysis.
Class d: Requires computer solution.

3.1[a]. Application of Raoult's law to a binary system

Repeat Example 3.1, but for a liquid concentration of 0.5 mol fraction of benzene and a temperature of 310 K.

Answer: $y_A = 0.758$

3.2[b]. Application of Raoult's law to a binary system

(a) Determine the composition of the liquid in equilibrium with a vapor containing 70 mol% benzene, 30 mol% toluene if the system exists in a vessel under a pressure of 2 atm. Predict the equilibrium temperature.

Answer: $x_A = 0.503$

(b) Determine the composition of the vapor in equilibrium with liquid containing 70 mol% benzene, 30 mol% toluene if the system exists in a vessel under a pressure of 2 atm. Predict the equilibrium temperature.

Answer: $y_A = 0.845$

3.3[a]. Application of Raoult's law to a binary system

Normal heptane, $n\text{-}C_7H_{16}$, and normal octane, $n\text{-}C_8H_{18}$, form ideal solutions. At 373 K, normal heptane has a vapor pressure of 106 kPa and normal octane of 47.1 kPa.

(a) What would be the composition of a heptane–octane solution that boils at 373 K under a pressure of 80 kPa?

Answer: $x_A = 0.559$

(b) What would be the composition of the vapor in equilibrium with the solution that is described in part (a)?

Answer: $y_A = 0.740$

Problems 205

3.4[a]. Henry's law: saturation of water with oxygen

A solution with oxygen dissolved in water containing 0.5 mg O_2/100 g of H_2O is brought in contact with a large volume of atmospheric air at 283 K and a total pressure of 1 atm. The Henry's law constant for the oxygen–water system at 283 K equals 32,700 atm/mol fraction.

(a) Will the solution gain or lose oxygen?

(b) What will be the concentration of oxygen in the equilibrium solution?

Answer: 11.4 mg O_2/L

3.5[b]. Material balances combined with equilibrium relations

Repeat Example 3.3, but assume that the ammonia, air, and water are brought into contact in a closed container. There is 10.0 m^3 of gas space over the liquid. Assuming that the gas-space volume and the temperature remain constant until equilibrium is achieved, modify the Mathcad program in Figure 3.2 to calculate:

(a) The total pressure at equilibrium.

Answer: 1.754 atm

(b) The equilibrium ammonia concentration in both phases.

3.6[c]. Isoelectric point of amino acid with neutral side chain

Consider the behavior of the amino acid alanine as a function of pH shown in the following sequence of reactions.

$$\underset{\text{(A)}}{CHCH_3NH_3^+COOH} \rightleftarrows \underset{\text{(B)}}{CHCH_3NH_3^+COO^-} + H^+ \tag{3-75}$$

$$\underset{\text{(B)}}{CHCH_3NH_3^+COO^-} \rightleftarrows \underset{\text{(C)}}{CHCH_3NH_2COO^-} + H^+ \tag{3-76}$$

The sequence shown starts at low pH (high hydrogen ion concentration) where alanine has a positive charge. Using equilibrium constants K_{a1} and K_{a2}, the coupled set of equations that result is (with the chemical species denoted as A, B, and C as illustrated above)

$$K_{a1} = \frac{[B][H^+]}{[A]} \qquad K_{a2} = \frac{[C][H^+]}{[B]} \tag{3-77}$$

Using $[B]_0$ to represent the initial molarity of the uncharged amino acid at neutral pH, we have

$$[B]_0 = [A] + [B] + [C] \tag{3-78}$$

Since the two equilibrium constants for amino acids are so different (for alanine, for example, pK_{a1} = 2.3, pK_{a2} = 9.9), it can be assumed that at low pH species C will not be present, while at high pH species A will not be present.

(a) Show that under these circumstances

$$K_{a1} \approx \frac{\left([B]_0 - [A]\right)[H^+]}{[A]} \qquad K_{a2} \approx \frac{[C][H^+]}{[B]_0 - [C]} \tag{3-79}$$

(b) Show that

$$\frac{[A]}{[B]_0} = \frac{10^{(pK_{a1} - pH)}}{1 + 10^{(pK_{a1} - pH)}} \qquad \frac{[C]}{[B]_0} = \frac{1}{1 + 10^{(pK_{a2} - pH)}} \tag{3-80}$$

c) The average charge of an amino acid as a function of pH, $f(pH)$, is given by

$$f(pH) = (+1) \times \frac{[A]}{[B]_0} + (-1) \times \frac{[C]}{[B]_0} \tag{3-81}$$

Combining equations (3-80) and (3-81) with the definition of the isoelectric point, show that, in this case,

$$pI = \frac{pK_{a1} + pK_{a2}}{2} \tag{3-82}$$

3.7^a. Isoelectric point of amino acid with neutral side chain

Serine ($C_3H_7NO_3$) is a nonessential amino acid that plays an important role in the catalytic function of many enzymes. The dissociation constants for serine are pK_{a1} = 2.21 and pK_{a2} = 9.15.

(a) Plot the charge of this amino acid as a function of pH.

(b) Calculate the isoelectric point of serine.

3.8^b. Mass-transfer resistances during absorption

In the absorption of component A (molecular weight = 60) from an airstream into an aqueous solution, the bulk compositions of the two adjacent streams at a point in the apparatus were analyzed to be $p_{A,G}$ = 0.1 atm and $c_{A,L}$ = 1.0 kmol of A/m^3 of solution. The total pressure was 2.0 atm; the density of the solution was 1100 kg/m^3. The Henry's law constant was 0.85 atm/mol fraction. The overall gas coefficient was K_G = 0.27 kmol/m^2·h·atm. If 60% of the total resistance to mass transfer resides in the gas film, determine

Problems

(a) The gas-film coefficient, k_G

(b) The liquid-film coefficient, k_L

Answer: 0.976 cm/h

(c) The concentration on the liquid side of the interface, $x_{A,i}$

Answer: 0.057 mol fraction

(d) The mass flux of A

3.9[b]. Mass-transfer resistances during absorption

For a system in which component A is transferring from the gas phase to the liquid phase, the equilibrium relation is given by $p_{A,i} = 0.80x_{A,i}$ where $p_{A,i}$ is the equilibrium partial pressure in atm and $x_{A,i}$ is the equilibrium liquid concentration in molar fraction. At one point in the apparatus, the liquid stream contains 4.5 mol% and the gas stream contains 9.0 mol% of A. The total pressure is 1 atm. The individual gas-film coefficient at this point is $k_G = 3.0$ mol/m^2-s-atm. Fifty percent of the overall resistance to mass transfer resides in the liquid phase. Evaluate

(a) The overall mass-transfer coefficient, K_y

Answer: 1.5 mol/m^2·s

(b) The molar flux of A

c) The liquid interfacial concentration of A

Answer: 7.88 mol%

3.10[d]. Absorption of ammonia by water: use of F-type mass-transfer coefficients

Modify the Mathcad program in Figure 3.6 to repeat Example 3.9, but with $y_{A,G} = 0.70$ and $x_{A,L} = 0.10$. Everything else remains constant.

Answer: $y_{A,i} = 0.588$

3.11[d]. Absorption of ammonia by water: use of F-type mass-transfer coefficients

Modify the Mathcad program in Figure 3.6 to repeat Example 3.9, but with $F_L = 5.0$ mol/m^2-s. Everything else remains constant.

Answer: $N_A = 0.62$ mol/m^2·s

3.12[b]. Mass-transfer resistances during absorption of ammonia

In the absorption of ammonia into water from an air–ammonia mixture at 300 K and 1 atm, the individual film coefficients were estimated to be $k_L = 6.3$ cm/h and $k_G = 1.17$ kmol/m^2·h·atm. The equilibrium relationship for very dilute solutions of ammonia in water at 300 K and 1 atm is $y_{A,i} = 1.64x_{A,i}$. Determine the following:

208 Interphase Mass Transfer

(a) k_y

(b) k_x

(c) K_y

Answer: 0.21 mol/m^2·s

(d) Fraction of the total resistance to mass transfer that resides in the gasphase.

Answer: 64.6%

3.13[b]. Mass-transfer resistances in hollow-fiber membrane contactors

For mass transfer across the hollow-fiber membrane contactors described in Example 2.14, the overall mass-transfer coefficient based on the liquid concentration, K_L, is given by (Yang and Cussler, 1986)

$$\frac{1}{K_L} = \frac{1}{k_L} + \frac{1}{mk_M} + \frac{1}{mk_c} \tag{3-83}$$

where k_L, k_M, and k_c are the individual mass-transfer coefficients in the liquid, across the membrane, and in the gas, respectively; and m is Henry's law constant, the gas equilibrium concentration divided by that in the liquid. The mass-transfer coefficient across a hydrophobic membrane is estimated from (Prasad and Sirkar, 1988)

$$k_M = \frac{D_{AB}\varepsilon_M}{\tau_M \delta} \tag{3-84}$$

where

D_{AB} = molecular diffusion coefficient in the gas filling the pores
ε_M = membrane porosity
τ_M = membrane tortuosity
δ = membrane thickness

For the modules of Example 2.14, $\varepsilon_M = 0.4$, $\tau_M = 2.2$, and $\delta = 25 \times 10^{-6}$ m (Prasad and Sirkar, 1988).

(a) Calculate the corresponding value of k_M.

Answer: 15.9 cm/s

(b) Using the results of part (a), Example 2.14, and Problem 2.27, calculate K_L, and estimate what fraction of the total resistance to mass transfer resides in the liquid film.

Problems 209

3.14^c. Combined use of F-type and k-type coefficients: absorption of low-solubility gases

During absorption of low-solubility gases, mass transfer from a highly concentrated gas mixture to a very dilute liquid solution frequently takes place. In that case, although it is appropriate to use a k-type mass-transfer coefficient in the liquid phase, an F-type coefficient must be used in the gas phase. Since dilute liquid solutions usually obey Henry's law, the interfacial concentrations during absorption of low-solubility gases are related through $y_{A,i} = mx_{A,i}$.

(a) Show that, under the conditions described above, the gas interfacial concentration satisfies the equation

$$\ln\left(1 - y_{A,i}\right) - \frac{k_x y_{A,i}}{mF_G} + \frac{k_x x_{A,L}}{F_G} - \ln\left(1 - y_{A,G}\right) = 0 \qquad (3\text{-}85)$$

(b) In a certain apparatus used for the absorption of SO_2 from air by means of water, at one point in the equipment the gas contained 30% SO_2 by volume and was in contact with a liquid containing 0.2% SO_2 by mol. The temperature was 303 K and the total pressure 1 atm. Estimate the interfacial concentrations and the local SO_2 molar flux. The mass-transfer coefficients were calculated as $F_G = 2.0$ mol/m^2·s and $k_x = 160$ mol/m^2·s. The equilibrium solubility data at 303 K are (Perry and Chilton, 1973)

kg SO_2/100 kg water	Partial pressure of SO_2, mm Hg
0.0	0
0.5	42
1.0	85
1.5	129
2.0	176
2.5	224

Answer: $N_A = 0.34$ mol/m^2·s

3.15^d. Distillation of a mixture of methanol and water in a packed tower: use of F-type mass-transfer coefficients

At a different point in the packed distillation column of Example 3.10, the methanol content of the bulk of the gas phase 76.2 mol%; that of the bulk of the liquid phase is 60 mol%. The temperature at that point in the tower is around 343 K. The packing characteristics and flow rates at that point are such that $F_G = 1.542$ mol/m^2·s and $F_L = 8.650$ mol/m^2·s. Calculate the interfacial compositions and the local methanol flux. To calculate the latent heats of vaporization at the new temperature, modify the values given in Example 3.6 using Watson's method (Smith et al., 1996):

Interphase Mass Transfer

$$\frac{\lambda_2}{\lambda_1} = \left[\frac{1 - T_{r2}}{1 - T_{r1}}\right]^{0.38} \tag{3-86}$$

For water, T_c = 647.1 K; for methanol, T_c = 512.6 K.

Answer: N_A = 0.123 mol/m^2·s

3.16[c, d]. Minimum flow rates: interpolated equilibrium data

Consider a countercurrent mass-transfer device for which the equilibrium distribution relation consists of a set of discrete values $\{X_{iD}, Y_{iD}\}$ instead of a continuous model such as Henry's law. An analysis similar to that presented in Example 3.13 is still possible using the *cubic spline* interpolation capabilities of Mathcad. Cubic spline interpolation passes a smooth curve through a set of points in such a way that the first and second derivatives are continuous across each point. Once the cubic spline describing a data set is assembled, it can be used as if it were a continuous model relating the two variables to predict accurately interpolated values. It can also be derived or integrated numerically as needed. For the case of transfer from phases V to L and a concave-downward equilibrium curve described by the data set $\{X_{iD}, Y_{iD}\}$, the procedure in Mathcad is as follows:

Define: vs: = cspline (X_{iD}, Y_{iD}) $Y_i(X_i)$: = interp (vs, X_{iD}, Y_{iD}, X_i)

Solve simultaneously equation (3-64) and:

$$\frac{L_S(\min)}{V_S} = \frac{dY_i(X_P)}{dX_P} \tag{3-87}$$

$$Y_P = Y_i(X_P) \tag{3-88}$$

[In equation (3-87), the first-derivative operator is from Operators list on the Math tab of Mathcad (Maxfield, 2014).]

Write a Mathcad program to solve equations (3-64), (3-87), and (3-88) and test it with data from the adsorber of Example 3.12.

Answer: $L_S'(\min)$ = 0.27 kg/s

3.17[b]. Material balances: adsorption of benzene vapor on activated carbon

Activated carbon is used to recover benzene from a nitrogen–benzene vapor mixture. The mixture, at 306 K and 1 atm containing 1.0% benzene by volume, is to be passed countercurrently at the rate of 1.0 m^3/s to a moving stream of activated carbon so as to remove 85% of the benzene from the gas in a continuous process. The entering activated carbon contains 15 cm^3 of benzene vapor (at STP) adsorbed per gram of the carbon. The temperature and total pressure are maintained at 306 K and 1 atm. Nitrogen is not adsorbed.

(a) Plot the equilibrium data given in the table below as X' = kg benzene/kg dry carbon, Y' = kg benzene/kg nitrogen for a total pressure of 1 atm.

Problems   211

(b) Calculate the minimum flow rate required of the entering carbon (remember that the entering carbon contains some adsorbed benzene).

Answer: 0.096 kg of activated carbon/s

(c) If the carbon flow is 20% above the minimum, what will be the concentrationof benzene adsorbed on the carbon leaving the process?

Answer: 0.295 kg benzene/kg activated carbon

(d) For the conditions of part (c), calculate the number of ideal stages required.

Answer: 3.32 stages

The equilibrium adsorption of benzene on this activated carbon at the temperature of 306 K is reported as follows:

Benzene vapor adsorbed, cm^3 (STP)/g of carbon	Partial pressure of benzene, mm Hg
15	0.55
25	0.95
40	1.63
50	2.18
65	3.26
80	4.88
90	6.22
100	7.83

3.18[b]. Material balances: desorption of benzene vapor from activated carbon

The activated carbon leaving the adsorber of Problem 3.17 is regenerated by countercurrent contact with steam at 380 K and 1 atm. The regenerated carbon is returned to the adsorber, while the mixture of steam and desorbed benzene vapors is condensed. The condensate separates into an organic and an aqueous phase and the two phases are separated by decantation. Due to the low solubility of benzene in water, most of the benzene will be concentrated in the organic phase, while the aqueous phase will contain only traces of benzene.

(a) Calculate the minimum steam flow rate required.

Answer: 0.035 kg/s

(b) For a steam flow rate 50% above the minimum, calculate the benzene concentration in the gas mixture leaving the desorber, and the number of ideal stages required.

Answer: 4.54 stages

The equilibrium adsorption of benzene on this activated carbon at the temperature of 380 K is reported as follows:

Benzene vapor adsorbed, kg benzene/100 kg of carbon	Partial pressure of benzene, kPa
2.9	1.0
5.5	2.0
12.0	5.0
17.1	8.0
20.0	10.0
25.7	15.0
30.0	20.0

3.19[b]. Material balances: adsorption of benzene vapor on activated carbon; cocurrent operation

If the adsorption process described in Problem 3.17 took place cocurrently, calculate the minimum flow rate of activated carbon required.

Answer: 0.543 kg/s

3.20[b]. Material balances in batch processes: drying of soap with air

It is desired to dry 10 kg of soap from 20% moisture by weight to no more than 6% moisture by contact with hot air. The wet soap is placed in a vessel containing 8.06 m^3 of air at 350 K and 1 atm, and a water-vapor partial pressure of 1.6 kPa. The system is allowed to reach equilibrium, and then the air in the vessel is entirely replaced by fresh air of the original moisture content and temperature. How many times must the process be repeated in order to reach the specified soap moisture content of no more than 6%? When this soap is exposed to air at 350 K and 1 atm, the equilibrium distribution of moisture between the air and the soap is as follows:

Wt% moisture in soap Partial pressure of	water, kPa
2.40	1.29
3.76	2.56
4.76	3.79
6.10	4.96
7.83	6.19
9.90	7.33
12.63	8.42
15.40	9.58
19.02	10.60

Answer: 5 times

Problems **213**

3.21[b]. Material balances in batch processes: extraction of an aqueous nicotine solution with kerosene

Nicotine in a water solution containing 2% nicotine is to be extracted with kerosene at 293 K. Water and kerosene are essentially insoluble in each other. Determine the percentage extraction of nicotine if 100 kg of the feed solution is extracted in a sequence of four batch ideal extractions using 49.0 kg fresh, pure kerosene each. The equilibrium data are given as follows (Claffey et al., 1950):

$X' \times 10^3$ kg nicotine/kg water	$Y' \times 10^3$ kg nicotine/kg kerosene
1.01	0.81
2.46	1.96
5.02	4.56
7.51	6.86
9.98	9.13
20.4	18.7

3.22[b]. Cross-flow cascade of ideal stages

The drying and liquid–liquid extraction operations described in Problems 3.20 and 3.21, respectively, are examples of a flow configuration called a *cross-flow cascade*. Figure 3.30 is a schematic diagram of a cross-flow cascade of ideal stages. Each stage is represented by a circle, and within each stage mass transfer occurs as if in cocurrent flow. The L phase flows from one stage to the next, being contacted in each stage by a fresh V phase. If the equilibrium-distribution curve of the cross-flow cascade is everywhere straight and of slope m, it can be shown that (Treybal, 1980)

$$N = \frac{\ln\left[\dfrac{X_0 - Y_0 / m}{X_N - Y_0 / m}\right]}{\ln\left(S + 1\right)} \tag{3-89}$$

where S is the *stripping factor*, mV_S/L_S, constant for all stages, and N is the total number of stages. Solve Problem (3.21) using equation (3-89) and compare the results obtained by both methods.

Answer: 78.1%

3.23[a]. Cross-flow cascade of ideal stages: nicotine extraction

Consider the nicotine extraction of Problems 3.21 and 3.22. Calculate the number of ideal stages required to achieve at least 95% extraction efficiency.

3.24[b]. Kremser equations: absorption of hydrogen sulfide

A scheme for the removal of H_2S from a flow of 1.0 std m^3/s of natural gas by scrubbing with water at 298 K and 10 atm is being considered. The initial composition of the feed gas is 2.5 mol% H_2S. A final gas stream containing only 0.1 mol%

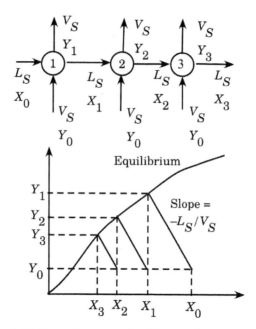

Figure 3.30 Cross-flow cascade of three ideal stages.

H$_2$S is desired. The absorbing water will enter the system free of H$_2$S. At the given temperature and pressure, the system will follow Henry's law, according to $Y_i = 48.3 X_i$, where X_i = mol H$_2$S/mol of water; Y_i = mol H$_2$S/mol of air.

(a) For a countercurrent absorber, determine the flow rate of water that is required if 1.5 times the minimum flow rate is used.

Answer: 54.5 kg water/s

(b) Determine the composition of the exiting liquid.

Answer: 0.67 g H$_2$S/L of water

(c) Calculate the number of ideal stages required.

Answer: 5.86 stages

3.25^b. Absorption with chemical reaction: H$_2$S scrubbing with MEA

As shown in Problem 3.24, scrubbing of hydrogen sulfide from natural gas using water is not practical since it requires large amounts of water due to the low solubility of H$_2$S in water. If a 2N solution of monoethanolamine (MEA) in water is used as the absorbent, however, the required liquid flow rate is reduced dramatically because the MEA reacts with the absorbed H$_2$S in the liquid phase, effectively increasing its solubility.

For this solution strength and a temperature of 298 K, the solubility of H$_2$S can be approximated by (de Nevers, 2000)

Problems

$$p_{H_2S,i} = \left(19.4 \text{ Pa}\right) \times \exp\left(195 x_{H_2S,i}\right)$$

Repeat the calculations of Problem 3.24 but using a 2N monoethanolamine solution as absorbent.

Answer: 0.76 kg MEA solution/s

3.26[b]. Kremser equations: absorption of sulfur dioxide

A flue gas flows at the rate of 10 kmol/s at 298 K and 1 atm with a SO_2 content of 0.15 mol%. Ninety percent of the sulfur dioxide is to be removed by absorption with pure water at 298 K. The design water flow rate will be 50% higher than the minimum. Under these conditions, the equilibrium line is (Benítez, 1993) $Y_i = 10X_i$, where Y_i = mol SO_2/mol air and X_i = mol SO_2/mol water.

(a) Calculate the water flow rate and the SO_2 concentration in the water leaving the absorber.

Answer: 2426 kg/s

(b) Calculate the number of ideal stages required for the specified flow rates and percentage SO_2 removal.

Answer: 4.01 stages

3.27[b]. Kremser equations: absorption of sulfur dioxide

An absorber is available to treat the flue gas of Problem 3.26, which is equivalent to 8.5 equilibrium stages.

(a) Calculate the water flow rate to be used in this absorber if 90% of the SO_2 is to be removed. Calculate also the SO_2 concentration in the water leaving the absorber.

Answer: 1819 kg/s

(b) What is the percentage removal of SO_2 that can be achieved with this absorber if the water flow rate used is the same as that calculated in Problem 3.26 (a)?

Answer: 97.9%

3.28[b]. Kremser equations: extraction of acetic acid with 3-heptanol

An aqueous acetic acid solution flows at the rate of 1000 kg/h. The solution is 1.1 wt% acetic acid. It is desired to reduce the concentration of this solution to 0.037 wt% acetic acid by extraction with 3-heptanol at 298 K. For practical purposes, water and 3-heptanol are immiscible. The inlet 3-heptanol contains 0.02 wt% acetic acid. An extraction column is available which is equivalent to a countercurrent cascade of 35 equilibrium stages. What solvent flow rate is required? Calculate the composition of the solvent phase leaving the column. For this system and range of concentrations, equilibrium is given by (Wankat, 1988):

216   Interphase Mass Transfer

wt ratio acetic acid in solvent = 0.828 × wt ratio acetic acid in water

Answer: 1247 kg/h

3.29^c. Countercurrent versus cross-flow extraction

A 1-butanol acid solution is to be extracted with pure water. The butanol solution contains 4.5 wt% of acetic acid and flows at the rate of 400 kg/h. A total water flow rate of 1005 kg/h is used. Operation is at 298 K and 1 atm. For practical purposes, 1-butanol and water are immiscible. At 298 K, the equilibrium data can be represented by $Y_{Ai} = 0.62X_{Ai}$, where Y_{Ai} is the weight ratio of acid in the aqueous phase and X_{Ai} is the weight ratio of acid in the organic phase.

(a) If the outlet butanol stream is to contain 0.10 wt% acid, how many equilibrium stages are required for a countercurrent cascade?

(b) If the water is split up equally among the same number of stages, but in a cross-flow cascade, what is the outlet acid concentration in the organic phase? (See Problem 3.22.)

Answer: 1.1%

3.30^c. Glucose sorption on an ion-exchange resin

Ching and Ruthven (1985) found that the equilibrium of glucose on an ion-exchange resin in the calcium form was linear for concentrations below 50 g/L. Their equilibrium expression at 303 K is $Y_{Ai} = 1.961X_{Ai}$, where X_{Ai} is the glucose concentration in the resin (g of glucose per liter of resin) and Y_{Ai} is the glucose concentration in solution (g of glucose per liter of solution).

(a) We wish to sorb glucose onto this ion-exchange resin at 303 K in a counter-current cascade of ideal stages. The concentration of the feed solution is 15 g/L. We want an outlet concentration of 1.0 g/L. The inlet resin contains 0.25 g of glucose/L. The feed solution flows at the rate of 100 L/min, while the resin flows at the rate of 250 L/min. Find the number of equilibrium stages required.

(b) If five equilibrium stages are added to the cascade of part (a), calculate the resin flow required to maintain the same degree of glucose sorption.

Answer: 216.3 L/min

REFERENCES

Benítez, J., *Process Engineering and Design for Air Pollution Control*, Prentice Hall, Englewood Cliffs, NJ (1993).

Ching, C. B., and D. M. Ruthven, *AIChE Symp. Ser.*, **81**, 242 (1985).

Claffey, J. B., et al., *Ind. Eng. Chem.*, **42**, 166 (1950).

de Nevers, N., *Air Pollution Control Engineering*, 2nd ed., McGraw-Hill, Boston (2000).

Appendix 3.1

Foust, A. S., et al., *Principles of Unit Operations*, 2nd ed., Wiley, New York (1980).

Himmelblau, D. M., *Basic Principles and Calculations in Chemical Engineering*, 5th ed., Prentice Hall, Englewood Cliffs, NJ (1989).

Kremser, A., *Natl. Petrol. News*, **22**, 42 (1930).

Lewis, W. K., and W. G. Whitman, *Ind. Eng. Chem.*, **16**, 1215 (1924).

Maxfield, B., *Essential PTC Mathcad Prime 3.0*, Academic Press, Amsterdam (2014).

Perchetti, G. A., et al., *Journal of Clinical Microbiology*, **58**, 8 (2020).

Perry, R. H., and C. H. Chilton (eds.), *Chemical Engineers' Handbook*, 5th ed., McGraw-Hill, New York (1973).

Prasad, R., and K. K. Sirkar, *AIChE J.*, **34**, 177 (1988).

Seader, J. D., E. J. Henley, and D. K. Roper, *Separation Process Principles*, 3rd ed., John Wiley and Sons, New York (2011).

Smith, J. M., H. C. Van Ness, and M. M. Abbott, *Introduction to Chemical Engineering Thermodynamics*, 5th ed., McGraw-Hill, New York (1996).

Treybal, R. E., *Mass-Transfer Operations*, 3rd ed., McGraw-Hill, New York (1980).

Wankat, P. C., *Equilibrium Staged Separations*, Elsevier, New York (1988).

Yang, M. C., and E. L. Cussler, *AIChE J.*, **32**, 1910 (1986).

APPENDIX 3.1

Python routine to solve Example 3.3.

```python
from scipy.optimize import fsolve
import numpy as np
def Example39a(p):
    yA = p[0]
    xA = p[1]
    gamA = p[2]
    return yA + 0.4*(0.88/(1 - xA))**1.75 - 1.0, yA - 10.51*xA*gamA,\
    gamA - 3.5858*xA**2 + 0.622*xA - 0.156
p = np.zeros(3,dtype=float)
# Initial estimates:
p[0] = 0.2
p[1] = 0.2
p[2] = 1.0
p = fsolve(Example39a,p)
print ('The gas-phase interfacial mol fraction yA = %5.3f' %p[0])
print('The liquid-phase interfacial mol fraction xA = %5.3f' %p[1])
print('The activity coefficient \u03B3A = %5.3f' %p[2])
```

218   **Interphase Mass Transfer**

Program results:

The gas-phase interfacial mol fraction $y_A = 0.214$
The liquid-phase interfacial mol fraction $x_A = 0.144$
The activity coefficient $\gamma_A = 0.141$
Ammonia in the liquid phase, in kmol, $L_A = 0.422$

APPENDIX 3.2

Python routine to solve Example 3.9a.

```python
from scipy.optimize import fsolve
import numpy as np
def Example33(p):
    yA = p[0]
    xA = p[1]
    gamA = p[2]
    LA = p[3]
    return xA - LA/(LA+2.5), yA - 10.51*xA*gamA,\
    gamA - 3.5858*xA**2 + 0.622*xA - 0.156,yA-(0.588-LA)/(1.197-LA)
p = np.zeros(4,dtype=float)
# Initial estimates:
p[0] = 0.5
p[1] = 0.5
p[2] = 1.0
p[3] = 0.5
p = fsolve(Example33,p)
print ('The gas-phase interfacial mol fraction yA = %5.3f' %p[0])
print('The liquid-phase interfacial mol fraction xA = %5.3f' %p[1])
print('The activity coefficient \u03B3A = %5.3f' %p[2])
print('Ammonia in the liquid phase, in kmoles, LA = %5.3f' %p[3])
```

Program results:
The gas-phase interfacial mol fraction $y_A = 0.494$
The liquid-phase interfacial mol fraction $x_A = 0.231$
The activity coefficient $\gamma_A = 0.204$

APPENDIX 3.3

Python routine to solve Example 3.10.

```python
from scipy.optimize import fsolve
import numpy as np
#Problem Parameters
a12 = 107.38; V1 = 40.73; V2 = 18.07; R = 1.987; a21 = 469.55; P = 101.3
Fg = 1.7; PsiAG = 5.155
def Example310(p):
    x1 = p[0]
    x2 = p[1]
    y1 = p[2]
    y2 = p[3]
    T = p[4]
    def La12(T):
        return (V2/V1)*np.exp(-a12/(R*T))
    def La21(T):
        return (V1/V2)*np.exp(-a21/(R*T))
    def P1(T):
        return np.exp(16.5938 - 3644.3/(T - 33))
    def P2(T):
        return np.exp(16.2620 - 3800.0/(T - 47))
    def gam1(T,x1,x2):
        return np.exp(-np.log(x1+x2*La12(T))+x2*(La12(T)/(x1+x2*La12(T))\
                                        -La21(T)/(x2+x1*La21(T))))
    def gam2(T,x1,x2):
        return np.exp(-np.log(x2+x1*La21(T))-x1*(La12(T)/(x1+x2*La12(T))\
                                        -La21(T)/(x2+x1*La21(T))))
    return x1+x2-1, y1+y2-1, y1*P-x1*gam1(T,x1,x2)*P1(T), y2*P-x2*gam2(T,\
                    x1,x2)*P2(T),y1+4.795*(4.955/(5.155-x1))**8.765-5.155
p = np.zeros(5,dtype=float)
# Initial estimates:
p[0] = 0.1; p[1] = 0.9; p[2] = 0.2; p[3] = 0.8; p[4] = 360
p = fsolve(Example310,p)
NA = PsiAG*Fg*np.log((PsiAG-p[2])/(PsiAG-0.36))

print ('x1 = %5.3f' %p[0]); print('x2 = %5.3f' %p[1]); print('y1 = %5.3f' %p[2])
print('y2 = %5.3f' %p[3]); print('Temperature (K) = %5.3f' %p[4])
print('The methanol molar flux, in mol/m^2-s, = %5.3f' %NA)
```

Program results:

$x_1 = 0.177$ $x_2 = 0.823$ $y_1 = 0.549$ $y_2 = 0.451$

Temperature = 356.46 K
Methanol molar flux = -0.352 mol/m^2-s

APPENDIX 3.4

Python routine to solve Example 3.13.

```python
# a) Absorber
from scipy.optimize import fsolve
import numpy as np
# Enter data
meq = 0.136; Vs = 38.0; Xin = 0.05; Yin = 0.08; Yout = 0.012
def Ex311a(p):
    YP = p[0]; XP = p[1]; Lsmin = p[2]
    return Lsmin/Vs -(YP - Yout)/(XP - Xin), Lsmin/Vs - meq/\
    (1 + (1 - meq)*XP)**2, YP - meq*XP/(1+(1-meq)*XP)
# Enter initial estimates of the variables:
p = np.zeros(3,dtype=float)
p[0] = (Yin + Yout)/2
p[1] = p[0]/meq
p[2] = meq*Vs
p = fsolve(Ex311a, p)
print ('a) Absorber Results')
print('Lsmin (mol/s) = %5.3f' %p[2])
print('YP = %5.3f' %p[0])
print('XP = %5.3f' %p[1])


# b) Stripper


# Enter data
meq = 1.770; Ls = 6.0; Xin = 0.48; Yin = 0.0; Xout = 0.05
def Ex311b(p):
    XP = p[0]; YP = p[1]; Vsmin = p[2]
    return Ls/Vsmin -(YP - Yin)/(XP - Xout), Ls/Vsmin - meq/\
    (1 + (1 - meq)*XP)**2, YP - meq*XP/(1+(1-meq)*XP)
# Enter initial estimates of the variables
p = np.zeros(3,dtype=float)
p[0] = (Xin + Xout)/2
p[1] = p[0]*meq
p[2] = Ls/meq
p = fsolve(Ex311b,p)
print('b) Stripper Results')
print('Vsmin (mol/s) = %5.3f' %p[2])
print('XP = %5.3f' %p[0])
print('YP = %5.3f' %p[1])
```

Appendix 3.5 221

Program results:
Absorber: $L_S(\text{min}) = 3.111$ mol/s
Stripper: $V_S(\text{min}) = 2.190$ mol/s

APPENDIX 3.5

Python routine to solve Example 3.16.

(a) Number of ideal stages for an absorber

```python
from scipy.optimize import fsolve
import numpy as np
#Enter data:
me = 0.136; Vs = 38.0; Ls = 6.0
X0 = 0.05; YNp1 = 0.08; Y1 = 0.012
def Yop(X):
    return Ls*(X - X0)/Vs + Y1

def Xeq(Y):
    return Y/(me + (me - 1)*Y)
#Calculate liquid exit concentration
XN = 1.0   # Initial estimate
def g(x):
    return Yop(x) - YNp1

XN = fsolve(g, XN)
print('The liquid exit concentration is XN = %5.3f' %XN)
# We are assuming that the number of ideal stages is lee than 20.
# If it is greater than 20, you get an error message.
X = np.zeros(20,dtype=float)
Y = np.zeros(20,dtype=float)
i = np.zeros(20,dtype=int)

def idealabs(XN, X0):
    i = 0
    Y[1] = Yop(X0)
    X[0] = X0
    while (X[i] < XN):
        i+=1
        X[i] = Xeq(Y[i])
        Y[i+1] = Yop(X[i])
    return i - 1, X[i-1], Y[i]

solution = idealabs(XN,X0)

N = solution[0]+(XN - solution[1])/(Xeq(solution[2]) - solution[1])
print('The number of ideal stages is = %5.3f' %N)
```

Program results:
Liquid exit concentration, $X_N = 0.481$
Number of ideal stages = 3.999

b) Number of ideal stages for a stripper

```python
from scipy.optimize import fsolve
import numpy as np
# Enter data:
me = 1.770; Vs = 3.3; Ls = 6.0; X0 = 0.48
YNp1 = 0.0; XN = 0.05
def Yop(X):
    return Ls*(X - XN)/Vs + YNp1

def Xeq(Y):
    return Y/(me + (me - 1)*Y)
# Calculate the gas exit concentration Y1
Y1 = Yop(X0)
print('The gas exit concentration is Y1 = %5.3f' %Y1)

# We assume that the number of stages is less than 20.

X = np.zeros(20,dtype=float)
Y = np.zeros(20,dtype=float)
i = np.zeros(20,dtype=int)

def idealstrip(XN, X0):
    i = 0
    Y[1] = Yop(X0)
    X[0] = X0
    while (X[i] > XN):
        i+=1
        X[i] = Xeq(Y[i])
        Y[i+1] = Yop(X[i])
    return i - 1, X[i-1], Y[i]
solution = idealstrip(XN, X0)

N = solution[0] + (XN - solution[1])/(Xeq(solution[2]) -\
                                    solution[1])

print('The number of ideal stages is = %5.3f' %N)
```

Appendix 3.5

Program results:
Gas exit concentration, $Y_1 = 0.782$
Number of ideal stages $= 5.16$

4

Equipment for Gas–Liquid Mass-Transfer Operations

4.1 INTRODUCTION

The purpose of the equipment used for mass-transfer operations is to provide intimate contact of the immiscible phases in order to permit interphase diffusion of the constituents. The rate of mass transfer is directly dependent upon the interfacial area exposed between the phases, and the nature and degree of dispersion of one phase into the other are therefore of prime importance.

The operations which include humidification and dehumidification, gas absorption and desorption, and distillation all have in common the requirement that a gas and a liquid phase be brought into contact for the purpose of diffusional interchange between them. The equipment for gas–liquid contact can be broadly classified according to whether its principal action is to disperse the gas or the liquid, although in many devices both phases become dispersed. In principle, at least, any type of equipment satisfactory for one of these operations is suitable for the others, and the major types are indeed used for all. For this reason, the main emphasis of this chapter is on equipment for gas–liquid operations.

4.2 GAS–LIQUID OPERATIONS: LIQUID DISPERSED

Your objectives in studying this section are to be able to:

1. Identify the most common types of tower packings.
2. Estimate gas-pressure drop in packed towers.
3. Estimate volumetric mass-transfer coefficients in packed towers.

Principles and Applications of Mass Transfer: The Design of Separation Processes for Chemical and Biochemical Engineering, Fourth Edition. Jaime Benitez.
© 2023 John Wiley & Sons, Inc. Published 2023 by John Wiley & Sons, Inc.

225

226 Equipment for Gas–Liquid Mass-Transfer Operations

This group includes devices in which the liquid is dispersed into thin films or drops, such as wetted-wall towers, sprays and spray towers, and packed towers. A thin film of liquid flowing down the inside of a vertical pipe, with gas flowing cocurrently or countercurrently, constitutes a *wetted-wall tower*. Such devices have been used for theoretical studies of mass transfer, as described in Chapter 2, because the interfacial surface is readily kept under control and is measurable. Industrially, they have been used as absorbers for hydrochloric acid, where absorption is accompanied by a very large evolution of heat (Hulswitt and Mraz, 1972). In this case, the wetted-wall tower is surrounded with rapidly flowing cooling water. Multitube devices have also been used for distillation, where the liquid film is generated at the top by partial condensation of the rising vapor. Gas-pressure drop in these towers is probably lower than in any other gas–liquid contacting device, for a given set of operating conditions.

The liquid can be sprayed into a gas stream by means of a nozzle which disperses the liquid into a fine spray of drops. The flow may be countercurrent, as in vertical spray towers with the liquid sprayed downward, or parallel, as in horizontal spray chambers. They are frequently used for adiabatic humidification–cooling operations with recirculating liquid. These devices have the advantage of low pressure drop for the gas, but also have a number of disadvantages. There is a relatively high pumping cost for the liquid, owing to the pressure drop through the spray nozzle. The tendency for entrainment of liquid by the gas leaving is considerable, and mist eliminators will almost always be necessary. Unless the diameter/length ratio is very small, the gas will be fairly thoroughly mixed by the spray and full advantage of countercurrent flow cannot be taken. Ordinarily, however, the diameter/length ratio cannot be made very small since then the spray would quickly reach the walls of the tower and become ineffective as a spray.

Packed towers, used for continuous contact between liquid and gas in both countercurrent and cocurrent flow, are vertical columns which have been filled with packings or devices of large surface, as in Figure 4.1. The liquid is distributed over, and trickles down through, the packed bed, exposing a large surface to contact the gas. The tower packing, or fill, should provide a large interfacial surface between liquid and gas per unit volume of packed space. It should possess desirable fluid-flow characteristics. This means that the fractional void volume ε in the packed bed should be large enough to permit passage of the volumes of gas and liquid to be processed at relatively high velocity, with low pressure drop for the gas. The packing should be chemically inert to the fluids being processed and should possess structural strength to permit easy handling and installation.

4.2.1 Types of Packing

Packings are of two major types, random and regular. *Random packings* are simply dumped into the tower during installation and allowed to fall at random. The Raschig ring, first patented by Dr. Fritz Raschig in Germany in 1907, was the first standardized packing. Until then, coke or broken glass and pottery were used as packings. Until the 1960s, packed columns were mostly filled with Raschig rings or Berl saddles, known as *first-generation packings*. Then development of more advanced packings with higher separation efficiency at low pressure drop accelerated. Today,

4.2 Gas–Liquid Operations: Liquid Dispersed

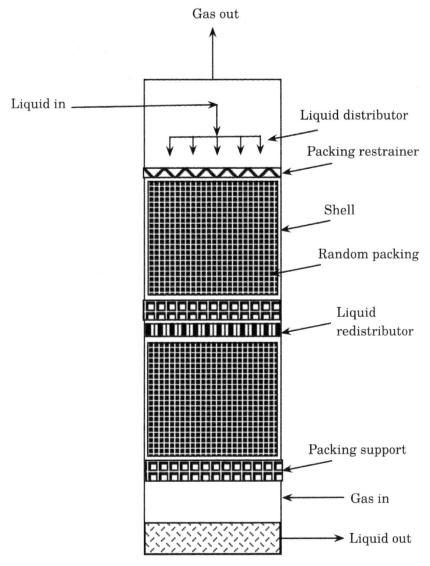

Figure 4.1 Schematic diagram of a packed tower.

Pall rings (*second-generation*) and exotically shaped saddles made of ceramics, metals, or plastics (*third-generation*) are widely used as packings.

Metal packings are lighter and resist breakage better than ceramic packings, making metal the choice for deep beds. Metal also lends itself to packing geometries that yield higher efficiencies than ceramic or plastic packing shapes. Compared to standard plastic packings, metal packings withstand higher temperatures and provide better wettability. Ceramic packings manufactured in chemical porcelain offer optimal corrosion resistance for applications, such as SO_2 and SO_3 absorption, mercaptan removal, natural gas or LPG sweetening, and corrosive distillation. Plastic packings offer the advantage of lightness in weight, but they must be chosen carefully since they may deteriorate rapidly with certain organic solvents and with oxygen-bearing gases at only slightly elevated temperatures.

228　　　　　　　　　**Equipment for Gas–Liquid Mass-Transfer Operations**

Raschig rings are hollow cylinders of diameters ranging from 6 mm to 100 mm or more. They may be made of metal, ceramic, or plastic. Pall rings (and as a variant, Hy-Pak), cylinders with partitions, are available in metal and plastic in five nominal sizes from 16 mm to 89 mm. The saddle-shaped packings, Intalox and Super Intalox saddles, are available in sizes from 6 mm to 75 mm, made of ceramic or plastic. The Super Intalox saddle has scalloped edges and holes for improved mass transfer and reduced settling and channeling.

More recently, through-flow packings of a lattice-work design have been developed. These packings, which include metal Intalox IMPT; metal, plastic, and ceramic Cascade Mini-Rings (CMR); metal Levapak; metal, plastic, and ceramic Hiflow rings; metal tri-packs; and plastic Nor Pac rings, exhibit even lower pressure drop per unit height of packing and even higher mass-transfer rates per unit volume of packing. Accordingly, they are called "high-efficiency" random packings. The IMTP packing is ideal in a wide range of mass-transfer services. It is used extensively in distillation towers and in absorbers and strippers. It is available in most metals and in six nominal sizes, ranging from 15 mm to 70 mm. Other, more exotic packing shapes are available, such as the plastic 38-mm Snowflakes.

Most recently, a *fourth-generation* of random packings, including VSP rings, Fleximax, and Raschig Super-Rings, have been developed. They feature a very open, undulating geometry that promotes even wetting, but with recurrent turbulence promotion. The result is lower pressure drop, but sustained mass transfer efficiency that may not decrease noticeably with increasing column diameter and may allow a greater depth of packing before a liquid redistributor is needed. Metal packings are usually preferred because of their superior strength and good wettability (Seader and Henley, 2006).

As a rough guide, packing sizes of 25 mm or larger are ordinarily used for gas rates of 0.25 m^3/s, and 50 mm or larger for gas rates of 1.0 m^3/s or more. As packing size increases, mass-transfer efficiency and pressure drop may decrease. Therefore, for a given column diameter, an optimal packing size exists that represent a compromise between these two factors, since low pressure drop and high mass-transfer rates are both desirable. However, to minimize channeling of liquid, the nominal size of the packing should be less than one-eighth of the column diameter. During installation, the packings are poured into the tower to fall at random, and in order to prevent breakage of ceramic packings, the tower may first be filled with water to reduce the velocity of fall (Treybal, 1980).

Regular or *structured packings* are of great variety. They offer the advantages of low pressure drop for the gas and greater possible flow rates, usually at the expense of more costly installation than random packing. Wood grids, or *hurdles*, are inexpensive and frequently used where large void volumes are required, as with cooling towers. Representative structured packings include the older corrugated sheets of metal gauze, such as Sulzer BX, Montz A, Gempak 4BG, and Intalox High-Performance Wire Gauze Packing. Newer and less-expensive structured packings are fabricated from sheet metal and plastics and may or may not be perforated, embossed, or surface roughened. They include metal and plastic Mellapak 250Y, metal Flexipac, metal and plastic Gempak 4A, metal Montz B1, and metal Intalox High-Performance Structured Packing.

4.2 Gas–Liquid Operations: Liquid Dispersed

Structured packings come with different size openings between adjacent corrugated layers and are stacked in the column. Although structured packings are considerably more expensive per unit volume than random packings, structured packings exhibit far less pressure drop per theoretical stage and have higher efficiency and capacity. *Static mixers* were originally designed as *line mixers* for mixing two fluids in cocurrent flow. There are several designs, but in general they consist of metal eggcrate-like devices installed in a pipe to cause a multitude of splits of cocurrently flowing fluids into left- and right-hand streams, breaking each stream down into increasingly smaller streams. The use of static mixers as alternative contacting devices for supercritical fluid separation processes has been proposed recently (Ruivo et al., 2006). The authors presented new data that addressed the hydrodynamic behavior and mass-transfer kinetics of a static mixer under high pressure conditions.

Packed tower shells may be of wood, metal, chemical stoneware, acidproof brick, plastic- or glass-lined metal, or other material, depending on corrosion conditions. For ease of construction and strength, they are usually circular in cross section. An open space at the bottom of the tower is necessary for ensuring good distribution of the gas into the packing. The support must, of course, be strong enough to carry the weight of a reasonable height of packing, and it must have ample free area to allow for flow of liquid and gas with a minimum of restriction. Specially designed supports which provide separate passageways for gas and liquid are available.

4.2.2 Liquid Distribution

Adequate initial distribution of the liquid at the top of the packing is of paramount importance. Otherwise, a significant portion of the packing near the top of the tower will remain dry. Dry packing, is of course, completely ineffective for mass transfer, and various devices are used for liquid distribution. The arrangement shown in Figure 4.1, or a ring of perforated pipe, can be used in small towers. For large diameters, special liquid distributors are available.

In the case of random packings, the packing density (i.e., the number of packing pieces per unit volume) is ordinarily less in the immediate vicinity of the tower walls, and this leads to a tendency of the liquid to segregate toward the walls and the gas to flow in the center of the tower (channeling). This tendency is much less pronounced when the diameter of the individual packing pieces (d_p) is smaller than one-eighth the tower diameter (D). It is recommended that the ratio $d_p/D = 1/15$ (Treybal, 1980). Even so, it is customary to provide for redistribution of the liquid at intervals varying from 3 to 10 times the tower diameter, but at least every 6 or 7 m.

Example 4.1 Oscillatory Variation of Void Fraction Near the Walls of Packed Beds

It is well known that the wall in a packed bed affects the packing density resulting in void fraction variations in the radial direction. Mueller (1992) developed the fol-

230　　　　　　　　　　**Equipment for Gas–Liquid Mass-Transfer Operations**

lowing equation to predict the radial void fraction distribution in a cylindrical tower packed with equal-sized spheres:

$$\varepsilon = \varepsilon_b + (1 - \varepsilon_b) J_0(\alpha r^*) \exp(-\beta r^*) \qquad (4\text{-}1)$$

where

$$\alpha = 7.45 - 3.15 \frac{d_p}{D} \qquad \text{for } 2.02 \leq \frac{D}{d_p} \leq 13.0$$

$$\alpha = 7.45 - 11.25 \frac{d_p}{D} \qquad \text{for } \frac{D}{d_p} \geq 13.0$$

$$\beta = 0.315 - 0.725 \frac{d_p}{D} \qquad \varepsilon_b = 0.365 + 0.220 \frac{d_p}{D}$$

$$r^* = \frac{r}{d_p} \quad 0 \leq r^* \leq \frac{D}{2 d_p}$$

(4-2)

J_0 = Bessel function of the first kind of order zero
r = radial distance measured from the wall

Consider a cylindrical vessel with a diameter of 305 mm packed with solid spheres with a diameter of 20 mm. Plot the radial void fraction fluctuations near the walls for this packed bed.

Solution
For this case, d_p = 20 mm, D = 305 mm, and D/d_p = 15.25. Substituting this ratio in the definitions of equation (4-2), α = 6.712, β = 0.267, and ε_b = 0.379.

Substituting in equation (4-1), yields

$$\varepsilon = 0.379 + 0.621 \times J_0(6.712 r^*) \exp(-0.267 r^*)$$

Figure 4.2 shows the result of the analysis. Notice the wide fluctuations of void fraction near the wall, decaying gradually as we move toward the center of the bed. According to Govindarao and Froment (1986), when D/d_p > 10, the radial void fraction approaches a constant asymptotic value (ε_b) at a distance from the wall of approximately 5 particle diameters, consistent with the results shown in Figure 4.2.

4.2.3 Liquid Holdup

For most random packings, the pressure drop suffered by the gas is influenced by the gas and liquid flow rates. At a fixed gas velocity, the gas-pressure drop increases with increased liquid rate, principally because of the reduced free cross section available for flow of gas resulting from the presence of the liquid. Below a certain limiting gas velocity, the *liquid holdup* (i.e., the quantity of liquid contained in the packed bed) is reasonably constant with changing gas velocity, although it increases with liquid rate. For a given liquid velocity, the upper limit to the gas

4.2 Gas–Liquid Operations: Liquid Dispersed

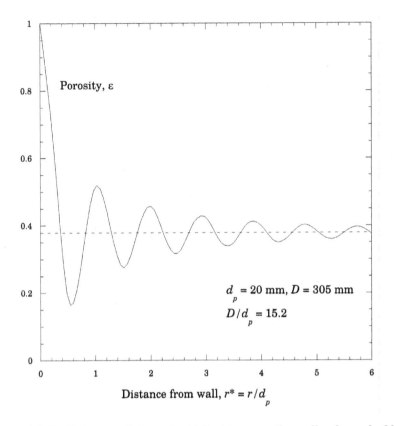

Figure 4.2 Oscillatory variation of void fraction near the walls of a packed bed.

velocity for a constant liquid holdup is termed the *loading point*. Below this point, the gas phase is the continuous phase. Above this point, liquid begins to accumulate or load the bed, replacing gas holdup and causing a sharp increase in pressure drop. Finally, a gas velocity is reached at which the liquid surface is continuous across the top of the packing and the column is flooded. At the *flooding point*, the pressure drop increases infinitely with increasing gas velocity.

The region between the loading point and the flooding point is the *loading region*; significant liquid entrainment is observed, liquid holdup increases sharply, mass-transfer efficiency decreases, and column operation is unstable. Typically, according to Billet (1989), the superficial gas velocity at the loading point is approximately 70% of that at the flooding point. Although a packed column can operate in the loading region, most packed columns are designed to operate below the loading point, in the *preloading region*.

To locate the loading point, Billet and Schultes (1999) developed the following equations to calculate the loading superficial gas velocity, $v_{G,S}$:

$$v_{G,S} = \sqrt{\frac{g}{\Psi_S}} \left[\frac{\varepsilon}{a^{1/6}} - a^{1/2} \left(\frac{12\mu_L v_{L,S}}{g\rho_L} \right)^{1/3} \right] \left(\frac{12\mu_L v_{L,S}}{g\rho_L} \right)^{1/6} \sqrt{\frac{\rho_L}{\rho_G}} \qquad (4\text{-}3)$$

Equipment for Gas–Liquid Mass-Transfer Operations

$$\Psi_S = \frac{9.8}{C_S^2}\left[X\left(\frac{\mu_L}{\mu_G}\right)^{0.4}\right]^{-2n_S} \tag{4-4}$$

$$v_{L,S} = \frac{\rho_G L'}{\rho_L V'}v_{G,S} \tag{4-5}$$

$$n_S = -0.326 \quad C_S = C_{S,T} \ \text{ for } X \le 0.4$$

$$n_S = -0.723 \quad C_S = 0.695 C_{S,T}\left(\frac{\mu_L}{\mu_G}\right)^{0.1588} \quad \text{for } X > 0.4 \tag{4-6}$$

where

X = flow parameter = $(L'/V')(\rho_G/\rho_L)^{0.5}$
$v_{L,S}$ = superficial liquid velocity at the loading point, m/s
g = acceleration of gravity
a = specific surface area of packing, m^2/m^3
L' and V' = liquid and gas flow rates, respectively, kg/s

Values of a and $C_{S,T}$ are characteristic of the particular type and size of packing, as listed together with packing void fraction, ε, and other packing constants in Table 4.1.

The specific liquid holdup (i.e., volume of liquid holdup/volume of packed bed) in the preloading region has been found from extensive experiments by Billet and Schultes (1995) for a wide variety of random and structured packings and for a number of gas–liquid systems to depend on packing characteristics, and the viscosity, density, and superficial velocity of the liquid according to the dimensionless expression

$$h_L = \left[\frac{12\text{Fr}_L}{\text{Re}_L}\right]^{1/3}\left(\frac{a_h}{a}\right)^{2/3} \tag{4-7}$$

where

h_L = specific liquid holdup, m^3/m^3 packed bed
Re_L = liquid Reynolds number = $v_{L,S}\rho_L/a\mu_L$
Fr_L = liquid Froude number = $(v_{L,S})^2 a/g$
a_h = hydraulic, or effective, specific area of packing

The ratio of specific areas is given by

$$\frac{a_h}{a} = C_h \text{Re}_L^{0.5} \text{Fr}_L^{0.1} \quad \text{for Re}_L < 0.5 \tag{4-8}$$

$$\frac{a_h}{a} = 0.85 C_h \text{Re}_L^{0.25} \text{Fr}_L^{0.1} \quad \text{for Re}_L \ge 0.5 \tag{4-9}$$

Values of C_h are characteristic of the particular type and size of packing and are listed in Table 4.1. Because the specific liquid holdup in the preloading region is constant, equation (4-7) does not involve gas-phase properties or gas velocity.

4.2 Gas–Liquid Operations: Liquid Dispersed

At low liquid velocities, liquid holdup can be so small that the packing is no longer completely wetted. When this happens, packing mass-transfer efficiency decreases dramatically, particularly for aqueous systems of high surface tension. To ensure complete wetting of packing, proven liquid distributors and redistributors should be used and superficial liquid velocities should exceed the following values (Seader and Henley, 1998):

Type of packing material	$v_{L, min}$, mm/s
Ceramic	0.15
Oxidized or etched metal	0.30
Bright metal	0.90
Plastic	1.20

To locate the flooding point, Billet and Schultes (1999) developed the following equations to calculate the flooding gas velocity, $v_{G,F}$:

$$v_{G,F} = \sqrt{\frac{2g\left(\varepsilon - h_{L,F}\right)^3 h_{L,F}\rho_L}{\Psi_F \varepsilon a \rho_G}} \qquad (4\text{-}10)$$

$$\Psi_F = \frac{9.8}{C_F^2}\left[X\left(\frac{\mu_L}{\mu_G}\right)^{0.2}\right]^{-2n_F} \qquad (4\text{-}11)$$

$$X = \frac{L'}{V'}\left(\frac{\rho_G}{\rho_L}\right)^{0.5} \qquad (4\text{-}12)$$

$$n_F = -0.194 \qquad C_F = C_{F,T} \qquad \text{for } X \leq 0.4$$

$$n_F = -0.708 \qquad C_F = 0.6244 C_{F,T}\left(\frac{\mu_L}{\mu_G}\right)^{0.1028} \qquad \text{for } X > 0.4 \qquad (4\text{-}13)$$

$$h_{L,F}^3\left(3h_{L,F} - \varepsilon\right) = \frac{6a^2 \varepsilon \mu_L L' \rho_G v_{G,F}}{g\rho_L^2 V'} \quad \text{for } \frac{\varepsilon}{3} \leq h_{L,F} \leq \varepsilon \qquad (4\text{-}14)$$

where values of $C_{F,T}$ are characteristic of the particular type and size of packing and are listed in Table 4.1.

Example 4.2 Loading and Flooding Points in Beds Packed with First- and Third-Generation Random Packings

Air containing 5 mol% NH_3 at a total flow rate of 20 kmol/h enters a packed column operating at 293 K and 1 atm where 90% of the ammonia is scrubbed by a countercurrent flow of 1500 kg/h of pure liquid water. Estimate the loading and flooding superficial gas velocities for two packing materials:

234   **Equipment for Gas–Liquid Mass-Transfer Operations**

(a) 25-mm ceramic Raschig rings
(b) 25-mm metal Hiflow rings

Solution

Because the superficial gas velocity is highest at the bottom of the column, calculations are made for conditions there. At this point, the liquid is a very dilute solution of NH_3 in water at 293 K. The properties of this solution can be approximated as those of pure water at 293 K ($\rho_L = 1000$ kg/m³, $\mu_L = 1$ cP). The density of the gas is from the ideal gas law. The viscosity of the gas can be approximated by that of pure air at 293 K ($\mu_G = 1.84 \times 10^{-5}$ kg/m·s).

Inlet gas:

$$M_G = 0.95 \times 29 + 0.05 \times 17 = 28.4$$

$$V' = \frac{20 \times 28.4}{3600} = 0.158 \text{ kg/s}$$

$$\rho_G = \frac{PM_G}{RT} = \frac{101.3 \times 28.4}{8.314 \times 293} = 1.181 \text{ kg/m}^3$$

$$Q_G = \frac{V'}{\rho_G} = 0.134 \text{ m}^3/\text{s}$$

Exiting liquid:

$$\text{Ammonia absorbed} = 20 \times 0.05 \times 0.9 \times 17$$

$$= 15.3 \text{ kg/h}$$

$$L' = \frac{1500 + 15.3}{3600} = 0.421 \text{ kg/s}$$

$$\rho_L = 1000 \text{ kg/m}^3$$

Then,

$$X = \frac{L'}{V'}\left(\frac{\rho_G}{\rho_L}\right)^{0.5} = \frac{0.421}{0.158}\left(\frac{1.181}{1000}\right)^{0.5} = 0.092$$

(a) From Table 4.1, for 25-mm ceramic Raschig rings (first-generation packing):

$$\varepsilon = 0.68 \qquad a = 190\,\text{m}^{-1} \qquad C_{S,T} = 2.454 \qquad C_{F,T} = 1.899$$

For $X = 0.092$, from equation (4-6), $n_S = -0.326$, $C_S = 2.454$. Equations (4-3), (4-4), and (4-5) are solved simultaneously using Mathcad to give $v_{G,S} = 0.954$ m/s. From equation (4-13), $n_F = -0.194$, $C_F = 1.899$. Equations (4-10), (4-11), (4-12), and (4-14) are solved simultaneously using Mathcad to give $v_{G,F} = 1.355$ m/s. Therefore, the ratio of loading to flooding velocity is $v_{G,S}/v_{G,F} = 0.704$, very close to the 70% value observed by Billet (1989).

(b) From Table 4.1, for 25-mm metal Hiflow rings (third-generation packing):

$$\varepsilon = 0.962 \qquad a = 202.9 \text{ m}^{-1} \qquad C_{S,T} = 2.918 \qquad C_{F,T} = 2.177$$

4.2 Gas–Liquid Operations: Liquid Dispersed

Table 4.1 Hydraulic Characteristics of Random Packings

Packing	Size	$C_{S,T}$	a, m^{-1}	ε	C_h	$C_{F,T}$
Berl saddle						
Ceramic	25 mm		260.0	0.680	0.620	
Ceramic	13 mm		545.0	0.65	0.833	
Bialecki ring						
Metal	50 mm	2.916	121.0	0.966	0.798	1.896
Metal	35 mm	2.753	155.0	0.967	0.787	1.885
Metal	25 mm	2.521	210.0	0.956	0.692	1.856
DIN-PAK ring						
Plastic	70 mm	2.970	110.7	0.938	0.991	1.912
Plastic	47 mm	2.929	131.2	0.923	1.173	1.991
Envi Pac ring						
Plastic	80, no. 3	2.846	60.0	0.955	0.641	1.522
Plastic	60, no. 2	2.987	98.4	0.961	0.794	1.864
Plastic	32, no. 1	2.944	138.9	0.936	1.039	2.012
Cascade mini-ring						
Metal	1.5" CMR;T		188.0	0.972	0.870	
Metal	1.5" CMR		174.9	0.974	0.935	
Metal	1.0" CMR		232.5	0.971	1.040	
Metal	0.5" CMR		356.0	0.955	1.338	
Hiflow ring						
Ceramic	75 mm		54.1	0.868		
Ceramic	50 mm	2.819	89.7	0.809		1.694
Ceramic	38 mm	2.840	111.8	0.788		1.930
Ceramic	20 mm, 6 stg.		265.8	0.776	0.958	
Ceramic	20 mm, 4 stg.	2.875	286.2	0.758	1.167	2.410
Metal	50 mm	2.702	92.3	0.977	0.876	1.626
Metal	25 mm	2.918	202.9	0.962	0.799	2.177
Plastic	90 mm		69.7	0.968		
Plastic	50 mm	2.894	118.4	0.925	1.038	1.871
Plastic	25 mm	2.841	194.5	0.918		1.989
IMTP						
Metal	# 25		225.8	0.966		
Metal	# 40		150.8	0.977		
Metal	# 50		100.0	0.980		
Metal	# 70		60.0	0.985		
Intalox saddle						
Ceramic	50 mm		114.6	0.761		
Plastic	50 mm		122.1	0.908		
NORPAC ring						
Plastic	50 mm	2.959	86.8	0.947	0.651	1.786
Plastic	35 mm	3.179	141.8	0.944	0.587	2.242
Plastic	25 mm	3.277	202.0	0.953	0.601	2.472
Nutter ring						
Metal	# 0.7		226.0	0.978		
Metal	# 1.0		167.0	0.978		
Metal	# 1.5		125.0	0.978		

Equipment for Gas–Liquid Mass-Transfer Operations

Table 4.1 (*continuation*)

Packing	Size	$C_{S,T}$	a, m^{-1}	ε	C_h	$C_{F,T}$
Metal	# 2.0		95.0	0.979		
Metal	# 2.5		82.0	0.982		
Metal	# 3.0		66.0	0.984		
Plastic	# 2.0		82.0	0.920		
Pall ring						
Ceramic	50 mm		116.5	0.783	1.335	
Metal	50 mm	2.725	112.6	0.951	0.784	1.580
Metal	35 mm	2.629	139.4	0.965	0.644	1.679
Metal	25 mm	2.627	223.5	0.954	0.719	2.083
Metal	15 mm		368.4	0.933	0.590	
Plastic	50 mm	2.816	111.1	0.919	0.593	1.757
Plastic	35 mm	2.654	151.1	0.906	0.718	1.742
Plastic	25 mm	2.696	225.0	0.887	0.528	2.064
Raschig ring						
Ceramic	25 mm	2.454	190.0	0.680	0.577	1.899
Ceramic	15 mm		312.0	0.690	0.648	
Ceramic	10 mm		440.0	0.650	0.791	
Ceramic	6 mm		771.9	0.620	1.094	
Metal	15 mm		378.4	0.917	0.455	
Raschig super-ring						
Metal	0.3 mm	3.560	315.0	0.960	0.750	2.340
Metal	0.5 mm	3.350	250.0	0.975	0.620	2.200
Metal	1.0 mm	3.491	160.0	0.980	0.750	2.200
Metal	2.0 mm	3.326	97.6	0.985	0.720	2.096
Metal	3.0 mm	3.260	80.0	0.982	0.620	2.100
Plastic	2.0 mm	3.326	100.0	0.960	0.720	2.096
Tellerette						
Plastic	25 mm	2.913	190.0	0.930	0.588	2.132
Top-Pak ring						
Aluminum	50 mm	2.528	105.5	0.956	0.881	1.579

Source: Data from Seader and Henley (1998) and Billet and Schultes (1999).

For $X = 0.092$, from equation (4-6), $n_S = -0.326$, $C_S = 2.918$. Equations (4-3), (4-4), and (4-5) are solved simultaneously using Mathcad to give $v_{G,S} = 1.772$ m/s.

From equation (4-13), $n_F = -0.194$, $C_F = 2.177$. Equations (4-10), (4-11), (4-12), and (4-14) are solved simultaneously using Mathcad to give $v_{G,F} = 2.528$ m/s. Therefore, the ratio of loading to flooding velocity is $v_{G,S}/v_{G,F} = 0.701$, again very close to the 70% value observed by Billet (1989).

Notice that the loading and flooding gas velocities for the third-generation packing are much higher than for the first-generation packing.

4.2 Gas–Liquid Operations: Liquid Dispersed

4.2.4 Pressure Drop

Most packed columns consist of cylindrical vertical vessels. The column diameter is determined so as to safely avoid flooding and operate in the preloading region. Usually, packed columns are designed based on either of two criteria: a fractional approach to flooding gas velocity or a maximum allowable gas-pressure drop. For given fluid flow rates and properties, and a given packing material, equations (4-10) to (4-14) are solved as illustrated in the previous example to compute the superficial gas velocity at flooding, $v_{G,F}$. Then, according to the first sizing criterion, a fraction of flooding, f, is specified (usually from 0.5 to 0.7), followed by calculation of the tower diameter from

$$D = \left[\frac{4Q_G}{\pi f v_{G,F}}\right]^{0.5} \tag{4-15}$$

According to the second sizing criterion, absorbers and strippers are usually designed with gas-pressure drops of 200 to 400 Pa/m of packed depth, atmospheric-pressure fractionators from 400 Pa/m to 600 Pa/m, and vacuum stills for 8 to 40 Pa/m (Kister, 1992). Theoretically based models to predict pressure drop in packed beds with countercurrent flows of gas and liquid have been presented by Stichlmair et al. (1989), who use a particle model, and Billet and Schultes (1991a, 1999), who use a channel model. Both models extend well-accepted equations for dry-bed pressure drop to account for the effect of liquid holdup. Based on extensive experimental studies using 54 different packing materials, including structured packings, Billet and Schultes (1991a, 1999) developed a correlation for dry-gas-pressure drop, ΔP_0. Their dimensionally consistent correlating equation is

$$\Delta P_0 = \Psi_0 \frac{a}{\varepsilon^3} \frac{\rho_G v_G^2}{2} \frac{1}{K_W} \tag{4-16}$$

where Z = packing height and K_W = wall factor. This wall factor can be important for columns with an inadequate ratio of effective particle diameter to inside column diameter, and is given by

$$\frac{1}{K_W} = 1 + \frac{2}{3}\left(\frac{1}{1-\varepsilon}\right)\frac{d_p}{D} \tag{4-17}$$

where the effective particle diameter, d_p, is given by

$$d_p = 6\left(\frac{1-\varepsilon}{a}\right) \tag{4-18}$$

The dry-packing resistance coefficient (a modified friction factor), ψ_0, is given by the empirical expression

$$\Psi_0 = C_p\left[\frac{64}{\mathrm{Re}_G} + \frac{1.8}{\mathrm{Re}_G^{0.08}}\right] \tag{4-19}$$

238   **Equipment for Gas–Liquid Mass-Transfer Operations**

where

$$\mathrm{Re}_G = \frac{v_G d_p \rho_G K_W}{(1-\varepsilon)\mu_G} \tag{4-20}$$

and C_p is a packing constant determined from experimental data and tabulated for a number of packings in Table 4.2.

When the packed bed is irrigated, the liquid holdup causes the pressure drop to increase. The experimental data in the preloading region are reasonably well correlated by (Billet and Schultes, 1999)

$$\frac{\Delta P}{\Delta P_0} = \left(\frac{\varepsilon}{\varepsilon - h_L}\right)^{1.5}\left(\frac{h_L}{h_{L,S}}\right)^{0.3}\exp\left[13{,}300\sqrt{\frac{\mathrm{Fr}_L}{a^3}}\right] \tag{4-21}$$

where a must be in units of $\mathrm{m}^2/\mathrm{m}^3$, $h_{L,S}$ is the liquid holdup at loading, and Fr_L was defined under equation (4-7). For operation in the preloading region, as usual, the liquid holdup can be estimated from equations (4-7) to (4-9).

Example 4.3 Pressure Drop in Beds Packed with First- and Third-Generation Random Packings

For the two packing materials considered in Example 4.2, calculate the column inside diameter and pressure drop of the gas for operation at 70% of the flooding superficial gas-velocity.

Solution
(a) For 25-mm ceramic Raschig rings:
From Example 4.2, 70% of the flooding gas-velocity corresponds to the loading point gas-velocity, $v_{G,S} = 0.954$ m/s, $Q_G = 0.134$ m³/s. Therefore, for operation at this point, the column inside diameter is

$$D = \left[\frac{4Q_G}{\pi f v_{G,F}}\right]^{0.5} = 0.423 \text{ m}$$

To calculate the gas-pressure drop, the liquid holdup must be estimated. The superficial liquid-mass velocity is from

$$G_x = \frac{4L'}{\pi D^2} = \frac{4 \times 0.421}{\pi \times (0.423)^2} = 3.0 \text{ kg/m}^2 \cdot \text{s}$$

Calculate Re_L and Fr_L:

$$\mathrm{Re}_L = \frac{v_{L,S}\rho_L}{a\mu_L} = \frac{G_x}{a\mu_L} = \frac{3.0}{190 \times 0.001} = 15.8$$

$$\mathrm{Fr}_L = \frac{v_{L,S}^2 a}{g} = \frac{G_x^2 a}{\rho_L^2 g} = \frac{(3.0)^2 \times 190}{(1000)^2 \times 9.8} = 1.744 \times 10^{-4}$$

4.2 Gas–Liquid Operations: Liquid Dispersed

239

From Table 4.1, C_h = 0.577. From equation (4-9),

$$\frac{a_h}{a} = 0.85 \times 0.577 \times \left(15.8\right)^{0.25} \times \left(1.744 \times 10^{-4}\right)^{0.1} = 0.412$$

From equation (4-7),

$$h_L = \left[\frac{12 \times 1.744 \times 10^{-4}}{15.8}\right]^{1/3} \times \left(0.412\right)^{2/3} = 0.0283 = h_{L,S}$$

From equation (4-18),

$$d_p = \frac{6 \times \left(1 - 0.68\right)}{190} = 0.0101 \text{ m}$$

From equation (4-17), $1/K_W$ = 1.049. The viscosity of the gas phase is basically that of air at 293 K and 1 atm, μ_G = 1.84 × 10^{-5} kg/m·s. Then, from equation (4-20),

$$\text{Re} = \frac{0.954 \times 0.0101 \times 1.181}{\left(1 - 0.68\right) \times 1.84 \times 10^{-5} \times 1.049} = 1842.4$$

From Table 4.2, C_p = 1.329. From equation (4-19), ψ_o = 1.357. The dry gas-pressure drop is from equation (4-16)

$$\frac{\Delta P_0}{Z} = \frac{1.357 \times 190 \times 1.181 \times \left(0.954\right)^2 \times 1.049}{2 \times \left(0.68\right)^2} = 462 \text{ Pa/m}$$

Correct the pressure drop for liquid holdup with equation (4-21)

$$\frac{\Delta P}{Z} = 462 \times \left(\frac{0.68}{0.68 - 0.0283}\right)^{1.5} \exp\left[13{,}300 \sqrt{\frac{1.744 \times 10^{-4}}{\left(190\right)^3}}\right]$$

$$= 521 \text{ Pa/m}$$

(b) For 25-mm metal Hiflow rings:
From Example 4.2, 70% of the flooding gas-velocity corresponds to the loading point gas-velocity, $v_{G,S}$ = 1.772 m/s. For operation at this point, the column diameter is D = 0.31 m. The resulting gas-pressure drop is 365 Pa/m.

Based on these results, the Hiflow rings have a much greater capacity than the Raschig rings, since the required column cross-sectional area is reduced by 50% while the resulting gas-pressure drop is also significantly lower.

4.2.5 Mass-Transfer Coefficients

In an extensive investigation, Billet and Schultes (1991b) measured and correlated mass-transfer coefficients for 31 different binary and ternary systems with 67 different types and sizes of packings in columns of diameter ranging from 6 cm to 1.4 m. The systems include some for which mass-transfer resistance resides

240   **Equipment for Gas–Liquid Mass-Transfer Operations**

mainly in the liquid phase and others for which resistance in the gas phase predominates. For the liquid-phase resistance, the proposed correlation is

$$k_L = 0.757 C_L \left[\frac{D_L a v_L}{\varepsilon h_L} \right]^{0.5} \tag{4-22}$$

where C_L is an empirical constant characteristic of the packing, as shown in Table 4.2. For the gas phase, they proposed

$$k_y = 0.1304 C_V \frac{D_G P}{RT} \frac{a}{\left[\varepsilon \left(\varepsilon - h_L \right) \right]^{0.5}} \left[\frac{Re_G}{K_W} \right]^{3/4} Sc_G^{1/3} \tag{4-23}$$

where C_V is an empirical constant included in Table 4.2 and Re_G is as defined in equation (4-20).

Example 4.4 Design of a Packed-Bed Ethanol Absorber

When molasses is fermented to produce a liquor containing ethanol, a CO_2-rich vapor containing a small amount of ethanol is evolved. The alcohol will be recovered by countercurrent absorption with water in a packed-bed tower. The gas will enter the tower at a rate of 180 kmol/h, at 303 K and 110 kPa. The molar composition of the gas is 98% CO_2 and 2% ethanol. The required recovery of the alcohol is 97%. Pure liquid water at 303 K will enter the tower at the rate of 151.5 kmol/h, which is 50% above the minimum rate required for the specified recovery (Seader and Henley, 1998). The tower will be packed with 50-mm metal Hiflow rings and will be designed for a maximum pressure drop of 100 Pa/m of packed height.

(a) Determine the column diameter for the design conditions.

(b) Estimate the fractional approach to flooding conditions.

(c) Estimate the gas and liquid volumetric mass-transfer coefficients, $k_y a_h$ and $k_L a_h$.

Solution

(a) Because the superficial gas velocity and pressure drop are highest at the bottom of the column, calculations are made for conditions there. The tower diameter that will result in the specified pressure drop of 100 Pa/m must be determined by an iterative procedure. As a first estimate of the diameter, assume operation at the loading point. Calculate the resulting pressure drop under those conditions by the Billet and Schultes (1999) correlation and compare to the specified value. Continue iterating, changing the column diameter and calculating the corresponding pressure drop, until convergence to the specified value is achieved.

4.2 Gas–Liquid Operations: Liquid Dispersed

Table 4.2 Other Parameters of Random Packings

Packing	Size	C_p	C_L	C_V
Berl saddle				
Ceramic	25 mm		1.246	0.387
Ceramic	13 mm		1.364	0.232
Bialecki ring				
Metal	50 mm	0.719	1.721	0.302
Metal	35 mm	1.011	1.412	0.390
Metal	25 mm	0.891	1.461	0.331
Envi Pac ring				
Plastic	80, no. 3	0.358	1.603	0.257
Plastic	60, no. 2	0.338	1.522	0.296
Plastic	32, no. 1	0.549	1.517	0.459
Hiflow ring				
Ceramic	50 mm	0.538	1.377	0.379
Ceramic	20 mm, 4stg.	0.628	1.744	0.465
Metal	50 mm	0.421	1.168	0.408
Metal	25 mm	0.689	1.641	0.402
Plastic	50 mm	0.327	1.487	0.345
Plastic	25 mm	0.741	1.577	0.390
NORPAC ring				
Plastic	50 mm	0.350	1.080	0.322
Plastic	35 mm	0.371	0.756	0.425
Plastic	25 mm, type B	0.397	0.883	0.366
Pall ring				
Ceramic	50 mm		1.227	0.415
Metal	50 mm	0.763	1.192	0.410
Metal	25 mm	0.957	1.440	0.336
Plastic	50 mm	0.698	1.239	0.368
Plastic	35 mm	0.927	0.856	0.380
Plastic	25 mm	0.865	0.905	0.446
Raschig ring				
Ceramic	25 mm	1.329	1.361	0.412
Ceramic	15 mm		1.276	0.401
Ceramic	10 mm		1.303	0.272
Raschig Super-Ring				
Metal	1.0 mm	0.500	1.290	0.440
Metal	2.0 mm	0.464	1.323	0.400
Metal	3.0 mm	0.430	0.850	0.300
Top-Pak ring				
Aluminum	50 mm	0.604	1.326	0.389
VSP ring				
Metal	50 mm, no. 2	0.773	1.222	0.420
Metal	25 mm, no. 1	0.782	1.376	0.405

Source: Data from Seader and Henley (1998) and Billet and Schultes (1999).

242   **Equipment for Gas–Liquid Mass-Transfer Operations**

Inlet gas:

$$M_G = 0.98 \times 0.44 + 0.02 \times 46 = 44.04$$

$$V' = \frac{180 \times 44.04}{3600} = 2.202 \text{ kg/s}$$

$$\rho_G = \frac{PM_G}{RT} = \frac{110 \times 44.04}{8.314 \times 303} = 1.923 \text{ kg/m}^3$$

$$\mu_G = 1.45 \times 10^{-5} \text{ Pa} \cdot \text{s}$$

$$Q_G = \frac{V'}{\rho_G} = 1.145 \text{ m}^3 / \text{s}$$

Exiting liquid:

$$\text{Ethanol absorbed} = 180 \times 0.02 \times 0.97 \times 46$$

$$= 160.6 \text{ kg/h}$$

$$L' = \frac{151.5 \times 18 + 160.6}{3600} = 0.804 \text{ kg/s}$$

$$\rho_L = 986 \text{ kg/m}^3 \qquad \mu_L = 0.631 \text{ cP}$$

$$X = \frac{0.804}{2.202}\left(\frac{1.923}{986}\right)^{0.5} = 0.016$$

For 50-mm metal Hiflow rings, from Tables 4.1 and 4.2:

$$\varepsilon = 0.977 \quad a = 92.3 \text{ m}^2/\text{m}^3 \quad C_{S,T} = 2.702 \quad C_{F,T} = 1.626 \quad C_h = 0.876$$

$$C_p = 0.421 \quad C_L = 1.168 \quad C_V = 0.408$$

Following a procedure similar to that presented in Example 4.2, the loading and flooding gas-velocities are found to be $v_{G,S}$ = 2.096 m/s and $v_{G,F}$ = 3.164 m/s, respectively ($v_{G,S}$ /$v_{G,F}$ = 0.663). Assuming operation at the loading point, an initial estimate of the column diameter is

$$D = \left[\frac{4 \times 1.145}{\pi \times 2.096}\right]^{0.5} = 0.834 \text{ m}$$

Following a procedure similar to that presented in Example 4.3, calculate the liquid holdup at loading, $h_{L,S}$, and the corresponding gas-pressure drop:

$$h_{L,S} = 0.015 \qquad \frac{\Delta P}{Z} = 168 \text{ Pa/m}$$

The resulting pressure drop is too high; therefore, we must increase the tower diameter to reduce the pressure drop. Appendix D presents a Mathcad computer program designed to iterate automatically until the pressure drop criterion is satisfied. Convergence is achieved, as shown in Appendix D, at a tower diameter of D = 0.938 m, at which $\Delta P/Z$ = 100 Pa/m of packed height.

4.3 Gas–Liquid Operations: Gas Dispersed 243

(b) For the tower diameter of D = 0.938 m, the following intermediate results were obtained from the computer program in Appendix D:

$$v_G = 1.656 \text{ m/s} \qquad v_L = 0.00118 \text{ m/s}, \qquad h_L = 0.013$$

$$a_h = 47.2 \text{ m}^2/\text{m}^3 \qquad Re_G = 13{,}650 \qquad Re_L = 19.96$$

The fractional approach to flooding, f, is

$$f = \frac{v_G}{v_{G,F}} = \frac{1.656}{3.164} = 0.523$$

(c) Next, we calculate the mass-transfer coefficients. Estimates of the diffusivities of ethanol in a dilute aqueous solution at 303 K (D_L) and in a gas mixture with CO_2 at 303 K and 110 kPa (D_G) are needed. For ethanol, V_c = 167.1 cm^3/mol, σ = 4.53 Å, ε/κ = 362.6 K; for CO_2, σ = 3.94 Å, ε/κ = 195.2 K (Reid et al., 1987).

From equation (1-48), V_{bA} = 60.9 cm^3/mol.
From equation (1-53), D_L = 1.91 $\times$ 10^{-5} cm^2/s.
From equation (1-49), D_G = 0.085 cm^2/s.
From equation (4-22), using the results obtained in (b), k_L = 1.141 $\times$ 10^{-4} m/s. Then, the volumetric liquid-phase mass-transfer coefficient is

$$k_L a_h = 1.141 \times 10^{-4} \times 47.2 = 5.39 \times 10^{-3} \text{ s}^{-1}$$

Calculate $Sc_G = \mu_G/\rho_G D_G$ = 0.887. From equation (4-23), k_y = 2.357 mol/m^2·s. The volumetric gas-phase mass-transfer coefficient is

$$k_y a_h = 2.357 \times 47.2 = 111.2 \text{ mol/m}^3 \cdot \text{s}$$

4.3 GAS–LIQUID OPERATIONS: GAS DISPERSED

Your objectives in studying this section are to be able to

1. Design bubble columns.
2. Design sieve-tray towers for gas–liquid operations.
3. Estimate stage efficiencies for sieve-tray towers.

In this group are included those devices, such as *sparged* and *agitated vessels* and various types of *tray towers*, in which the gas phase is dispersed into bubbles or foams. Tray towers are the most important of the group, since they produce countercurrent, multistage contact, but the simple vessel contactors have many applications (Treybal, 1980).

244   **Equipment for Gas–Liquid Mass-Transfer Operations**

Gas and liquid can conveniently be contacted, with gas dispersed as bubbles, in agitated vessels whenever multistage, countercurrent effects are not required. This is particularly the case when a chemical reaction between the dissolved gas and a constituent of the liquid is required. The carbonation of a lime slurry, the chlorination of paper stock, the hydrogenation of vegetable oils, the aeration of fermentation broths as in the production of penicillin, the production of citric acid from beet sugar by the action of microorganisms, and the aeration of activated sludge for biological oxidation are all examples (Treybal, 1980). The gas–liquid mixture can be mechanically agitated, or agitation can be accomplished by the gas itself in *sparged vessels*. The operation may be batch, semibatch with continuous flow of gas and a fixed quantity of liquid, or continuous with flow of both phases.

4.3.1 Sparged Vessels (Bubble Columns)

A *sparger* is a device for introducing into a liquid a stream of gas in the form of small bubbles. If the vessel diameter is small, the sparger, located at the bottom of the vessel, may simply be an open tube through which the gas issues into the liquid. For vessels of diameter greater than 0.3 m, it is better to use several orifices for introducing the gas to ensure good gas distribution. In that case, the orifices may be holes from 1.5 mm to 3.0 mm in diameter drilled in a pipe distributor placed horizontally at the bottom of the vessel. The purpose of the sparging may be contacting the sparged gas with the liquid, or it may simply be a device for agitation.

The size of gas bubbles depends upon the rate of flow through the orifices, the orifice diameter, the fluid properties, and the extent of turbulence prevailing in the liquid. What follows is for cases where turbulence in the liquid is solely that generated by the rising bubbles and when orifices are horizontal and sufficiently separated to prevent bubbles from adjacent orifices from interfering with each other (at least $3d_p$ apart). For air–water, the following correlations can be used to estimate the size of the bubbles, d_p, as they leave the orifices of the sparger (Leibson et al., 1956):

$$d_p = 0.0287 d_o^{1/2} \text{Re}_o^{1/3} \qquad \text{for Re}_o \leq 2100 \qquad (4\text{-}24)$$

$$d_p = 0.0071 \text{Re}_o^{-0.05} \qquad \text{for } 10{,}000 \leq \text{Re}_o \leq 50{,}000 \qquad (4\text{-}25)$$

where the bubble size, d_p, and the orifice diameter, d_o, are in meters and $\text{Re}_o = 4V'/\pi d_o \mu_G$ where V' is the mass flow rate of the gas. For the transition range ($2100 < \text{Re}_o < 10{,}000$), there is no correlation of data. It is suggested that d_p for air–water can be approximated by log–log interpolation between the points given by equation (4-24) at $\text{Re}_o = 2100$ and by equation (4-25) at $\text{Re}_o = 10{,}000$ (Treybal, 1980).

The volume fraction of the gas–liquid mixture in the vessel which is occupied by the gas is called the *gas holdup*, ϕ_G. If the superficial gas velocity in the vessel is v_G, then v_G/ϕ_G is the true gas velocity relative to the vessel walls. If the liquid flows

4.3 Gas–Liquid Operations: Gas Dispersed 245

upward, cocurrently with the gas, at a velocity relative to the vessel walls $v_L/(1 - \phi_G)$, the relative velocity of gas and liquid, or *slip velocity*, is

$$v_S = \frac{v_G}{\phi_G} - \frac{v_L}{1 - \phi_G} \qquad (4\text{-}26)$$

Equation (4-26) will also give the slip velocity for countercurrent flow of liquid if v_L for the downward liquid flow is assigned a negative sign.

The holdup for sparged vessels can be correlated through the slip velocity by the following equation developed from data presented by Hughmark (1967):

$$\ln\left[\frac{v_G}{v_S}\right] = -0.847 + 0.0352 \ln V_G - 0.1835 \left(\ln V_G\right)^2 - 0.01348 \left(\ln V_G\right)^3$$

$$V_G = v_G \left[\frac{\rho_W \sigma_{AW}}{\rho_L \sigma}\right]^{1/3} \qquad (4\text{-}27)$$

Knowing v_G, and v_L, equations (4-26) and (4-27) can be combined to estimate the gas holdup and slip velocity. The following restrictions apply when using equation (4-27), satisfactory for no liquid flow ($v_L = 0$), cocurrent liquid flow up to $v_L = 0.1$ m/s, and also for small countercurrent flow:

1) Vessel diameter, D = above 0.1 m
2) Liquid density, ρ_L = 770 to 1700 kg/m^3
3) Liquid viscosity, μ_L = 0.0009 to 0.152 Pa-s
4) Density of water, ρ_W = 1000 kg/m^3
5) Air–water surface tension, σ_{AW} = 0.072 N/m
6) Surface tension, σ = 0.025 to 0.076 N/m

If a unit volume of a gas–liquid mixture contains a gas volume ϕ_G made up of n bubbles of diameter d_p, then $n = 6\phi_G/\pi d_p^3$. If the interfacial area per unit volume is a, then $n = a/(\pi d_p^2)$. Equating the two expressions for n provides the specific area

$$a = \frac{6\phi_G}{d_p} \qquad (4\text{-}28)$$

For low liquid velocities, the bubble size may be taken as that produced at the orifices of the sparger—according to equation (4-24) or (4-25)—corrected as necessary for pressure. For high liquid velocities, the bubble size may be altered by turbulent breakup and coalescence of bubbles. For example, for air–water in the range $\phi_G = 0.1$ to 0.4 and $v_L/(1 - \phi_G) = 0.15$ to 15 m/s, the bubble size is approximated by (Petrick, 1962)

$$d_p = \frac{0.002344}{\left[\dfrac{v_L}{1 - \phi_G}\right]^{0.67}} \qquad (4\text{-}29)$$

246   **Equipment for Gas–Liquid Mass-Transfer Operations**

where d_p is in meters and v_L in m/s.

In practically all the gas-bubble liquid systems, the liquid-phase mass-transfer resistance is strongly controlling, and gas-phase coefficients are not needed. The liquid-phase coefficients are correlated by (Hughmark, 1967)

$$\text{Sh}_L = \frac{F_L d_p}{c \text{D}_L} = 2 + b' \text{Re}_G^{0.779} \text{Sc}_L^{0.546} \left(\frac{d_p g^{1/3}}{\text{D}_L^{2/3}} \right)^{0.116} \tag{4-30}$$

where

$$b' = \begin{cases} 0.0610 & \text{single gas bubbles} \\ 0.0187 & \text{swarms of bubbles} \end{cases}$$

The gas-bubble Reynolds number must be calculated with the slip velocity: $\text{Re}_G = d_p v_S \rho_L / \mu_L$.

The power supplied to the vessel contents, which is responsible for the agitation and creation of large interfacial area, is derived from the gas flow. Bernoulli's equation, a mechanical energy balance written for the gas between location o (just above the sparger orifices) and location s (at the liquid surface), is

$$\frac{v_s^2 - v_o^2}{2 g_c} + \left(Z_s - Z_o \right) \frac{g}{g_c} + \int_o^s \frac{dP}{\rho_G} + W + h_f = 0 \tag{4-31}$$

The friction loss, h_f, and the velocity at the liquid surface, v_s, can be neglected, and the gas density can be described by the ideal gas law, whereupon

$$W = \frac{v_o^2}{2 g_c} + \frac{P_o}{\rho_{G,o}} \ln \frac{P_o}{P_s} + \left(Z_o - Z_s \right) \frac{g}{g_c} \tag{4-32}$$

W is the work done by the gas on the vessel contents, per unit mass of gas. Work for the gas compressor will be larger to account for friction losses in the piping and orifices and compressor inefficiency.

Example 4.5 Stripping Chloroform from Water by Sparging with Air

Chlorinating drinking water kills microbes but produces trace amounts of a group of potentially harmful substances called *trihalomethanes* (THMs), the most prevalent of which is chloroform. Federal regulations in the United States limit total trihalomethanes (TTHMs) levels in drinking water to 80 µg/L. Sparging with ambient air in a bubble column is an effective method to strip the water of TTHMs.

A vessel 1.0 m in diameter is to be used for stripping chloroform from water by sparging with air at 298 K. The water will flow continuously downward at the rate of 10.0 kg/s at 298 K. The water contains 240 µg/L of chloroform. It is desired to remove 90% of the chloroform in the water using an airflow that is 50% higher than the minimum required. At these low concentrations, chloroform–water solutions follow Henry's law ($y_i = m x_i$) with m = 220. The sparger is in the form of a ring located at the bottom of the vessel, 50 cm in diameter,

4.3 Gas–Liquid Operations: Gas Dispersed 247

containing 90 orifices, each 3 mm in diameter. Estimate the depth of the water column required to achieve the specified 90% removal efficiency. Estimate the power required to operate the air compressor if the mechanical efficiency of the system is 60%.

Solution

Following the procedure discussed in Chapter 3 for countercurrent flow and dilute solutions, it is determined that the air flow rate corresponding to 50% above the minimum is 0.10 kg/s. The mass flow rate through each orifice is $V' = 0.10/90 = 0.0011$ kg/s. The viscosity of air at 298 K is $\mu_G = 1.8 \times 10^{-5}$ kg/m·s. The orifice diameter is $d_o = 0.003$ m. Then, $\mathrm{Re}_o = 25{,}940$. From equation (4-25), the bubble diameter at the orifice outlet is $d_p = 0.00427$ m.

To estimate the gas holdup, the density of the gas at the average pressure in the vessel must be estimated. Since the water column height is not known at this point, an iterative procedure must be implemented. Assume an initial value of the column height (say, $Z = 0.5$ m) to estimate the average pressure, as a first step in calculating an improved estimate of the column height. With the refined estimate of Z, calculate a new value of average pressure, and continue iterating until convergence is achieved. For $Z = 0.5$ m, $\rho_L = 1000$ kg/m^3, and $P_s = 101.3$ kPa, $P_o = 101.3 + 1000 \times 9.8 \times 0.5/1000 = 106.2$ kPa. Then, the average pressure is $P_{av} = (P_s + P_o)/2 = 103.8$ kPa. From the ideal gas law, the gas density at this average pressure and 298 K is $\rho_G = 1.215$ kg/m^3. The cross-sectional area of the 1.0-m diameter vessel is 0.785 m^2; therefore, the gas superficial velocity is $v_G = 0.10/(1.215 \times 0.785) = 0.105$ m/s. In this case, $\rho_L = \rho_W$, $\sigma = \sigma_{AW}$. Then, from equation (4-27), $V_G = v_G$, $v_G/v_S = 0.182$, and $v_S = 0.577$ m/s. Taking into consideration that the water flows downward, the superficial liquid velocity is $v_L = -10/(1000 \times 0.785) = -0.0127$ m/s. Substituting in equation (4-26) gives us

$$0.577 = \frac{0.105}{\phi_G} + \frac{0.0127}{1 - \phi_G}$$

Solving, the gas holdup is $\phi_G = 0.187$.

To estimate the interfacial area and the liquid-phase mass-transfer coefficient, the average bubble size along the water column must be estimated. In this case,

$$\frac{v_L}{1 - \phi_G} = \frac{0.0127}{1 - 0.187} = 0.0156 \ll 0.15 \text{ m/s}$$

Then, the bubble diameter may be taken as that produced at the orifices of the sparger, corrected for pressure. Therefore, the bubble diameter at the average column pressure, calculated from d_p at the orifice, is from the ideal gas law

$$d_p = \left[(0.00427)^3 \frac{106.2}{103.8} \right]^{1/3} = 0.0043 \text{ m}$$

Then, from equation (4-28), $a = 261$ m^{-1}. Next, an estimate of the diffusivity of chloroform in dilute aqueous solution at 298 K, D_L, is needed. Use the corresponding Hayduk–Minhas correlation, equation (1-53). The molar volume for chloroform,

248   **Equipment for Gas–Liquid Mass-Transfer Operations**

estimated from its critical volume, is V_{bA} = 88.6 cm^3/mol (Reid et al., 1987). The viscosity of water at 298 K is approximately 0.9 cP. Then, D_L = 1.08 × 10^{-9} m^2/s and Sc_L = 833. Calculate

$$\frac{d_p g^{1/3}}{D_L^{2/3}} = \frac{0.0043 \times 9.8^{1/3}}{\left(1.08 \times 10^{-9}\right)^{2/3}} = 8742$$

$$Re_G = \frac{d_p v_S \rho_L}{\mu_L} = \frac{0.0043 \times 0.577 \times 1000}{0.9 \times 10^{-3}} = 2760$$

From equation (4-30),

$$Sh_L = 2 + 0.0187 \times 2760^{0.779} \times 833^{0.546} \times 8742^{0.116} = 1012$$

For dilute aqueous solutions, $x_{B,M} \approx 1.0$, $c \approx 55.5$ kmol/m^3. Then, for $N_B = 0$,

$$k_x \approx \frac{Sh_L c D_L}{d_p} = \frac{1012 \times 55.5 \times 1.08 \times 10^{-9}}{0.0043} = 0.0141 \text{ kmol/m}^2 \cdot \text{s}$$

The volumetric mass-transfer coefficient is

$$k_x a = 0.0141 \times 261 = 3.68 \text{ kmol/m}^3 \cdot \text{s}$$

Next, we estimate the height of the water column required for 90% removal of the chloroform. If we assume that all of the resistance to mass transfer in this case resides in the liquid phase, the mass-transfer circumstances are analogous to those of the dissolved-oxygen stripping of boiler feed water described in Example 2.14. Therefore, equation (2-99) applies, modified here as

$$Z = \frac{G_{Mx}}{k_x a (1 - A)} \ln \left[\frac{x_{in}}{x_{out}} (1 - A) + A \right] \tag{4-33}$$

where G_{Mx} is the liquid molar velocity and A is the absorption factor, $A = L/mV$. For this case, L = 10/18 = 0.556 kmol/s; G_{Mx} = 0.556/0.785 = 0.708 kmol/m^2·s; V = 0.10/29 = 0.00344 kmol/s; A = 0.732. Substituting in equation (4-33) for x_{in}/x_{out} = 10, Z = 0.90 m.

With this new estimate of Z, calculate again the average pressure in the column of water: P_o = 110.1 kPa, P_{av} = 105.7 kPa. The gas density at this pressure is ρ_G = 1.237 kg/m^3, and the average gas velocity is v_G = 0.103 m/s. This is very close to the value used in the first iteration (v_G = 0.105 m/s); therefore, there will be only a small change in the estimate of the column height. Convergence is achieved in three iterations at the value of Z = 0.904 m.

Use equation (4-32) to calculate the work done by the gas on the vessel contents per unit mass of gas, W. Calculate each of the individual terms in equation (4-32) as follows:

$$\rho_{Go} = \frac{P_o}{RT} = 1.289 \text{ kg/m}^3$$

4.3 Gas–Liquid Operations: Gas Dispersed

$$v_o = \frac{4V'}{\pi d_o^2 \rho_{Go}} = \frac{4 \times 0.0011}{\pi \times (0.003)^2 \times 1.289} = 121.9 \text{ m/s}$$

$$\frac{v_o^2}{2g_c} = \frac{(121.9)^2}{2 \times 1.0} = 7431 \text{ J/kg}$$

$$\frac{P_o}{\rho_{Go}} \ln\left[\frac{P_o}{P_s}\right] = \frac{RT}{M} \ln\left[\frac{P_o}{P_s}\right] = \frac{8314 \times 298}{29} \ln\left[\frac{110.1}{101.3}\right] = 7165 \text{ J/kg}$$

$$\left(Z_o - Z_s\right)\frac{g}{g_c} = \frac{0.904 \times 9.8}{1.0} = 9.0 \text{ J/kg}$$

Adding all the contributions, $W = 14{,}605$ J/kg of air. For a total air flow rate of 0.1 kg/s and 60% mechanical efficiency, the air-compressor power is

$$\dot{W} = \frac{14{,}605 \times 0.1}{0.6} = 2434 \text{ W} = 2.434 \text{ kW}$$

4.3.2 Tray Towers

Tray towers are vertical cylinders in which the liquid and gas are contacted in stepwise fashion on trays or plates. The liquid enters at the top and flows downward by gravity. On the way, it flows across each tray and through a downspout to the tray below. The gas passes upward through openings of one sort or another in the tray, then bubbles through the liquid to form a froth, disengages from the froth, and passes on to the next tray above. The overall effect is a multiple countercurrent contact of gas and liquid, although each tray is characterized by a cross flow of the two. Each tray of the tower is a stage, since on the tray the fluids are brought into intimate contact, interphase diffusion occurs, and the fluids are then separated.

The number of equilibrium stages (theoretical trays) in a tower is dependent only upon the difficulty of the separation to be carried out and is determined solely from material balances and equilibrium considerations. The stage efficiency, and therefore the number of real trays, is determined by the mechanical design used and the conditions of operation. The diameter of the tower, on the other hand, depends upon the quantities of gas and liquid flowing through the tower per unit time. Once the number of theoretical plates, or equilibrium stages, required has been determined, the principal problem in the design of the tower is to choose dimensions and arrangements which will represent the best compromise between several opposing tendencies, since it is generally found that conditions leading to high tray efficiencies will ultimately lead to operational problems (Treybal, 1980).

High efficiencies require deep pools of liquid on the tray (long contact time) and relatively high gas velocities (large interfacial contact areas and mass-transfer coefficients). These conditions, however, lead to a number of difficulties. One is the mechanical entrainment of liquid droplets in the rising gas stream. At high gas velocities, when the gas disengages from the froth, small droplets of liquid are carried by the gas to the tray above. Liquid carried up the tower in this manner reduces mass transfer and consequently affects the tray efficiency adversely.

Furthermore, great liquid depths on the tray and high gas velocities both result in high pressure drop for the gas in flowing through the tray. For absorbers and humidifiers, high pressure drop results in high fan power to force the gas through the tower, and consequently, high operating costs. In the case of distillation, high pressure at the bottom of the tower results in high boiling temperatures, which in turn may lead to heating difficulties and possibly damage to heat-sensitive compounds.

Ultimately, purely mechanical difficulties arise. High pressure drop may lead directly to flooding. With a large pressure difference in the space between trays, the level of liquid leaving a tray at relatively low pressure and entering one of high pressure must necessarily assume an elevated position in the downspouts, as shown in Figure 4.3. As the pressure difference increases due to an increased rate of flow of the gas, the level in the downspout will rise further to permit the liquid to enter the lower tray. Ultimately, the liquid level in the downcomer may reach that on the tray above and the liquid will fill the entire space between the trays. The tower is then flooded, the tray efficiency falls to a very low value, the flow of gas is erratic, and liquid may be forced out of the gas exit pipe at the top of the tower.

For liquid–gas combinations which tend to foam excessively, high gas velocities may lead to a condition of *priming*, which is also an inoperative situation. Here, the foam persists throughout the space between trays, and a great deal of liquid is carried by the gas from one tray to the tray above. This is an exaggerated condition of entrainment. The liquid so carried recirculates between trays, and the added liquid-handling load increases the gas-pressure drop sufficiently to lead to flooding.

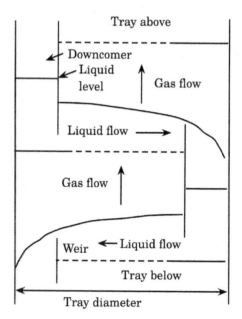

Figure 4.3 Cross-flow tray tower.

4.3 Gas–Liquid Operations: Gas Dispersed

If liquid rates are too low, the gas rising through the openings of the trays may push the liquid away (*coning*) and contact of the gas and liquid is poor. If the gas rate is too low, much of the liquid may rain down through the openings of the tray (*weeping*), thus failing to obtain the benefit of complete flow over the tray; and at very low gas rates, none of the liquid reaches the downspouts (*dumping*). The relations between these conditions are shown schematically in Figure 4.4, and all types of trays are subject to these difficulties in some form. The various arrangements, dimensions, and operating conditions chosen for design are those which experience have proved to be reasonably good compromises. The general design procedure involves a somewhat empirical application of them, following by computational check to ensure that pressure drop and flexibility to handle varying flow quantities are satisfactory.

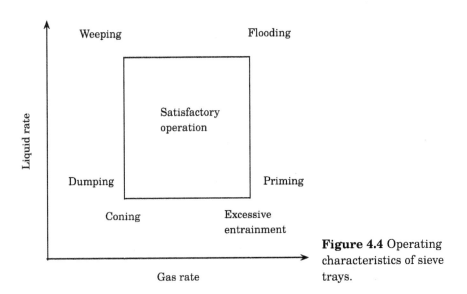

Figure 4.4 Operating characteristics of sieve trays.

A great variety of tray designs have been and are being used. The simplest is a sieve tray—it has a perforated tray deck with a uniform hole diameter of from less than a millimeter to about 25 mm. Trays with valves, which can be fixed or floating, are also very common. During the first half of the twentieth century, practically all towers were fitted with bubble-cap trays, but new installations now use either sieve trays or one of the proprietary designs which have proliferated since 1950. Figure 4.5 shows samples of each family of tray-deck design.

Because of their simplicity and low cost, sieve (perforated) trays are now the most important of tray devices. In the design of sieve trays, the diameter of the tower must be chosen to accommodate the flow rates, the details of the tray layout must be selected, estimates must be made of the gas-pressure drop and approach to flooding, and assurance against excessive weeping and entrainment must be established.

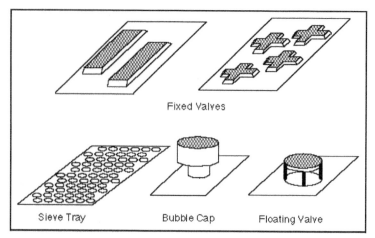

Figure 4.5 Types of tray-deck mass-transfer devices. Adapted from Bennett and Kovak (2000).

4.3.3 Tray Diameter

The tower diameter and, consequently, its cross-sectional area must be sufficiently large to handle the gas and liquid rates within the satisfactory region of Figure 4.4. For a given type of tray at flooding, the superficial velocity of the gas v_{GF} (volumetric rate of gas flow Q_G per net cross-sectional area for gas flow in the space between trays, A_n) is related to the fluid densities by

$$v_{GF} = C \left[\frac{\rho_L - \rho_G}{\rho_G} \right]^{1/2} \tag{4-34}$$

The net cross-sectional area A_n is the tower cross-sectional area A_t minus the area taken up by the downspout A_d, in the case of a cross-flow tray as in Figure 4.3. The value of the empirical constant C depends on the tray design, tray spacing, flow rates, liquid surface tension, and foaming tendency. It is estimated according to the empirical relationship (Fair, 1961; Seader and Henley, 1998)

$$C = F_{ST} F_F F_{HA} C_F \tag{4-35}$$

where F_{ST} = surface tension factor = $(\sigma/20)^{0.2}$
σ = liquid surface tension, dyn/cm
F_F = foaming factor = 1.0 for nonfoaming systems; for many absorbers may be 0.75 or even less (Kister, 1992)
F_{HA} = 1.0 for $A_h/A_a \geq 0.10$, and $5(A_h/A_a) + 0.5$ for $A_h/A_a < 0.1$
A_h/A_a = ratio of vapor hole area to tray active area

$$C_F = \alpha \log \frac{1}{X} + \beta \tag{4-36}$$

$$\alpha = 0.0744t + 0.01173$$
$$\beta = 0.0304t + 0.015 \tag{4-37}$$

4.3 Gas–Liquid Operations: Gas Dispersed

X = flow parameter = $(L'/G')(\rho_G/\rho_L)^{0.5}$

> If the value of X is in the range 0.01 to 0.10, use $X = 0.10$ in equation (4-36).

t = tray spacing, m

Typically, the column diameter D is based on a specified fractional approach to flooding f. Then,

$$D = \left[\frac{4Q_G}{f v_{GF} \pi \left(1 - A_d / A_t \right)} \right]^{0.5} \tag{4-38}$$

Oliver (1966) suggests that A_d/A_t must be chosen, based on the value of X, as

$$\frac{A_d}{A_t} = \begin{cases} 0.1 & \text{for } X < 0.1 \\ 0.1 + \dfrac{X - 0.1}{9} & \text{for } 0.1 \leq X \leq 1.0 \\ 0.2 & \text{for } X > 1.0 \end{cases} \tag{4-39}$$

In most sieve trays, the holes are placed in the corners of equilateral triangles at distances between centers (pitch, p') of from 2.5 to 5 hole diameters, d_o, as shown in Figure 4.6a. For such an arrangement, it is easily shown that

$$\frac{A_h}{A_a} = \frac{\pi}{4 \sin(60°)} \left[\frac{d_o}{p'} \right]^2 = 0.907 \left[\frac{d_o}{p'} \right]^2 \tag{4-40}$$

The downcomer geometry is shown in Figure 4.6b. From this and geometric relationships, the weir length, L_w, and the distance of the weir from the center of the tower, r_w, can be estimated for a specified value of A_d/A_t:

$$\frac{A_d}{A_t} = \frac{\theta - \sin \theta}{2\pi} \quad \theta \text{ in radians}$$
$$\frac{L_w}{D} = \sin\left[\frac{\theta}{2}\right] \quad \frac{r_w}{D} = \frac{1}{2}\cos\left[\frac{\theta}{2}\right] \tag{4-41}$$

Because of the need for internal access to columns with trays, a packed column is generally used if the diameter calculated from equation (4-38) is less than 60 cm. Tray spacing must be specified to compute column diameter, as shown by equation (4-37). On the other hand, recommended values of tray spacing depend on column diameter, as summarized in Table 4.3. Therefore, calculation of the tower diameter involves an iterative procedure: (a) an initial value of tray spacing is chosen (usually, $t = 0.6$ m); (b) the tower diameter is calculated based on this value of tray spacing; (c) the value of t is modified as needed according to the recommendations of Table 4.3 and the current estimate of D; (d) the procedure is repeated until convergence is achieved.

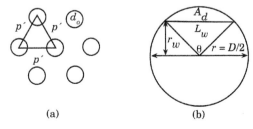

Figure 4.6 Tray geometry: (a) equilateral-triangular arrangement of holes in a sieve tray; (b) downcomer area geometry.

254   **Equipment for Gas–Liquid Mass-Transfer Operations**

Table 4.3 Recommended Tray Spacing

Tower diameter, D, m	Tray spacing, t, m
1 or less	0.50
1–3	0.60
3–4	0.75
4–8	0.90

Source: Data from Treybal (1980).

Example 4.6 Design of a Sieve-Tray Column for Ethanol Absorption

Design a sieve-tray column for the ethanol absorber of Example 4.4. For alcohol absorbers, Kister (1992) recommends a foaming factor $F_F = 0.9$. The liquid surface tension is estimated as $\sigma = 70$ dyn/cm. Take $d_o = 5$ mm on an equilateral-triangular pitch 15 mm between hole centers, punched in stainless steel sheet metal 2 mm thick. Design for an 80% approach to the flood velocity.

Solution
From Example 4.4, $X = 0.016$. From equation (4-40), $A_h/A_a = 0.907(5/15)^2 = 0.101$. Assume that $t = 0.5$ m; then, from equation (4-37), $\alpha = 0.0489$, $\beta = 0.0302$. Since $X < 0.1$, using $X = 0.1$ in equation (4-36), $C_F = 0.0791$ m/s. Since $A_h/A_a > 0.1$, then $F_{HA} = 1$. Also,

$$F_{ST} = \left[\frac{\sigma}{20}\right]^{0.2} = \left[\frac{70}{20}\right]^{0.2} = 1.285 \text{ and } F_F = 0.90$$

From equation (4-35), $C = 0.091$ m/s. From Example 4.4, $\rho_G = 1.923$ kg/m³, $\rho_L = 986$ kg/m³, $Q_G = 1.145$ m³/s; then, from equation (4-34), $v_{GF} = 2.07$ m/s. Since $X < 0.1$ m, equation (4-39) recommends $A_d/A_t = 0.1$. For an 80% approach to flooding, equation (4-38) yields

$$D = \left[\frac{4 \times 1.145}{0.8 \times 2.07 \times (1-0.1) \times \pi}\right]^{0.5} = 0.989 \text{ m}$$

At this point, the assumed value of tray spacing ($t = 0.5$ m) must be checked against the recommended values of Table 4.3. Since the calculated value of $D < 1.0$ m, $t = 0.5$ m is the recommended tray spacing, and no further iteration is needed.

Next, we calculate some further details of the tray design. From equation (4-41), for $A_d/A_t = 0.1$ and $D = 0.989$ m, $\theta = 1.627$ rad; $L_w = 0.727 \times 0.989 = 0.719$ m, and $r_w = 0.3435 \times 0.989 = 0.34$ m.

Summarizing, the details of the sieve-tray design are as follows:

Diameter $D = 0.989$ m
Tray spacing $t = 0.5$ m

4.3 Gas–Liquid Operations: Gas Dispersed 255

Total cross-sectional area $A_t = 0.768$ m^2
Downcomer area $A_d = 0.077$ m^2
Active area over the tray $A_a = A_t - 2A_d = 0.615$ m^2
Weir length $L_w = 0.719$ m
Distance from tray center to weir $r_w = 0.34$ m
Total hole area $A_h = 0.101A_a = 0.062$ m^2
Hole arrangement: 5-mm diameter on an equilateral-triangular pitch 15 mm between hole centers, punched in stainless steel sheet metal 2 mm thick

---

4.3.4 Tray Gas-Pressure Drop

Typical tray pressure drop for flow of vapor in a tower is from 0.3 kPa/tray to 1.0 kPa/tray. Pressure drop (expressed as head loss) for a sieve tray is due to friction for vapor flow through the tray perforations, holdup of the liquid on the tray, and a loss due to surface tension:

$$h_t = h_d + h_l + h_\sigma \tag{4-42}$$

where

h_t = total head loss/tray, cm of liquid
h_d = dry tray head loss, cm of liquid
h_l = equivalent head of clear liquid on tray, cm of liquid
h_σ = head loss due to surface tension, cm of liquid

The dry sieve-tray pressure drop is given by a modified orifice equation (Ludwig, 1979),

$$h_d = 0.0051 \left[\frac{v_o}{C_o}\right]^2 \rho_G \left[\frac{\rho_W}{\rho_L}\right]\left[1 - \left(\frac{A_h}{A_d}\right)^2\right] \tag{4-43}$$

where v_o is the hole velocity, in m/s, ρ_W is the density of liquid water at the liquid temperature, and the orifice coefficient C_o can be determined by the following equation (Wankat, 1988):

$$C_o = 0.85032 - 0.04231\left(\frac{d_o}{l}\right) + 0.00117954\left(\frac{d_o}{l}\right)^2 \quad \text{for} \left(\frac{d_o}{l}\right) \geq 1.0 \tag{4-44}$$

where l is the tray thickness.

The equivalent height of clear liquid holdup on a tray depends on weir height, h_W, liquid and vapor densities and flow rates, and downcomer weir length, as given by the following empirical expression developed from experimental data (Bennett et al., 1983):

$$h_l = \phi_e\left[h_w + C_l\left(\frac{q_L}{L_w\phi_e}\right)^{2/3}\right] \tag{4-45}$$

256 Equipment for Gas–Liquid Mass-Transfer Operations

where

h_w = weir height, cm (typical values are from 2.5 cm to 7.5 cm)

ϕ_e = effective relative froth density (height clear liquid/froth height)

= exp $(-12.55\,K_s^{0.91})$

K_s = capacity parameter, m/s

$$K_s = v_a \left[\frac{\rho_G}{\rho_L - \rho_G} \right]^{1/2} \tag{4-46}$$

v_a = superficial gas velocity based on tray active area, m/s

q_L = liquid flow rate across tray, m^3/s

$C_l = 50.12 + 43.89\,$exp$\,(-1.378h_w)$

As the gas emerges from the tray perforations, the bubbles must overcome surface tension. The pressure drop due to surface tension is given by the difference between the pressure inside the bubble and that of the liquid according to the theoretical relation

$$h_\sigma = \frac{6\sigma}{g\rho_L d_o} \tag{4-47}$$

where it is assumed that the maximum bubble size may be taken as the perforation diameter d_o.

Methods for estimating gas-pressure drop for bubble-cap trays and valve trays are discussed by Kister (1992).

Example 4.7 Gas-Pressure Drop in a Sieve-Tray Ethanol Absorber

Estimate the tray gas-pressure drop for the ethanol absorber of Examples 4.4 and 4.6. Use a weir height h_w of 50 mm.

Solution

Calculate first the dry tray pressure drop. From Example 4.4, $Q_G = 1.145\ m^3$/s; from Example 4.6, $A_h = 0.062\ m^2$. Then, $v_o = 1.145/0.062 = 18.48$ m/s. From Example 4.6, $d_o/l = 5\,/2 = 2.5$, $A_h/A_a = 0.101$, $\rho_G = 1.923\ kg/m^3$, and $\rho_L = 986\ kg/m^3$. The density of liquid water at 303 K is $\rho_W = 995\ kg/m^3$. From equation (4-44), $C_o = 0.756$. Then, from equation (4-43), $h_d = 5.85$ cm.

Calculate the equivalent head of clear liquid on the tray: from Example 4.6, $A_a = 0.615\ m^2$, then $v_a = 1.145/0.615 = 1.863$ m/s. From equation (4-46), $K_s = 0.082$ m/s; $\phi_e = 0.274$. From Example 4.4, $q_L = 0.804/986 = 0.000815\ m^3$/s; From Example 4.6, $L_w = 0.719$ m. For $h_w = 5.0$ cm, $C_l = 50.16$. Substituting in equation (4-45), $h_l = 1.73$ cm.

Calculate head loss due to surface tension: from Example 4.6, the surface tension is 70 dyne/cm = 0.07 N/m. Substituting in equation (4-47), $h_\sigma = 0.0087$ m = 0.87 cm. Calculate the total head loss/tray: Substituting in equation (4-42) gives $h_t = 8.45$ cm of clear liquid/tray. For a liquid density of 986 kg/m^3, this is equivalent to a gas-pressure drop of $\Delta P_G = 0.0845 \times 986 \times 9.8 = 817$ Pa/tray.

4.3 Gas–Liquid Operations: Gas Dispersed

4.3.5 Weeping and Entrainment

For a tray to operate at high efficiency, weeping of liquid through the tray perforations must be small compared to flow over the outlet weir and into the down-comer, and entrainment of liquid by the gas must not be excessive. Weeping occurs at low vapor velocities and/or high liquid rates. Bennett et al. (1997) report that weeping did not appear to substantially degrade tray performance as long as the orifice Froude number is greater or equal than 0.5:

$$\mathrm{Fr}_o = \left[\frac{\rho_G v_o^2}{\rho_L g h_l}\right]^{0.5} \geq 0.5 \qquad (4\text{-}48)$$

At high vapor rates, entrainment becomes significant and decreases tray performance. To take this into account, Bennett et al. (1997) defined the fractional entrainment E as

$$E = \frac{\text{liquid entrainment mass flow rate}}{\text{upward gas mass flow rate}} \qquad (4\text{-}49)$$

and developed the correlation

$$E = 0.00335 \left(\frac{h_{2\phi}}{t}\right)^{1.1} \left(\frac{\rho_L}{\rho_G}\right)^{0.5} \left(\frac{h_l}{h_{2\phi}}\right)^{\kappa} \qquad (4\text{-}50)$$

where

t = tray spacing

$h_{2\phi}$ = height of two-phase region on the tray, given by

$$h_{2\phi} = \frac{h_l}{\phi_e} + 7.79 \left[1 + 6.9 \left(\frac{d_o}{h_l}\right)^{1.85}\right] \frac{K_s^2}{\phi_e g \left(A_h / A_a\right)} \qquad (4\text{-}51)$$

κ = constant defined by (Bennett et al., 1995)

$$\kappa = 0.5 \left[1 - \tanh\left[1.3 \ln\left(\frac{h_l}{d_o}\right) - 0.15\right]\right] \qquad (4\text{-}52)$$

Example 4.8 Weeping and Entrainment in a Sieve-Tray Ethanol Absorber

Estimate the entrainment flow rate for the ethanol absorber of Examples 4.4, 4.6, and 4.7. Determine whether significant weeping occurs.

Solution
First, check whether weeping is significant. From equation (4-48), using information from the previous examples, we have

$$\mathrm{Fr}_o = \left[\frac{1.923}{986} \times \frac{18.48^2}{9.8 \times 0.0173}\right]^{0.5} = 1.98 > 0.5$$

258   **Equipment for Gas–Liquid Mass-Transfer Operations**

It can be concluded from this result that weeping is not a problem under these circumstances. Next, estimate the entrainment flow rate. From equation (4-52),

$$\kappa = 0.5\left[1 - \tanh\left(1.3 \times \ln\frac{17.3}{5} - 0.15\right)\right] = 0.0511$$

From equation (4-51),

$$h_{2\phi} = \frac{0.0173}{0.2740} + 7.79\left[1 + 6.9\times\left(\frac{5}{17.3}\right)^{1.85}\right] \times \frac{0.082^2}{0.274\times9.8\times0.101} = 0.394 \text{ m}$$

From equation (4-50),

$$E = 0.00335\left(\frac{0.394}{0.500}\right)^{1.1}\left(\frac{986}{1.923}\right)^{0.5}\left(\frac{0.0173}{0.394}\right)^{0.0511} = 0.05$$

From Example 4.4, the gas mass flow rate is $V' = 2.202$ kg/s. From equation (4-49), the entrainment mass flow rate L_e' is

$$L_e' = E \times V' = 0.050 \times 2.202 = 0.110 \text{ kg/s}$$

4.3.6 Tray Efficiency

The graphical and algebraic methods for determining stage requirements for absorption and stripping discussed in Chapter 3 assume that the streams leaving each stage are in equilibrium with respect to both heat and mass transfer. The assumption of thermal equilibrium is reasonable, but the assumption of equilibrium with respect to mass transfer is seldom justified. To determine the actual number of stages required for a given separation, the number of equilibrium stages must be adjusted with a *stage efficiency* (or *tray efficiency*).

> Stage efficiency concepts are applicable only to devices in which the phases are contacted and then separated, that is, when discrete stages can be identified. This is not the case for packed columns or other continuous contact devices. For these, the efficiency is already embedded into the design equation.

Tray efficiency is the fractional approach to an equilibrium stage which is attained by a real tray. Ultimately, we require a measure of approach to equilibrium of all the vapor and liquid from the tray, but since the conditions at various locations on the tray may differ, we begin by considering the local or *point efficiency* of mass transfer at a particular place on the tray surface.

Figure 4.7 is a schematic representation of one tray of a multitray tower. The tray n is fed from tray $n - 1$ above by liquid of average composition x_{n-1}, and it delivers liquid of average composition x_n to the tray below. At the place under consideration, a pencil of gas of composition $y_{n+1,\text{local}}$ rises from below and, as a result of mass transfer, leaves with a concentration $y_{n,\text{local}}$. At the place in question, it is assumed that the local liquid concentration x_{local} is constant in the vertical direction. The point efficiency is then defined by

4.3 Gas–Liquid Operations: Gas Dispersed

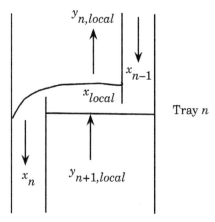

Figure 4.7 Point efficiency.

$$\mathbf{E}_{OG} = \frac{y_{n,local} - y_{n+1,local}}{y^*_{local} - y_{n+1,local}} \qquad (4\text{-}53)$$

Here, y^*_{local} is the concentration in equilibrium with x_{local}, and equation (4-53) then represents the change in gas concentration that actually occurs as a fraction of that which would occur if equilibrium were established. The subscript G signifies that gas concentrations are used, and the O emphasizes that $\mathbf{E}_{OG}$ is a measure of the overall resistance to mass transfer for both phases.

Consider that the gas rises at a rate G_M mol/(area)(time). Let the interfacial surface between gas and liquid be a (area/volume of liquid–gas froth). As the gas rises a differential height dh_l, the area of contact is $a\,dh_l$ per unit active area of the tray. If, while of concentration y, it undergoes a concentration change dy in this height, and if the total quantity of gas is assumed to remain essentially constant, the rate of solute transfer is $G_M\,dy$:

$$G_M\,dy = K_y a \left(y^*_{local} - y \right) dh_l \qquad (4\text{-}54)$$

Then,

$$\int_{y_{n+1,local}}^{y_{n,local}} \frac{dy}{y^*_{local} - y} = \frac{K_y a h_l}{G_M} \qquad (4\text{-}55)$$

Since y^*_{local} is constant for constant x_{local}

$$-\ln \frac{y^*_{local} - y_{n,local}}{y^*_{local} - y_{n+1,local}} = -\ln\left(1 - \frac{y_{n,local} - y_{n+1,local}}{y^*_{local} - y_{n+1,local}}\right)$$

$$= -\ln(1 - \mathbf{E}_{OG}) = \frac{K_y a h_l}{G_M} \qquad (4\text{-}56)$$

Therefore,

$$\mathbf{E}_{OG} = 1 - \exp\left(-\frac{K_y a h_l}{G_M}\right) \qquad (4\text{-}57)$$

260 Equipment for Gas–Liquid Mass-Transfer Operations

A method to estimate the overall volumetric mass-transfer coefficient at any point over the tray, $K_y a$, is needed in order to use equation (4-57). Alternatively, Bennett et al. (1997) proposed the following correlation for estimating sieve-tray point efficiency:

$$\mathbf{E}_{OG} = 1 - \exp\left[-\frac{0.029}{1 + m\dfrac{c_G}{c_L}\sqrt{\dfrac{D_G\left(1-\phi_e\right)}{D_L\left(A_h / A_a\right)}}} \mathrm{Re}_{Fe}^{a1}\left(\frac{h_l}{d_o}\right)^{a2}\left(\frac{A_h}{A_a}\right)^{a3} \right] \quad (4\text{-}58)$$

where

$$\mathrm{Re}_{Fe} = \frac{\rho_G v_o h_l}{\mu_G \phi_e} \quad (4\text{-}59)$$

$a1 = 0.4136$
$a2 = 0.6074$
$a3 = -0.3195$
m = local slope of the equilibrium curve
c_G, c_L = molar concentration of gas and liquid, respectively
D_G, D_L = diffusivities of gas and liquid, respectively

The bulk-average concentration of all the local pencils of gas of Figure 4.7 are y_{n+1} and y_n. The efficiency of the entire tray, also known as the *Murphree efficiency*, is then

$$\mathbf{E}_{MG} = \frac{y_n - y_{n+1}}{y^* - y_{n+1}} \quad (4\text{-}60)$$

where y^* is in equilibrium with the liquid leaving the tray of concentration x_n. The relationship between $\mathbf{E}_{MG}$ and $\mathbf{E}_{OG}$ can then be derived by integrating the local $\mathbf{E}_{OG}$ over the surface of the tray. Four cases will be considered, depending upon the degree of mixing of the vapor and liquid phases:

1. If the liquid on the tray is completely backmixed, everywhere of uniform concentration x_n, and the vapor entering the tray is perfectly mixed, then $\mathbf{E}_{MG} = \mathbf{E}_{OG}$.
2. If the movement of the liquid on the tray is in plug flow, with no mixing, while the vapor entering the tray is perfectly mixed (Kister, 1992),

$$\mathbf{E}_{MG} = \frac{1}{\lambda}\left[\exp\left(\lambda \mathbf{E}_{OG}\right) - 1\right] \quad (4\text{-}61)$$

where $\lambda = mV/L$.
3. If the entering vapor is well mixed but the liquid is only partially mixed (Gerster et al., 1958),

$$\frac{\mathbf{E}_{MG}}{\mathbf{E}_{OG}} = \frac{1 - \exp\left[-\left(\eta + \mathrm{Pe}_L\right)\right]}{\left(\eta + \mathrm{Pe}_L\right)\left[1 + \dfrac{\eta + \mathrm{Pe}_L}{\eta}\right]} + \frac{\exp\left(\eta\right) - 1}{\eta\left[1 + \dfrac{\eta + \mathrm{Pe}_L}{\eta}\right]}$$

$$\eta = \frac{\mathrm{Pe}_L}{2}\left[\left[1 + \frac{4\lambda \mathbf{E}_{OG}}{\mathrm{Pe}_L}\right]^{0.5} - 1\right] \quad (4\text{-}62)$$

4.3 Gas–Liquid Operations: Gas Dispersed

For $\lambda < 3.0$, equation (4-62) can be approximated to within an average error of about 0.4% by (Bennett et al., 1997):

$$\mathbf{E}_{MG} = \frac{\left[1 + \dfrac{\lambda \mathbf{E}_{OG}}{N}\right]^N - 1}{\lambda} \qquad (4\text{-}63)$$

where

$$N = \frac{\mathrm{Pe}_L + 2}{2}$$

$$\mathrm{Pe}_L = \frac{4 q_L r_w^2}{A_a h_l \mathrm{D}_{E,L}} \qquad (4\text{-}64)$$

$\mathrm{D}_{E,L}$ = liquid eddy diffusivity, estimated from (Bennett et al., 1997)

$$\mathrm{D}_{E,L} = 0.1 \left(g h_{2\phi}^3 \right)^{0.5} \qquad (4\text{-}65)$$

4. If both the vapor and liquid are only partially mixed, for cross-flow trays and $\lambda < 3.0$ (Bennett et al., 1997),

$$\mathbf{E}_{MG} = \frac{\left\{\left[1 + \dfrac{\lambda \mathbf{E}_{OG}}{N}\right]^N - 1\right\}}{\lambda} \times \left[1 - 0.0335 \lambda^{1.073} \mathbf{E}_{OG}^{2.518} \mathrm{Pe}_L^{0.175}\right] \qquad (4\text{-}66)$$

The degree of vapor mixing can be estimated from Katayama and Imoto (1972), who report that the vapor can be considered unmixed if the vapor-Peclet number, $\mathrm{Pe}_G > 50$, or if $h_{2\phi}/t \geq 1$, where

$$\mathrm{Pe}_G = \frac{4 Q_G r_w^2}{A_a \left(t - h_{2\phi}\right) \mathrm{D}_{E,G}} \quad \text{for} \quad \frac{h_{2\phi}}{t} < 1.0 \qquad (4\text{-}67)$$

Lockett (1986) recommends a value of gas eddy diffusivity $\mathrm{D}_{E,G} = 0.01 \ \mathrm{m}^2/\mathrm{s}$.

A further correction of the tray efficiency is required for the damage done by entrainment, a form of backmixing which acts to destroy the concentration changes produced by the trays. Bennett et al. (1997) suggest

$$\mathbf{E}_{MGE} = \mathbf{E}_{MG} \left[1 - \frac{0.8 \mathbf{E}_{OG} \lambda^{1.543}}{m}\right] \qquad (4\text{-}68)$$

where $\mathbf{E}_{MGE}$ is the Murphree tray efficiency corrected for the effect of entrainment. Examples 4.9 and 4.10 illustrate the use of these equations.

Example 4.9 Murphree Efficiency of a Sieve-Tray Ethanol Absorber

Estimate the entrainment-corrected Murphree tray efficiency for the ethanol absorber of Examples 4.4, 4.6, 4.7, and 4.8.

Solution
From equation (4-59), $\mathrm{Re}_{Fe} = 1.543 \times 10^5$. The molar densities are $c_G = \rho_G/M_G = 0.044$ kmol/m^3 and $c_L = \rho_L/M_L = 52.93$ kmol/m^3. For the low concentrations prevailing

262   **Equipment for Gas–Liquid Mass-Transfer Operations**

in the liquid phase, the ethanol–water solution at 303 K obeys Henry's law, and the slope of the equilibrium curve is $m = 0.57$ (Seader and Henley, 1998). Equation (4-58) yields a point efficiency of $\mathbf{E}_{OG} = 0.809$. Next, check the degree of mixing of the vapor phase. Equation (4-67) yields $Pe_G = 810 > 50$. Therefore, the vapor is unmixed. From equation (4-65), $D_{E,L} = 0.077$ m^2/s; from equation (4-64), $Pe_L = 0.459$, $N = 1.229$. Next, calculate $\lambda = mG/L = 0.66$. Equation (4-66) yields a Murphree tray efficiency of $\mathbf{E}_{MG} = 0.836$. From Example 4.8, the fractional entrainment is $E = 0.05$. Substituting in equation (4-68), the tray efficiency corrected for entrainment is $\mathbf{E}_{MGE} = 0.811$.

Example 4.10 Sieve-Tray Distillation Column for Ethanol Purification

In the development of a new process, it will be necessary to fractionate 910 kg/h of an ethanol–water solution containing 0.3 mol fraction ethanol. It is desired to produce a distillate containing 0.80 mol fraction ethanol, with negligible loss of ethanol in the residue, using a sieve-tray distillation column at atmospheric pressure. Preliminary calculations show that the maximum vapor-volumetric flow rate occurs at a point in the column just above the feed tray, where the liquid and the vapor contain 0.5 and 0.58 mol fraction of ethanol, respectively. The temperature at that point in the column is about 353 K; the local slope of the equilibrium curve is $m = 0.42$. The liquid and vapor molar flow rates are 39 and 52 kmol/h, respectively.

(a) Specify a suitable sieve-tray design for this application. Check your design for vapor-pressure drop, weeping, and excessive entrainment. Estimate the corresponding Murphree tray efficiency, corrected for entrainment.

(b) Check your design for the conditions prevailing at the bottom of the tower where the vapor molar flow rate remains the same, but the liquid molar flow rate is 73 kmol/h. The temperature is 373 K, and both the liquid and vapor phases are virtually pure water.

Solution

a) For the given concentrations of the liquid and vapor phases, the average molecular weights are $M_G = 34.2$ and $M_L = 32.0$. The corresponding mass flow rates are $V' = 0.494$ kg/s and $L' = 0.347$ kg/s. The density of the vapor is from the ideal gas law, $\rho_G = 1.18$ kg/m^3. The viscosity of the vapor, as estimated from the Lucas' method, is $\mu_G = 105$ µP; the diffusivity is from the Wilke–Lee equation, $D_G = 0.158$ cm^2/s. Properties of the liquid phase are (Treybal, 1980) $\rho_L = 791$ kg/m^3, $D_L = 2.07 \times 10^{-5}$ cm^2/s, $\sigma = 21$ dyn/cm, and $m = 0.42$. The density of liquid water at 353 K is $\rho_W = 970$ kg/m^3. The foaming factor for alcohol distillation is $F_F = 0.9$ (Kister, 1992). Use stainless steel, 2-mm-thick sheet metal with 4.5-mm perforations on an equilateral-triangular pitch 12 mm between hole centers. Set the weir height at 5 cm, and design for an 80% fractional approach to flooding.

   Appendix E presents Mathcad and Python sieve-tray design computer programs. For a given set of data on properties of the gas and liquid streams, tray geometry, and fractional approach to flooding, the programs calculate the tray diameter,

4.3 Gas–Liquid Operations: Gas Dispersed

weir length and location, and recommended tray spacing; estimate the vapor-pressure drop per tray; check the design for weeping; calculate fractional liquid entrainment; and estimate the tray Murphree efficiency, corrected for entrainment.

For the conditions specified at a point in the column, just above the feed tray of the ethanol fractionator, the results are

1. Tray diameter = 0.631 m
2. Recommended tray spacing = 0.5 m
3. Weir length = 0.458 m
4. Weir location = 0.217 m from center of tray
5. Gas-pressure drop = 358 Pa/tray
6. Orifice Froude number = 1.11 > 0.5; no excessive weeping
7. Fractional liquid entrainment = 0.024
8. Point efficiency = 0.760
9. Murphree tray efficiency = 0.795
10. Murphree tray efficiency, corrected for entrainment = 0.783

(b) At the bottom of the tower the liquid and vapor phases are very dilute solutions of ethanol in water; therefore, $M_G = M_L = 18$. The corresponding mass flow rates are $V' = 0.260$ kg/s and $L' = 0.365$ kg/s. The density of the vapor and of the liquid are, from the Steam Tables (Smith et al., 1996), for saturated water at 373 K, $\rho_L = \rho_W = 958$ kg/m^3 and $\rho_G = 0.600$ kg/m^3. The viscosity of the gas is estimated by Lucas' method, $\mu_G = 1.21 \times 10^{-5}$ kg/m-s; the surface tension of the liquid is estimated as $\sigma = 63$ dyn/cm (Reid et al., 1987). The gas-phase diffusivity is from the Wilke–Lee correlation, $D_G = 0.177$ cm^2/s; the liquid-phase diffusivity is from the Hayduk–Minhas correlation for aqueous solutions, $D_L = 5.54 \times 10^{-5}$ cm^2/s. The slope of the equilibrium curve for very dilute solutions can be derived from the modified Raoult's law, equation (3-2), written here as

$$m = \lim_{x_A \to 0} \frac{\gamma_A P_A}{P} = \frac{\gamma_A^{\infty} P_A \left(@373 \text{ K}\right)}{P} \tag{4-69}$$

The activity coefficient at infinite dilution for ethanol in water at 373 K is 5.875 (Reid et al., 1987); the vapor pressure of ethanol at 373 K is 226 kPa (Smith et al., 1996). For a total pressure of 101.3 kPa, equation (4-64) yields $m = 13.1$.

Next, enter the data presented above into the Mathcad or Python sieve-tray design computer programs of Appendix E. Since the dimensions of the tray are known, the fractional approach to flooding is adjusted until the tray design coincides with the tray dimensions determined in part (a) of this example. Convergence is achieved at a value of $f = 0.431$. This means that at the bottom of the distillation column the gas velocity is only 43.1% of the flooding velocity. Other important results obtained from the program are as follows:

1. Gas-pressure drop = 434 Pa/tray
2. Orifice Froude number = 0.65 > 0.5; no excessive weeping
3. Fractional liquid entrainment = 0.019
4. Point efficiency = 0.500

264   **Equipment for Gas–Liquid Mass-Transfer Operations**

5. Murphree tray efficiency = 0.807
6. Murphree tray efficiency, corrected for entrainment = 0.792

Notice that even though the point efficiency for the conditions prevailing at the bottom of the column is significantly smaller than the point efficiency just above the feed tray, the entrainment-corrected Murphree tray efficiencies at both points are very similar.

PROBLEMS

The problems at the end of each chapter have been grouped into four classes (designated by a superscript after the problem number).

Class a: Illustrates direct numerical application of the formulas in the text.
Class b: Requires elementary analysis of physical situations, based on the subject material in the chapter.
Class c: Requires somewhat more mature analysis.
Class d: Requires computer solution.

4.1[a]. Void fraction near the walls of packed beds

Consider a cylindrical vessel with a diameter of 610 mm packed with solid spheres with a diameter of 50 mm.

(a) From equation (4-1), calculate the asymptotic porosity of the bed.

Answer: 38.3%

(b) Estimate the void fraction at a distance of 100 mm from the wall.

Answer: 42.3%

4.2[b]. Void fraction near the walls of packed beds

Because of the oscillatory nature of the void-fraction radial variation of packed beds, there are a number of locations close to the wall where the local void fraction is exactly equal to the asymptotic value (see Figure 4.4). For the bed described in Example 4.1, calculate the distance from the wall to the first five such locations.

Answer: 7.2, 16.4, 25.8, 35.1, 44.5 mm

4.3[c, d]. Void fraction near the walls of packed beds

(a) Show that the radial location of the maxima and minima of the function described by equation (4-1) are the roots of the equation

$$J_0\left(\alpha r^*\right) + \frac{\alpha}{\beta} J_1\left(\alpha r^*\right) = 0 \qquad (4\text{-}70)$$

Problems   **265**

(b) For the packed bed of Example 4.1, calculate the radial location of the first
five maxima and of the first five minima; calculate the amplitude of the void
fraction oscillations at those points.

(c) Calculate the distance from the wall at which the absolute value of the porosity
fluctuations has been dampened to less than 10% of the asymptotic bed porosity.

Answer: 69 mm

(d) What fraction of the cross-sectional area of the packed bed is characterized by
porosity fluctuations which are within ± 10% of the asymptotic bed porosity?

Answer: 35.6%

(e) Show that the average void fraction of a bed packed with equal-sized spheres
ε_{av} is given by

$$\varepsilon_{av} = \frac{2}{\left(R^*\right)^2} \int_0^{R^*} \left(R^* - r^*\right)\varepsilon\left(r^*\right)dr^* \quad \text{where } R^* = \frac{D}{2d_p} \qquad (4\text{-}71)$$

(f) For the packed bed of Example 4.1, estimate the average void fraction by
numerical integration of equation (4-71) and estimate the ratio $\varepsilon_{av}/\varepsilon_b$.

Answer: $\varepsilon_{av}/\varepsilon_b = 1.064$

4.4^c. Void fraction near the walls of annular packed beds

Annular packed beds (APBs) involving the flow of fluids are used in many
technical and engineering applications, such as in chemical reactors, heat
exchangers, and fusion reactor blankets. It is well known that the wall in a packed
bed affects the radial void fraction distribution. Since APBs have two walls that can
simultaneously affect the radial void fraction distribution, it is essential to include
this variation in transport models. A correlation for this purpose was recently formulated (Mueller, 1999). The correlation is restricted to randomly packed beds in
annular cylindrical containers of outside diameter D_o, inside diameter D_i, equivalent
diameter $D_e = D_o - D_i$, consisting of equal-sized spheres of diameter d_p, with diameter aspect ratios of $4 \le D_e/d_p \le 20$. The correlation is

$$\varepsilon = \varepsilon_b + \left(1 - \varepsilon_b\right)\left\{J_0\left(\alpha r^*\right)\exp\left(\beta r^*\right) + J_0\left(\alpha\right)\exp\left[\beta\left(R^* - r^*\right)\right]\right\} \qquad (4\text{-}72)$$

where

$$r^* = \frac{r}{d_p} \qquad R^* = \frac{D_e}{2d_p} \qquad 0 \le r^* \le R^* \qquad (4\text{-}73)$$

$r =$ radial position measured from the outer wall

$$\alpha = 6.64 - 6.1\exp\left(-R^*\right)$$
$$\beta = -\left(0.69 + 0.015R^*\right) \qquad (4\text{-}74)$$
$$\varepsilon_b = 0.456 - 0.3\exp\left(-R^*\right)$$

266   **Equipment for Gas–Liquid Mass-Transfer Operations**

Consider an APB with outside diameter of 140 mm, inside diameter of 40 mm, packed with identical 10-mm-diameter spheres.

(a) Estimate the void fraction at a distance from the outer wall of 25 mm.

Answer: 42.2%

(b) Plot the void fraction, as predicted by equation (4-72), for $r*$ from 0 to $R*$.

(c) Show that the average porosity for an APB is given by

$$\varepsilon_{av} = \frac{2}{\left(R_o^*\right)^2 - \left(R_i^*\right)^2} \int_0^{R*} \left(R_o^* - r*\right)\varepsilon(r*)dr*$$

$$R_o^* = \frac{D_o}{2d_p} \qquad R_i^* = \frac{D_i}{2d_p}$$

(4-75)

(d) Estimate the average porosity for the APB described above.

Answer: 48.7%

4.5ᵃ. Minimum liquid mass velocity for proper wetting of packing

A 1.0-m-diameter bed used for absorption of ammonia with pure water at 298 K is packed with 25-mm plastic Intalox saddles. Calculate the minimum water flow rate, in kg/s, needed to ensure proper wetting of the packing surface.

Answer: 0.94 kg/s

4.6ᵃ. Minimum liquid mass velocity for proper wetting of packing

Repeat Problem 4.5 using ceramic instead of plastic Intalox saddles.

Answer: 0.12 kg/s

4.7ᵇ. Loading and flooding in second-generation random packing

Repeat Example 4.2 using 25-mm metal Pall rings as packing material.

Answer: $v_{G,S} = 1.514$ m/s

4.8ᵇ. Loading and flooding in structured packing

Repeat Example 4.2 but using Montz metal B1-200 structured packing (very similar to the one shown in Figure 4.3). For this packing, $a = 200$ m^{-1}, $\varepsilon = 0.979$, $C_{S,T} = 3.116$, and $C_{F,T} = 2.339$ (Billet and Schultes, 1999).

Answer: $v_{G,F} = 2.8184$ m/s

4.9ᵇ. Specific liquid holdup and void fraction in first-generation random packing

A tower packed with 25-mm ceramic Raschig rings is to be used for absorbing benzene vapor from a dilute mixture with an inert gas using a wash oil at 300 K.

The viscosity of the oil is 2.0 cP and its density is 840 kg/m³. The liquid mass velocity is $G_x = 2.71$ kg/m²-s. Estimate the liquid holdup, the operational void fraction, and the hydraulic specific area of the packing.

Answer: $(\varepsilon - h_L) = 0.646$, $a_h = 65$ m⁻¹

4.10[b, d]. Pressure drop in beds packed with fourth-generation random packings

Since its introduction in the market in 1995, numerous mass-transfer columns have been packed with Raschig Super-Rings—a new fourth-generation packing shown in Figure 4.8—in various chemical processes, petrochemical, refining, and environmental applications (Schultes, 2003).

Figure 4.8 Fourth-generation Raschig Super-Ring.

Repeat Example 4.3 using metal Raschig Super-Rings No. 2 as packing material.

Answer: $\Delta P = 277$ Pa/m, $D = 0.259$ m

4.11[b, d]. Pressure drop and approach to flooding in structured packing

Repeat Example 4.3 using Montz metal B1-200 structured packing (very similar to the one shown in Figure 4.3). For this packing, $C_{S,T} = 3.116$, $C_{F,T} = 2.339$, $a = 200$ m⁻¹, $\varepsilon = 0.979$, $C_h = 0.547$, and $C_p = 0.355$ (Seader and Henley, 1998; Billet and Schultes, 1999).

Answer: $\Delta P = 214$ Pa/m, $D = 0.295$ m

4.12[b, d]. Pressure drop and mass-transfer coefficients in beds packed with fourth-generation random packings

A packed tower is to be designed for the countercurrent contact of a benzene–nitrogen gas mixture with kerosene to wash out the benzene from the gas. The gas enters the tower at the rate of 1.5 m³/s, measured at 110 kPa and 298 K, containing 5 mol% benzene. Essentially, all the benzene is absorbed by the kerosene. The liquid enters the tower at the rate of 4.0 kg/s; the liquid density is 800 kg/m³ and the viscosity is 2.3 cP. The packing will be metal Raschig Super-Rings No. 2, and the tower diameter will be chosen to produce a gas-pressure drop of 275 Pa/m of irrigated packing. The gas viscosity, from Lucas method, is 1.66×10^{-5} kg/m-s.

(a) Calculate the tower diameter to be used, and the resulting fractional approach to flooding.

Answer: $f = 0.82$, $D = 0.885$ m

268   Equipment for Gas–Liquid Mass-Transfer Operations

(b) Estimate the volumetric mass-transfer coefficients for the gas and liquid phases. Assume that $D_L = 5.0 \times 10^{-10}$ m^2/s, $D_G = 8.85 \times 10^{-6}$ m^2/s.

Answer: $k_L a_h = 0.00621$ s^{-1}

(c) Assume that for the diameter chosen, the irrigated packed height will be 5 m and that 1 m of unirrigated packing will be placed over the liquid inlet to act as entrainment separator. The blower-motor combination to be used at the gas inlet will have an overall mechanical efficiency of 60%. Calculate the power required to blow the gas through the packing.

Answer: 3.87 kW

4.13[b, d]. Pressure drop in beds packed with structured packings

Redesign the packed bed of Problem 4.12 using Montz metal B1-200 structured packing. Estimate the corresponding mass-transfer coefficients. For this packing, $C_{S,T} = 3.116$, $C_{F,T} = 2.339$, $a = 200$ m^{-1}, $\varepsilon = 0.979$, $C_h = 0.547$, $C_p = 0.355$, $C_L = 0.971$, and $C_V = 0.390$ (Seader and Henley, 1998; Billet and Schultes, 1999).

Answer: $D = 0.961$ m; $k_y a_h = 0.250$ kmol/m^3·s

4.14[c, d]. Air stripping of wastewater in a packed column

A wastewater stream of 0.038 m^3/s, containing 10 ppm (by weight) of benzene, is to be stripped with air in a packed column operating at 298 K and 2 atm to reduce the benzene concentration to 0.005 ppm. The packing specified is 50-mm plastic Pall rings. The airflow rate to be used is five times the minimum. Henry's law constant for benzene in water at this temperature is 0.6 kPa·m^3/mol (Davis and Cornwell, 1998). Calculate the tower diameter if the gas-pressure drop is not to exceed 250 Pa/m of packed height. Estimate the corresponding mass-transfer coefficients. The diffusivity of benzene vapor in air at 298 K and 1 atm is 0.096 cm^2/s; the diffusivity of liquid benzene in water at infinite dilution at 298 K is 1.02×10^{-5} cm^2/s (Cussler, 1997).

4.15[b]. Stripping chloroform from water by sparging with air

Repeat Example 4.5 using an airflow rate that is twice the minimum required.

Answer: $W = 4.09$ kW

4.16[b]. Stripping chloroform from water by sparging with air

Repeat Example 4.5 using the same airflow rate used in Problem 4.15 and specifying a chloroform removal efficiency of 99%.

Answer: $Z = 1.30$ m

4.17[b]. Stripping chlorine from water by sparging with air

A vessel 2.0 m in diameter and 2.0-m deep (measured from the gas sparger at the bottom to liquid overflow at the top) is to be used for stripping chlorine from water by sparging with air. The water will flow continuously downward at the rate of 7.5 kg/s with an initial chlorine concentration of 5 mg/L. Airflow will be 0.22 kg/s

Problems 269

at 298 K. The sparger is in the form of a ring, 25 cm in diameter, containing 200 orifices, each 3.0 mm in diameter. The Henry's law constant for chlorine in water at this temperature is 0.11 kPa·m³/mol (Perry and Chilton, 1973). The diffusivity of chlorine at infinite dilution in water at 298 K is 1.25×10^{-5} cm²/s (Cussler, 1997).

(a) AnswerAssuming that all the resistance to mass transfer resides in the liquid phase, estimate the chlorine removal efficiency achieved.

Answer: 99.34%

(b) Estimate the power required to operate the air compressor if the mechanical efficiency of the system is 65%.

Answer: 7.17 kW

4.18[c, d]. Batch wastewater aeration using spargers

In the treatment of wastewater, undesirable gases are frequently stripped or desorbed from the water, and oxygen is adsorbed into the water when bubbles of air are dispersed near the bottom of aeration tanks or ponds. As the bubbles rise, solute can be transferred from the gas to the liquid or from the liquid to the gas, depending upon the concentration driving force.

For batch aeration in a constant-volume tank, an oxygen mass balance can be written as

$$\frac{dc_A}{dt} = K_L a \left(c_A{}^* - c_A \right) \qquad (4\text{-}76)$$

where $c_A{}^*$ is the oxygen saturation concentration. Integrating between the time limits zero and t and the corresponding dissolved oxygen concentration limits $c_{A,0}$ and $c_{A,t}$, and assuming that $c_A{}^*$ remains essentially constant and that all the resistance to mass transfer resides in the liquid phase, we obtain

$$\ln \left[\frac{c_A{}^* - c_{A,0}}{c_A{}^* - c_{A,t}} \right] = k_L a t \qquad (4\text{-}77)$$

In aeration tanks, where air is released at an increased liquid depth, the solubility of oxygen is influenced both by the increasing pressure of the air entering the aeration tank and by the decreasing oxygen partial pressure in the air bubble as oxygen is absorbed. For these cases, the use of a mean saturation value corresponding to the aeration tank mid-depth is suggested (Eckenfelder, 2000):

$$c_A{}^* = \frac{c_s}{2} \left(\frac{P_o}{P_s} + \frac{O_t}{20.9} \right) \qquad (4\text{-}78)$$

where

c_s = saturation dissolved oxygen concentration in fresh water exposed to atmospheric air at 101.3 kPa containing 20.9% oxygen

P_o = absolute pressure at the depth of air release

P_s = atmospheric pressure

O_t = molar oxygen percent in the air leaving the aeration tank

270 **Equipment for Gas–Liquid Mass-Transfer Operations**

The molar oxygen percent in the air leaving the aeration tank is related to the oxygen transfer efficiency, O_{eff}, through

$$O_t = \frac{21(1-O_{eff})}{79+21(1-O_{eff})} \times 100 \qquad (4\text{-}79)$$

where

$$O_{eff} = \frac{\text{mass of oxygen absorbed by the water}}{\text{total mass of oxygen supplied}} \qquad (4\text{-}80)$$

Consider a 567-m^3 aeration pond aerated with 15 spargers, each using compressed air at a rate of 0.01 kg/s. Each sparger is in the form of a ring 100 cm in diameter, containing 20 orifices, each 3.0 mm in diameter. The spargers will be located 5 m below the surface of the pond. The water temperature is 298 K; atmospheric conditions are 298 K and 101.3 kPa. Under these conditions, c_s = 8.38 mg/L (Davis and Cornwell, 1998).

(a) Estimate the volumetric mass-transfer coefficient for these conditions from equations (4-28) and (4-30).

Answer: k_La = 2.83 h^{-1}

(b) Estimate the time required to raise the dissolved oxygen concentration from 0.5 mg/L to 6.0 mg/L and calculate the resulting oxygen transfer efficiency.

Answer: 17.9 min

(c) Estimate the power required to operate the 15 spargers if the mechanical efficiency of the compressor is 60%.

Answer: 8.64 kW

4.19[c, d]. Batch wastewater aeration using spargers; effect of liquid depth

Consider the situation described in Problem 4.18. According to Eckenfelder (2000), for most types of bubble-diffusion aeration systems the volumetric mass-transfer coefficient will vary with liquid depth Z according to the relationship

$$\frac{k_La @ Z_1}{k_La @ Z_2} = \left[\frac{Z_1}{Z_2}\right]^n \qquad (4\text{-}81)$$

where the exponent n has a value near 0.7 for most systems. For the aeration pond of Problem 4.18, calculate k_La at values of Z = 3 m, 4 m, 6 m, and 7 m. Estimate the corresponding value of n from regression analysis of the results. *Hint*: Remember that the total volume of the pond must remain constant; therefore, the cross-sectional area of the pond must change as the water depth changes.

4.20[c]. Flooding conditions in a packed cooling tower

A cooling tower, 2 m in diameter, packed with 75-mm ceramic Hiflow rings, is fed with water at 316 K at a rate of 25 kg/m^2·s. The water is contacted with air, at

Problems

300 K and 101.3 kPa essentially dry, drawn upward countercurrently to the water flow. Neglecting evaporation of the water and changes in the air temperature, estimate the volumetric rate of airflow, in m^3/s, which would flood the tower.

4.21[c, d]. Design of a sieve-tray column for ethanol absorption

Repeat the calculations of Examples 4.6, 4.7, 4.8, and 4.9 for a column diameter corresponding to 50% of flooding.

Answer: $D = 1.176$ m

4.22[c, d]. Design of a sieve-tray column for aniline stripping

A sieve-tray tower is to be designed for stripping an aniline (C_6H_7N)–water solution with steam. The circumstances at the top of the tower, which are to be used to establish the design, are

Temperature = 371.5 K $\qquad$ Pressure = 100 kPa

Liquid:
Rate = 10.0 kg/s Composition = 7.00 wt% aniline
Density = 961 kg/m^3 Viscosity = 0.3 cP
Surface tension = 58 dyn/cm
Diffusivity = 4.27×10^{-5} cm^2/s (est.) $\qquad$ Foaming factor = 0.90

Vapor:
Rate = 5.0 kg/s $\qquad$ Composition = 3.6 mol% aniline
Density = 0.670 kg/m^3 $\qquad$ Viscosity = 118 μP (est.)
Diffusivity = 0.116 cm^2/s (est.)

The equilibrium data at this concentration indicate that $m = 0.0636$ (Treybal, 1980).

(a) Design a suitable cross-flow sieve tray for such a tower. Take $d_o = 5.5$ mm on an equilateral-triangular pitch 12 mm between hole centers, punched in stainless steel sheet metal 2 mm thick. Use a weir height of 40 mm. Design for a 75% approach to the flood velocity. Report details respecting tower diameter, tray spacing, weir length, gas-pressure drop, and entrainment in the gas. Check for excessive weeping.

Answer: $D = 1.93$ m

(b) Estimate the tray efficiency corrected for entrainment for the design reported in part (a).

Answer: $E_{MGE} = 0.72$

4.23[c, d]. Design of a sieve-tray column for aniline stripping

Repeat Problem 4.22, but for a 45% approach to flooding. Everything else remains the same as in Problem 4.22.

Answer: $D = 2.49$ m; $E_{MGE} = 0.67$; excessive weeping

272 Equipment for Gas–Liquid Mass-Transfer Operations

4.24[c, d]. Design of a sieve-tray column for methanol stripping

A dilute aqueous solution of methanol is to be stripped with steam in a sieve-tray tower. The conditions chosen for design are

Temperature = 368 K   Pressure = 101.3 kPa

Liquid:
   Rate = 0.25 kmol/s   Composition = 15.0 mass% methanol
   Density = 961 kg/m^3 Viscosity = 0.3 cP
   Surface tension = 40 dyn/cm
   Diffusivity = 5.70 $_x$ 10^{-5} cm^2/s (est.)   Foaming factor = 1.0

Vapor:
   Rate = 0.1 kmol/s   Composition = 18 mol% methanol
   Viscosity = 125 $_\mu$P (est.)   Diffusivity = 0.213 cm^2/s (est.)

The equilibrium data at this concentration indicate that m = 2.5 (Perry and Chilton, 1973).

(a) Design a suitable cross-flow sieve tray for such a tower. Take d_o = 6.0 mm on an equilateral-triangular pitch 12 mm between hole centers, punched in stainless-steel sheet metal 2 mm thick. Use a weir height of 50 mm. Design for an 80% approach to the flood velocity. Report details respecting tower diameter, tray spacing, weir length, gas-pressure drop, and entrainment in the gas. Check for excessive weeping.

Answer: D = 1.174 m; $_\Delta P$ = 386 Pa/tray

(b) Estimate the tray efficiency corrected for entrainment for the design reported in part (a).

Answer: E_{OG} = 0.60

4.25[c, d]. Sieve-tray column for methanol stripping; effect of hole size

Repeat Problem 4.24, but changing the perforation size to 4.5 mm, keeping everything else constant.

Answer: $_\Delta P$ = 686 Pa/tray; E_{OG} = 0.793

4.26[c, d]. Design of a sieve-tray column for butane absorption

A gas containing methane, propane, and n-butane is to be scrubbed countercurrently in a sieve-tray tower with a hydrocarbon oil to absorb principally the butane. It is agreed to design a tray for the circumstances existing at the bottom of the tower, where the conditions are

Temperature = 310 K   Pressure = 350 kPa
Liquid:
   Rate = 0.50 kmol/s   Average molecular weight = 150
   Density = 850 kg/m^3 Viscosity = 1.6 cP

Problems 273

Surface tension = 25 dyn/cm
Diffusivity = 1.14×10^{-5} cm^2/s (est.)   Foaming factor = 0.9
Vapor:
Rate = 0.3 kmol/s   Composition = 86% CH_4, 12% C_3H_8, 2% C_4H_{10}
Viscosity = 113 µP (est.)   Diffusivity = 0.035 cm^2/s (est.)

The system obeys Raoult's law; the vapor pressure of n-butane at 310 K is 3.472 bar (Reid et al., 1987).

(a) Design a suitable cross-flow sieve tray for such a tower. According to Bennett and Kovak (2000), the optimal value of the ratio A_h/A_a is that which yields an orifice Froude number, $Fr_o = 0.5$. Design for the optimal value of d_o on an equilateral-triangular pitch 12 mm between hole centers, punched in stainless-steel sheet metal 2 mm thick. Use a weir height of 50 mm. Design for a 75% approach to the flood velocity. Report details respecting tower diameter, tray spacing, weir length, gas-pressure drop, and entrainment in the gas.

Answer: $D = 2.73$ m

(b) Estimate the tray efficiency corrected for entrainment for the design reported in part (a).

4.27[c, d]. Design of a sieve-tray column for ammonia absorption

A process for making small amounts of hydrogen by cracking ammonia is being considered, and residual uncracked ammonia is to be removed from the resulting gas. The gas will consist of H_2 and N_2 in the molar ratio 3:1, containing 3% NH_3 by volume. The ammonia will be removed by scrubbing the gas countercurrently with pure liquid water in a sieve-tray tower. Conditions at the bottom of the tower are

Temperature = 303 K   Pressure = 200 kPa
Liquid:
Rate = 6.0 kg/s   Average molecular weight = 18
Density = 996 kg/m^3 Viscosity = 0.9 cP
Surface tension = 68 dyn/cm
Diffusivity = 2.42×10^{-5} cm^2/s (est.)   Foaming factor = 1.0
Vapor:
Rate = 0.7 kg/s
Viscosity = 113 µP (est.)   Diffusivity = 0.230 cm^2/s (est.)

For dilute solutions, NH_3–H_2O follows Henry's law, and at 303 K the slope of the equilibrium curve is $m = 0.85$ (Treybal, 1980).

(a) Design a suitable cross-flow sieve tray for such a tower. Take $d_o = 4.75$ mm on an equilateral-triangular pitch 12.5 mm between hole centers, punched in stainless-steel sheet metal 2 mm thick. Use a weir height of 40 mm. Design for an 80% approach to the flood velocity. Report details respecting tower diameter, tray spacing, weir length, gas-pressure drop, and entrainment in the gas. Check for excessive weeping.

Answer: $D = 0.775$ m

274 Equipment for Gas–Liquid Mass-Transfer Operations

(b) Estimate the tray efficiency corrected for entrainment for the design reported in part (a).

Answer: E_{MGE} = 0.847

4.28[c,d]. Design of a sieve-tray column for toluene–methylcyclohexane distillation

A sieve-tray tower is to be designed for distillation of a mixture of toluene and methylcyclohexane. The circumstances which are to be used to establish the design are

Temperature = 380 K   Pressure = 98.8 kPa
Liquid:
    Rate = 4.8 mol/s   Composition = 48.0 mol% toluene
    Density = 726 kg/m^3 Viscosity = 0.22 cP
    Surface tension = 16.9 dyn/cm
    Diffusivity = 7.08 × 10^{-5} cm^2/s (est.)   Foaming factor = 0.80

Vapor:
    Rate = 4.54 mol/s   Composition = 44.6 mol% toluene
    Density = 2.986 kg/m^3 Viscosity = 337 μP (est.)
    Diffusivity = 0.0386 cm^2/s (est.)

The equilibrium data at this concentration indicate that m = 1.152.

(a) Design a suitable cross-flow sieve tray for such a tower. Take d_o = 4.8 mm on an equilateral-triangular pitch 12.7 mm between hole centers, punched in stainless-steel sheet metal 2 mm thick. Use a weir height of 50 mm. Design for a 60% approach to the flood velocity. Report details respecting tower diameter, tray spacing, weir length, gas-pressure drop, and entrainment in the gas. Check for excessive weeping.

Answer: D = 0.6 m

(b) Estimate the tray efficiency corrected for entrainment for the design reported in part (a).

Answer: E_{MGE} = 0.71

REFERENCES

Bennett, D. L., R. Agrawal, and P. J. Cook, *AIChE J.*, **29**, 434–442 (1983).
Bennett, D. L., A. S. Kao, and L. W. Wong, *AIChE J.*, **41**, 2067–2082 (1995).
Bennett, D. L., and K. W. Kovak, *Chem. Eng. Progr.*, **96** (5), 20 (2000).
Bennett, D. L., D. N. Watson, and M. A. Wiescinski, *AIChE J.*, **43**, 1611–1626 (1997).
Billet, R. *Packed Column Analysis and Design*, Ruhr-University at Bochum, Germany (1989).
Billet, R., and M. Schultes, *Chem. Eng. Technol.*, **14**, 89–95 (1991a).

References

Billet, R., and M. Schultes, *Beitrage zur Verfahrens–und Umwelt–Technik*, Ruhr-University at Bochum, Germany, 88–106 (1991b).

Billet, R., and M. Schultes, *Packed Towers in Processing and Environmental Technology*, VCH, New York (1995).

Billet, R., and M. Schultes, *Trans IChemE*, **77**, 498–504 (1999).

Cussler, E. L., *Diffusion*, 2nd ed., Cambridge University Press, New York (1997).

Davis, M. L., and D. A. Cornwell, *Introduction to Environmental Engineering*, 3rd ed. McGraw-Hill, New York (1998).

Eckenfelder, W. W., Jr., *Industrial Water Pollution Control*, 3rd ed., McGraw-Hill, Boston (2000).

Fair, J. R., *Petro/Chem. Eng.*, **33**, 211–218 (1961).

Gerster, J. A. et al., Tray Efficiencies in Distillation Columns, *Final Report from the University of Delaware*, AIChE, New York (1958).

Govindarao, V. M. H., and G. F. Froment, *Chem. Eng. Sci.*, **41**, 533 (1986).

Hughmark, G. A., *Ind. Eng. Chem. Process Des. Dev.*, **6**, 218 (1967).

Hulswitt, C., and J. A. Mraz, *Chem Eng.*, **79**, 80 (1972).

Katayama, H., and T. Imoto, *J. Chem. Soc. Japan*, **9**, 1745 (1972).

Kister, H. Z., *Distillation Design*, McGraw-Hill, New York (1992).

Leibson, I. et al., *AIChE J.*, **2**, 296 (1956).

Lockett, M. J., *Distillation Tray Fundamentals*, Cambridge University Press, New York (1986).

Ludwig, E. E., *Applied Process Design for Chemical and Petrochemical Plants*, vol. **2**, 2nd ed., Gulf Publishing Company, Houston, TX (1979).

Mueller, G. E., *Powder Technol.*, **72**, 269 (1992).

Mueller, G. E., *AIChE J.*, **45**, 2458–2460 (1999).

Oliver, E. D., *Diffusional Separation Processes: Theory, Design, and Evaluation*, Wiley, New York (1966).

Perry, R. H., and C. H. Chilton (eds.), *Chemical Engineers Handbook*, 5th ed., McGraw-Hill, New York (1973).

Petrick, M., *U. S. Atomic Energy Comm.*, **ANL-658, 1** (1962).

Reid, R. C., J. M. Prausnitz, and B. E. Poling, *The Properties of Gases and Liquids*, 4th ed., McGraw-Hill, Boston (1987).

Ruivo, R., A. Paiva, and P. C. Simoes, *Chem. Eng. Process*, **45**, 224–231 (2006).

Schultes, M., *Chem. Eng. Res. Design*, **81**, 48–57 (2003).

Seader, J. D., and E. J. Henley, *Separation Process Principles*, 2nd ed.,Wiley, New York (2006).

Seader, J. D., and E. J. Henley, *Separation Process Principles*, Wiley, New York (1998).

Smith, J. M., H. C. Van Ness, and M. M. Abbott, *Introduction to Chemical Engineering Thermodynamics*, 5th ed., McGraw-Hill, New York (1996).

Stichlmair, J., J. L. Bravo, and J. R. Fair, *Gas Sep. Purif.*, **3**, 19–28 (1989).

Treybal, R. E., *Mass-Transfer Operations*, 3rd ed., McGraw-Hill, New York (1980).

Wankat, P. C. *Equilibrium Staged Separations*, Elsevier, New York (1988).

5

Absorption and Stripping

5.1 INTRODUCTION

Gas absorption is an operation in which a gas mixture is contacted with a liquid for the purpose of preferentially dissolving one or more components of the gas into the liquid. For example, the gas from by-product coke ovens is washed with water to remove ammonia, and again with an oil to remove benzene and toluene vapors. Objectionable hydrogen sulfide is removed from such a gas, or from naturally occurring hydrocarbon gases by washing with various alkaline solutions in which it is absorbed. Acid-rain precursor sulfur dioxide is removed from stack gases by scrubbing the gaseous effluent with an alkaline aqueous solution.

The operations described above require mass transfer of a substance from the gas stream to the liquid. When mass transfer occurs in the opposite direction (i.e., from the liquid to the gas) the operation is called *desorption, or stripping*. For example, the benzene and toluene are removed from the absorption oil mentioned above by contacting the liquid solution with steam, whereupon the aromatic vapors enter the gas stream and are carried away, and the absorption oil can be used again. Since the principles of both absorption and desorption are the same, we can study both operations simultaneously.

Absorption and stripping are usually conducted in packed columns or in trayed towers. Packed columns are preferred when (1) the required column diameter is less than 60 cm; (2) the pressure drop must be low, as for vacuum service; (3) corrosion considerations favor the use of ceramic or polymeric materials; and/or (4) low liquid holdup is desirable. Trayed towers are preferred when (1) the liquid/gas ratio is very low and (2) frequent cleaning is required. If there is no overriding consideration, cost is the major factor to be considered when choosing between packed columns and trayed towers for absorption or stripping.

Principles and Applications of Mass Transfer: The Design of Separation Processes for Chemical and Biochemical Engineering, Fourth Edition. Jaime Benitez.
© 2023 John Wiley & Sons, Inc. Published 2023 by John Wiley & Sons, Inc.

278   **Absorption and Stripping**

Absorption and stripping are technically mature separation operations. Design procedures are well developed for both packed columns and tray towers, and commercial processes are common. In most applications, the solutes are contained in gaseous effluents from chemical reactors. Passage of strict environmental standards with respect to air pollution by emission of noxious gases from industrial sources has greatly increased the use of gas absorbers (also known as *scrubbers*) in the past decades.

As was shown in Chapter 3, the fraction of a component absorbed in a countercurrent cascade depends on the number of equilibrium stages and the absorption factor, $A = L_S/mV_S$, for that component. If the value of A is greater than 1.0, any degree of absorption can be achieved: the larger the value of A, the fewer the number of stages required to absorb a desired fraction of the solute. However, very large values of A can correspond to absorbent flow rates that are larger than necessary. From an economic standpoint, the value of A for the main species to be absorbed should be in the range of 1.25 to 2.0, with 1.4 being a frequently recommended value. For a stripper, the stripping factor, $S = 1/A$, is crucial. The optimum value of S is also in the vicinity of 1.4 (Seader and Henley, 1998).

5.2 COUNTERCURRENT MULTISTAGE EQUIPMENT

Your objectives in studying this section are to be able to:

1. Estimate the number of real trays required for a sieve-tray absorber or stripper by graphical stepwise construction on an xy diagram, incorporating in the analysis the Murphree tray efficiency.
2. Estimate the number of real trays required for a sieve-tray absorber or stripper by algebraic methods, for the case of dilute solutions and equilibrium described by Henry's law.

5.2.1 Graphical Determination of the Number of Ideal Trays

Tray towers and similar devices bring about stepwise contact of the liquid and gas and are, therefore, countercurrent multistage cascades. On each tray of a sieve-tray tower, for example, the gas and liquid are brought into intimate contact and separated—somewhat in the manner of Figure 3.24—and the tray thus constitutes a stage. Few of the tray devices described in Chapter 4 actually provide the parallel flow on each tray as shown in Figure 3.24. Nevertheless, it is convenient to use the latter as an arbitrary standard for design and for assessment of the performance of actual trays regardless of their method of operation. For this purpose, a *theoretical* or *ideal* tray is defined as one where the average composition of all the gas leaving the tray is in equilibrium with the average composition of all the liquid leaving the tray.

5.2 Countercurrent Multistage Equipment

The number of ideal trays required to bring about a given change in composition of the liquid or the gas, for either absorbers or strippers, can be determined graphically on an *XY* diagram in the manner of Figure 3.25, with each step representing an ideal stage. The nearer the operating line is to the equilibrium curve, the more steps are required. Should the two curves touch at any time, corresponding to a minimum L_S/V_S ratio, the number of steps will be infinite. The construction for strippers is the same, except, of course, that the operating line lies below the equilibrium curve. The steps can be equally constructed on diagrams plotted in terms of any concentration units, such as mol fractions or partial pressures.

5.2.2 Tray Efficiencies and Real Trays by Graphical Methods

Methods for estimating the Murphree tray efficiency corrected for entrainment, E_{MGE}, for sieve trays were discussed in detail in Chapter 4. For a given stripper or absorber, these allow estimation of the tray efficiency as a function of fluid compositions and flow rates, as they vary from one end of the tower to the other. Usually, it is sufficient to make such computations at only three or four locations and then proceed as in Figure 5.1. The dashed line is drawn on an *xy* diagram between the operating line and the equilibrium curve at a fractional vertical distance from the operating line equal to the prevailing Murphree gas efficiency, E_{MGE}. Thus, the value of E_{MGE} for the bottom tray is the ratio of the line segments AB/AC. Since the dashed line then represents the real effluent compositions from the trays, it is used instead of the equilibrium curve to complete the tray construction, which now provides the number of real trays. Notice that on the *xy* diagram, the operating line is usually curved.

5.2.3 Dilute Mixtures

Where both operating line and equilibrium curve are straight, the number of ideal trays can be determined without recourse to graphical methods. This will frequently be the case for relatively dilute gas and liquid mixtures, such as those described in Example 5.1. In this case, Henry's law usually applies; therefore, the equilibrium relation in terms of mol fractions is a straight line. If the quantity of

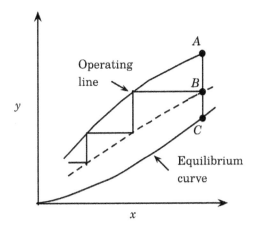

Figure 5.1 Use of Murphree efficiencies for an absorber.

280 Absorption and Stripping

gas absorbed is small, the total flow of liquid entering and leaving the absorber remains substantially constant; therefore, $L_0 \approx L_N \approx L$, and, similarly, the total flow of gas remains relatively constant at V. An operating line plotted in terms of mol fractions then will be virtually straight. For such cases, the absorption factor, $A = L_S/mV_S \approx L/mV$. Small variations in A from one end of the tower to the other due to changing L/V as a result of absorption or stripping can be roughly allowed for by using the geometric average of the values of A at top and bottom (Hachmuth, 1951). The following modified versions of the Kremser equations can be used to estimate the number of equilibrium stages required for a given separation.

For transfer from L to V (stripping of L)

$A = L/mV = 1.0$

$$N = \frac{x_0 - x_N}{x_N - \dfrac{y_{N+1}}{m}} \tag{5-1}$$

$A \neq 1.0$

$$N = \frac{\ln\left[\dfrac{x_0 - \left(y_{N+1}/m\right)}{x_N - \left(y_{N+1}/m\right)}\left(1 - A\right) + A\right]}{\ln\left(1/A\right)} \tag{5-2}$$

For transfer from V to L (absorption into L)

$A = L/mV = 1.0$

$$N = \frac{y_{N+1} - y_1}{y_1 - mx_0} \tag{5-3}$$

$A \neq 1.0$

$$N = \frac{\ln\left[\dfrac{y_{N+1} - mx_0}{y_1 - mx_0}\left(1 - \dfrac{1}{A}\right) + \dfrac{1}{A}\right]}{\ln\left(A\right)} \tag{5-4}$$

When the Murphree efficiency is constant for all trays, and under conditions such that the operating line and equilibrium curve are straight, the overall tray efficiency can be computed and the number of real trays can be determined analytically from

$$\mathbf{E}_O = \frac{\text{equilibrium trays}}{\text{real trays}} = \frac{\ln\left[1 + \mathbf{E}_{MGE}\left(\dfrac{1}{A} - 1\right)\right]}{\ln\left(\dfrac{1}{A}\right)} \tag{5-5}$$

5.2 Countercurrent Multistage Equipment

Example 5.1 Number of Real Sieve Trays in an Absorber

A sieve-tray tower is being designed for a gas absorption process. The entering gas contains 1.8 mol% of A, the component to be absorbed. The gas should leave the tower containing no more than 0.1 mol% of A. The liquid to be used as an absorbent initially contains 0.01 mol% of A. The system obeys Henry's law with $m = y_i / x_i = 1.41$. At the bottom of the tower, the molar liquid to gas ratio is $L/V = 2.115$, while at the other extreme $L/V = 2.326$. For these conditions, it has been found that the Murphree efficiency is constant at $\mathbf{E}_{MGE} = 0.65$.

(a) Estimate the number of trays required.

(b) Based on criteria presented in Chapter 4, the tower diameter must be 1.5 m. Estimate the tower height.

Solution

(a) Calculate the absorption factor from the molar liquid-to-gas ratio given at the extremes of the column and the slope of the equilibrium curve: $A_1 = 2.115/1.41 = 1.50$; $A_2 = 2.326/1.41 = 1.65$. As recommended above, a geometric average is calculated: $A = (A_1 A_2)^{0.5} = 1.573$. From the concentration data given, $y_{N+1} = 0.018$, $y_1 = 0.001$, $x_0 = 0.0001$. From equation (5-4), $N = 4.647$ ideal stages. From equation (5-5), $\mathbf{E}_O = 0.596$. Therefore, the number of real trays required = $4.647/0.596 = 7.79$. Use 8 trays since it is not possible to specify a fractional number of trays. We can back-calculate the actual concentration of the gas leaving the tower when 8 trays are used. For an overall efficiency of 0.596 and an actual number of trays of 8, equation (5-5) yields $N = 4.784$ ideal trays. Solving equation (5-4), $y_1 = 0.094\%$, which satisfies the requirement that the gas exit concentration should not exceed 0.1%.

(b) For a tower diameter of 1.5 m, Table 4.3 recommends a plate spacing of 0.6 m. Therefore, the tower height will be $Z = (8)(0.6) = 4.8$ m.

Example 5.2 Sieve-Tray Absorber for Recovery of Benzene Vapors

Consider the absorber of Example 3.11. Assume that the wash oil is n-tetradecane ($C_{14}H_{30}$). The absorber will be a cross-flow sieve-tray tower with $d_o = 4.5$ mm on an equilateral-triangular pitch 12 mm between hole centers, punched in stainless steel sheet metal 2 mm thick, with a weir height of 50 mm. Estimate the number of real trays required, the dimensions of the absorber, and the power required to pump the gas and the liquid through the tower. Design for a 65% approach to the flooding velocity.

Solution

To determine the tower diameter, consider the conditions at the bottom of the absorber where the maximum gas and liquid flow rates are found. From Examples 3.11 and 3.15:

282   **Absorption and Stripping**

Gas in:

Rate = 1.0 m³/s = 0.0406 kmol/s at 300 K and 1 atm
Benzene content: $y_{N+1} = 0.074$ mol fraction, $Y_{N+1} = 0.080$ kmol/kmol
Average molecular weight, $M_G = 32.63$ kg/kmol
Density, $\rho_G = 1.325$ kg/m³ (from the ideal gas law)
Mass flow rate, $V' = 1.325$ kg/s
Viscosity, $\mu_G = 166$ µP (estimated by the Lucas method)
Diffusivity, $D_G = 0.096$ cm²/s (Appendix A)
Slope of equilibrium curve = 0.136

Liquid out:

Rate = 0.00888 kmol/s (ideal solution of benzene in tetradecane)
Benzene content: $x_N = 0.324$ mol fraction, $X_N = 0.48$ kmol/kmol
Average molecular weight, $M_L = 159.12$ kg/kmol
Mass flow rate, $L' = 1.413$ kg/s
Density, $\rho_L = 780$ kg/m³ (estimated from the pure-component densities)
Viscosity, $\mu_L = 1.41$ cP (estimated from the pure-component viscosities)
Diffusivity, $D_L = 0.93 \times 10^{-5}$ cm²/s [estimated from equation (1-56) applied to
an ideal solution, $\Gamma = 1.0$]
Surface tension, $\sigma = 24$ dyne/cm (estimated)
Foaming factor, $F_F = 0.9$

Run the Sieve-Tray Design Program of Appendix E using these data. The program
converges to a tower diameter $D = 1.036$ m, and a tray spacing $t = 0.6$ m.
When convergence is achieved, the results at the bottom of the tower are

$$f = 65\% \qquad \Delta P_G = 375 \text{ Pa/tray}$$

$$\text{Fr}_o = 0.966 \qquad E = 0.0169$$

$$E_{OG} = 0.705 \qquad E_{MGE} = 0.77$$

Consider now the conditions at the top of the 1.036-m diameter tower:

Gas out:

Rate = 0.0381 kmol/s at 300 K and 1 atm
Benzene content; $y_1 = 0.0119$ mol fraction, $Y_1 = 0.012$ kmol/kmol
Average molecular weight, $M_G = 29.58$ kg/kmol
Density, $\rho_G = 1.201$ kg/m³ (from the ideal gas law)
Mass flow rate, $V' = 1.126$ kg/s
Viscosity, $\mu_G = 180$ µP (estimated by the Lucas method)
Diffusivity, $D_G = 0.096$ cm²/s (Appendix A)

Liquid in:

Rate = 0.00630 kmol/s (ideal solution of benzene in tetradecane)
Benzene content; $x_0 = 0.0476$ mol fraction, $X_0 = 0.05$ kmol/kmol
Average molecular weight, $M_L = 192.7$ kg/kmol

5.2 Countercurrent Multistage Equipment 283

Mass flow rate, $L' = 1.214$ kg/s
Density, $\rho_L = 765$ kg/m^3 (estimated from the pure-component densities)
Viscosity, $\mu_L = 1.98$ cP (estimated from the pure-component viscosities)
Diffusivity, $D_L = 0.83 \times 10^{-5}$ cm^2/s [estimated from equation (1-56) applied to an ideal solution, $\Gamma = 1.0$]
Surface tension, $\sigma = 22$ dyne/cm (estimated)
Foaming factor, $F_F = 0.9$

Run the Sieve-Tray Design Program of Appendix E using these data. Try different values of the fractional approach to flooding, f, until the program converges to the specified tower diameter $D = 1.036$ m. When convergence is achieved, the results at the top of the tower are

$f = 59.6\%$ $\qquad \Delta P_G = 348$ Pa/tray

$\text{Fr}_o = 0.847$ $\qquad E = 0.0146$

$E_{OG} = 0.656$ $\qquad E_{MGE} = 0.730$

From these results, we conclude that the plate design is appropriate from the point of view of pressure drop, entrainment, weeping, and plate efficiency. The next step in the design is to determine the number of real stages required. Since the plate efficiency does not change much from one end of the tower to the other, we will assume a constant value taken as the arithmetic average of the efficiency values at the tower ends: $E_{MGE} = (0.77 + 0.73)/2 = 0.75$.

To determine the number of real stages, we must generate the operating line and the equilibrium distribution curve on an xy diagram. For this system, Raoult's law applies; therefore the equilibrium relation is straight on the xy diagram ($y_i = 0.136\, x_i$). However, the operating line will not be straight since we are dealing with fairly concentrated solutions. From Example 3.11, the operating line can be written in terms of mol ratios as

$$\frac{Y_{N+1} - Y}{X_N - X} = \frac{L_S}{V_S}$$

After some algebraic manipulation, this equation can be written in terms of mol fractions as

$$y = \frac{C + (R - C)\, x}{(1 + C) + (R - C - 1)\, x} \quad \text{where } R = \frac{L_S}{V_S} \quad C = Y_{N+1} - R X_N$$

Substituting numerical values gives

$$y = \frac{0.004 + 0.154x}{1.004 - 0.846x} \quad \text{operating line}$$

Figure 5.2 shows the xy diagram for this example. Notice how the operating line curves upward. The broken line is a pseudoequilibrium curve located between the operating and the equilibrium lines, at 75% (the average value of $\mathbf{E}_{MGE}$) of the distance between the lines, measured from the operating line. The pseudoequilibrium curve is used to complete the tray construction, which now yields the number of real trays. As Figure 5.2 shows, in this case five real trays are not enough to achieve the desired separation, while six real trays accomplish more than the specified 85% benzene recovery. We choose six trays. For a tray spacing of 0.6 m, the total height of the mass-transfer region of the absorber is calculated as $Z = 3.6$ m.

A conservative estimate of the total gas-pressure drop can be obtained assuming that the pressure drop through each tray remains constant at the value estimated for the bottom of the absorber where the gas and liquid velocities are the highest. Therefore, the total gas-pressure drop is about $\Delta P_G = 375 \times 6 = 2250$ Pa. Assuming a mechanical efficiency of the motor-fan system $E_m = 60\%$, the power required to force the gas through the tower is approximately

$$W_G = \frac{Q_G \Delta P_G}{E_m} = \frac{1.0 \times 2250}{0.60} = 3750 \text{ W}$$

The liquid must be pumped to the top of the absorber, and gravity will force it through the tower. The power required to pump the liquid to the top of the absorber is

$$W_L = \frac{L_0' g Z}{E_m} = \frac{1.214 \times 9.8 \times 3.6}{0.60} = 71.4 \text{ W}$$

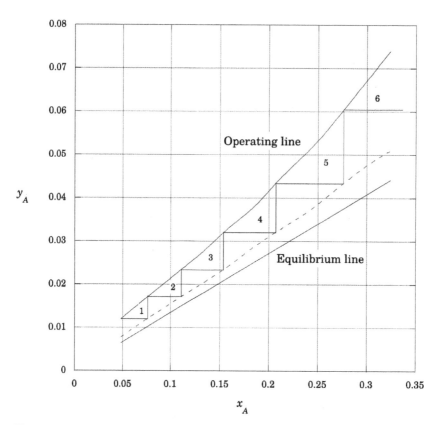

Figure 5.2 Example 5.2: absorber real stages on xy diagram.

5.3 COUNTERCURRENT CONTINUOUS-CONTACT EQUIPMENT

Your objectives in studying this section are to be able to:

1. Define the concepts: number of mass-transfer units (NTU), and height of a mass-transfer unit (HTU).
2. Estimate the required packed height for a continuous-contact absorber or stripper in terms of the number and height of mass-transfer units.

Countercurrent packed towers operate in a different manner from plate towers in that the fluids are in contact continuously in their path through the tower, rather than intermittently. Thus, in a packed bed the liquid and gas compositions change continuously with height of packing. Every point on an operating line therefore represents conditions found somewhere in the tower; whereas for tray towers, only the isolated points on the operating line corresponding to the trays have real significance.

The development of the design equation for a countercurrent packed tower absorber or stripper begins with a differential mass balance of component A in the gas phase, in a manner similar to that of Example 2.12. However, this time we do not restrict the analysis to dilute solutions nor to constant molar velocity. If only component A is transferred ($\Psi_{A,G} = \Psi_{A,L} = 1.0$), considering the fact that the gas-phase molar velocity will change along the column, and that F-type mass-transfer coefficients are required for concentrated solutions, the mass balance is

$$\frac{d(Vy)}{Sdz} = F_G a_h \ln\left[\frac{1-y_i}{1-y}\right] \tag{5-6}$$

where

S = cross-sectional area of the tower

a_h = effective specific area of the packing

Both V and y vary from one end of the tower to the other, but V_S does not. Therefore,

$$d(Vy) = d\left(\frac{V_S y}{1-y}\right) = \frac{V_S dy}{(1-y)^2} = \frac{Vdy}{1-y} \tag{5-7}$$

Substituting in equation (5-6), rearranging, and integrating yields

$$Z = \int_{y_2}^{y_1} \frac{G_{My} dy}{F_G a_h (1-y) \ln\left[\frac{1-y_i}{1-y}\right]} \tag{5-8}$$

where Z is the total packed height and G_{My} is the gas-phase molar velocity. The value of y_i can be found by the methods of Chapter 3. For any value of (x, y) on the

Absorption and Stripping

operating curve plotted in terms of mol fractions, a value of (x_i, y_i) on the equilibrium curve can be found using

$$\frac{1-y_i}{1-y} = \left(\frac{1-x}{1-x_i}\right)^{F_L/F_G} \tag{5-9}$$

This provides the local y and y_i for use in equation (5-8). Equation (5-8) can then be integrated numerically or graphically after plotting the integrand as ordinate vs. y as abscissa.

However, it is more customary to proceed as follows. Since

$$y - y_i = \left(1-y_i\right)-\left(1-y\right) \tag{5-10}$$

the numerator and denominator of the integral of equation (5-8) can be multiplied respectively by the right- and left-hand sides of equation (5-10) to provide

$$Z = \int_{y_2}^{y_1} \frac{G_{My}\left(1-y\right)_{iM} dy}{F_G a_h \left(1-y\right)\left(y-y_i\right)} \tag{5-11}$$

where $(1-y)_{i,M}$ is the logarithmic mean of $(1-y)$ and $(1-y_i)$.

It has been observed that the ratio $G_{My}/F_G a_h$ is very much more constant along the tower than either G_{My} or $F_G a_h$, and in many cases may be considered constant within the accuracy of the available data. Define the *height of a gas-phase transfer unit* H_{tG} (HTU) as

$$H_{tG} \equiv \text{HTU} = \frac{G_{My}}{F_G a_h} \tag{5-12}$$

(Since there will be some variation of the molar gas velocity, the mass-transfer coefficients, and the effective specific area along the tower, estimate these values at both ends of the tower and use arithmetic averages to estimate the HTU and the local values of y_i.) Equation (5-11) becomes

$$Z = H_{tG}\int_{y_2}^{y_1} \frac{\left(1-y\right)_{iM} dy}{\left(1-y\right)\left(y-y_i\right)} = H_{tG}N_{tG} = \text{HTU} \times \text{NTU} \tag{5-13}$$

where the integral is called the *number of gas transfer units, N_{tG}* (NTU). Equation (5-13) can be further simplified by substituting the arithmetic average for the logarithmic average

$$\left(1-y\right)_{iM} \simeq \frac{\left(1-y_i\right)+\left(1-y\right)}{2} \tag{5-14}$$

which involves little error. Then

$$N_{tG} \equiv \text{NTU} \simeq \int_{y_2}^{y} \frac{dy}{y-y_i} + \frac{1}{2}\ln\left[\frac{1-y_2}{1-y_1}\right] \tag{5-15}$$

5.3 Countercurrent Continuous-Contact Equipment 287

which makes for simpler numerical or graphical integration.

The above relationships all have their counterparts in terms of liquid concentrations, derived in exactly the same way

$$Z = H_{tL}N_{tL} \tag{5-16}$$

$$H_{tL} = \frac{G_{Mx}}{F_L a_h} \tag{5-17}$$

$$N_{tL} \simeq \int_{x_2}^{x_1} \frac{dx}{x_i - x} + \frac{1}{2}\ln\left[\frac{1 - x_1}{1 - x_2}\right] \tag{5-18}$$

where

G_{Mx} = the liquid-phase molar velocity
H_{tL} = the *height of a liquid-phase transfer unit* (HTU)
N_{tL} = the *number of liquid-phase transfer units* (NTU)

Example 5.3 Packed-Tower Absorber for Recovery of Benzene Vapors

Consider the absorber of Example 3.11 and Example 5.2. The adsorber will be packed with 50-mm metal Hiflow rings. Estimate the dimensions of the absorber, and the power required to pump the gas and the liquid through the tower. Design for a gas-phase pressure drop of 150 Pa/m of packed height.

Solution

To determine the tower diameter, consider the conditions at the bottom of the absorber where the maximum gas and liquid flow rates are found. Run the Packed-Tower Design Program of Appendix D using the data from Example 5.2 and packing parameters from Chapter 4. From Table 4.1 for 50-mm metal Hiflow rings:

$$C_{S,T} = 2.702 \qquad a = 92.3 \text{ m}^2/\text{m}^3 \qquad \varepsilon = 0.977 \qquad C_h = 0.876 \qquad C_{F,T} = 1.626$$

From Table 4.2: $C_L = 1.168 \quad C_V = 0.408 \quad C_p = 0.421$

For a pressure drop of 150 Pa/m, the program converges to a tower diameter $D = 0.768$ m. When convergence is achieved, the results at the bottom of the tower are

$$f = 79.4\%, \ a_h = 62.45 \text{ m}^{-1}, \ G_{My} = 87.6 \text{ mol/m}^2 \cdot \text{s}$$

$$k_y = 2.372 \text{ mol/m}^2 \cdot \text{s}, \ k_L = 9.10 \times 10^{-5} \text{ m/s}$$

From equations (2-6) and (2-11),

$$\begin{aligned} F_G &= k_y y_{B,M} = k_y \left(1 - y\right)_{i,M} \\ F_L &= k_x x_{B,M} = k_x \left(1 - x\right)_{i,M} = k_L c_L \left(1 - x\right)_{i,M} \end{aligned} \tag{5-19}$$

288   **Absorption and Stripping**

We do not know yet the interfacial concentrations along the column; therefore, we cannot evaluate the terms $(1 - y)_{i,M}$ and $(1 - x)_{i,M}$. However, a good estimate of these terms is

$$\left(1-y\right)_{iM} \approx \left(1-y_1\right) = 0.926$$
$$\left(1-x\right)_{iM} \approx \left(1-x_1\right) = 0.676 \tag{5-20}$$

The liquid molar concentration is $c_L = \rho_L/M_L = 780/159.12 = 4.902$ kmol/m^3. Substituting in equation (5-19), $F_G = 2.200$ mol/m^2·s, $F_L = 0.302$ mol/m^2·s:

$$H_{tG} = \frac{G_{My}}{F_G a_h} = \frac{87.6}{2.20 \times 62.45} = 0.638 \text{ m}$$

Now, we consider the conditions at the top of the absorber. Run the Packed-Tower Design Program of Appendix D using the data for that part of the column from Example 5.2. Try different values of the pressure drop per unit packed height until the program converges to the tower diameter of 0.768 m already selected. Convergence is achieved at a gas-pressure drop of 116 Pa/m. The results at the top of the tower are

$$f = 71.9\%, \quad a_h = 53.8 \text{ m}^{-1}, \quad G_{My} = 82.2 \text{ mol/m}^2 \cdot \text{s}$$

$$k_y = 2.095 \text{ mol/m}^2 \cdot \text{s}, \quad k_L = 8.15 \times 10^{-5} \text{ m/s}$$

The liquid molar concentration is $c_L = \rho_L/M_L = 765/192.7 = 3.970$ kmol/m^3. Substituting in equation (5-19), $F_G = 2.070$ mol/m^2·s, $F_L = 0.308$ mol/m^2·s:

$$H_{tG} = \frac{G_{My}}{F_G a_h} = \frac{82.2}{2.07 \times 53.80} = 0.738 \text{ m}$$

The average HTU value is $H_{tG} = 0.688$ m. The average mass-transfer coefficients are $F_G = 2.135$ mol/m^2·s and $F_L = 0.305$ mol/m^2·s.

The next step is to estimate the number of gas transfer units from (5-15). The operating curve equation for this system in terms of mol fractions was developed in Example 5.2 and is

$$y = \frac{0.004 + 0.154x}{1.004 - 0.846x} \quad \text{operating line}$$

Figure 5.3 presents a Mathcad program to estimate N_{tG}. Appendix 5.1 presents the corresponding Python solution. A vector of $(n+1)$ x-values from $x_2 = 0.0476$ to $x_1 = 0.324$ is created. The operating curve equation is used to generate the corresponding y vector. The average values of F_G and F_L are supplied to the program. For each point (x, y) on the operating curve, equation (5-9) is solved for the interfacial gas-phase composition y_i (notice that $x_i = y_i/m$). The y and y_i-vectors are

5.3 Countercurrent Continuous-Contact Equipment

Enter data

$$L_S := 6.0 \, \frac{\text{mol}}{\text{s}} \qquad V_S := 38 \, \frac{\text{mol}}{\text{s}} \qquad m_e := 0.136 \qquad X_2 := 0.05 \quad Y_2 := 0.012$$

$$Y_1 := 0.08 \qquad X_1 := 0.481 \qquad y_1 := \frac{Y_1}{1+Y_1} = 0.074 \qquad y_2 := \frac{Y_2}{1+Y_2} = 0.012$$

$$x_2 := \frac{X_2}{1+X_2} = 0.048 \qquad x_1 := \frac{X_1}{1+X_1} = 0.325 \qquad R := \frac{L_S}{V_S} = 0.158 \qquad n := 20$$

$$C := Y_1 - R \cdot X_1 = 0.004 \qquad \Delta x := \frac{x_1 - x_2}{n} = 0.014 \quad j := 0 \dots n \qquad x_j := x_2 + j \cdot \Delta x$$

Generate the operating line

$$FG := 2.135 \, \frac{\text{mol}}{\text{m}^2 \cdot \text{s}} \qquad FL := 0.305 \, \frac{\text{mol}}{\text{m}^2 \cdot \text{s}} \qquad \overrightarrow{y := \frac{C+(R-C)\cdot x}{(1+C)+(R-C-1)\cdot x}}$$

Generate the vector of interfacial concentrations

Guess Values
$$y_{int} := \left(\frac{y_1 + y_2}{2} \right)$$

Constraints
$$\frac{1 - y_{int}}{1 - y} = \left[\frac{1 - x}{1 - \dfrac{y_{int}}{m_e}} \right]^{\frac{FL}{FG}}$$

Solver
$$yi(x, y) := \text{Find}\,(y_{int})$$

$$yint_j := yi\left(x_j, y_j\right)$$

Calculate the number of transfer units using cubic spline interpolation and numerical integration.

$$f := \overrightarrow{\frac{1}{y - yint}} \qquad vs := \text{cspline}\,(y, f) \qquad N_{tG} := \int_{y_0}^{y_n} \text{interp}\,(vs, y, f, \theta) \, d\theta = 9.023$$

Figure 5.3 Example 5.3: calculation of N_{tG}.

combined to generate the integrand in equation (5-15), $(y - y_i)^{-1}$. An interpolation formula relating the integrand to y is generated in the form of a cubic spline, which is then used to evaluate the integral in equation (5-15) numerically. The result is $N_{tG} = 9.02$ units. The total packed height is from equation (5-13),

$$Z = N_{tG} H_{tG} = 9.02 \times 0.688 = 6.21 \text{ m}$$

290   **Absorption and Stripping**

A conservative estimate of the total gas-pressure drop can be obtained assuming that the pressure drop through the tower remains constant at the value estimated for the bottom of the absorber where the gas and liquid velocities are the highest. Therefore, the total gas-pressure drop is approximately $\Delta P_G = 150 \times 6.21 = 932$ Pa. Assuming a mechanical efficiency of the motor-fan system $E_m = 60\%$, the power required to force the gas through the tower is

$$W_G = \frac{Q_G \Delta P_G}{E_m} = \frac{1.0 \times 932}{0.60} = 1553 \text{ W}$$

The liquid must be pumped to the top of the absorber, and gravity will force it through the tower. The power required to pump the liquid to the top of the absorber is

$$W_L = \frac{L_2' g Z}{E_m} = \frac{1.214 \times 9.8 \times 6.21}{0.6} = 123 \text{ W}$$

5.3.1 Dilute Solutions; Henry's Law

For dilute solutions and when Henry's law applies, the equilibrium curve and the operating lines are straight lines. Overall mass-transfer coefficients are convenient in this case. The expression for the height of packing can be written as

$$Z = H_{tOG} N_{tOG} \tag{5-21}$$

where
    Z = the total packed height
    H_{tOG} = the *overall height of a gas-phase transfer unit* (HTU)
    N_{tOG} = the *overall number of gas-phase transfer units* (NTU)

$$H_{tOG} = \frac{G_{My}}{K_y a_h} = H_{tG} + \frac{H_{tL}}{A} \tag{5-22}$$

$$N_{tOG} = \int_{y_2}^{y_1} \frac{dy}{y - y^*} + \frac{1}{2} \ln \left[\frac{1 - y_2}{1 - y_1} \right] \tag{5-23}$$

In equation (5.22), K_y is the gas-phase overall mass-transfer coefficient as defined in equation (3-24). For dilute solutions in both phases and equilibrium described by Henry's law, equation (5-23) becomes (Treybal, 1980)

$$N_{tOG} = \frac{\ln \left[\dfrac{y_1 - mx_2}{y_2 - mx_2} \left(1 - \dfrac{1}{A} \right) + \dfrac{1}{A} \right]}{1 - 1/A} \tag{5-24}$$

5.3 Countercurrent Continuous-Contact Equipment   291

The corresponding expressions for strippers are

$$Z = H_{tOL}N_{tOL} \qquad (5\text{-}25)$$

$$H_{tOL} = \frac{G_{Mx}}{K_x a_h} \qquad (5\text{-}26)$$

$$N_{tOL} = \frac{\ln\left[\dfrac{x_2 - y_1/m}{x_1 - y_1/m}(1-A)+A\right]}{1-A} \qquad (5\text{-}27)$$

where G_{Mx} is the liquid-phase molar velocity. Example 5.4 illustrates the application of these concepts to the design of an absorber.

Example 5.4 Packed Height of an Ethanol Absorber

Consider the ethanol absorber of Example 4.4, packed with 50-mm metal Hiflow rings. Estimate the packed height required to recover 97% of the alcohol, using pure water at a rate 50% above the minimum, and with a gas-pressure drop of 100 Pa/m. The system ethanol–CO_2–water at the given temperature and pressure obeys Henry's law with $m = 0.57$.

Solution
The conditions specified correspond to those of Example 4.4 where the tower diameter ($D = 0.938$ m) and the volumetric mass-transfer coefficients for both phases were evaluated. The amount of ethanol absorbed is $(180)(0.02)(0.97) = 3.5$ kmol/h. The inlet gas molar velocity is

$$G_{My1} = 180(4)/[(3600)\,(\pi)(0.938)^2] = 0.0724 \text{ kmol/m}^2\text{·s}$$

The outlet gas velocity is

$$G_{My2} = (180 - 3.5)(4)/[(3600)\,(\pi)(0.938)^2] = 0.0709 \text{ kmol/m}^2\text{-s}$$

The average molar gas velocity is

$$G_{My} = (G_{My1} + G_{My2})/2 = 0.0717 \text{ kmol/m}^2\text{·s}$$

The inlet liquid molar velocity is

$$G_{Mx2} = 151.5(4)/[(3600)(\pi)(0.938)^2] = 0.0607 \text{ kmol/m}^2\text{·s}$$

The outlet liquid molar velocity is

$$G_{Mx1} = (151.5 + 3.5)(4)/[(3600)\,(\pi)(0.938)^2] = 0.0625 \text{ kmol/m}^2\text{·s}$$

Calculate the absorption factor at both ends of the column:

$$A_1 = 0.0625 / \left[(0.57)(0.0724) \right] = 1.514$$

$$A_2 = 0.0607 / \left[(0.57)(0.0709) \right] = 1.502$$

Calculate the geometric average,

$$A = (A_1 \times A_2)^{1/2} = 1.508$$

For 97% removal of the ethanol, $y_2 \approx (0.03)(0.02) = 0.0006$. Since pure water is used, $x_2 = 0$. Substituting in equation (5-24), $N_{tOG} = 7.35$.

From Example 4.4, $k_y a_h = 0.111$ kmol/m^3·s; $k_L a_h = 0.00539$ s^{-1}. The total molar concentration in the liquid phase is

$$c \approx \rho_L / M_L = 986 / 18 = 54.78 \text{ kmol/m}^3.$$

Therefore, $k_x a_h = k_L a_h c = 0.295$ kmol/m^3-s.

The overall volumetric mass-transfer coefficient is given by

$$K_y a_h = \left[k_y a_h^{\ -1} + m / k_x a_h \right]^{-1} = 0.0914 \text{ kmol/m}^3 \cdot \text{ s}$$

From equation (5-22),

$$H_{tOG} = 0.0717 / 0.0914 = 0.785 \text{ m}$$

The packed height is given by equation (5-21),

$$Z = (7.35)(0.785) = 5.77 \text{ m}$$

5.4 THERMAL EFFECTS DURING ABSORPTION AND STRIPPING

Your objectives in studying this section are to be able to:

1. Combine material and energy balances with equilibrium considerations to determine the number of equilibrium stages required by a tray absorber or stripper when heat effects are important.
2. Combine material and energy balances with mass-transfer considerations to determine the height required by a packed-bed absorber or stripper when heat effects are important.

5.4 Thermal Effects During Absorption and Stripping 293

Many absorbers and strippers deal with dilute gas mixtures and liquid solutions, and it is satisfactory in these cases to assume that the operation is isothermal. But actually absorption operations are usually exothermic, and when large quantities of solute gas are absorbed to form concentrated solutions, the thermal effects cannot be ignored. If by absorption the temperature of the liquid is raised to a considerable extent, the equilibrium solubility of the solute will be appreciably reduced and the capacity of the absorber decreased (or else much larger flow rates of liquid will be required). For stripping, an endothermic process, the temperature of the liquid tends to fall. To consider thermal effects during absorption and stripping, energy balances must be combined with the material balances presented in Chapter 3.

Consider *adiabatic operation* of a countercurrent tray absorber represented by a cascade of ideal stages such as illustrated in Figure 3.24. The temperature of the streams leaving the absorber will generally be higher than the entering temperatures owing to the heat of solution. The design of such absorbers may be done numerically, calculating tray by tray from the bottom (gas entrance) to the top. The principle of an ideal tray, that the effluent streams from the tray are in equilibrium both with respect to temperature and composition, is utilized for each tray. Thus, total and solute balances around an envelope from the bottom of the absorber up to tray n are

$$L_n + V_{N+1} = L_N + V_{n+1} \tag{5-28}$$

$$L_n x_n + V_{N+1} y_{N+1} = L_N x_N + V_{n+1} y_{n+1} \tag{5-29}$$

from which L_n and x_n are computed. An energy balance around the same envelope is

$$L_n H_{L,n} + V_{N+1} H_{V,N+1} = L_N H_{L,N} + V_{n+1} H_{V,n+1} \tag{5-30}$$

where H represents in each case the molal enthalpy of the stream at its particular thermodynamic state.

The temperature of stream L_n can be obtained from equation (5-30). Stream V_n is then at the same temperature as L_n and in composition equilibrium with it. Equations (5-28) to (5-30) are then applied to tray $n-1$, and so forth. To get started, since usually only the temperatures of the entering streams L_0 and V_{N+1} are known, it is necessary to make an initial estimate of the temperature T_1 of the gas leaving V_1. An energy balance over the entire absorber,

$$L_0 H_{L,0} + V_{N+1} H_{V,N+1} = L_N H_{L,N} + V_1 H_{V,1} \tag{5-31}$$

is used to compute T_N, the temperature of the liquid leaving at the bottom of the tower. The estimate of T_1 is checked when the calculations reach the top tray, and if necessary, the entire computation is repeated. The method is best illustrated by Example 5.5.

294   **Absorption and Stripping**

Example 5.5 Tray Tower for Adiabatic Pentane Absorption

One kmol/s of a gas consisting of 75% methane and 25% n-pentane at 300 K and 1 atm is to be scrubbed with 2 kmol/s of a nonvolatile paraffin oil entering the absorber free of pentane at 308 K. Estimate the number of ideal trays for adiabatic absorption of 98.6% of the pentane. Neglect the solubility of methane in the oil and assume operation to be at constant pressure. The pentane forms ideal solutions with the paraffin oil. The average molecular weight of the oil is 200, heat capacity is 1.884 kJ/kg·K. The heat capacity of methane over the range of temperatures to be encountered is 35.6 kJ/kmol·K; for liquid pentane is 177.5 kJ/kmol·K; for pentane vapor is 119.8 kJ/kmol·K. The latent heat of vaporization of n-pentane at 273 K is 27.82 MJ/kmol (Treybal, 1980).

Solution

It is convenient to refer all enthalpies to the condition of pure liquid solvent, pure diluent gas, and pure solute at some base temperature T_0, with each substance assigned zero enthalpy for its normal state of aggregation at T_0 and 1 atm pressure. In this case, we use a base temperature T_0 = 273 K. Enthalpies referred to 273 K, liquid pentane, liquid paraffin oil, and gaseous methane, are then

$$H_L = (1-x)(1.884)(200)(T_L - 273) + x(177.5)(T_L - 273) + \Delta H_S$$

where ΔH_S is the heat of solution. Since pentane forms ideal solutions with the paraffin oil, ΔH_S = 0 and simplifying we obtain

$$H_L = (T_L - 273)(376.8 - 199.3x) \text{ kJ/kmol liquid solution.}$$

$$H_V = (1-y)(35.6)(T_G - 273) + y\left[(119.8)(T_G - 273) + 27,820\right]$$
$$= (T_G - 273)(35.6 + 84.2y) + 27,820y \text{ kJ/kmol gas—vapor mixture}$$

In this case, equilibrium is described by Raoult's law, $y_i = P_A x_i / P = m x_i$. The vapor pressure for n-pentane as a function of temperature can be estimated from the Antoine equation, in this case (Smith et al., 1996)

$$P_A = \exp\left[13.8183 - \frac{2477.07}{T - 40}\right]$$

where P_A is in kPa. Then, the slope of the equilibrium curve, m, as a function of temperature is given by

$$m(T) = \frac{\exp\left[13.8183 - \dfrac{2477.07}{T - 40}\right]}{101.3}$$

Calculate the flow rate and composition of the two streams leaving the tower. From material balances for the required 98.6% pentane recovery:

5.4 Thermal Effects During Absorption and Stripping 295

Pentane entering with the incoming gas = (1 kmol/s)(0.25) = 0.25 kmol/s
Pentane absorbed = (0.986)(0.25) = 0.2465 kmol/s
Methane in the outgoing gas = methane in the incoming gas = 0.75 kmol/s

$$V_1 = 0.75 + (0.014)(0.25) = 0.7535 \text{ kmol/s} \qquad y_1 = (0.014)(0.25)/0.7535 = 0.0046$$

$$L_N = 2.0 + 0.2465 = 2.2465 \text{ kmol/s} \qquad x_N = 0.2465/2.2465 = 0.1097$$

Calculate the enthalpy of the two streams entering the tower:

$$H_{L,0} = 13,190 \text{ kJ/kmol} \quad H_{V,N+1} = 8480 \text{ kJ/kmol}$$

Assume $T_1 = 308.5$ K (to be checked later), and calculate $H_{V,1} = 1405$ kJ/kmol. Substitute into the energy balance for the entire absorber, equation (5-31):

$$2.0 \times 13,190 + 1.0 \times 8480 = 2.2465 H_{L,N} + 0.7535 \times 1405$$

Then,

$$H_{L,N} = 15,050 = \left(T_N - 273\right)\left[376.8 - \left(199.3 \times 0.1097\right)\right]$$

$$T_N = 315.4 \text{ K}$$

$$m\left(T_N\right) = 1.228 \quad y_N = m\left(T_N\right)x_N = \left(1.228\right)\left(0.1097\right) = 0.135$$

$$V_N = V_S / \left(1 - y_N\right) = 0.75 / \left(1 - 0.135\right) = 0.867 \text{ kmol/s}$$

From equation (5-28) with $n = N - 1$,

$$L_{N-1} = L_N + V_N - V_{N+1} = 2.2465 + 0.867 - 1.000 = 2.114 \text{ kmol/s}$$

From equation (5-29) with $n = N - 1$, $x_{N-1} = 0.0521$.
From equation (5-30) with $n = N - 1$, $H_{L,N-1} = 14,300$ kJ/kmol, $T_{N-1} = 312$ K.

The computations are continued upward through the tower in this manner until the gas composition falls at least to $y = 0.0046$. The results are

n = tray number	T_n, K	x_n	y_n
$N = 4$	315.3	0.1091	0.1340
$N - 1 = 3$	312.0	0.0521	0.0568
$N - 2 = 2$	309.8	0.0184	0.0187
$N - 3 = 1$	308.5	0.0046	0.0045

The required $y_1 = 0.0046$ occurs at about 4 ideal trays, and the temperature on the top tray is essentially that assumed. Had this not been so, a new assumed value of T_1 and a new stage-by-stage calculation would have been required.

296   **Absorption and Stripping**

5.4.1 Adiabatic Operation of a Packed-Bed Absorber

During absorption in a packed bed, release of energy at the interface due to latent heat and heat of solution raises the interface temperature above that of the bulk liquid. This changes physical properties, mass-transfer coefficients, and equilibrium concentrations. Because of the possibility of substantial temperature effects, there is the possibility that the solvent evaporates in the warm parts of the tower and recondenses in the cooler parts. This has been confirmed experimentally (Raal and Khurani, 1973).

For this problem, therefore, the components are defined as follows:

A = principal transferred solute, present in both gas and liquid phases
B = carrier gas, not dissolving in the liquid, present only in the gas
C = principal liquid solvent; can evaporate and condense; present in both phases

Consider material and energy balances over a differential section of the packed tower. The mass and energy fluxes are considered positive in the direction gas to liquid, negative in the opposite direction. Looking at the gas phase: solute A and solvent vapor C may transfer, but carrier gas B will not ($N_B = 0$):

$$N_A a_h dz = \Psi_A F_{G,A} \ln \left[\frac{\Psi_A - y_{A,i}}{\Psi_A - y_A} \right] a_h dz = -G_{M,B} dY_A \qquad (5\text{-}32)$$

$$N_C a_h dz = \Psi_C F_{G,C} \ln \left[\frac{\Psi_C - y_{C,i}}{\Psi_C - y_C} \right] a_h dz = -G_{M,B} dY_C \qquad (5\text{-}33)$$

where Y_A = mol A/mol B, Y_C = mol C/mol B. Since the fluxes N_A and N_C may have either sign, the Ψ's can be greater or less than 1.0, positive or negative. However, $\Psi_A + \Psi_C = 1.0$. Neglecting the liquid heat-transfer resistance ($T_L = T_i$), the heat transfer rate for the gas is

$$q_G a_h dz = h_{G,c} a_h \left(T_G - T_L \right) dz \qquad (5\text{-}34)$$

where $h_{G,c}$ is the gas-phase convection heat-transfer coefficient corrected for mass transfer (Treybal, 1980). Similarly, for the liquid,

$$N_A a_h dz = \Psi_A F_{L,A} \ln \left[\frac{\Psi_A - x_A}{\Psi_A - x_{A,i}} \right] a_h dz \qquad (5\text{-}35)$$

$$N_C a_h dz = \Psi_C F_{L,C} \ln \left[\frac{\Psi_C - x_C}{\Psi_C - x_{C,i}} \right] a_h dz \qquad (5\text{-}36)$$

At steady-state, with no chemical reaction, a mass balance on the differential section yields (for countercurrent flow)

$$dL = dV \qquad (5\text{-}37)$$

5.4 Thermal Effects During Absorption and Stripping

Since

$$V = V_B \left(1 + Y_A + Y_C \right) \tag{5-38}$$

Then

$$dL = V_B \left(dY_A + dY_C \right) \tag{5-39}$$

The concentration gradients along the column are provided by equations (5-32) and (5-33) written here as

$$\frac{dY_A}{dz} = -\frac{\Psi_A F_{G,A} a_h}{G_{M,B}} \ln \left[\frac{\Psi_A - y_{A,i}}{\Psi_A - y_A} \right] \tag{5-40}$$

$$\frac{dY_C}{dz} = -\frac{\Psi_C F_{G,C} a_h}{G_{M,B}} \ln \left[\frac{\Psi_C - y_{C,i}}{\Psi_C - y_C} \right] \tag{5-41}$$

The gas-phase temperature gradient is from an energy balance over the gas phase in the differential section:

$$\frac{dT_G}{dz} = -\frac{h_{G,c} a_h \left(T_G - T_i \right)}{G_{M,B} \left(C_B + Y_A C_A + Y_C C_C \right)} \tag{5-42}$$

The interface concentration of A is obtained from equations (5-32) and (5-35)

$$y_{A,i} = \Psi_A - \left(\Psi_A - y_A \right) \left[\frac{\Psi_A - x_A}{\Psi_A - x_{A,i}} \right]^{F_{L,A}/F_{G,A}} \tag{5-43}$$

Similarly, for component C,

$$y_{C,i} = \Psi_C - \left(\Psi_C - y_C \right) \left[\frac{\Psi_C - x_C}{\Psi_C - x_{C,i}} \right]^{F_{L,C}/F_{G,C}} \tag{5-44}$$

These equations are solved simultaneously with their respective equilibrium-distribution curves, as described in Chapter 3. The calculation is by trial-and-error: Ψ_A is assumed, whence $\Psi_C = 1 - \Psi_A$. The correct Ψ's are those for which $x_{A,i} + x_{B,i} = 1.0$.

Heat-transfer coefficients can be estimated, if not available otherwise, through the heat-mass-transfer analogies discussed in Chapter 2. For the gas phase, the heat-transfer coefficient must be corrected for mass transfer (Treybal, 1980):

$$h_{G,c} a_h = -\frac{G_{M,B} \left(C_A \dfrac{dY_A}{dz} + C_C \dfrac{dY_C}{dz} \right)}{1 - \exp \left[\dfrac{G_{M,B}}{h_G a_h} \left(C_A \dfrac{dY_A}{dz} + C_C \dfrac{dY_C}{dz} \right) \right]} \tag{5-45}$$

298   **Absorption and Stripping**

Effective diffusivities for the three-component gas mixtures are estimated through equation (1-58), which reduces in this case to

$$D_{A,eff} = \frac{\Psi_A - y_A}{\Psi_A \left(\dfrac{y_B}{D_{AB}} + \dfrac{y_A + y_C}{D_{AC}} \right) - \dfrac{y_A}{D_{AC}}} \qquad (5\text{-}46)$$

$$D_{C,eff} = \frac{\Psi_C - y_C}{\Psi_C \left(\dfrac{y_B}{D_{CB}} + \dfrac{y_A + y_C}{D_{AC}} \right) - \dfrac{y_C}{D_{AC}}} \qquad (5\text{-}47)$$

In the design of an absorption tower, the cross-sectional area and hence the mass velocities of gas and liquid are established through pressure drop consider-ations as described in Chapter 4. Assuming that entering flow rates, compositions, and temperatures, pressure of absorption, and percentage absorption (or stripping) of one component are specified, the packed height is then fixed. The problem is therefore to estimate the packed height and the conditions (temperature and com-position) of the outlet streams. Fairly extensive trial-and-error is required, for which the relations outlined above can best be solved by a digital computer (Taylor and Krishna, 1993; Seader and Henley, 1998).

Example 5.6 Packed Bed for Adiabatic Ammonia Absorption (Treybal, 1980)

A gas consisting of 41.6% ammonia (A) and 58.4% air (B) at 20 °C and 1 atm is to be scrubbed countercurrently and adiabatically with liquid water (C) entering at 20 °C to remove 99% of the ammonia. The gas enters the absorber at the rate of 33.9 mol/ m²-s while the water enters at the rate of 271 mol/m²-s. The absorber is to be packed with 38-mm ceramic Berl saddles. Estimate the packed height and the out-let gas and liquid temperatures.

Solution

The dissolution of gaseous ammonia into liquid water is highly exothermic so we can expect that adiabatic operation of this absorber will result in significant increases in the temperature of both the gas and liquid phases. The trial-and-error solution of this problem is presented in some detail in Treybal (1980). The procedure begins assuming an outlet gas temperature (24 °C) and that the outlet gas is satu-rated with water vapor. From material balances, the rate and composition of the outlet is calculated. From an energy balance over the entire adiabatic absorber, the outlet liquid temperature is estimated (41.3 °C). Using equilibrium data from the literature for the system ammonia-air-water, the interfacial temperature and com-position are calculated. The concentration gradients of components A and C are calculated. A small increment in ΔY_A (−0.05) is taken up the tower and ΔZ is

5.4 Thermal Effects During Absorption and Stripping

calculated. The procedure is repeated until the specified gas outlet concentration is reached, whereupon the assumed outlet gas temperature and water concentration can be checked. The latter are adjusted as necessary and the entire computation repeated until convergence is achieved. The packed depth required is the sum of the final ΔZ's.

The final results are shown in Figure 5.4 illustrating the bulk gas and bulk liquid temperature profiles. The required packed depth is 1.3 m. The bulk liquid temperature increases continuously as it moves down the column until it reaches an outlet temperature of 42 °C. The gas bulk temperature increases initially until it reaches a maximum value of about 32 °C at a height of 0.3 m and then diminishes until it reaches a temperature of about 22 °C at the exit. Water is stripped in the lower part of the tower and reabsorbed in the upper portion.

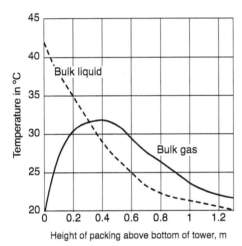

Figure 5.4 Temperature profiles in the adiabatic absorber of Example 5.6.

PROBLEMS

The problems at the end of each chapter have been grouped into four classes (designated by a superscript after the problem number).

Class a: Illustrates direct numerical application of the formulas in the text.
Class b: Requires elementary analysis of physical situations, based on the subject material in the chapter.
Class c: Requires somewhat more mature analysis. Class d: Requires computer solution.

5.1[a]. Overall tray efficiency

Consider a tray absorber with a constant Murphree efficiency $\mathbf{E}_{MGE} = 0.75$, and an average absorption factor $A = 1.25$.

(a) Estimate the overall tray efficiency.

Answer: 71%

(b) If the absorber requires 6.34 equilibrium stages, calculate the number of real trays.

Answer: 9 trays

5.2[c]. Relation between Murphree tray efficiency and overall efficiency

Derive equation (5-5). *Hint*: Start with the definition of $\mathbf{E}_{MGE}$ and locate the pseudoequilibrium line which could be used together with the operating line for graphically constructing steps representing real trays. Then use the Kremser equation (5-3) with the pseudoequilibrium line by moving the origin of the xy diagram to the intercept of the pseudoequilibrium line with the x axis.

5.3[a]. Ammonia stripping from a wastewater in a tray tower

A tray tower providing six equilibrium stages is used for stripping ammonia from a wastewater stream by means of countercurrent air at 1 atm and 300 K. Calculate the concentration of ammonia in the exit water if the inlet liquid concentration is 0.1 mole% ammonia, the inlet air is free of ammonia, and 1.873 standard cubic meter of air are fed to the tower per kilogram of wastewater. The equilibrium data for this system, in this range of concentrations and 300 K, can be represented by $y_{A,i} = 1.414\, x_{A,i}$ (King, 1971).

Answer: 5.74×10^{-4} mol%

Problems 301

5.4^a. Ammonia stripping from a wastewater in a tray tower

The Murphree plate efficiency for the ammonia stripper of Problem 5.3 is constant at 53.5%. Estimate the number of real trays required.

Answer: 8 trays

5.5^b. Ammonia stripping from a wastewater in a tray tower

If the air flow rate to the absorber of Problems 5.3 and 5.4 is reduced to 1.0 standard m^3/kg of water, calculate concentration of ammonia in the exit water if 8 real trays are used and the Murphree efficiency remains constant at 53.5%.

Answer: 1.4×10^{-2} mol%

5.6^b. Absorption of an air pollutant in a tray tower

A heavy-oil stream at 320 K is used in an absorber to remove dilute quantities of pollutant A from an air stream. The heavy oil is then recycled back to the process where A is stripped. The process is being run on a pilot plant basis, and information for scale-up is desired. The current absorber is a 16-plate sieve-tray column. Pilot plant data are as follows:

Liquid flow rate = 5.0 mol/h

Gas flow rate = 2.5 mol/h

$$y_{A,in} = 0.04 \qquad y_{A,out} = 0.0001 \qquad x_{A,in} = 0.0001$$

Equilibrium for A is given as $y_{A,i} = 0.7x_{A,i}$. Find the overall column efficiency, and the Murphree plate efficiency. Assume that the liquid and gas flow rates are roughly constant.

Answer: $E_{MGE} = 0.53$

5.7^b. Absorption of ammonia in a laboratory-scale tray tower

An absorption column for laboratory use has been carefully constructed so that it has exactly 4 equilibrium stages and is being used to measure equilibrium data. Water is used as the solvent to absorb ammonia from air. The system operates isothermally at 300 K and 1 atm. The inlet water is pure distilled water. The ratio of $L/V = 1.2$, inlet gas concentration is 0.01 mol fraction ammonia; the measured outlet gas concentration is 0.0027 mol fraction ammonia. Assuming that Henry's law applies, calculate the slope of the equilibrium line.

Answer: $m = 1.413$

302   **Absorption and Stripping**

5.8[c, d]. Absorption of ammonia in a sieve-tray tower

A process for making small amounts of hydrogen by cracking ammonia is being considered and residual, uncracked ammonia is to be removed from the resulting gas. The gas will consist of H_2 and N_2 in the molar ratio 3:1, containing 3% NH_3 by volume, at a pressure of 2 bars and a temperature of 303 K.

There is available a sieve-tray tower, 0.75 m diameter, containing 6 cross-flow trays at 0.5 m tray spacing. The perforations are 4.75 mm in diameter, arranged in triangular pitch on 12.5 mm centers, punched in sheet metal 2 mm thick. The weir height is 40 mm. Assume isothermal scrubbing with pure water at 303 K. The water flow rate to be used should not exceed 50% of the maximum recommended for cross-flow sieve trays, which is 0.015 m^3/s·m of tower diameter (Treybal, 1980). The gas flow rate should not exceed 80% of the flooding value.

(a) Estimate the gas flow rate that can be processed in the column under the circumstances described above.

(b) Calculate the concentration of the gas leaving the absorber in part (a).

Answer: 3.16×10^{-5} mol fraction

(c) Estimate the total gas-pressure drop and check the operation for excessive weeping and entrainment.

Answer: $\Delta P = 3.7$ kPa

Data:

Liquid density = 996 kg/m^3 Surface tension = 0.068 N/m

Foaming factor = 1.0   Liquid diffusivity = 2.42×10^{-9} m^2/s

Gas viscosity = 1.122×10^{-5} kg/m·s   Gas diffusivity = 0.23 cm^2/s

Slope of the equilibrium line, $m = 0.85$

5.9[b]. Absorption of carbon dioxide in a bubble-cap tray tower

A plant manufacturing dry ice will burn coke in air to produce a flue gas which, when cleaned and cooled, will contain 15% CO_2, 6% O_2, and 79% N_2. The gas will be blown into a bubble-cap tower scrubber at 1.2 atm and 298 K, to be scrubbed countercurrently with a 30 wt% monoethanolamine (C_2H_7ON) aqueous solution entering at 298 K. The scrubbing liquid, which is recycled from a stripper, will contain 0.058 mol CO_2/mol solution. The gas leaving the scrubber is to contain 2% CO_2. A liquid-to-gas ratio of 1.2 times the minimum is specified. Assume isothermal operation. At 298 K and 1.2 atm, the equilibrium mol fraction of carbon dioxide over aqueous solutions of monoethanolamine (30 wt%) is given by

$$y_{A,i} = 7.5503 - 272.01 x_{A,i} + 2452.4 x_{A,i}^2 \quad \text{for } 0.055 \le x_{A,i} \le 0.065$$

Problems 303

where $x_{A,i}$ is the mol fraction of CO_2 in the liquid solution.

(a) Calculate the kilograms of solution entering the tower per cubic meter of entering gas.

Answer: 21.0 kg

(b) Determine the number of theoretical trays required for part (a).

Answer: 2.5

(c) The monoethanolamine solution has a viscosity of 6.0 cP and a density of 1012 kg/m^3. Estimate the overall tray efficiency for the absorber, and the number of real trays required. Seader and Henley (1998) proposed the following empirical correlation to estimate the overall efficiency of absorbers and strippers using bubble-cap trays (it has also been used to obtain rough estimates for sieve-tray towers):

$$\log \mathbf{E}_O = -0.773 - 0.415 \log\left(\frac{mM_L\mu_L}{\rho_L}\right) - 0.0896\left[\log\left(\frac{mM_L\mu_L}{\rho_L}\right)\right]^2$$

where
$\mathbf{E}_O$ = overall fractional efficiency
m = slope of equilibrium curve
μ_L = liquid viscosity, in cP
ρ_L = liquid density, in kg/m^3

Hint: In this problem, the equilibrium-distribution curve is not a straight line; therefore, m is not constant. Estimate the average value of m in the range of liquid concentrations along the operating line and use the average value in the correlation given above.

Answer: 20 trays

5.10[b]. Design equation of a boiler feed water deaerator

Example 2.14 introduced the design equation for a countercurrent hollow-fiber boiler feed water deaerator when all the resistance to mass transfer resides on the liquid phase and the entering gas contains no oxygen:

$$V_T = \frac{L}{k_L a c \left(1-A\right)} \ln\left[\frac{x_{in}}{x_{out}}\left(1-A\right)+A\right] \qquad (2\text{-}99)$$

Derive equation (2-99) from equations (5-25), (5-26), and (5-27).

5.11[c, d]. Absorption of carbon disulfide in a sieve-tray tower

Carbon disulfide, CS_2, used as a solvent in a chemical plant, is evaporated from the product in a dryer into an inert gas (essentially N_2) in order to avoid an explosion hazard. The CS_2–N_2 mixture is to be scrubbed with an absorbent

304   **Absorption and Stripping**

hydrocarbon oil (octadecane, $C_{18}H_{38}$). The gas will flow at the rate of 0.4 m³/s at 297 K and 1 atm. The partial pressure of CS_2 in the original gas is 50 mm Hg, and the CS_2 concentration in the outlet gas is not to exceed 0.5%. The oil enters the absorber essentially pure at a rate 1.5 times the minimum, and solutions of oil and CS_2 follow Raoult's law. Design a sieve-tray tower for this process. Design for a gas velocity which is 70% of the flooding velocity. Assuming isothermal operation, determine

(a) Liquid flow rate, kg/s.

Answer: 2.24 kg/s

(b) The tower diameter and plate spacing.

(c) The details of the tray design.

(d) Number of real trays required.

(e) Total gas-pressure drop.

Data (at 297 K):

Oil average molecular weight = 254   Oil viscosity = 4 cP

Oil density = 810 kg/m³   Surface tension = 0.030 N/m

Foaming factor = 0.9 CS_2 vapor pressure = 346 mm Hg

Gas viscosity = 1.7×10^{-5} kg/m·s   Gas diffusivity = 0.114 cm²/s

Liquid diffusivity = 0.765×10^{-5} cm²/s

5.12[c, d]. Absorption of carbon disulfide in a sieve-tray tower

Repeat Problem 5.11, but design for a liquid rate 2.0 times the minimum and 60% of the flooding velocity.

Answer: 7 real trays

5.13[b]. Steam-stripping of benzene in a sieve-plate column

A straw oil used to absorb benzene from coke-oven gas is to be steam-stripped in a sieve-plate column at atmospheric pressure to recover the dissolved benzene, C_6H_6. Equilibrium conditions at the operating temperature are approximated by Henry's law such that, when the oil phase contains 10 mol% benzene, the equilibrium benzene partial pressure above the oil is 5.07 kPa. The oil may be considered nonvolatile. It enters the stripper containing 8 mol% benzene, 75% of which is to be removed. The steam leaving contains 3 mol% C_6H_6.

(a) How many theoretical stages are required?

Answer: 2.51 ideal stages

Problems 305

(b) How many mol of steam are required per 100 mol of the oil–benzene mixture?

Answer: 194 mol

(c) If 85% of the benzene is to be recovered with the same steam and oil rates, how many theoretical stages are required?

Answer: 5 ideal stages

5.14[b]. Relation between N and N_{tOG} for constant absorption factor

With the help of the Kremser equation (5-3) and equation (5-24), derive the relation between N and N_{tOG} for constant absorption factor. Establish the condition for which $N = N_{tOG}$.

5.15[c, d]. Absorption of carbon disulfide in a random-packed tower

Design a tower packed with 50-mm ceramic Hiflow rings for the carbon disulfide scrubber of Problem 5.11. Assume isothermal operation and use a liquid rate of 1.5 times the minimum and a gas-pressure drop not exceeding 175 Pa/m of packing. Calculate the tower diameter, packed height, and total gas-pressure drop. Assume that C_h for the packing is 1.0.

Answer: $D = 0.623$ m

5.16[b, d]. Absorption of sulfur dioxide in a random-packed tower

It is desired to remove 90% of the sulfur dioxide in a flue gas stream at 298 K and 1 atm by countercurrent absorption with pure water at the same temperature, using a packed tower that is 0.7 m in diameter. The tower is packed with 35-mm plastic NORPAC rings. The average gas-pressure drop is 200 Pa/m. Equilibrium is described by Henry's law with $y_i = 8.4x_i$. If the liquid flow is adjusted so that the driving force $(y - y^*)$ is constant, calculate the height of the packed section. Flue gases usually contain less than 1 mol% of SO_2, an air pollutant regulated by law. Assume that the properties of the liquid are similar to those of pure water, and that the properties of the flue gases are similar to those of air. *Hint*: For the driving force $(y - y^*)$ to remain constant, the operating and equilibrium lines must be parallel, which means that $A = 1$.

5.17[c]. Benzene vapor recovery system

Benzene vapor in the gaseous effluent of an industrial process is scrubbed with a wash oil in a countercurrent packed absorber. The resulting benzene–wash oil solution is then heated to 398 K and stripped in a tray tower, using steam as the stripping medium. The stripped wash oil is then cooled and recycled to the absorber. Some data relevant to the operation follow:

306   **Absorption and Stripping**

Absorption:

Benzene in entering gas = 1.0 mol%

Operating pressure of absorber = 800 mm Hg

Oil circulation rate = 2 m³/1000 m³ of gas at STP

Oil specific gravity = 0.88   Molecular weight = 260

Henry's law constant = 0.095 at 293 K;

= 0.130 at 300 K

N_{tOG} = 5 transfer units

Stripping:

Pressure = 1 atm Steam at 1 atm, 398 K

Henry's law constant = 3.08 at 398 K

Number of equilibrium stages = 5

(a) In the winter, it is possible to cool the recycled oil to 293 K, at which temperature the absorber then operates. Under these conditions 72.0 kg of steam is used in the stripper per 1000 m³ of gas at STP entering the absorber. Calculate the percent benzene recovery in the winter.

Answer: 92.5%

(b) In the summer, it is impossible to cool the recycled wash oil to lower than 300 K with the available cooling water. Assuming that the absorber then operates at 300 K, with the same oil and steam rates, and that N_{tOG} and equilibrium stages remain the same, what summer recovery of benzene can be expected?

Answer: 86.7%

(c) If the oil rate cannot be increased, but the steam rate in the summer is increased by 50% over the winter value, what summer recovery of benzene can be expected?

Answer: 87.9%

5.18[b]. Overall transfer units for cocurrent absorption

For dilute mixtures and when Henry's law applies, prove that the number of overall transfer units for cocurrent gas absorption in packed towers is given by

$$N_{tOG} = \frac{A}{A+1} \ln\left|\frac{y_1 - mx_1}{y - mx_2}\right| \qquad (5\text{-}48)$$

where subscript 1 indicates the top (where gas and liquid enter) and subscript 2 indicates the bottom of the absorber.

Problems 307

5.19[c, d]. Absorption of germanium tetrachloride used for optical fibers

Germanium tetrachloride ($GeCl_4$) and silicon tetrachloride ($SiCl_4$) are used in the production of optical fibers. Both chlorides are oxidized at high temperature and converted to glasslike particles. However, the $GeCl_4$ oxidation is quite incomplete, and it is necessary to scrub the unreacted $GeCl_4$ from its air carrier in a packed column operating at 298 K and 1 atm with a dilute caustic solution. At these conditions, the dissolved $GeCl_4$ has no vapor pressure, and mass transfer is controlled by the gas phase. Thus, the equilibrium curve is a straight line of zero slope. The entering gas flows at the rate of 23,850 kg/day of air containing 288 kg/day of $GeCl_4$. The air also contains 540 kg/day of Cl_2, which, when dissolved, also will have no vapor pressure.

It is desired to absorb at least 99% of both $GeCl_4$ and Cl_2 in an existing 0.75-m-diameter column that is packed to a height of 3.0 m with 20-mm ceramic Hiflow rings. The liquid rate should be set so that the column operates at 75% of flooding. Because the solutions are very dilute, it can be assumed that both gases are absorbed independently. For the two diffusing species, take $D_{GeCl4} = 0.06$ cm^2/s; $D_{Cl2} = 0.13$ cm^2/s. Determine

(a) Liquid flow rate, in kg/s.

(b) The percentage of absorption of $GeCl_4$ and Cl_2 based on the available 3.0 m of packing.

5.20[c, d]. Absorption of carbon disulfide in a structured-packed tower

Redesign the absorber of Problem 5.15 using metal Montz B2-300 structured packing. The characteristics of this packing are (Billet and Schultes, 1999)

$$C_{S,T} = 3.098 \qquad a = 300 \text{ m}^{-1} \qquad \varepsilon = 0.93 \qquad C_h = 0.482$$

$$C_p = 0.295 \qquad C_L = 1.165 \qquad C_V = 0.422 \qquad C_{F,T} = 2.464$$

5.21[c, d]. Absorption of ammonia in a random-packed tower

It is desired to reduce the ammonia content of 0.05 m^3/s of an ammonia–air mixture (300 K and 1 atm) from 5.0% to 0.04% by volume by water scrubbing. There is available a 0.3-m-diameter tower packed with 25-mm ceramic Raschig rings to a depth of 3.5 m. Is the tower satisfactory, and if so, what water rate should be used? At 300 K, ammonia–water solutions follow Henry's law up to 5 mol% ammonia in the liquid, with $m = 1.414$.

308   Absorption and Stripping

Data

Liquid:

Density $= 998$ kg/m^3 Viscosity $= 0.8$ cP

Diffusivity $= 1.64 \times 10^{-5}$ cm^2/s

Gas:

Viscosity $= 1.84 \times 10^{-5} Pa \cdot s$ Diffusivity $= 0.28$ cm^2 / s

5.22[c, d]. Absorption of sulfur dioxide in a structured-packed tower

A tower packed with metal Montz B2-300 structured packing is to be designed to absorb SO_2 from air by scrubbing with water. The entering gas, at an SO_2-free flow rate of 37.44 mol/m^2-s of bed cross-sectional area, contains 20 mol% of SO_2. Pure water enters at a flow rate of 1976 mol/m^2-s of bed cross-sectional area. The exiting gas is to contain only 0.5 mol% SO_2. Assume that neither air nor water will transfer between the phases and that the tower operates isothermally at 2 atm and 303 K. Equilibrium data for solubility of SO_2 in water at 303 K and 1 atm have been fitted by least-squares to the equation (Seader and Henley, 1998):

$$y_i = 12.697x_i + 3148x_i^2 - 4.724 \times 10^5 x_i^3 + 3.001 \times 10^7 x_i^4 - 6.524 \times 10^8 x_i^5 \tag{5-49}$$

(a) Derive the following operating line equation for the absorber:

$$x = \frac{0.01895\left(\dfrac{y}{1-y}\right) - 9.5213 \times 10^{-5}}{1 + 0.01895\left(\dfrac{y}{1-y}\right)} \tag{5-50}$$

(b) If the absorber is to process 1.0 m^3/s (at 2 atm and 303 K) of the entering gas, calculate the water flow rate, the tower diameter, and the gas-pressure drop per unit of packing height at the bottom of the absorber.

Answer: ΔP = 167 Pa/m

(c) Calculate N_{tG}, H_{tG}, and the height of packing required.

Answer: Z = 1.04 m

Data:

Liquid:

Density $= 998$ kg / m^3 Viscosity $= 0.7$ cP

Diffusivity $= 1.61 \times 10^{-5}$ cm^2/s

Problems 309

Gas:

$$\text{Viscosity } (\text{in}) = 1.75 \times 10^{-5} \text{Pa} \cdot \text{s} \quad \text{Diffusivity} = 0.067 \text{ cm}^2 / \text{s}$$

$$\text{Viscosity } (\text{out}) = 1.87 \times 10^{-5} \text{ Pa} \cdot \text{s}$$

Packing characteristics are given in Problem 5.20.

5.23[c, d]. Absorption of sulfur dioxide in a random-packed tower

Repeat Problem 5.22 using 50-mm ceramic Pall rings as packing material. Compare your results to those of Problem 5.22.

5.24[c, d]. Isothermal absorption of methanol in a random-packed tower

A system for recovering methanol from a solid product wet with methanol involves evaporation of the alcohol into a stream of inert gas, essentially nitrogen. In order to recover the methanol from the nitrogen, an absorption scheme involving washing the gas countercurrently with water in a packed tower is being considered. The resulting water–methanol solution is then to be distilled to recover the methanol.

The absorption tower will be filled with 50-mm ceramic Pall rings. Design for a gas-pressure drop not to exceed 400 Pa/m of packed depth. Assume cooling coils will allow isothermal operation at 300 K. The gas will enter the column at the rate of 1.0 m³/s at 300 K and 1 atm. The partial pressure of methanol in the inlet gas is 200 mm Hg (Sc_G = 0.783). The partial pressure of methanol in the outlet gas should not exceed 15 mm Hg. Pure water enters the tower at the rate of 0.50 kg/s at 300 K. Neglecting evaporation of water, calculate the diameter and packed depth of the absorber.

Solutions of methanol and water follow Wilson equation. The corresponding parameters can be found in Smith et al. (1996).

5.25[c, d]. Isothermal absorption of oxygen in a random-packed tower

A tower 0.6 m in diameter, packed with 50-mm ceramic Raschig rings to a depth of 1.2 m, is to be used for producing a solution of oxygen in water for certain pollution control operation. The packed space will be connected to a gas cylinder containing pure oxygen and maintained at an absolute pressure of 5 atm. There is no gas outlet, and gas will enter from the cylinder only to maintain pressure, as oxygen dissolves in the water. Water will flow down the packing continuously at the rate of 1.5 kg/s. The temperature is to be 298 K. Assuming that the entering water is oxygen-free, what concentration of oxygen can be expected in the water effluent? Neglect the water-vapor content of the gas.

310   **Absorption and Stripping**

Data:

$$D_L = 2.5 \times 10^{-5} \text{ cm}^2/\text{s} \quad \mu_L = 0.894 \text{ cP}$$

Henry's law for O_2 in water: p_A (atm) = 4.38 × $10^4 x_A$

5.26[b, d]. Isothermal absorption of acetone in a random-packed tower

Air at 300 K is used to dry a plastic sheet. The solvent wetting the plastic is acetone. At the end of the dryer, the air leaves containing 2.0 mol% acetone. The acetone is to be recovered by absorption with water in a packed tower. The gas composition is to be reduced to 0.05 mol% acetone. The absorption will be isothermal because of cooling coils inside the tower. For the conditions of the absorber, the equilibrium relationship is $y_i = 1.8x_i$. The rich gas enters the tower at a rate of 0.252 kg/s, and pure water enters the top at the rate of 0.34 kg/s. The tower is packed with 25-mm metal Hiflow rings. Determine the dimensions of the tower. Design for a gas-pressure drop not to exceed 150 Pa/m.

5.27[b]. Isothermal absorption of acetone in a random-packed tower: dilute solutions

You are asked to design a packed column to recover acetone from air continuously by absorption with pure water at 289 K. The air contains 3.0 mol% acetone, and a 97% recovery is desired. The gas flow rate is 1.42 m^3/min at 289 K and 1 atm. The maximum-allowable gas superficial velocity is 0.73 m/s. It may be assumed that in the range of operation, $Y_i = 1.75X_i$, where Y_i and X_i are mol ratios for acetone. Calculate

(a) The minimum water-to-air molar flow rate ratio.

Answer: 1.7 mol water/mol air

(b) The maximum acetone concentration possible in the aqueous solution.

Answer: 1.74 mol%

(c) The overall number of gas-phase transfer units for a molar flow rate ratio of 1.4 times the minimum.

(d) The packing height, assuming that $K_y a_h$ = 3.2 kmol/(m^3·min).

(e) The packing height as a function of the molar flow rate ratio, assuming that V and H_{tOG} remain constant.

REFERENCES

Billet, R., and M. Schultes, *Trans IChemE*, **77**, 498–504 (1999).

Hachmuth, K. H., *Chem. Eng. Prog.*, **47**, 523, 621 (1951).

King, C. J., *Separation Processes*, McGraw-Hill, New York (1971).

Raal, J. D., and M. K. Khurani, *Can. J. Chem. Eng.*, **51**, 162 (1973).

Seader, J. D., and E. J. Henley, *Separation Process Principles*, Wiley, New York (1998).

Smith, J. M., H. C. Van Ness, and M. M. Abbott, *Introduction to Chemical Engineering Thermodynamics*, 5th ed., McGraw-Hill, New York (1996).

Taylor, R., and R. Krishna, *Multicomponent Mass Transfer*, Wiley, New York (1993).

Treybal, R. E., *Mass-Transfer Operations*, 3rd ed., McGraw-Hill, New York (1980

312   Absorption and Stripping

APPENDIX 5.1

Python solution of Example 5.3.

```python
import numpy as np
from scipy.integrate import quad
from scipy.optimize import fsolve
# Enter data:
LS = 6.0; VS = 38.0; me = 0.136; FG = 2.135; FL = 0.305
X2 = 0.05; Y2 = 0.012; X1 = 0.481; Y1 = 0.08; n = 20
# Convert molar ratios to molar fractions:
x1 = X1/(1 + X1); x2 = X2/(1 + X2); y1 = Y1/(1 + Y1)
y2 = Y2/(1 + Y2)
R = LS/VS; C = Y1 - R*X1; delx = (x1 - x2)/n
# Generate the operating line:
x = np.zeros(21,dtype=float)
y = np.zeros(21,dtype=float)
n = 20
delx = (x1 - x2)/n

for i in range(21):
    x[i] = x2 + i*delx
    y[i] = (C + (R - C)*x[i])/((1 + C)+(R - C - 1)*x[i])
    i = i + 1
#Generate the vector of interfacial concentrations:
yint = np.zeros(21,dtype=float)
yint0 = (y1 + y2)/2  # Initial estimate
for i in range(21):
    def g(w):
        return (1-w)/(1-y[i]) - ((1 - x[i])/\
                       (1 - w/me))**(FL/FG)
    yint[i] = fsolve(g, yint0)
    i+= 1
# Estimate the number of transfer units by cubic spline
# interpolation and numerical integration:
f = np.zeros(21,dtype=float)
for i in range(21):
    f[i] = 1/(y[i] - yint[i])
    i+= 1
from scipy.interpolate import CubicSpline
from scipy.integrate import quad
cs = CubicSpline(y, f)
def interp(y):
```

Appendix 5.1

Appendix 5.1 Python solution of Example 5.3 (continuation).

```
        return cs(y)
INTG = quad(interp, y[0], y[20])
NTG = INTG[0] + 0.5*np.log((1 - y2)/(1 - y1))
print('NTG = %5.3f' %NTG)
import matplotlib.pyplot as plt
plt.plot(y, f)
plt.ylabel('f = 1/(y - yint)')
plt.xlabel('y')
plt.show()
```

Program results: The number of transfer units, NTG = 9.023.

The program also generates the following plot that illustrates the function that must be integrated numerically to estimate NTG.

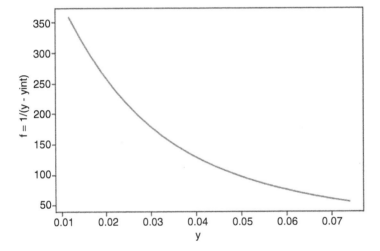

6

Distillation

6.1 INTRODUCTION

Distillation is a method of separating the components of a solution which depends upon the distribution of the substances between a liquid and a gas phase, applied to cases where all components are present in both phases (Treybal, 1980). Instead of introducing a new substance into the mixture in order to provide the second phase, as is done in gas absorption or desorption, the new phase is created from the original solution by vaporization or condensation.

The advantages of distillation as a separation method are clear. In distillation, the new phases differ from the original by their heat content, but heat is readily added or removed, although, of course, the cost of doing this must inevitably be considered. Absorption or desorption operations, on the other hand, which depend upon the introduction of a foreign substance, result in a new solution which in turn may have to be separated by one of the diffusional operations unless it happens that the new solution is directly useful.

During the first quarter of the twentieth century, the application of distillation expanded from a tool for enhancing the alcohol content of beverages into the prime separation technique in the chemical industry. During most of the century, distillation was by far the most widely used method for separating liquid mixtures of chemical components (Seader and Henley, 1998). Despite the emergence in recent years of many new separation techniques (e.g., membranes), distillation retains its position of supremacy among chemical engineering unit operations (Taylor and Krishna, 1993). Unfortunately, distillation is a very energy-intensive technique, especially when the relative volatility of the components being separated is low. This is a major drawback of distillation in our times, characterized by spiraling costs of energy.

Distillation is most frequently carried out in multitray columns, although packed columns have long been the preferred alternative when pressure drop is an important consideration. In recent years, the development of highly efficient structured packings has led to increased use of packed columns in distillation.

Principles and Applications of Mass Transfer: The Design of Separation Processes for Chemical and Biochemical Engineering, Fourth Edition. Jaime Benitez.
© 2023 John Wiley & Sons, Inc. Published 2023 by John Wiley & Sons, Inc.

315

6.2 SINGLE-STAGE OPERATION—FLASH VAPORIZATION

Your objectives in studying this section are to be able to:

1. Define the concepts of flash vaporization and partial condensation.
2. Relate—through material and energy balances, and equilibrium considerations—operating temperature and pressure during flash vaporization to the fraction of the feed vaporized, and to the concentrations of the liquid and vapor products.
3. Calculate the heat requirements of flash vaporization.

Flash vaporization is a single-stage operation wherein a liquid mixture is partially vaporized, the vapor is allowed to come to equilibrium with the residual liquid, and the resulting vapor and liquid phases are separated and removed from the apparatus. It may be batch or continuous.

A typical flowsheet is shown schematically in Figure 6.1 for continuous operation. The liquid feed is heated in a conventional tubular heat exchanger. The pressure is reduced, vapor forms at the expense of the liquid adiabatically, and the mixture is introduced into a vapor–liquid separating vessel. The liquid portion of the mixture leaves at the bottom of the separating vessel while the vapor rises and leaves at the top. The vapor may then pass to a condenser, not shown in the figure. The product D—richer in the more volatile—is, in this case, entirely a vapor. The material and energy balances for the process are

$$F = D + W \tag{6-1}$$

$$Fz_F = Dy_D + Wx_W \tag{6-2}$$

$$FH_F + Q = DH_D + WH_W \tag{6-3}$$

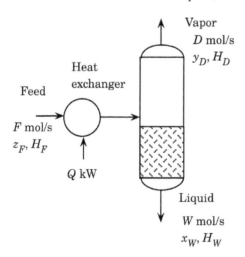

Figure 6.1 Schematic of the continuous flash vaporization process.

6.2 Single-Stage Operation—Flash Vaporization

In equation (6-2), z_F is used to represent the feed composition because the feed could be a liquid or a vapor. Solved simultaneously, equations (6-1) to (6-3) yield

$$-\frac{W}{D} = \frac{y_D - z_F}{x_W - z_F} = \frac{H_D - \left(H_F + \dfrac{Q}{F}\right)}{H_W - \left(H_F + \dfrac{Q}{F}\right)} \quad (6\text{-}4)$$

The two left-hand members of equation (6-4) represent the usual single-stage operating line on distribution coordinates, of negative slope as for all single-stage cocurrent operations (see Chapter 3). It passes through compositions representing the influent and effluent streams, points F and M on Figure 6.2. If the effluent streams were in equilibrium, the device would be an ideal stage and the products would be on the equilibrium curve at N on Figure 6.2. In that case, for a given pressure in the separator and a specified degree of vaporization, y_D, x_W, and the separator temperature can be calculated solving simultaneously equation (6-4) and the *vapor–liquid equilibrium* (VLE) *relationships* for the system. Once the concentrations of the products are known, H_D and H_W can be evaluated. Since the enthalpy of the feed, H_F, is known, the two right-hand members of equation (6-4) can be used to determine the thermal load of the heat exchanger for the degree of vaporization of the feed specified.

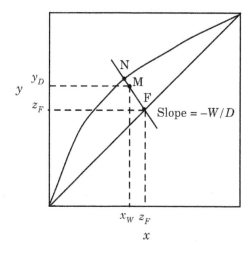

Figure 6.2 Flash vaporization process on the xy diagram.

The equations developed above apply equally well to the case where the feed is a vapor and Q, the heat removed in the heat exchanger to produce partial condensation, is taken as negative.

Example 6.1 Flash Vaporization of a Heptane–Octane Mixture

A liquid mixture containing 50 mol% n-heptane (A), 50 mol% n-octane (B), at 303 K, is to be continuously flash-vaporized at a pressure of 1 atm to vaporize 60 mol% of the feed. What will be the composition of the vapor and liquid and the temperature in the separator if it behaves as an ideal stage? Calculate the

318   **Distillation**

amount of heat to be added per mole of feed. n-Heptane and n-octane form ideal solutions.

Solution
The VLE relationship that applies under these conditions is Raoult's law. The vapor pressure of the two components (P_j) is related to temperature through Wagner's equation (Reid et al., 1987)

$$\ln\left(\frac{P_j}{P_{c,j}}\right) = \frac{A_j \chi_j + B_j \chi_j^{1.5} + C_j \chi_j^3 + D_j \chi_j^6}{1 - \chi_j}$$

$$\chi_j = \frac{T}{T_{c,j}}$$

(6-5)

The parameters in equation (6-5) for n-heptane and n-octane are

Component	T_c, K	P_c, bar	A	B	C	D
n-heptane	540.3	27.4	−7.675	1.371	−3.536	−3.202
n-octane	568.8	24.9	−7.912	1.380	−3.804	−4.501

Figure 6.3 shows a Mathcad computer program to perform flash vaporization calculations. Appendix 6.1 presents the corresponding Python program. Data supplied to the program include the parameters of the Wagner equation for both components, the separator pressure, feed concentration, and the desired fractional vaporization. Initial estimates are given to the program of the separator temperature (320 K) and the vapor and liquid concentrations ($y_D = 0.6$, $x_W = 0.4$). The "solve block" includes three relations between the variables: equation (6-4), and Raoult's law for components A and B. Convergence is achieved at a separator temperature of 385.9 K, and equilibrium compositions of $y_D = 0.576$, $x_W = 0.386$.

The next step is to determine the heat required to vaporize 60% of the feed. The feed is initially at 303 K and 1 atm. To determine its state of aggregation under these conditions, estimate the bubble point of the feed. The Mathcad program of Figure 6.3 can be used to determine both the bubble point (D approaching zero; $T_{BP} = 382.8$ K) and the dew point of the mixture ($D = 1.0$; $T_{DP} = 387.9$ K). Therefore, at 303 K and 1 atm the feed is a subcooled liquid. Since n-heptane and n-octane form ideal solutions, molar liquid enthalpies with respect to a reference temperature T_0 can be calculated from

$$H_L = \left[x_A C_{L,A} + \left(1 - x_A\right)C_{L,B}\right]\left(T_L - T_0\right)$$

(6-6)

Molar enthalpies for vapors are from

$$H_G = \left[y_A C_{p,A} + \left(1 - y_A\right)C_{p,B}\right]\left(T_G - T_0\right) + y_A \lambda_A + \left(1 - y_A\right)\lambda_B$$

(6-7)

where λ_A and λ_B are the latent heats of vaporization at T_0.

6.2 Single-Stage Operation—Flash Vaporization

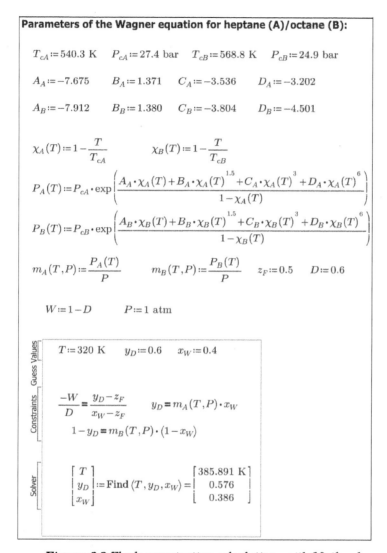

Figure 6.3 Flash vaporization calculations with Mathcad.

The following data were obtained from *Chemical Engineers' Handbook* (1973):

1. Latent heats of vaporization at $T_0 = 298$ K:

$$\lambda_A = 36.5 \text{ kJ/mol} \qquad \lambda_B = 41.4 \text{ kJ/mol}$$

2. Heat capacities of liquids, average in the specified temperature range:

$$C_{L,A} = 218 \text{ J/mol·K} (298-303 \text{ K}) \quad C_{L,A} = 241 \text{ J/mol·K} (298-386 \text{ K})$$
$$C_{L,B} = 253 \text{ J/mol·K} (298-303 \text{ K}) \quad C_{L,B} = 268 \text{ J/mol·K} (298-386 \text{ K})$$

320   **Distillation**

3. Heat capacities of gases, average in the range 298–386 K (Smith et al., 1996):

$$C_{p,A} = 187 \text{ J/mol·K} \quad C_{p,B} = 247 \text{ J/mol·K}$$

Substituting in equation (6-6), H_F = 1.18 kJ/mol of feed, H_W = 22.64 kJ/mol of liquid residue. From equation (6-7), H_D = 57.12 kJ/mol of vapor. From equation (6-4), Q = 42.15 kJ of heat added/mol of feed.

Example 6.2 Flash Vaporization of a Ternary Mixture

A liquid containing 50 mol% benzene (A), 25 mol% toluene (B), and 25 mol% o-xylene (C) is flash-vaporized at 1 atm and 373 K. Compute the amounts of liquid and vapor products and their composition. These components form ideal mixtures. The vapor pressures of the three components at 373 K are P_A = 178.8 kPa, P_B = 73.6 kPa, and P_C = 26.3 kPa. Therefore, for a total pressure of 1 atm and a temperature of 373 K, m_A = 1.765, m_B = 0.727, m_C = 0.259.

Solution
Equation (6-4) and Raoult's law apply for each of the three components in the mixture. Also, the mol fractions must add up to 1.0 in each of the two phases leaving the separator. The Mathcad program of Figure 6.3 is easily modified to include an additional component and solve for D, W, and the vapor and liquid compositions (see Problem 6.3). The results are D = 0.297, y_{AD} = 0.719, y_{BD} = 0.198, y_{CD} = 0.083, x_{AW} = 0.408, x_{BW} = 0.272, x_{CW} = 0.320.

6.3 DIFFERENTIAL DISTILLATION

Your objectives in studying this section are to be able to:

1. Understand the concept of differential distillation, and how it differs from flash vaporization.
2. Derive and apply *Rayleigh equation* to relate the compositions of the residue solution and composited distillate to the fraction of the feed differentially-distilled.

If during an infinite number of flash vaporizations of a liquid only an infinitesimal portion of the liquid were flashed each time, the net result would be equivalent to a *differential distillation*.

In practice, this can only be approximated. A batch of liquid is charged to a kettle or still fitted with some sort of heating device as in Figure 6.4. The charge is boiled slowly and as rapidly as they form the vapors are withdrawn to a condenser where they are liquefied. The condensate (distillate) is collected in a receiver. The first portion of the distillate would be the richest in the more volatile substance. As

6.3 Differential Distillation

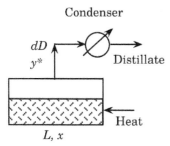

Figure 6.4 Schematic of the simple differential distillation process.

the distillation proceeds, the vaporized product becomes leaner. The distillate can, therefore, be collected in separate batches, called *cuts*, to give a series of distilled products of various purities. Thus, for example, if a ternary mixture contained a small amount of a very volatile substance A, a majority of substance B of intermediate volatility, and a small amount of substance C of low volatility, the first cut—which would be small—would contain the majority of A. A large second cut would contain the majority of B reasonably pure, and the residue left in the kettle would be largely C. While all three cuts would contain all three substances, some separation would have been achieved.

In the simple differential distillation process, the vapor product is in equilibrium with the liquid in the reboiler at any given time but changes continuously in composition. The mathematical approach must therefore be differential. Assume that at any time during the course of the distillation there are L mol of liquid in the still of composition x mol fraction of A and that an amount dD mol of distillate is vaporized, of mole fraction y^* in equilibrium with the liquid. Then we have the following differential material balances:

$$dL = -dD \tag{6-8}$$

$$0 - y^* dD = d(Lx) = L dx + x dL \tag{6-9}$$

Combining these equations

$$y^* dL = L dx + x dL \tag{6-10}$$

Rearranging and integrating

$$\int_W^F \frac{dL}{L} = \ln \frac{F}{W} = \int_{x_W}^{x_F} \frac{dx}{y^* - x} \tag{6-11}$$

where F is the mol of charge of composition x_F and W the mol of residual liquid of composition x_W. This is known as the *Rayleigh equation*, after Lord Rayleigh, who first derived it (Treybal, 1980). It can be used to determine F, W, x_F, or x_W when three of these are known. The *composited* distillate composition, $y_{D,av}$ can be determined by a simple material balance,

$$y_{D,av} = \frac{F x_F - W x_W}{D} \tag{6-12}$$

Example 6.3 Differential Distillation of a Heptane–Octane Mixture

Suppose the liquid of Example 6.1 [50 mol% n-heptane (A), 50 mol% n-octane (B)] were subjected to a batch distillation at atmospheric pressure, with 60 mol% of the liquid distilled. Compute the composition of the composited distillate and the residue.

Solution

Take as basis for calculation F = 100 mol. Then, D = 60 mol, W = 40 mol, x_F = 0.50. Substituting in equation (6-11) yields

$$\ln\frac{100}{40} = 0.916 = \int_{x_w}^{0.50} \frac{dx}{y^* - x} \qquad (6\text{-}13)$$

The equilibrium–distribution data for this system can be generated using the Mathcad program of Figure 6.3, calculating the liquid composition ($x = x_W$) at the dew point (D = 1.0) for different feed compositions ($y^* = z_F$). For example, for $y^* = z_F = 0.5$, $x = 0.317$. The following table is generated in that manner.

y^*	0.500	0.55	0.60	0.65	0.686	0.70	0.75
x	0.317	0.361	0.409	0.460	0.500	0.516	0.577
$(y^* - x)^{-1}$	5.464	5.291	5.236	5.263	5.376	5.435	5.780

A cubic spline interpolation formula is generated with these data, which is then integrated according to equation (6-13) to determine the residue concentration, x_W, when 60% of the feed has been distilled. Figure 6.5 shows the details of the calculations. Appendix 6.2 is the Python solution.

6.4 CONTINUOUS RECTIFICATION—BINARY SYSTEMS

Continuous distillation, or fractionation, is a multistage, countercurrent distillation operation. For a binary solution, with certain exceptions, it is ordinarily possible by this method to separate the solution into its components, recovering each in any state of purity desired.

Consider the general countercurrent, multistage, binary distillation operation shown in Figure 6.6. The operation consists of a column containing the equivalent of N theoretical stages arranged in a two-section cascade; a total condenser in which the overhead vapor leaving the top stage is totally condensed to give a liquid distillate product and liquid reflux that is returned to the top stage; a partial reboiler in which liquid from the bottom stage is partially vaporized to give a liquid bottoms product and vapor boilup that is returned to the bottom stage; and an intermediate feed stage.

6.4 Continuous Rectification—Binary Systems

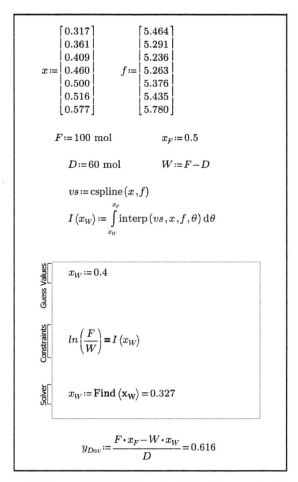

Figure 6.5 Example 6.3: differential distillation calculations.

The feed, which contains a more volatile component—the *light key*, LK—and a less volatile component—the *heavy key*, HK—enters the column at a feed stage, f. At the feed-stage pressure, the feed may be liquid, vapor, or a mixture of liquid and vapor, with its overall mol-fraction composition with respect to the light component denoted z_F. Vapor rising in the section above the feed (called the *enriching* or *rectifying* section) is washed with liquid to remove or absorb the heavy key. In the section below the feed (*stripping* or *exhausting* section), the liquid is stripped of the light key by the rising vapor. The mol fraction of the light key in the distillate is x_D, while the mol fraction of the light component in the bottoms product is x_B. Inside the tower, the liquids and vapors are always at their bubble points and dew points, respectively, so that the highest temperatures are at the bottom, the lowest at the top. The entire device is called a *fractionator*.

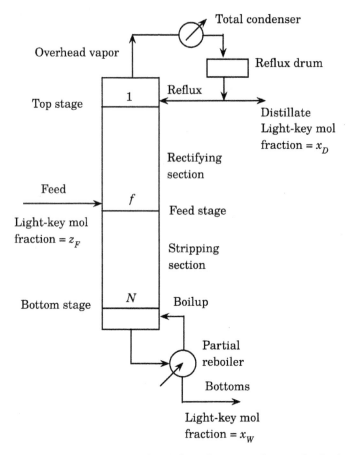

Figure 6.6 Fractionator with total condenser and partial reboiler.

6.5 McCABE–THIELE METHOD FOR TRAYED TOWERS

Your objectives in studying this section are to be able to:

1. Understand the fundamentals of the McCabe–Thiele graphical method to analyze binary distillation in trayed towers.
2. Apply the McCabe–Thiele graphical method to determine the number of equilibrium stages required for a given separation, and the optimal location along the cascade for introduction of the feed.

McCabe and Thiele (1925) developed an approximate graphical method for combining the equilibrium–distribution curve for a binary system with operating-line curves for the rectifying and stripping sections of a fractionator to estimate, for a given feed mixture and column operating pressure, the number of equilibrium

6.5 McCabe-Thiele Method for Trayed Towers

stages and the amount of reflux required for a desired degree of separation of the feed. The graphical nature of the McCabe–Thiele method greatly facilitates the visualization of many of the important aspects of multistage distillation, and therefore, the effort required to learn the method is well justified (Seader and Henley, 1998).

The McCabe–Thiele method hinges upon the fact that, as an approximation, the operating lines on the xy diagram can be considered straight for each section of a fractionator between points of addition or withdrawal of streams. This is true only if the total molar flow rates of liquid and vapor do not vary from stage to stage in each section of the column. This is the case if

1. The two components have equal and constant molar latent heats of vaporization.
2. Sensible enthalpy changes and heat of mixing are negligible compared to latent heats of vaporization.
3. Heat losses are negligible.
4. The pressure is uniform throughout the column (negligible pressure drop).

These assumptions are referred to as the *McCabe–Thiele assumptions* leading to the condition of *constant molar overflow*. For constant molar overflow, the analysis of a distillation column is greatly simplified because it is not necessary to consider energy balances in either the rectifying or stripping sections; only material balances and a VLE curve are required.

6.5.1 Rectifying Section

As shown in Figure 6.6, the rectifying section extends from the top stage, 1, to just above the feed stage, f. Consider a top portion of the rectifying stages, including the total condenser. A material balance for the light key over the envelope shown in Figure 6.7a for the total condenser and stages 1 to n is as follows:

$$y_{n+1} = \frac{L_n}{V_{n+1}} x_n + \frac{D}{V_{n+1}} x_D \tag{6-14}$$

For constant molar overflow, $V_{n+1} = V$ = constant; $L_n = L$ = constant and equation (6-14) can be written as

$$y_{n+1} = \frac{L}{V} x_n + \frac{D}{V} x_D \tag{6-15}$$

Equation (6-15) is the operating line for the rectifying section of the fractionator.

The liquid entering the top stage is the external reflux rate, L_0, and its ratio to the distillate rate, L_0/D, is the *reflux ratio*, R. For the case of a total condenser, with reflux returned to the column at its bubble point, $L_0 = L$ and $R = L/D$, a constant in the rectifying section. Also, since from a total molar balance $V = L + D$, then

$$\frac{L}{V} = \frac{L}{L+D} = \frac{L/D}{L/D+1} = \frac{R}{R+1} \tag{6-16}$$

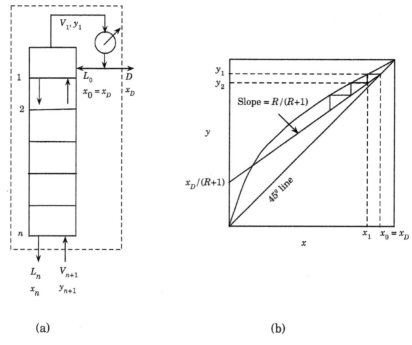

(a) (b)

Figure 6.7 McCabe–Thiele operating line for the rectifying section.

$$\frac{D}{V} = \frac{D}{L+D} = \frac{1}{L/D+1} = \frac{1}{R+1} \qquad (6\text{-}17)$$

Combining equations (6-15), (6-16), and (6-17) produces the most useful form of the operating line for the rectifying section:

$$y = \frac{R}{R+1} x + \frac{1}{R+1} x_D \qquad (6\text{-}18)$$

If values of R and x_D are specified, equation (6-18) plots as a straight line on the xy diagram with intersection at $y_1 = x_D$ on the 45° line, slope $= R/(R+1)$, and intersection at $y = x_D/(R+1)$ for $x = 0$, as shown in Figure 6.7b. The equilibrium stages are stepped off in the manner described in Chapter 3.

6.5.2 Stripping Section

As shown in Figure 6.6, the stripping section extends from the feed to the bottom stage. Consider a bottom portion of the stripping stages, including the partial reboiler. A material balance for the light key over the envelope shown in Figure 6.8a for the partial reboiler and stages N to $(m+1)$ is as follows:

$$y_{m+1} = \frac{L_{st}}{V_{st}} x_m - \frac{W}{V_{st}} x_W \qquad (6\text{-}19)$$

6.5 McCabe-Thiele Method for Trayed Towers

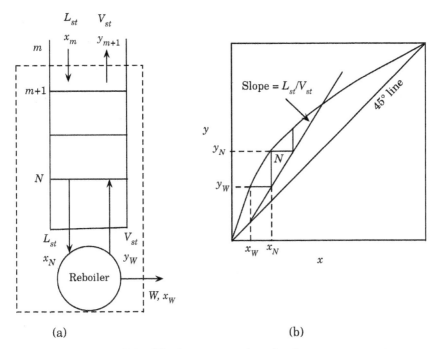

Figure 6.8 McCabe–Thiele operating line for the stripping section.

or

$$y = \frac{L_{st}}{V_{st}} x - \frac{W}{V_{st}} x_W \qquad (6\text{-}20)$$

where L_{st} and V_{st} are the total molar flows in the stripping section which, by the constant-molar-overflow assumption, remain constant from stage to stage. The vapor leaving the partial reboiler is assumed to be in equilibrium with the liquid bottoms product. Thus, the partial reboiler acts as an additional equilibrium stage. The vapor rate leaving it is called the boilup, V_{st}, and its ratio to the bottoms product rate, $V_B = V_{st}/W$, is the *boil-up ratio*. Since $L_{st} = V_{st} + W$,

$$\frac{L_{st}}{V_{st}} = \frac{V_{st} + W}{V_{st}} = \frac{V_B + 1}{V_B} \qquad (6\text{-}21)$$

Similarly

$$\frac{W}{V_{st}} = \frac{1}{V_B} \qquad (6\text{-}22)$$

Combining equations (6-20), (6-21), and (6-22), the operating line for the stripping section becomes

$$y = \left(\frac{V_B + 1}{V_B}\right) x - \frac{x_W}{V_B} \qquad (6\text{-}23)$$

328   **Distillation**

If values of V_B and x_W are known, equation (6-23) can be plotted on the xy diagram as a straight line with intersection at $y = x_W$ on the 45° line and slope = $(V_B + 1)/V_B$, as shown in Figure 6.8. The equilibrium stages are stepped off in a manner similar to that described for the rectifying section. The very last stage at the bottom corresponds to the partial reboiler.

6.5.3 Feed Stage

Thus far, the McCabe–Thiele construction has not considered the feed to the column. In determining the operating lines for the rectifying and stripping sections, it is very important to note that although x_D and x_W can be selected independently, R and V_B are related by the feed phase condition.

Consider the section of the column at the tray where the feed is introduced. The quantities of the liquid and vapor streams change abruptly at this tray since the feed may consist of liquid, vapor, or a mixture of both (fraction vaporized = V_F/F). If, for example, the feed is a saturated liquid, L_{st} will exceed L by the amount of the added feed liquid. To establish a general relationship, an overall material balance around the feed plate is

$$F + L + V_{st} = V + L_{st} \tag{6-24}$$

and an energy balance,

$$FH_F + LH_{L,f-1} + V_{st}H_{G,f+1} = VH_{G,f} + L_{st}H_{L,f} \tag{6-25}$$

The vapors and liquids inside the tower are all saturated, and the molal enthalpies of all saturated vapors at this section are essentially identical since the temperature and composition changes over one tray are small. Then, $H_{G,\,f+1} = H_{G,\,f}$. The same is true of the molal enthalpies of the saturated liquids; therefore, $H_{L,\,f-1} = H_{L,\,f}$. equation (6-25) then becomes

$$\left(L_{st} - L\right)H_{L,f} = \left(V_{st} - V\right)H_{G,f} + FH_F \tag{6-26}$$

Combining this with equation (6-24):

$$\frac{L_{st} - L}{F} = \frac{H_{G,f} - H_F}{H_{G,f} - H_{L,f}} = q \tag{6-27}$$

Since there are only small composition changes across the feed plate, $H_{G,\,f}$ is basically the molal enthalpy that the feed would have if it were a saturated vapor, while $H_{L,\,f}$ is basically the molal enthalpy that the feed would have if it were a saturated liquid. Therefore, the quantity q in equation (6-27) is the energy required to convert 1 mol of feed from its condition H_F to a saturated vapor, divided by the molal latent heat of evaporation ($H_{G,\,f} - H_{L,\,f}$).

The feed may be introduced under five different thermal conditions ranging from a liquid well below its bubble point to a superheated vapor. For each condition, the value of q will be different, as the next table shows:

6.5 McCabe-Thiele Method for Trayed Towers 329

Feed condition	q-Value
Subcooled liquid	> 1
Bubble-point liquid	1
Partially vaporized	$L_F/F = 1 - V_F/F$
Dew-point vapor	0
Superheated vapor	< 0

Combining equations (6-24) and (6-27) yields

$$V_{st} = V + F\left(q - 1\right) \qquad (6\text{-}28)$$

which provides a convenient method for determining V_{st}.

The point of intersection of the two operating lines will help locate the exhausting-section operating line. This can be established as follows. Subtracting equation (6-20) from (6-15) gives

$$y\left(V - V_{st}\right) = L\left(L - L_{st}\right)x + Dx_D + Wx_W \qquad (6\text{-}29)$$

Further, by an overall material balance

$$Fz_F = Dx_D + Wx_W \qquad (6\text{-}30)$$

Combining equations (6-27) to (6-30) gives

$$y = \frac{q}{q-1}x - \frac{z_F}{q-1} \qquad (6\text{-}31)$$

Equation (6-31), representing the locus of intersection of operating lines (the q-line), is a straight line on the xy diagram of slope $= q/(q-1)$ and it passes through the point $x = y = z_F$ on the 45° line. Figure 6.9 shows the graphical interpretation of the q-line for typical cases. Here the operating-lines intersection is shown for a particular case of feed as a mixture of vapor and liquid. Following the placement of the rectifying-section operating line and the q-line, the stripping-section operating line is located by drawing a straight line from the point $x = y = x_W$ on the 45° line to and through the point of intersection of the q-line and the rectifying-section operating line as shown in Figure 6.9. The point of intersection must lie somewhere between the equilibrium curve and the 45° line. It is clear from this analysis that, for a given feed condition, fixing the reflux ratio automatically establishes the liquid/vapor ratio in the stripping section and the reboiler heat load as well.

As q changes from a value greater than 1 (subcooled liquid) to a value less than 0 (superheated vapor) the slope of the q-line changes from positive to negative, and back to positive as shown in Figure 6.9. For a saturated liquid feed, the q-line is vertical; for a saturated vapor, it is horizontal.

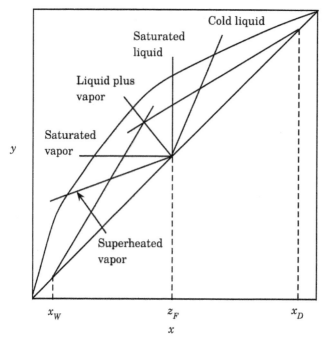

Figure 6.9 Location of the q-line for typical feed conditions.

6.5.4 Number of Equilibrium Stages and Feed-Stage Location

The q-line is useful in simplifying the graphical location of the stripping-section line, but the point of intersection of the two operating lines does not necessarily establish the demarcation between the stripping and rectifying sections of the fractionator. Rather, it is the introduction of the feed which governs the change from one operating line to the other. At least in the design of a new column some latitude in the introduction of the feed is available to the designer.

Consider the separation shown in Figure 6.10, for example. For a given feed, z_F and the q-line are fixed. For particular overhead and residue products, x_D and x_W are fixed. If the reflux ratio is specified, the location of the rectifying line is fixed and the stripping line must pass through the q-line at E. If the feed is introduced upon the fourth stage from the top (Figure 6.10a) the rectifying line is used for stages 1 through 3, and, beginning with stage 4, the stripping line must be used. The total number of ideal stages required is approximately 6.5, including the partial reboiler. If, on the other hand, the feed is introduced upon the sixth stage from the top (Figure 6.10b) the rectifying line is used for stages 1 through 5. The total number of ideal stages required this time is approximately 7.3. If the feed is introduced upon the third stage from the top (Figure 6.10c) the rectifying line is used for stages 1 and 2, and, beginning with stage 3, the stripping line must be used, for a total of approximately 7.2 stages. The least total number of trays will result if the steps on the diagram are kept as large as possible, a condition accomplished if the transition from one line to the other is made at the first opportunity after passing the operating-line intersection at E, as shown in Figure 6.10a. In the design of a new column, this is the practice to be followed.

6.5 McCabe-Thiele Method for Trayed Towers

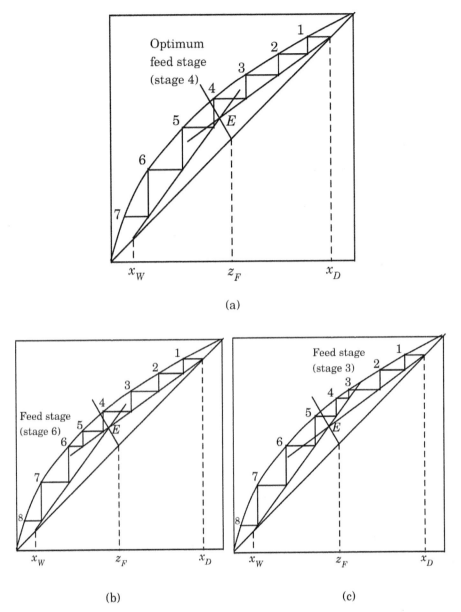

Figure 6.10 Location of feed stage: (a) optimum location; (b) location below optimum stage; (c) location above optimum stage.

In the adaptation of an existing column to a new separation, the point of introducing the feed is limited to the location of existing nozzles in the column wall. The slope of the operating lines and the product compositions to be realized must then be determined by trial and error, in order to obtain numbers of theoretical stages in the two sections of the fractionator consistent with the number of real trays in each section and the expected tray efficiency.

6.5.5 Limiting Conditions

For a given specification, a reflux ratio can be selected anywhere from the minimum, R_{min}, to an infinite value (total reflux) where all the overhead vapor is condensed and returned to the top stage (thus, no distillate is withdrawn!). The minimum reflux corresponds to an infinite number of stages, while an infinite reflux ratio corresponds to the minimum number of stages.

As the reflux ratio increases, the operating lines of both sections of the tower move toward the 45° line until, eventually, at total reflux they will coincide. Because the operating lines are located as far away as possible from the equilibrium curve, a minimum number of stages is required, as shown in Figure 6.11.

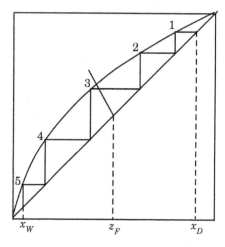

Figure 6.11 Total reflux and minimum stages.

As the reflux ratio decreases from the limiting case of total reflux, the intersection of the two operating lines and the q-line moves from 45° line toward the equilibrium curve. The number of ideal stages required increases because the operating lines move closer and closer to the equilibrium curve, thus requiring more and more steps to move from the top of the column to the bottom.

Finally, a limiting condition is reached when the point of intersection is on the equilibrium curve, as shown in Figure 6.12. For binary mixtures that are not highly nonideal, the typical case is shown in Figure 6.12a where the intersection, P, is at the feed stage. To reach that stage from either the rectifying section or the stripping section, an infinite number of stages is required. The point P is called a *pinch point*.

For a highly nonideal binary mixture, the pinch may occur at a stage above or below the feed stage. A pinch point above the feed stage is illustrated in Figure 6.12b, where the operating line for the rectifying section is tangent to the equilibrium curve at point P, well before the feed stage is reached. The slope of this tangent operating line cannot be reduced any further because it would then cross over the equilibrium curve, as shown on Figure 6.12b for the operating line through point K. The line through K clearly represents too small a reflux ratio. Because of the interdependence of the liquid/vapor ratios in the two sections of the column, a tangent operating line in the exhausting section may also set the minimum reflux ratio (see Problem 6.9).

6.5 McCabe-Thiele Method for Trayed Towers

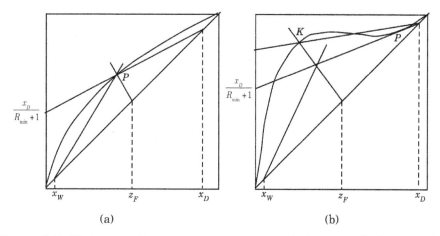

Figure 6.12 Minimum reflux ratio construction: (a) ideal or near ideal system; (b) nonideal system, pinch point above the feed stage.

6.5.6 Optimum Reflux Ratio

Any reflux ratio between the minimum and infinity will provide the desired separation, with the corresponding number of theoretical trays required varying from infinity to the minimum number. The reflux ratio to be used for a new design should be the optimum, the one for which the total cost of the operation will be the least. At the minimum reflux ratio, the column requires an infinite number of stages and, consequently, the fixed cost is infinite, but the operating costs (heat for the reboiler, condenser cooling water, power for the reflux pump) are least. As R increases, the number of trays rapidly decreases, but the column diameter increases owing to the larger quantities of recycled liquid and vapor per unit quantity of feed. The condenser, reboiler, and reflux pump must also be larger. The fixed costs therefore fall through a minimum value and rise to infinity again at total reflux. On the other hand, the operating costs increase almost directly with reflux ratio. The total cost, which is the sum of the fixed and operating costs, must therefore pass through a minimum at the optimum reflux ratio, as Figure 6.13 shows. This frequently occurs at a value of $R = R_{opt}$ in the range of $1.2 R_{min}$ to $1.5 R_{min}$ (Treybal, 1980).

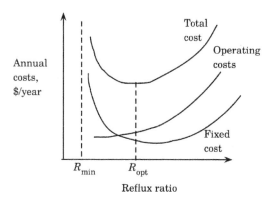

Figure 6.13 Optimum reflux ratio.

334   **Distillation**

Example 6.4 Rectification of a Benzene–Toluene Mixture

A trayed tower operating at 1 atm is to be designed to continuously distill 200 kmol/h (55.6 mol/s) of a binary mixture of 60 mol% benzene, 40 mol% toluene. A liquid distillate and a liquid bottoms product of 95 mol% and 5 mol% benzene, respectively, are to be produced. Before entering the column, the feed—originally at 298 K—is flash-vaporized at 1 atm to produce an equimolal vapor–liquid mixture ($V_F/F = L_F/F = 0.5$). A reflux ratio 30% above the minimum is specified. Calculate: (a) quantity of the products; (b) minimum number of theoretical stages, N_{min}; (c) minimum reflux ratio; (d) number of equilibrium stages and the optimal location of the feed stage for the reflux ratio specified; and (e) thermal load of the condenser, reboiler, and feed preheater.

Solution
(a) Calculate D and W. An overall material balance on benzene gives

$$0.60\left(200\right)=0.95D+0.05W$$

A total balance gives

$$200=D+W$$

Combining these two balances gives $D = 122.2$ kmol/h, $W = 77.8$ kmol/h.

(b) Benzene and toluene form ideal solutions; therefore the VLE data for this system at 1 atm are generated in the manner illustrated in Example 6.1. The parameters in equation (6-5) for benzene and toluene are (Reid et al., 1987):

Component	T_c, K	P_c, bar	A	B	C	D
Benzene	562.2	48.9	−6.983	1.332	−2.629	−3.333
Toluene	591.8	41.0	−7.286	1.381	−2.834	−2.792

Applying Raoult's law at 1 atm, the following VLE data are generated:

x	0.10	0.20	0.30	0.40	0.50	0.60	0.70	0.80	0.90
y^*	0.21	0.37	0.51	0.64	0.72	0.79	0.86	0.91	0.96
T	379.4	375.5	371.7	368.4	365.1	362.6	359.8	357.7	355.3

In Figure 6.14, where y and x refer to benzene—the light key—with $x_D = 0.95$ and $x_W = 0.05$, the minimum number of equilibrium stages is stepped off between the equilibrium curve and the 45° line, starting from the top, giving $N_{min} = 6.7$ stages.

6.5 McCabe-Thiele Method for Trayed Towers 335

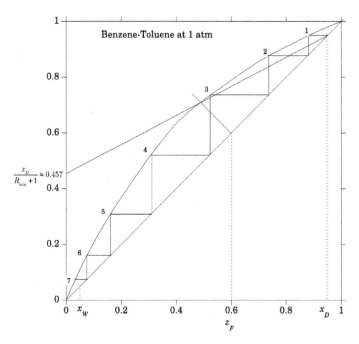

Figure 6.14 Determination of minimum stages and minimum reflux ratio for Example 6.4.

(c) Calculate the slope of the q-line: For this example, $q = L_F/F = 0.5$. From equation (6-31), the slope of the q-line $= q/(q-1) = -1.0$. In Figure 6.14, a q-line is drawn that has a slope of -1.0 and passes through the point $x = y = z_F = 0.6$ on the 45° line. For the minimum reflux ratio, an operating line for the rectifying section passes through the point $x = y = x_D = 0.95$ on the 45° line and through the point of intersection of the q-line and the equilibrium curve, to a y-intercept of 0.457. Then,

$$\frac{x_D}{R_{\min}+1} = \frac{0.95}{R_{\min}+1} = 0.457$$

Therefore, $R_{\min} = 1.079$ mol reflux/mol distillate.

(d) For $R = 1.3 R_{\min} = 1.403$, the y-intercept of the rectifying-section operating line is $(0.95)/(1.403 + 1) = 0.395$. Figure 6.15 shows the location of both operating lines. The operating line for the stripping section is drawn to pass through the point $x = y = x_W = 0.05$ on the 45° line and the point of intersection of the q-line and the rectifying-section operating line. The number of equilibrium stages is stepped off between first, the rectifying-section operating line and the equilibrium curve, and then, the equilibrium curve and the stripping-section operating line. For the optimal feed-stage location, the transition from one operating line to the other occurs at the first opportunity after passing the operating-line intersection. Therefore, Figure 6.15 shows that the feed is to be introduced on the sixth ideal stage from the top. A total of 13 equilibrium stages, including the reboiler, is required, and the tower must then contain 12 ideal stages.

336   **Distillation**

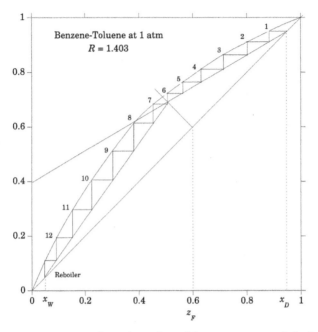

Figure 6.15 Determination of number of equilibrium stages and feed-stage location for Example 6.4.

(e) Calculate the molal flows of liquid and vapor throughout the column:

$$L = L_0 = RD = (1.403)(122.2) = 171.4 \text{ kmol/h}$$

$$V = L + D = 171.4 + 122.2 = 293.6 \text{ kmol/h}$$

From equation (6-27),

$$L_{st} = L + qF = 171.4 + (0.5)(200) = 271.4 \text{ kmol/h}$$

From equation (6-28),

$$V_{st} = V + (q-1)F = 293.6 \ (0.5)(200) = 193.6 \text{ kmol/h}$$

Consider the feed preheater, a flash-vaporization unit, shown schematically in Figure 6.16. For 50% vaporization of the feed (z_F = 0.60), from calculations similar to those illustrated in Example 6.1, the separator temperature is T_F = 365.6 K and the equilibrium compositions are y_F = 0.707, x_F = 0.493. Since benzene (A) and toluene (B) form ideal solutions, molar liquid and vapor enthalpies with respect to a reference temperature T_0 can be calculated from equations (6-6) and (6-7). The following data were obtained from Perry and Chilton (1973):

6.5 McCabe-Thiele Method for Trayed Towers

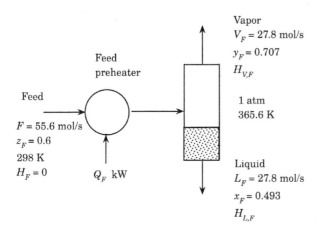

Figure 6.16 Feed preheater for Example 6.4.

Latent heats of vaporization at $T_0 = 298$ K:

$$\lambda_A = 33.9 \text{ kJ/mol}, \quad \lambda_B = 38.0 \text{ kJ/mol}$$

Heat capacities of liquids (298 – 366 K):

$$C_{L,A} = 0.147 \text{ kJ/mol K}, \quad C_{L,B} = 0.174 \text{ kJ/mol K}$$

Heat capacities of gases, average in the range 298 – 366 K (Smith et al., 1996):

$$C_{p,A} = 0.094 \text{ kJ/mol K}, \quad C_{p,B} = 0.118 \text{ kJ/mol K}$$

Substituting in equation (6-6), $H_F = 0$,

$$H_{LF} = \left[x_F C_{L,A} + (1 - x_F) C_{L,B}\right](T_F - T_0) = 10.86 \text{ kJ/mol of liquid feed}$$

From equation (6-7),

$$H_{VF} = \left[y_F C_{p,A} + (1 - y_F) C_{p,B}\right](T_F - T_0) + y_F \lambda_A + (1 - y_F) \lambda_B$$
$$= 41.93 \text{ kJ/mol of vapor feed}$$

From equation (6-3),

$$Q_F = V_F H_{VF} + L_F H_{LF} - F H_F = 1{,}450 \text{ kW}$$

Consider now the total condenser. Saturated vapor containing 95 mol% benzene at a dew-point temperature of 355.8 K will enter the condenser at the rate of 293.6 kmol/h (81.6 mol/s). It will leave the condenser as saturated liquid of the same composition and a bubble-point temperature of 354.3 K. An energy balance around the condenser is as follows:

$$Q_c = V\left(H_{V,1} - H_{L,0}\right)$$

From equation (6-7),

$$H_{V,1} = [y_1 C_{p,A} + (1-y_1)C_{p,B}](T_1 - T_0) + y_1 \lambda_A + (1-y_1)\lambda_B$$
$$= 40.0 \text{ kJ/mol of vapor}$$

From equation (6-6),

$$H_{L,0} = [x_0 C_{L,A} + (1-x_0)C_{L,B}](T_{L,0} - T_0)$$
$$= 8.3 \text{ kJ/mol of liquid}$$

$$Q_c = V(H_{V,1} - H_{L,0}) = 2{,}590 \text{ kW}$$

Consider now the partial reboiler shown schematically in Figure 6.17. Saturated liquid leaving the last equilibrium stage in the tower enters the reboiler at a rate of 271.4 kmol/h (75.4 mol/s). Saturated vapor leaves the reboiler and returns to the column at the rate of 193.6 kmol/h (53.8 mol/s), while the liquid residue is withdrawn as the bottoms product at the rate of 77.8 kmol/h (21.6 mol/s). The bottoms product is a saturated liquid with a composition of 5 mol% benzene. A flash-vaporization calculation is done in which the fraction vaporized is known (53.8/75.4 = 0.714) and the concentration of the liquid residue is fixed at $x_W = 0.05$. The calculations yield the following results: $T_R = 381.6$ K, $x_{12} = 0.093$, $y_{13} = 0.111$. The liquid entering the reboiler is at its bubble point, which is $T_{12} = 379.7$ K. An energy balance around the reboiler is

$$Q_R = V_{st} H_{V,13} + W H_{L,W} - L_{st} H_{L,12}$$

From equation (6-7),

$$H_{V,13} = [y_{13} C_{p,A} + (1-y_{13})C_{p,B}](T_R - T_0) + y_{13}\lambda_A + (1-y_{13})\lambda_B$$
$$= 47.19 \text{ kJ/mol of vapor feed}$$

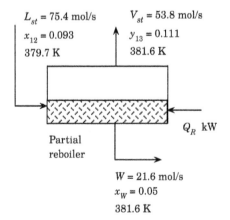

Figure 6.17 Partial reboiler for Example 6.4.

6.5 McCabe-Thiele Method for Trayed Towers 339

From equation (6-6),

$$H_{L,\,12} = \left[x_{12}C_{L,A} + \left(1 - x_{12}\right)C_{L,B}\right]\left(T_{12} - T_0\right)$$
$$= 14.01 \text{ kJ/mol of liquid}$$

$$H_{L,W} = \left[x_W C_{L,A} + \left(1 - x_W\right)C_{L,B}\right]\left(T_R - T_0\right)$$
$$= 14.43 \text{ kJ/mol of liquid}$$

$$Q_R = \left(53.8\right)\left(47.19\right) + \left(21.6\right)\left(14.43\right) - \left(75.4\right)\left(14.01\right) = 1{,}790 \text{ kW}$$

6.5.7 Large Number of Stages

The McCabe–Thiele graphical construction is difficult to apply when conditions of *relative volatility* and/or product purities are such that a large number of stages must be stepped off. In that event, one of the following techniques can be used to determine the stage requirements.

1. Separate plots of expanded scales and/or larger dimensions are used for stepping off stages at the ends of the xy diagram. For example, the additional plots may cover just the regions (1) 0.95 to 1.0 and (2) 0 to 0.05.
2. The stages are determined by combining the McCabe–Thiele graphical construction, for a suitable region in the middle, with the Kremser equations for the low and/or high ends, where absorption and stripping factors are almost constant.
3. If the equilibrium data are given in analytical form, a McCabe–Thiele computer program can be used. Appendix F is an example of Mathcad programs to implement the McCabe–Thiele method (Hwalek, 2001). They generate the required VLE data from Antoine equation for vapor pressure and the NRTL equation for liquid-phase activity coefficients. Appendix F-1 is for column feed as saturated liquid; Appendix F-2 is for column feed as saturated vapor.

Example 6.5 Rectification of a Methanol–Water Solution

A methanol (A)–water (B) solution containing 36 mol% methanol is to be continuously rectified at 1 atm pressure at a rate of 216.8 kmol/h (60.22 mol/s) to provide a distillate containing 99.9 mol% methanol and a residue containing 0.1 mol% methanol. The feed to the column is to be preheated to its bubble point. The distillate is to be totally condensed and the reflux returned at the bubble point. A reflux ratio of 1.5 times the minimum will be used. Determine the number of theoretical stages required if the feed is introduced at the optimal location:

(a) using the Mathcad program of Appendix F-1,
(b) combining the McCabe–Thiele graphical method with the Kremser equations.

340 **Distillation**

Solution

(a) The Antoine constants for methanol (1) and water (2) are (Smith et al., 1996): A_1 = 16.5938, B_1 = 3644.3 K, C_1 = 239.76 K; A_2 = 16.2620, B_2 = 3799.89 K, C_2 = 226.35 K. The NRTL constants are b_{12} = 253.88 cal/mol, b_{21} = 845.21 cal/mol, α = 0.2994 (Smith et al., 1996). Table 6.1 presents the VLE data for the system generated by the computer program in Appendix F-1.

A summary of the results obtained in Appendix F-1 for R = 1.5 R_{min} is as follows:

Distillate flow rate = 21.66 mol/s   Bottoms flow rate = 38.56 mol/s
Minimum reflux ratio = 0.84   Reflux ratio = 1.26
N = 22 ideal stages   Feed introduced in stage 16 from top
L = 27.29 mol/s V = 48.95 mol/s
L_{st} = 87.51 mol/s V_{st} = V= 48.95 mol/s

(b) Figure 6.18 shows the McCabe–Thiele construction for the region of x from 0.017 to 0.961, where the stages have been stepped off in two directions starting from the feed stage. Notice that, because of the high purity of the products, the operating lines originate basically from the corners of the xy diagram where the 45° line ends. In this middle region, eight stages are stepped off above the feed stage and four below it, for a total of 13 stages including the feed stage. The Kremser equations can now be applied to determine the remaining stages needed to achieve the desired high purities for the distillate and bottoms.

Let now us consider the number of additional ideal stages required in the stripping section (N_S) to achieve the desired bottoms concentration x_W = 0.001. This portion of the column behaves like a methanol stripper with very dilute solutions of

Table 6.1 VLE for the System Methanol–Water at 1 atm

CH_3OH mol fraction in liquid, x	CH_3OH mol fraction in gas, y^*	Temp. (K)
0.000	0.000	373.1
0.100	0.416	361.2
0.200	0.576	355.3
0.300	0.667	351.5
0.400	0.731	348.8
0.500	0.784	346.5
0.600	0.832	344.5
0.700	0.876	342.7
0.800	0.919	340.9
0.900	0.960	339.3
1.000	1.000	337.7

Source: Computer program in Appendix F-1. Data from Hwalek (2001).

6.5 McCabe-Thiele Method for Trayed Towers

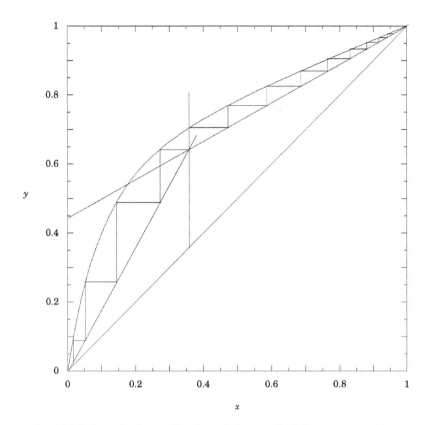

Figure 6.18 McCabe–Thiele construction for Example 6.5, from $x = 0.017$ to $x = 0.961$.

methanol in water. From the construction in Figure 6.18, the liquid entering the stripper contains 1.7 mol% of methanol, while the liquid leaving the section contains only 0.1 mol% methanol. Since the operating line for the stripping section of the distillation column ends on the 45° line at $y = x = x_W$, the stripper operates as if the vapor entering it had a concentration $y = x_W = 0.001$. To calculate the absorption factor for the "stripper," we must estimate the slope of the equilibrium curve in the limit as x_A tends to zero (m_{st}). In that portion of the column, the temperature is very close to the normal boiling point of pure water, then $T_{st} = 373.1$ K. From the modified Raoult's law,

$$m_{st} = \frac{P_A(T_{st})\gamma_A^\infty(T_{st})}{P} \tag{6-32}$$

The infinite-dilution activity coefficients for liquid solutions described by the NRTL equation are given by the following equations (Smith et al., 1996)

$$\ln \gamma_A^\infty(T) = \tau_{21}(T) + \tau_{12}(T)\exp\left[-\alpha\tau_{12}(T)\right] \tag{6-33}$$

$$\ln \gamma_B^\infty(T) = \tau_{12}(T) + \tau_{21}(T)\exp\left[-\alpha\tau_{21}(T)\right] \tag{6-34}$$

342   **Distillation**

where

$$\tau_{12}(T) = \frac{b_{12}}{RT} \qquad \tau_{21}(T) = \frac{b_{21}}{RT} \tag{6-35}$$

Substituting in equations (6-32) to (6-35), $m_{st} = 7.452$. Then, the absorption factor for the stripper is $A_{st} = L_{st}/(m_{st}V_{st}) = 87.51/(7.542 \times 48.95) = 0.24$. Substituting in equation (5-1), $N_S = 1.9$ stages.

Let us consider now the number of additional ideal stages required in the rectifying section (N_R) to achieve the desired distillate concentration $x_D = 0.001$. This portion of the column behaves like a water absorber with very dilute solutions of water in methanol. From the construction in Figure 6.18, the vapor entering the absorber contains 2.3 mol% water (97.7 mol% methanol), while the vapor leaving it contains only 0.1 mol% water. The liquid entering it has a concentration $x = x_D = 0.001$. To calculate the absorption factor for the absorber, we must estimate the slope of the equilibrium curve in the limit as x_B tends to zero (m_{ab}). In that portion of the column, the temperature is very close to the normal boiling point of pure methanol, then $T_{ab} = 337.7$ K. From the modified Raoult's law,

$$m_{ab} = \frac{P_B(T_{ab})\gamma_B^\infty(T_{ab})}{P} \tag{6-36}$$

Substituting in equations (6-34) to (6-36), $m_{ab} = 0.393$. Then, the absorption factor for the "absorber" is $A_{ab} = L/(m_{ab}V) = 27.29/(0.393 \times 48.95) = 1.418$. Substituting in equation (5-3), $N_R = 7.0$ stages.

Combining these results with those obtained by the McCabe–Thiele graphical construction for the intermediate portion of the column, we have $8 + 7 = 15$ stages above the feed stage, and $4 + 1.9 = 5.9$ stages below the feed stage, for a total of 21.9 ideal stages, and feed introduced in stage 16 from the top. These results agree with those obtained in part a using the computer program in Appendix F-1.

6.5.8 Use of Open Steam

Ordinarily, heat is applied at the base of the fractionator by means of a reboiler. However, when an aqueous solution is distilled to give the nonaqueous solute as the distillate and the water is removed as the bottoms product, the heat required may be provided by the use of open steam at the bottom of the tower. The reboiler is then dispensed with. For a given reflux ratio and overhead composition, however, more trays will be required in the tower.

Figure 6.19 shows the effects of using open steam instead of a reboiler. Overall material balances are equation (6-30) and

$$F + V_{st} = D + W \tag{6-37}$$

where V_{st} is the molal flow of steam used, assumed saturated at the fractionator pressure. The rectifying operating line is located as usual, and the slope of the stripping line, L_{st}/V_{st}, is related to L/V and the feed conditions in the same manner as before. However, the stripping line now ends at the point $y = 0$, $x = x_W$, as shown

6.5 McCabe-Thiele Method for Trayed Towers

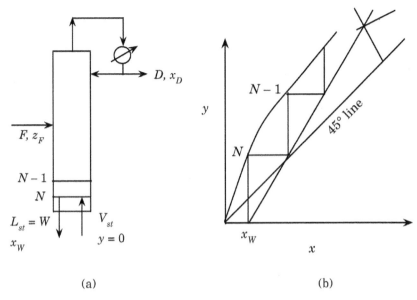

Figure 6.19 Use of open steam instead of a reboiler.

in Figure 6.19b. The graphical tray construction must therefore be continued to the x-axis of the diagram.

If the steam entering the tower, V_{N+1}, is superheated, it will vaporize liquid on tray N to the extent necessary to bring it to saturation. Then,

$$V_{st} = V_{N+1}\left[1 + \frac{H_{G,N+1} - H_{G,sat}}{\lambda_B}\right]$$

$$L_{st} = V_{st} - V_{N+1} + W$$

(6-38)

where $H_{G,sat}$ is the molar enthalpy of steam saturated at the column pressure, and λ_B is the molar heat of vaporization of water at the saturation temperature corresponding to the column pressure.

6.5.9 Tray Efficiencies

Methods for estimating tray efficiencies were discussed in Chapter 4. Murphree vapor efficiencies are most simply used graphically on the xy diagram. Overall efficiencies $\mathbf{E}_O$ strictly have meaning only when the Murphree efficiency of all trays is the same, and the equilibrium and operating line are both straight over the concentration range considered. Nevertheless, empirical correlations to estimate $\mathbf{E}_O$ are easy to use and are useful for rough estimates, if not for final designs.

The most widely used empirical approach to estimate $\mathbf{E}_O$ is the O'Connell correlation (illustrated in Example 6.7), although it has been pointed out that it

344 **Distillation**

usually underpredicts the overall efficiency in the distillation of water solutions (Wankat, 1988; Kister, 1992):

$$\mathbf{E}_O = 0.52782 - 0.27511\log_{10}\left(\alpha\mu_L\right) + 0.044923\left[\log_{10}\left(\alpha\mu_L\right)\right]^2 \quad (6\text{-}39)$$

$$0.1 \text{ cP} \leq \alpha\mu_L \leq 10 \text{ cP}$$

where

μ_L = viscosity of the feed as liquid at the average temperature of the tower, expressed in cP

α = average relative volatility. The relative volatility is defined, for a binary system, as the ratio of the equilibrium concentration ratio of A and B in one phase to that in the other; then

$$\alpha = \frac{y^*\left(1-x\right)}{x\left(1-y^*\right)} \quad (6\text{-}40)$$

Example 6.6 Rectification of an Ethanol–Water Solution

In the development of a new process, it will be necessary to fractionate 34.43 kmol/h of an ethanol–water solution containing 30 mol% ethanol, available at the bubble point. It is desired to recover 99.9% of the ethanol in a distillate containing 80 mol% ethanol. The tower will be designed for operation at 1 atm, with a reflux ratio twice the minimum. Open steam at 1 atm and 523 K is available for heating. Calculate

(a) Product rates, kmol/h, and rate of steam use, kg/h.
(b) Number of theoretical trays required.
(c) For a sieve-tray tower of "conventional design," determine the tower diameter for the gas velocity not to exceed 70% of the flooding velocity.
(d) Estimate the number of real trays combining the Murphree efficiency and the McCabe–Thiele method and calculate the total tower height.

Solution

(a) Table 6.2 shows the VLE data for the system ethanol–water at 1 atm. A minimum-boiling azeotropic mixture is formed at 89.43 mol% ethanol. The distillate concentration of 80 mol% is safely below the azeotropic composition. To recover 99.9% of the alcohol in the overhead product,

$$D = \frac{0.999Fz_F}{x_D} = \frac{0.999 \times 34.43 \times 0.3}{0.8} = 12.90 \text{ kmol/h}$$

Figure 6.20 is the xy diagram for this example. Since the feed is a saturated liquid, the q-line is vertical ($q = 1.0$). The pinch point that determines the minimum reflux ratio is above the feed stage. From the y-intercept of the pinch-point tangent rectifying operating line, $x_D/(R_{min} + 1) = 0.387$, $R_{min} = 1.067$. For a reflux ratio that is

6.5 McCabe-Thiele Method for Trayed Towers

Table 6.2 VLE for the System Ethanol–Water at 1 atm

Ethanol mol fraction in liquid, x	Ethanol mol fraction in gas, y^*	Temp. (K)
0.000	0.000	373.10
0.0721	0.3891	362.10
0.1238	0.4704	358.40
0.2377	0.5445	355.80
0.2608	0.5580	355.40
0.3965	0.6122	353.80
0.5198	0.6599	352.80
0.5732	0.6841	352.40
0.6763	0.7385	351.84
0.7472	0.7815	351.51
0.8943	0.8943	351.25
1.000	1.000	351.40

Source: Data from Wankat (1988).

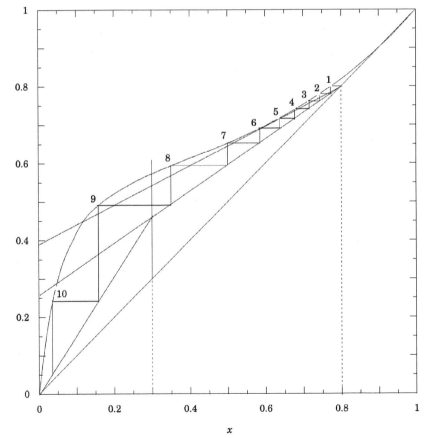

Figure 6.20 Ideal stages for the fractionator of Example 6.6.

346   **Distillation**

twice the minimum, $R = 2.134$. Calculate the molal flows of liquid and vapor throughout the column:

$$L = L_0 = RD = (2.134)(12.90) = 27.53 \text{ kmol/h}$$

$$V = L + D = 27.53 + 12.90 = 40.43 \text{ kmol/h}$$

$$L_{st} = L + qF = 27.53 + (1.0)(34.43) = 61.96 \text{ kmol/h}$$

$$V_{st} = V + (q-1)F = V = 40.43 \text{ kmol/h}$$

From the Steam Tables (Smith et al., 1996), the enthalpy of saturated steam at 101.3 kPa is $H_{G,sat} = 2676$ kJ/kg; the enthalpy of superheated steam at 101.3 kPa and 523 K is $H_{G,N+1} = 2825.8$ kJ/kg; the latent heat of vaporization of water at 373.1 K is $\lambda_B = 2256.9$ kJ/kg. Substituting in equation (6-38), the flow rate of open steam required is $V_{N+1} = 37.93$ kmol/h (or 682.74 kg/h), and the flow rate of the bottoms product is $W = 59.46$ kmol/h. Since 0.1% of the ethanol in the feed will be lost in the bottoms product,

$$x_W = \frac{0.001 F z_F}{W} = \frac{0.001 \times 34.43 \times 0.3}{59.46} = 1.74 \times 10^{-4}$$

(b) For $R = 2.134$, the rectifying operating line y-intercept in Figure 6.20 is $x_D/(R+1) = 0.255$. The operating line for the stripping section, on this scale of plot, for all practical purposes passes through the origin. Ideal stages are stepped off as usual starting at the top of the column. The feed should be introduced at ideal stage number 9 from the top. Below ideal stage 10, Kremser equations should be used to determine the additional number of trays required to reach the bottoms concentration, N_S. From Figure 6.20, the concentration of the liquid leaving ideal stage 10 is $x = 0.025$, and from this concentration down to $x_W = 1.74 \times 10^{-4}$ the equilibrium curve is essentially straight ($m = 8.95$). The absorption factor $A = L_{st}/mV_{st} = 61.96/[(8.95)(40.43)] = 0.171$. When open steam is used, the concentration of the entering vapor is $y = 0$. Substituting in equation (5-1), $N_S = 2.71$ stages. Therefore, the total number of ideal stages required is 12.71.

(c) For a single-feed, two-product column, there is generally a need to carry out the column sizing calculations for the top tray, bottom tray, tray just above the feed, and tray just below the feed. The column is then designed for the more severe conditions (Kister, 1992). Consider first the bottom tray. In this part of the tower both the liquid and the vapor are virtually pure water ($M_G = M_L = 18$). Therefore, the liquid mass flow rate is $L'_{st} = (61.96)(18)/3600 = 0.3098$ kg/s; the vapor mass flow rate is $V'_{st} = 0.2022$ kg/s. The gas density, ρ_G, is from the ideal gas law,

$$\rho_G = \frac{P M_G}{RT} = \frac{101.3 \times 18}{8.314 \times 373.1} = 0.588 \text{ kg/m}^3$$

The density of the liquid is from the Steam Tables, $\rho_L = 958$ kg/m^3. The viscosity of the vapor is $\mu_G = 1.28 \times 10^{-5}$ kg/m-s (Holman, 1990). The foaming factor for aqueous solutions of alcohols is 1.0 (Kister, 1992).

The surface tension of liquid water at 293 K is 72.8 dyn/cm. It can be corrected to a different temperature from

6.5 McCabe-Thiele Method for Trayed Towers

$$\sigma_2 = \sigma_1 \left[\frac{1-T_{r2}}{1-T_{r1}} \right]^{1.22}$$

(6-41)

For water, T_c = 647.3 K; therefore, the surface tension at 373.1 K is

$$\sigma_2 = 72.8 \left[\frac{1-\left(\dfrac{293}{647.3}\right)}{1-\left(\dfrac{373.13}{647.3}\right)} \right]^{1.22} = 53.3 \text{ dyn/cm}$$

The slope of the equilibrium curve is m = 8.95. The gas-phase diffusivity is from the Wilke–Lee equation, D_G = 0.177 cm^2/s. The liquid-phase diffusivity is from the Hayduk–Minhas correlation for aqueous solutions, D_L = 5.54 × 10^{-5} cm^2/s. Take d_o = 4.5 mm on an equilateral-triangular pitch 12.0 mm between hole centers, punched in stainless steel sheet metal 2 mm thick. Use a weir height of 50 mm. Design for a 70% approach to the flood velocity. From the Mathcad program in Appendix E, we get the following results for conditions at the bottom tray:

$$D=0.423 \text{ m} \qquad t=0.5 \text{ m} \qquad \mathbf{E}_{MGE}=0.567$$

Consider now conditions at the top tray. Both the liquid entering the tray and the vapor leaving it contain 80 mol% ethanol. The temperature there is around 351.4 K (see Table 6.2). The average molecular weight of both streams is 40.5 kg/kmol. Therefore, the liquid mass flow rate is L' = (27.53)(40.5)/3600 = 0.310 kg/s; the vapor mass flow rate is V' = (40.43)(40.5)/3600 = 0.455 kg/s. For the conditions at the top tray, the following data are available from Wankat (1988):

$$\rho_G=1.393 \text{ kg/m}^3 \qquad \rho_L=772 \text{ kg/m}^3 \qquad \sigma=18.2 \text{ dyn/cm}$$
$$m=0.63 \qquad\qquad D_L=3.64\times10^{-5}\text{cm}^2/\text{s}$$

From the Lucas method, μ_G = 1.04 × 10^{-5} kg/m·s. From the Wilke–Lee equation, D_G = 0.157 cm^2/s. From the Mathcad program in Appendix E, we get the following results for conditions at the top tray:

$$D=0.6 \text{ m} \qquad\qquad t=0.5 \text{ m} \qquad\qquad \mathbf{E}_{MGE}=0.80$$
$$\Delta P=336 \text{ Pa/tray} \qquad \text{Fr}_o=1.03 \qquad\qquad E=0.0193$$

Similar calculations for the trays just above and just below the feed tray show that the conditions at the top tray govern the design of the tower. Therefore, a tower diameter of 0.6 m is specified, with tray spacing of 0.5 m.

(d) The Murphree vapor efficiency corrected for entrainment remains relatively constant at $\mathbf{E}_{MGE}$ = 0.80 for the rectifying section. For the stripping section, the efficiency is recalculated for the specified tower diameter D = 0.60 m. For this diameter, flow conditions at the bottom tray correspond to only a 34% approach to flooding, and $\mathbf{E}_{MGE}$ = 0.78. Therefore, it is appropriate to assume that the Murphree vapor efficiency for this problem remains basically constant throughout the column at an average value of $\mathbf{E}_{MGE}$ = 0.79.

348   **Distillation**

Figure 6.21 shows the xy diagram for this problem in which a pseudoequilibrium curve (dotted-line) has been added such that the vertical distance from the operating lines to the pseudoequilibrium curve is 79% of the distance from the operating lines to the original equilibrium curve. Real stages are stepped off between the operating lines and the pseudoequilibrium curve. Feed is introduced in tray 11 from the top. The liquid leaving tray 14 contains 0.017 mol fraction of ethanol. From calculations similar to those in part (b), 2.50 additional ideal stages are required to reduce the liquid concentration from 0.017 mol fraction to $x_W = 1.74 \times 10^{-4}$. In that part of the column, $\mathbf{E}_{MGE} = 0.78$, $A = 0.171$. The overall efficiency for that part of the column can be estimated from equation (5-5):

$$\mathbf{E}_O = \frac{\ln\left[1 + \mathbf{E}_{MGE}\left(A^{-1} - 1\right)\right]}{\ln\left(A^{-1}\right)} = 0.886$$

Therefore, the number of additional real stages is 2.5/0.886 = 2.81 stages, which must be rounded up to the next integer, in this case, 3 additional stages. A total of 17 real stages is required, with the feed introduced in stage 11 from the top. In this case, all 17 stages must be in the column since open steam is used instead of a partial reboiler. For a tray spacing of 0.5 m, the total height between the bottom and top tray is 8.5 m. Add 1 m above the top tray for entrainment separation and add 3 m beneath the bottom tray for bottoms surge capacity (Seader and Henley, 1998). The total column height is then 12.5 m.

Example 6.7 Overall Efficiency of a Benzene–Toluene Fractionator

A binary distillation operation separates 78.1 mol/s of a mixture of 46 mol% benzene and 54 mol% toluene. The purpose of the sieve-tray column (equivalent to 20 theoretical stages and a partial reboiler) is to separate the feed into a liquid distillate of 99 mol% benzene and a liquid bottoms product of 98 mol% toluene. The pressure in the reboiler is 141 kPa. In this range of pressure, benzene and toluene form ideal solutions with a relative volatility of 2.26 at the bottom tray (395 K) and 2.52 at the top tray (360 K). The feed to the column is a mixture of vapor and liquid; 23.4 mol% of the feed is vaporized. A reflux ratio 30% above the minimum is used. The diameter of the column is 1.53 m, which corresponds to 84% of flooding at the top tray, 81% of flooding at the bottom tray, and an average gas-pressure drop of 700 Pa/tray.

(a) Estimate the overall tray efficiency using the O'Connell correlation.

(b) Calculate the number of real trays and the total tower height.

(c) Estimate the total gas-pressure drop through the column.

Solution

(a) The average temperature in the column is (365 + 390)/2 = 377.5 K. The average relative volatility is $\alpha = (2.26 + 2.52)/2 = 2.39$. The viscosity of an ideal binary solution can be estimated in terms of the viscosities of the pure components at the solution temperature and the solution composition from (Reid et al., 1987)

6.5 McCabe-Thiele Method for Trayed Towers

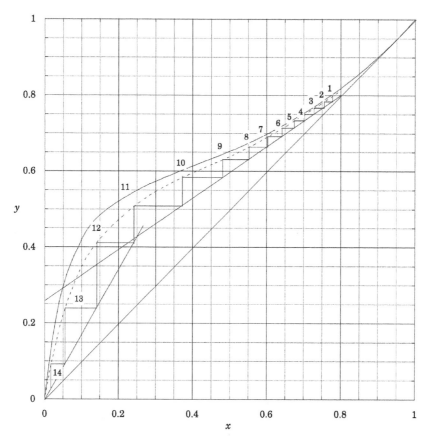

Figure 6.21 Real stages for the fractionator of Example 6.6.

$$\mu_L = \mu_A^{x_A} \mu_B^{x_B} \tag{6-42}$$

The viscosity of liquid benzene and liquid toluene as functions of temperature are given by (Reid et al., 1987)

$$\ln \mu = A + \frac{B}{T} + CT + DT^2 \tag{6-43}$$

μ in cP; T in K

where

Component	A	B	C	D × 10^5
Benzene	4.612	148.9	−0.0254	2.222
Toluene	−5.878	1287	0.00458	−0.450

Source: Data from Reid et al. (1987).

350   **Distillation**

At the average column temperature of 377.5 K, $\mu_A = 0.242$ cP, $\mu_B = 0.251$ cP, $\mu_L = 0.247$ cP. Then, $\alpha\mu_L = 0.5903$; $\log_{10}(\alpha\mu_L) = -0.229$. From the O'Connell correlation, $E_O = 0.593$.

(b) Number of real trays = number of ideal trays/E_O = 33.7 or call it 34 trays. For a tower diameter of 1.53 m, Table 4.3 recommends a tray spacing of 0.6 m. Adding 1 m over the top tray as entrainment separator and 3 m beneath the bottom tray for bottoms surge capacity, the total column height is 4 + (34)(0.6) = 24.4 m.

(c) Total gas-pressure drop = (0.700 kPa/tray)(34 trays) = 23.8 kPa.

6.6 BINARY DISTILLATION IN PACKED TOWERS

Your objectives in studying this section are to be able to:

1. Understand the advantages of using packed versus trayed towers for certain distillation applications.
2. Combine material-balances information from the McCabe–Thiele graphical method with interphase mass-transfer considerations to determine the packed height of the rectifying and stripping sections of a continuous-contact binary distillation tower.

With the availability of economical and efficient packings, packed towers are finding increasing use in new distillation processes and for retrofitting existing trayed towers. They are particularly useful in applications where pressure drop must be low, as in low-pressure distillation, and where liquid holdup must be small, such as when distilling heat-sensitive materials whose exposure to high temperatures must be minimized.

As in the case of packed absorbers, the changes in concentration with height produced by these towers are continuous rather than stepwise as for tray towers, and the computation procedure must take this into consideration. Figure 6.22a shows a schematic diagram of a packed-tower fractionator. Like tray towers, it must be provided with a reboiler at the bottom (or open steam may be used if an aqueous residue is produced), a condenser, means of returning reflux and reboiled vapor, as well as means for introducing feed. The last can be accomplished by providing a short, unpacked section at the feed entry, with adequate distribution of liquid over the top of the exhausting section.

The operating diagram, Figure 6.22b, is determined exactly as for tray towers using the McCabe–Thiele method. Equations for operating lines already derived for trays are also applicable, except that tray-number subscripts are omitted. The operating lines are then simply the relation between x and y, the bulk liquid and gas compositions, prevailing at each horizontal section of the tower. As before, the change from rectifying- to stripping-section operating lines is made at the point where the feed is actually introduced, and for new designs a shorter column results, for a given reflux ratio, if this is done at the intersection of the operating lines. In what follows, this practice is assumed.

6.6 Binary Distillation in Packed Towers

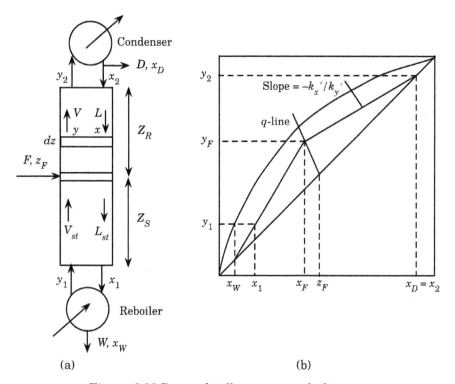

Figure 6.22 Binary distillation in a packed tower.

By material balance over an incremental section of packed height in the rectifying section, assuming equimolar counterdiffusion ($N_B = -N_A$)

$$Vdy = k'_y a_h (y_i - y) S dz = L dx = k'_x a_h (x - x_i) S dz \qquad (6\text{-}44)$$

Integrating over the entire rectifying section

$$Z_R = \int_{y_F}^{y_2} \frac{V dy}{k'_y a_h S(y_i - y)} = \int_{x_F}^{x_D} \frac{L dx}{k'_x a_h S(x - x_i)} \qquad (6\text{-}45)$$

Integrating over the stripping section

$$Z_S = \int_{y_1}^{y_F} \frac{V_{st} dy}{k'_y a_h S(y_i - y)} = \int_{x_1}^{x_F} \frac{L_{st} dx}{k'_x a_h S(x - x_i)} \qquad (6\text{-}46)$$

For any point (x, y) on the operating lines, the corresponding point (x_i, y_i) on the equilibrium curve is obtained at the intersection with a line of slope $-k'_x/k'_y$ drawn from (x, y), as shown in Figure 6.22b. For $k'_x > k'_y$, so that the principal resistance to mass transfer lies within the vapor, $(y_i - y)$ is more accurately read than $(x - x_i)$. The middle integral of equations (6-44) and (6-45) is then best used. For $k'_x < k'_y$, it is better to use the last integral. In this manner, variations in V, L, the mass-transfer coefficients, and the interfacial area with location on the operating lines are readily dealt with.

352   Distillation

Example 6.8 Benzene–Toluene Fractionator Using a Structured Packed Tower

Determine suitable dimensions of packed sections of a tower for the benzene–toluene separation of Example 6.4, using Montz B2-300 metal structured packing. Design for a gas-pressure drop not to exceed 400 Pa/m. The hydraulic and mass-transfer characteristics for this packing are (Billet and Schultes, 1999)

$$C_{S,T}=3.098 \quad a=300 \text{ m}^{-1} \quad \varepsilon=0.930 \quad C_h=0.482$$
$$C_p=0.295 \quad C_L=1.165 \quad C_{F,T}=2.464 \quad C_V=0.422$$

Solution

From Example 6.4, L = 171.4 kmol/h, V = 293.6 kmol/h, L_{st} = 271.4 kmol/h, and V_{st} = 193.6 kmol/h. Vapor and liquid mass-flow rates throughout the tower are

Rectifying Section

x	T_L, K	y	T_G, K	M_G	V', kg/s	M_L	L', kg/s
0.900	355.4	0.922	357.1	79.2	6.459	79.5	3.785
0.800	357.6	0.861	359.8	80.0	6.524	80.9	3.852
0.700	360.0	0.802	362.1	80.9	6.598	82.3	3.918
0.600	362.6	0.743	364.3	81.7	6.663	83.7	3.985
0.505	365.2	0.691	366.1	82.4	6.720	85.0	4.047

Stripping Section

x	T_L, K	y	T_G, K	M_G	V_{st}', kg/s	M_L	L_{st}', kg/s
0.505	365.2	0.691	366.1	82.4	4.431	85.0	6.408
0.400	368.4	0.5401	370.9	84.5	4.544	86.5	6.521
0.300	371.7	0.401	374.4	86.5	4.652	87.9	6.627
0.200	375.4	0.261	378.2	88.4	4.802	89.3	6.732
0.100	379.4	0.120	381.4	90.4	4.862	90.7	6.838

The x and y values are from the operating lines (Figure 6.15). The temperatures are bubble and dew points for the liquids and vapors, respectively. The operating lines intersect at $x = x_F = 0.505$, $y = y_F = 0.691$, the dividing point between the rectifying and stripping sections.

    The tower diameter will be set by the conditions at the bottom of the rectifying section, right above the feed entrance, because of the large vapor flow at this point. The gas and liquid properties at this point are estimated as follows. The gas density is from the ideal gas law

$$\rho_G = \frac{PM_G}{RT_G} = \frac{101.3 \times 82.4}{8.314 \times 366.1} = 2.742 \text{ kg/m}^3$$

6.6 Binary Distillation in Packed Towers

Since benzene and toluene form virtually ideal solutions, to estimate the liquid density only the molar fractions and the densities of the pure components as liquids at the solution temperature and pressure are required. Use the Rackett equation to estimate the molar volumes of the pure components as saturated liquids (Smith et al., 1996):

$$V^{sat} = V_c Z_c^{(1-T_r)^{0.2857}}$$ (6-47)

The values predicted by equation (6-47) can be used, with no further correction for pressure, as good estimates of the molar volumes of the pure components at the solution temperature and pressure. Then, V_A = 98.4 cm³/mol; V_B = 114.5 cm³/mol;

$$V_L = (0.505)(98.4) + (0.495)(114.5) = 106.4 \text{ cm}^3/\text{mol}$$

$$\rho_L = M_L / V_L = 799.4 \text{ kg/m}^3$$

The liquid viscosity is estimated from equations (6-42) and (6-43), μ_L = 0.274 cP. The gas viscosity is from the method of Lucas, μ_G = 8.92 × 10⁻⁶ kg/m·s. Gas diffusivity is from the Wilke–Lee equation, D_G = 0.0456 cm²/s. The liquid diffusivity is from equation (1-56), with Γ = 1.0, using the corresponding Hayduk–Minhas correlation to estimate the infinite dilution diffusion coefficients: D_L = 5.43 × 10⁻⁵ cm²/s.

The Mathcad program in Appendix D is used to determine the tower diameter that satisfies the gas-pressure drop criterion at the bottom of the rectifying section. The result is D = 1.41 m, which corresponds to a 67% approach to flooding. The effective specific interfacial area is a_h = 130.2 m²/m³. The mass-transfer coefficients are k'_x = 3.705 mol/m²·s; k'_y = 2.125 mol/m²·s.

To determine the conditions at the interface for this point in the tower, solve simultaneously the equilibrium relationship (Raoult's law, in this case) and the interface mass-transfer condition given by an equation similar to (3-22):

$$\frac{y - y_i}{x - x_i} = -\frac{k'_x}{k'_y}$$ (6-48)

$$y_i = \frac{P_A(T)x_i}{P} \qquad 1 - y_i = \frac{P_B(T)(1-x_i)}{P}$$ (6-49)

These equations are easily solved using the "solve block" capabilities of Mathcad. The solution is T = 365.5 K, x_i = 0.495, y_i = 0.709. Values at other concentrations in the rectifying section are

y	k'_y mol/m²·s	k'_x mol/m²·s	a_h m⁻¹	y_i	$[k'_y a_h(y_i-y)]^{-1}$ m³·s/mol	z m	dP/dz Pa/m
0.691	2.125	3.705	130.2	0.709	0.201	0	400
0.743	2.119	3.7461	128.7	0.777	0.108	0.393	393
0.802	2.115	3.804	126.8	0.842	0.093	0.691	385
0.861	2.107	3.853	125.1	0.900	0.097	0.981	376
0.922	2.095	3.914	123.4	0.952	0.131	1.335	368
0.950	2.097	3.944	122.5	0.975	0.158	1.546	364

354 Distillation

The height, z, for a given gas composition along the rectifying section is calculated from equation (6-45), written here for constant molar overflow as

$$z = \int_{y_F}^{y} \frac{V dy}{k'_y a_h S \left(y_i - y\right)} = G_{My} \int_{y_F}^{y} \frac{dy}{k'_y a_h \left(y_i - y\right)}$$

$$Z_R = G_{My} \int_{y_F}^{y_g} \frac{dy}{k'_y a_h \left(y_i - y\right)}$$

(6-50)

For the rectifying section, $G_{My} = V/(\pi D^2/4) = 52.25$ mol/m^2·s. The integrals in equation (6-50) are estimated numerically using Mathcad, as Example 5.3 illustrated while calculating N_{tG}. The total height of the rectifying section is $Z_R = 1.546$ m. The gas-pressure drop per unit packed height (dP/dz), estimated at different positions along the section, is integrated in a similar manner to yield the total pressure drop for the rectifying section, $\Delta P_R = 591$ Pa.

As Figure 6.17 shows, conditions at the bottom of the tower are $y_1 = 0.111$, $x_1 = 0.093$. This is one extreme of the stripping-section operating line. The other extreme is at the point right underneath the introduction of the feed: $x = x_F = 0.505$, $y = y_F = 0.691$. For the rectifying section, $G_{My} = V_{st}/(\pi D^2/4) = 34.45$ mol/m^2·s. Calculations of mass-transfer coefficients, interfacial area, interfacial gas and liquid concentrations, and gas-pressure drop at different points on the stripping-section operating line yield the following results:

y	k'_y mol/m^2·s	k'_x mol/m^2·s	a_h m^{-1}	y_i	$[k'_y a_h(y_i-y)]^{-1}$ m^3·s/mol	z m	dP/dz Pa/m
0.111	1.601	4.192	170.1	0.160	0.098	0	240
0.120	1.602	4.197	170.1	0.172	0.093	0.030	240
0.261	1.608	4.151	167.4	0.333	0.068	0.395	234
0.401	1.587	4.113	165.2	0.475	0.068	0.723	221
0.540	1.577	4.079	162.8	0.599	0.087	1.069	212
0.691	1.565	4.032	160.2	0.711	0.261	1.869	202

The total height of the stripping section is $Z_S = 1.859$ m; the total pressure drop for the stripping section is $\Delta P_S = 409$ Pa. Therefore, the total packed height is $Z = Z_R + Z_S = 3.415$ m. The total gas-pressure drop is $\Delta P = \Delta P_R + \Delta P_S = 1.0$ kPa.

6.7 MULTICOMPONENT DISTILLATION

Your objectives in studying this section are to be able to:

1. Explain why analysis of multicomponent distillation problems is always by trial-and-error.
2. Make appropriate assumptions and solve the overall material balances.

6.7 Multicomponent Distillation

Many of the distillations of industry involve more than two components. While the principles established for binary solutions generally apply to such distillations, new problems of design are introduced which require special consideration.

An important principle to be emphasized is that a single fractionator cannot separate more than one component in reasonably pure form from a multicomponent solution, and that a total of $C - 1$ fractionators will be required for complete separation of a system of C components. Consider, for example, the continuous separation of a ternary solution consisting of components A, B, and C whose relative volatilities are in that order (A most volatile). In order to obtain the three substances in substantially pure form, the following two-column scheme can be used. The first column is used to separate C as a residue from the rest of the solution. This residue is necessarily contaminated with a small amount of B and an even smaller amount of A. The distillate, which is necessarily contaminated with a small amount of C, is then fractionated in the second column to give nearly pure A and B.

Another significant difference between multicomponent and binary distillation problems arises from a degree-of-freedom analysis around the column (Wankat, 1988). Assuming constant pressure and negligible heat losses in the column, the number of degrees of freedom is $C + 6$. For ternary distillation, then, there are 9 variables that can be specified by the designer; the most likely are listed in Table 6.3.

In multicomponent distillation, neither the distillate nor the bottoms composition is completely specified because there are not enough degrees of freedom to allow complete specification. This inability to completely specify the distillate and bottoms compositions has major effects on the calculation procedure. The components that do have their distillate and bottoms fractional recoveries specified (such as component 1 in the distillate and component 2 in the bottoms in Table 6.3) are called *key components*. The most volatile of the keys is called the *light key* (LK), and the least volatile the *heavy key* (HK). The other components are *nonkeys* (NK). If a nonkey is more volatile than the light key, it is a *light nonkey* (LNK); if it is less volatile than the heavy key, it is a *heavy nonkey* (HNK).

Table 6.3 Specified Design Variables: Ternary Distillation

Number of variables	Variable
1	Feed rate, F
2	Feed composition, z_1, z_2
1	Feed quality, q (or H_F or T_F)
1	Distillate, $x_{1, D}$ (or D or one fractional recovery)
1	Bottoms, $x_{2, W}$ (or W or one fractional recovery)
1	Reflux ratio
1	Saturated liquid reflux or T_0
1	Optimum feed plate
9	Total

Source: Data from Wankat (1988).

356   **Distillation**

Consider the overall mass balances around a ternary distillation column. They are

$$Fz_i = Wx_{i,W} + Dx_{i,D} \qquad i = 1,2,3 \qquad (6\text{-}51)$$

$$\sum_{i=1}^{3} x_{i,D} = 1.0 \qquad \sum_{i=1}^{3} x_{i,W} = 1.0 \qquad (6\text{-}52)$$

The unknowns are six: D, W, $x_{2,\,D}$, $x_{3,\,D}$, $x_{1,\,W}$, and $x_{3,\,W}$. There are only five independent equations. Additional equations (energy balances and equilibrium expressions) always add additional variables, so we cannot start out by solving the overall mass and energy balances.

Can we do the internal stage-by-stage calculations first and then solve the overall balances? To begin the stage-by-stage calculation procedure in a distillation column, we need to know all the compositions at one end of the column. For ternary systems with the variables specified as in Table 6.3, these compositions are unknown. To begin the analysis, we would have to assume that one of them is known. Therefore, internal calculations for multicomponent distillation problems are necessarily trial-and-error. In a ternary system, once an additional composition is assumed, both the overall and internal calculations are easily done. The results can then be compared and the assumed composition modified as needed until convergence is achieved.

Fortunately, in many cases it is easy to make an excellent first guess. If a sharp separation of the keys is required, then almost all of the heavy nonkeys will appear only in the bottoms, and almost all of the light nonkeys will appear only in the distillate. If there are only light nonkeys or only heavy nonkeys, then an accurate first guess of compositions can be made as illustrated in Example 6.9.

Example 6.9 Overall Mass Balances Using Fractional Recovery

We wish to distill 2000 kmol/h of a saturated liquid feed of composition 45.6 mol% propane, 22.1 mol% n-butane, 18.2 mol% n-pentane, and 14.1 mol% n-hexane at a total pressure of 101.3 kPa. A fractional recovery of 99.4% of propane is desired in the distillate and 99.7% of the n-butane in the bottoms. Estimate distillate and bottoms compositions and flow rates.

Solution

This appears to be a straightforward application of overall material balances, except that there are two variables too many. Thus, we will have to assume the recoveries or concentrations of two of the components. The normal boiling point of propane is 231.1 K, that of n-butane is 272.7 K, 309.2 for n-pentane, and 341.9 for n-hexane. Thus, the order of volatilities is propane > n-butane > n-pentane > n-hexane. This makes propane the light key, n-butane the heavy key, and n-pentane and n-hexane the heavy nonkeys (HNKs). Since the recoveries of the keys are quite high, it is reasonable to assume that all of the HNKs appear only in the bottoms.

6.8 Fenske–Underwood–Gilliland Method

Based on this assumption, the overall material balances yield: $D = 907.9$ kmol/h, $W = 1092.1$ kmol/h. The estimated compositions are as follows:

Component	Mol fraction in D	Mol fraction in W
Propane	0.9985	0.0060
n-Butane	0.0015	0.4030
n-Pentane	0.0000	0.3330
n-Hexane	0.0000	0.2580

The composition calculated in the bottoms is quite accurate. Thus, in this case we can step off stages from the bottoms upward and be confident that the results are reliable. On the other hand, if only light nonkeys are present, the distillate composition can be estimated with high accuracy, and stage-by-stage calculations should proceed from the top downward. When both light and heavy nonkeys are present, stage-by-stage calculation methods are difficult and other design procedures should be used.

6.8 FENSKE–UNDERWOOD–GILLILAND METHOD

Your objectives in studying this section are to be able to:

1. Derive the Fenske equation and use it to determine the number of stages required at total reflux and the splits of nonkey components.
2. Use the Underwood equations to determine the minimum reflux ratio for multicomponent distillation.
3. Use the Gilliland correlation to estimate the actual number of stages in a multicomponent column, and the optimum feed-stage location.

Although rigorous computer methods are available for solving multicomponent separation problems, approximate methods continue to be used in practice for various purposes, including preliminary design, parametric studies to establish optimum design conditions, and process synthesis studies to determine optimal separation sequences (Seader and Henley, 1998). A widely used approximate method is commonly referred to as the *Fenske–Underwood–Gilliland* (FUG) method.

6.8.1 Total Reflux: Fenske Equation

Fenske (1932) derived a rigorous solution for binary and multicomponent distillation at total reflux. The derivation assumes that the stages are equilibrium stages. Consider a multicomponent distillation column operating at total reflux. For an equilibrium partial reboiler, for any two components A and B,

$$\left[\frac{y_{A,R}}{y_{B,R}}\right] = \alpha_R \left[\frac{x_{A,W}}{x_{B,W}}\right] \tag{6-53}$$

Equation (6-53) is just the definition of the relative volatility applied to the conditions in the reboiler. Material balances for these components around the reboiler are

$$V_R y_{A,R} = L_N x_{A,N} - W x_{A,W} \tag{6-54a}$$

$$V_R y_{B,R} = L_N x_{B,N} - W x_{B,W} \tag{6-54b}$$

However, at total reflux, $W = 0$ and $L_N = V_R$. Thus, the mass balances become

$$y_{A,R} = x_{A,N} \qquad y_{B,R} = x_{B,N} \quad \text{(at total reflux)} \tag{6-55}$$

For a binary system this means, naturally, that the operating line is the $y = x$ line. Combining equations (6-53) and (6-55),

$$\left[\frac{x_{A,N}}{x_{B,N}} \right] = \alpha_R \left[\frac{x_{A,W}}{x_{B,W}} \right] \tag{6-56}$$

If we now move up the column to stage N, combining the equilibrium equation and the mass balances, we obtain

$$\left[\frac{x_{A,N-1}}{x_{B,N-1}} \right] = \alpha_N \left[\frac{x_{A,N}}{x_{B,N}} \right] \tag{6-57}$$

Then equations (6-56) and (6-57) can be combined to give

$$\left[\frac{x_{A,N-1}}{x_{B,N-1}} \right] = \alpha_N \alpha_R \left[\frac{x_{A,W}}{x_{B,W}} \right] \tag{6-58}$$

We can repeat this procedure until we reach the top stage. The result is

$$\left[\frac{x_{A,D}}{x_{B,D}} \right] = \alpha_1 \alpha_2 \alpha_3 \cdots \alpha_{N-1} \alpha_N \alpha_R \left[\frac{x_{A,W}}{x_{B,W}} \right] \tag{6-59}$$

If we define α_{AB} as the geometric average relative volatility,

$$\alpha_{AB} = \left[\alpha_1 \alpha_2 \alpha_3 \cdots \alpha_{N-1} \alpha_N \alpha_R \right]^{1/N_{\min}} \tag{6-60}$$

Equation (6-59) becomes

$$\left[\frac{x_{A,D}}{x_{B,D}} \right] = \alpha_{AB}^{N_{\min}} \left[\frac{x_{A,W}}{x_{B,W}} \right] \tag{6-61}$$

6.8 Fenske–Underwood–Gilliland Method

Solving equation (6-61) for $N_{\min}$ yields

$$N_{\min} = \frac{\ln\left[\dfrac{x_{A,D}x_{B,W}}{x_{B,D}x_{A,W}}\right]}{\ln\alpha_{AB}} \tag{6-62}$$

which is one form of the Fenske equation. In this equation, $N_{\min}$ is the number of equilibrium stages required at total reflux, including the partial reboiler.

An alternative form of the Fenske equation that is very convenient for multi-component calculations is easily derived. Equation (6-62) can also be written as

$$N_{\min} = \frac{\ln\left[\dfrac{(Dx_{A,D})(Wx_{B,W})}{(Dx_{B,D})(Wx_{A,W})}\right]}{\ln\alpha_{AB}} \tag{6-63}$$

The amount of substance A recovered in the distillate is $(Dx_{A,D})$ and is also equal to the fractional recovery of A in the distillate, $FR_{A,D}$, times the amount of A in the feed

$$Dx_{A,D} = \left(FR_{A,D}\right)\left(Fz_A\right) \tag{6-64}$$

From the definition of the fractional recovery,

$$Wx_{A,W} = \left(1 - FR_{A,D}\right)\left(Fz_A\right) \tag{6-65}$$

Substituting equations (6-64) and (6-65) and the corresponding equations for component B into equation (6-63) gives

$$N_{\min} = \frac{\ln\left[\dfrac{\left(FR_{A,D}\right)\left(FR_{B,W}\right)}{\left(1 - FR_{A,D}\right)\left(1 - FR_{B,W}\right)}\right]}{\ln\alpha_{AB}} \tag{6-66}$$

For multicomponent systems, calculations with the Fenske equation are straightforward if fractional recoveries of the two keys, A and B, are specified. If the relative volatility is not constant, the average defined in equation (6-60) can be approximated by

$$\alpha_{AB} = \left(\alpha_R \alpha_D\right)^{1/2} \tag{6-67}$$

where α_D is determined at the distillate composition as illustrated in Example 6.10. Once $N_{\min}$ is known, the fractional recovery of the nonkeys can be found by writing equation (6-66) for a nonkey component, C, and either key component. Then, solve the resulting equation for $FR_{C,W}$ or $FR_{C,D}$, depending on the key component chosen. If the key component chosen is B, the result is

$$FR_{C,D} = \frac{\alpha_{CB}^{N_{\min}}}{\dfrac{FR_{B,W}}{1 - FR_{B,W}} + \alpha_{CB}^{N_{\min}}} \tag{6-68}$$

360   **Distillation**

Example 6.10 Use of Fenske Equation for Ternary Distillation

A distillation column with a partial reboiler and a total condenser is being used to separate a mixture of benzene, toluene, and 1,2,3-trimethylbenzene. The feed, 40 mol% benzene, 30 mol% toluene, and 30 mol% 1,2,3-trimethylbenzene, enters the column as a saturated vapor. We desire 95% recovery of the toluene in the distillate and 95% of the 1,2,3-trimethylbenzene in the bottoms. The reflux is returned as a saturated liquid, and constant molar overflow can be assumed. The column operates at a pressure of 1 atm. Find the number of equilibrium stages required at total reflux, and the recovery fraction of benzene in the distillate. Solutions of benzene, toluene, and 1,2,3-trimethylbenzene are ideal.

Solution

In this case, toluene (A) is the light key, 1,2,3-trimethylbenzene (B) is the heavy key, and benzene (C) is the light nonkey. Since the liquid solutions are ideal, and the gases are ideal at the pressure of 1 atm, the VLE can be described by Raoult's law. The relative volatilities will be simply the ratio of the vapor pressures. To estimate the relative volatilities, we need an initial estimate of the conditions at the bottom and top of the column. For a 95% recovery of the LK in the distillate, we may assume initially that virtually all of the LNK will be recovered in the distillate. Therefore, the bottoms will be almost pure 1,2,3-trimethylbenzene, and the reboiler temperature will be close to the normal boiling point of 1,2,3-trimethylbenzene, 449.3 K. For a 95% recovery of the HK in the bottoms, we may assume initially that the distillate will contain all of the LK, all of the LNK, and virtually none of the HK. Therefore, a first estimate of the distillate composition is $x_C = 40/70 = 0.571$; $x_A = 30/70 = 0.429$; $x_B = 0.0$. The bubble point temperature for this solution is 390 K, a good preliminary estimate of the conditions at the top of the column.

    The vapor pressures of the pure components as functions of temperature are given by equation (6-5). The corresponding parameters for benzene and toluene are given in Example 6.4. For 1,2,3-trimethylbenzene, they are: $T_c = 664.5$ K, $P_c = 34.5$ bar, $A = -8.442$, $B = 2.922$, $C = -5.667$, and $D = 2.281$ (Reid et al., 1987). At the estimated reboiler temperature of 449.3 K, the vapor pressures are $P_A = 3.389$ atm, $P_B = 1.0$ atm, $P_C = 6.274$ atm. The relative volatilities are $\alpha_{AB} = 3.389$, $\alpha_{CB} = 6.274$. At the estimated distillate temperature of 390 K, the vapor pressures are $P_A = 0.779$ atm, $P_B = 0.173$ atm, $P_C = 1.666$ atm. The relative volatilities at this temperature are $\alpha_{AB} = 0.779/0.173 = 4.503$, $\alpha_{CB} = 1.666/0.173 = 9.63$. The geometric-average relative volatilities are $\alpha_{AB} = 3.91$, $\alpha_{CB} = 7.77$. Equation (6-66) gives

$$N_{min} = \frac{\ln\left[\dfrac{0.95 \times 0.95}{0.05 \times 0.05}\right]}{\ln(3.91)} = 4.32$$

Equation (6-67) gives the desired benzene fractional recovery in the distillate

$$FR_{C,D} = \frac{7.77^{4.32}}{\dfrac{0.95}{1-0.95} + 7.77^{4.32}} = 0.997$$

6.8 Fenske–Underwood–Gilliland Method

Thus, the assumption that virtually all of the LNK will be recovered in the distillate is justified.

6.8.2 Minimum Reflux: Underwood Equations

For binary distillation at minimum reflux, most of the stages are crowded into a constant-composition zone that bridges the feed stage. In this zone, all liquid and vapor streams have compositions essentially identical to those of the flashed feed. This zone constitutes a single pinch point as shown in Figure 6.12. This is also true for multicomponent distillation when all the components in the feed distribute to both the distillate and bottoms product. When this occurs, an analytical solution for the limiting flows can be derived (King, 1980). Unfortunately, for multicomponent systems, there will be separate pinch points in both the stripping and rectifying sections if one or more of the components appear in only one of the products. In this case, an alternative analysis procedure developed by Underwood is used to find the minimum reflux ratio (Wankat, 1988).

If there are nondistributing heavy nonkeys present, a pinch point of constant composition will occur at minimum reflux in the rectifying section above where the heavy nonkeys are fractionated out. With nondistributing light nonkeys present, a pinch point will occur in the stripping section. Consider the case where the pinch point is in the rectifying section. The mass balance for component i around the top portion of the rectifying section as illustrated in Figure 6.7a is

$$V_{\min} y_{i,n+1} = L_{\min} x_{i,n} + D x_{i,D} \qquad (6\text{-}69)$$

At the pinch point where compositions are constant

$$x_{i,n-1} = x_{i,n} = x_{i,n+1} \quad \text{and} \quad y_{i,n-1} = y_{i,n} = y_{i,n+1} \qquad (6\text{-}70)$$

The equilibrium expression can be written as

$$y_{i,n+1} = m_i x_{i,n+1} \qquad (6\text{-}71)$$

Combining equations (6-69) to (6-71), we obtain a simplified balance valid in the region of constant composition

$$V_{\min} y_{i,n+1} = \frac{L_{\min} y_{i,n+1}}{m} + D x_{i,D} \qquad (6\text{-}72)$$

Defining the relative volatility $\alpha_i = m_i / m_{HK}$ and combining terms in equation (6-72) yields

$$V_{\min} y_{i,n+1} \left[1 - \frac{L_{\min}}{V_{\min} \alpha_i m_{HK}} \right] = D x_{i,D} \qquad (6\text{-}73)$$

Rearranging,

$$V_{\min} y_{i,n+1} = \frac{\alpha_i D x_{i,D}}{\left[\alpha_i - \dfrac{L_{\min}}{V_{\min} m_{HK}} \right]} \qquad (6\text{-}74)$$

Equation (6-74) can be summed over all components to give the total vapor flow in the enriching section at minimum reflux:

$$V_{\min} = \sum_i V_{\min} y_{i,n+1} = \sum_i \dfrac{\alpha_i D x_{i,D}}{\left[\alpha_i - \dfrac{L_{\min}}{V_{\min} m_{HK}}\right]} \tag{6-75}$$

In the stripping section, a similar analysis can be used to derive

$$-V_{st,\min} = \sum_i \dfrac{\alpha_{i,st} W x_{i,W}}{\left[\alpha_{i,st} - \dfrac{L_{st,\min}}{V_{st,\min} m_{HK,st}}\right]} \tag{6-76}$$

Defining

$$\phi = \dfrac{L_{\min}}{V_{\min} m_{HK}} \quad \text{and} \quad \phi_{st} = \dfrac{L_{st,\min}}{V_{st,\min} m_{HK,st}} \tag{6-77}$$

equations (6-75) and (6-76) become polynomials in ϕ and ϕ_{st} and have C roots. The equations are now

$$V_{\min} = \sum_i \dfrac{\alpha_i D x_{i,D}}{\alpha_i - \phi} \quad \text{and} \quad -V_{st,\min} = \sum_i \dfrac{\alpha_{i,st} W x_{i,W}}{\alpha_{i,st} - \phi_{st}} \tag{6-78}$$

Assuming constant molar overflow and constant relative volatilities, Underwood showed there are common values of $\phi = \phi_{st}$ that satisfy both equations. Adding both equations in (6-78), we have

$$\Delta V_{feed} = V_{\min} - V_{st,\min} = \sum_i \left[\dfrac{\alpha_i D x_{i,D}}{\alpha_i - \phi} + \dfrac{\alpha_i W x_{i,W}}{\alpha_i - \phi}\right] \tag{6-79}$$

where ΔV_{feed} is the change in vapor flow at the feed stage, and α_i is now an average relative volatility. equation (6-79) is easily simplified, combining it with the overall column mass balance for component i to give

$$\Delta V_{feed} = \sum_i \left[\dfrac{\alpha_i F z_i}{\alpha_i - \phi}\right] \tag{6-80}$$

If q is known

$$\Delta V_{feed} = F\left(1 - q\right) \tag{6-81}$$

Combining equations (6-80) and (6-81),

$$1 - q = \sum_i \left[\dfrac{\alpha_i z_i}{\alpha_i - \phi}\right] \tag{6-82}$$

6.8 Fenske–Underwood–Gilliland Method

Equation (6-82) is known as the first Underwood equation. It can be used to calculate appropriate values of ϕ. Equation (6-78) is known as the second Underwood equation and is used to calculate $V_{\min}$. Once $V_{\min}$ is known, $L_{\min}$ is calculated from the mass balance:

$$L_{\min} = V_{\min} - D \qquad (6\text{-}83)$$

The exact method for using the Underwood equations depends on what can be assumed about the distillation process. Three cases will be considered.

Case A. Assume that none of the nonkeys distribute. In this case, the amounts of nonkeys in the distillate are

$$Dx_{HNK,D} = 0 \ \text{ and } \ Dx_{LNK,D} = Fz_{LNK}$$

while the amounts for the keys are

$$Dx_{LK,D} = FR_{LK,D}Fz_{LK}$$

$$Dx_{HK,D} = \left[1 - FR_{HK,W}\right]Fz_{HK}$$

Equation (6-82) can now be solved for the one value of ϕ between the relative volatilities of the two keys, $\alpha_{HK} < \phi < \alpha_{LK}$. This value of ϕ can now be substituted into equation (6-78) to immediately calculate $V_{\min}$. Then,

$$D = \sum_{i=1}^{C} \left(Dx_{i,D}\right) \qquad (6\text{-}84)$$

and $L_{\min}$ is found from mass balance equation (6-83).

Case B. Assume that the distribution of the nonkeys determined from the Fenske equation at total reflux are also valid at minimum reflux. In this case, the $Dx_{HNK,D}$ values are obtained from the Fenske equation as described before. The rest of the procedure is similar to the one described for Case A.

Case C. Exact solution without further assumptions. Equation (6-82) is a polynomial with C roots. Solve this equation for all values of ϕ lying between the relative volatilities of all components. This gives $C - 1$ valid roots. Now write equation (6-78) $C - 1$ times, once for each value of ϕ. There are now $C - 1$ equations in $C - 1$ unknowns ($V_{\min}$ and $Dx_{i,D}$ for all nonkeys). Solve these simultaneous equations; calculate D from equation (6-84); calculate $L_{\min}$ from equation (6-83). This is illustrated in Example 6.12.

Example 6.11 Underwood Equations for Ternary Distillation

For the distillation problem of Example 6.10, find the minimum reflux ratio. Use a basis of 100 kmol/h of feed.

364 **Distillation**

Solution
This problem fits into *Case B*. Since the feed is a saturated vapor, $q = 0$ and equation (6-82) becomes

$$1 = \frac{3.91 \times 0.3}{3.91 - \phi} + \frac{1.0 \times 0.3}{1.0 - \phi} + \frac{7.77 \times 0.4}{7.77 - \phi}$$

Solving for the value of ϕ between 1 and 3.91, we obtain $\phi = 2.2085$. From the problem statement (A = toluene, B = 1,2,3-trimethylbenzene, C = benzene),

$$Dx_{A,D} = (100)(0.3)(0.95) = 28.5 \text{ kmol/h}$$

$$Dx_{B,D} = (100)(0.3)(0.05) = 1.5 \text{ kmol/h}$$

From the results of Example 6.10,

$$Dx_{C,D} = (100)(0.4)(0.997) = 39.9 \text{ kmol/h}$$

Summing the three distillate flows, $D = 69.9$ kmol/h. Equation (6-78) becomes

$$V_{min} = \frac{3.91 \times 28.5}{3.91 - 2.2085} + \frac{1.0 \times 1.5}{1.0 - 2.2085} + \frac{7.77 \times 39.9}{7.77 - 2.2085}$$
$$= 120 \text{ kmol/h}$$

From the mass balance, $L_{min} = 120.0 - 69.9 = 50.1$ kmol/h. The minimum reflux ratio is $R_{min} = L_{min}/D = 0.717$.

Example 6.12 Underwood Equations for a Depropanizer

The feed to a depropanizer is 66% vaporized at the column inlet. The feed composition and average relative volatilities are given in the table below. It is required that 98% of the propane in the feed is recovered in the distillate, and 99% of the pentane is to be recovered in the bottoms product. Calculate the minimum reflux ratio for this case using Underwood's method.

Component	Mol fraction in feed	Relative volatility
Methane (C)	0.26	39.47
Ethane (D)	0.09	10.00
Propane (A)	0.25	4.08
Butane (E)	0.17	2.11
Pentane (B)	0.11	1.00
Hexane (F)	0.12	0.50

Solution
In this case, propane and pentane are the light key and heavy key, respectively. Methane and ethane are LNK, hexane is an HNK, while butane is a "sandwich component,"

6.8 Fenske–Underwood–Gilliland Method 365

meaning that it has a volatility intermediate between the keys. This problem fits into *Case C*, where the distribution of the nonkeys at minimum reflux must be determined, simultaneously with the minimum reflux ratio. Shiras et al. (1950) developed the following equation to determine whether or not a component is distributed at minimum reflux

$$D_{i,R} = \frac{\alpha_i - 1}{\alpha_{LK} - 1} FR_{LK,D} + \frac{\alpha_{LK} - \alpha_i}{\alpha_{LK} - 1} FR_{HK,D} \qquad (6\text{-}85)$$

where the relative volatilities are based on a reference value of 1.0 for the heavy key component. The Shiras et al. criterion applies at minimum reflux as follows:

$D_{i,R} > 1.0$ Component is nondistributed; contained entirely in distillate.
$0 < D_{i,R} < 1.0$ Component is distributed.
$D_{i,R} < 0$ Component is nondistributed; contained entirely in bottoms.

We will determine which of the nonkey components distribute according to the Shiras et al. criterion. From the statement of the problem, $FR_{LK,D}$ = 0.98, $FR_{HK,D}$ = 0.01. For methane,

$$D_{C,R} = \frac{39.47 - 1}{4.08 - 1} \times 0.98 + \frac{4.08 - 39.47}{4.08 - 1} \times 0.01 = 12.13$$

For ethane,

$$D_{D,R} = \frac{10.0 - 1}{4.08 - 1} \times 0.98 + \frac{4.08 - 10.0}{4.08 - 1} \times 0.01 = 2.84$$

For butane,

$$D_{E,R} = \frac{2.11 - 1}{4.08 - 1} \times 0.98 + \frac{4.08 - 2.11}{4.08 - 1} \times 0.01 = 0.36$$

For hexane,

$$D_{F,R} = \frac{0.50 - 1}{4.08 - 1} \times 0.98 + \frac{4.08 - 0.50}{4.08 - 1} \times 0.01 = -0.15$$

Then, according to the Shiras et al. criterion, methane and ethane appear only in the distillate, hexane appears only in the bottoms, and butane is the only distributed nonkey.

Equation (6-82) is now solved for two values of ϕ such that $1.0 < \phi_1 < 2.11$, and $2.11 < \phi_2 < 4.08$. Since the feed is 66% vaporized, $1 - q = 0.66$. The two values of ϕ in the given intervals are $\phi_1 = 1.263$, $\phi_2 = 2.846$. For the purpose of calculating the minimum reflux ratio, choose a basis and calculate the corresponding distillate flows:

Basis: 100 mol of feed

$$Dx_{A,D} = (100)(0.25)(0.98) = 24.5 \text{ mol } (\text{propane})$$

$$Dx_{B,D} = (100)(0.11)(0.01) = 0.11 \text{ mol } (\text{pentane})$$

$$Dx_{C,D} = (100)(0.26) = 26 \text{ mol } (\text{methane})$$

$$Dx_{D,D} = (100)(0.09) = 9 \text{ mol } (\text{ethane})$$

$$Dx_{E,D} = \text{unknown } (\text{butane})$$

$$Dx_{F,D} = (100)(0.12)(0) = 0 \text{ moles } (\text{hexane})$$

366   **Distillation**

Applying equation (6-78) for each value of ϕ yields

$$V_{\min} = \frac{39.47 \times 26.0}{39.47 - 1.263} + \frac{10.0 \times 9.0}{10.0 - 1.263} + \frac{4.08 \times 24.5}{4.08 - 1.263}$$
$$+ \frac{2.11 \times \left(Dx_{E,D}\right)}{2.11 - 1.263} + \frac{1.0 \times 0.11}{1.0 - 1.263}$$

$$V_{\min} = \frac{39.47 \times 26.0}{39.47 - 2.846} + \frac{10.0 \times 9.0}{10.0 - 2.846} + \frac{4.08 \times 24.5}{4.08 - 2.846}$$
$$+ \frac{2.11 \times \left(Dx_{E,D}\right)}{2.11 - 2.846} + \frac{1.0 \times 0.11}{1.0 - 2.846}$$

or

$$V_{\min} = 72.243 + 2.494 \times \left(Dx_{E,D}\right)$$
$$V_{\min} = 121.614 - 2.863 \times \left(Dx_{E,D}\right)$$

Solving simultaneously, $V_{\min}$ = 95.23 mol, $Dx_{E,D}$ = 9.22 mol. Summing the five distillate flows, D = 68.83 mol. From the mass balance, $L_{\min}$ = 95.23 – 68.83 = 26.40 mol, and the minimum reflux ratio is $R_{\min} = L_{\min}/D$ = 26.40/68.83 = 0.384.

6.8.3 Gilliland Correlation for Number of Stages at Finite Reflux

A general shortcut method for determining the number of stages required for a multicomponent distillation at finite reflux ratios would be extremely useful. Unfortunately, such a method has not been developed. However, Gilliland (1940) noted that he could empirically relate the number of stages N at a finite reflux ratio L/D to the minimum number of stages and to the minimum reflux ratio. Gilliland did a series of accurate stage-by-stage calculations and found that he could correlate the variable

$$Y = \frac{N - N_{\min}}{N + 1} \tag{6-86}$$

with the variable

$$X = \frac{R - R_{\min}}{R + 1} \tag{6-87}$$

The original Gilliland correlation was graphical. Molkanov et al. (1972) fit the Gilliland correlation to the equation

$$Y = 1 - \exp\left[\frac{1 + 54.4X}{11 + 117.2X}\left(\frac{X - 1}{X^{0.5}}\right)\right] \tag{6-88}$$

Implicit in the application of the Gilliland correlation is the specification that the theoretical stages be distributed optimally between the rectifying and stripping

6.8 Fenske–Underwood–Gilliland Method

sections. A reasonably good approximation of optimum feed-stage location, according to Seader and Henley (1998), can be made by employing the empirical equation of Kirkbride (1944):

$$\frac{N_R}{N_S} = \left[\frac{z_{HK}}{z_{LK}}\left(\frac{x_{LK,W}}{x_{HK,D}}\right)^2 \frac{W}{D}\right]^{0.206} \tag{6-89}$$

where N_R and N_S are the number of stages in the rectifying and stripping sections, respectively. Application of equation (6-89) requires knowledge of the distillate and bottoms composition at the specified reflux ratio. Seader and Henley (1998) suggest that the distribution of the nonkey components at a finite reflux ratio is close to that estimated by the Fenske equation at total reflux conditions. The procedure is illustrated in Example 6.13.

Example 6.13 Application of the Gilliland Correlation

Estimate the total number of equilibrium stages and the optimum feed-stage location for the distillation problem presented in Examples 6.10 and 6.11 if the actual reflux ratio is set at $R = 1.0$.

Solution
From Example equation (6.10, $N_{min} = 4.32$; from Example 6.11, $R_{min} = 0.717$. For $R = 1.0, X = (1.0 - 0.717)/(1 + 1) = 0.142$. Substituting in equation (6-88), $Y = 0.513$. From equation (6-86), $N = 9.93$ equilibrium stages.

To use the Kirkbride equation to determine the feed-stage location, we must first estimate the composition of the distillate and bottoms from the results of Example 6.10. We found there that, at total reflux, 99.7% of the LNK (benzene) is recovered in the distillate, 95% of the light key is in the distillate, and 95% of the heavy key is in the bottoms. For a basis of 100 mol of feed, the material balances for the three components are

Component	Distillate		Bottoms	
	Mol	Molar fraction	Mol	Molar faction
Benzene (LNK)	39.88	0.571	0.120	0.004
Toluene (LK)	28.50	0.408	1.50	0.050
Trimethylbenzene (HK)	1.50	0.021	28.50	0.946
Total	69.88 = D		30.12 = W	

From the problem statement, $z_{LK} = z_{HK} = 0.30$. Substituting in equation (6-89), $N_R/N_S = 1.202$. On the other hand, $N = N_R + N_S = 9.93$. Solving simultaneously, $N_R = 5.42$, $N_S = 4.51$. Rounding the estimated equilibrium stage requirement leads to 1 stage as a partial reboiler, 4 stages below the feed, and 5 stages above the feed.

368   **Distillation**

6.9 RIGOROUS CALCULATION PROCEDURES FOR MULTICOMPONENT DISTILLATION

Your objectives in studying this section are to be able to:

1. Understand the importance of the introduction of computers as tools for rigorous analysis and design of multicomponent distillation equipment.
2. Understand the difference between the equilibrium-efficiency approach and the rate-based approach to multicomponent distillation problems.
3. Mention some of the computer programs available for implementation of the equilibrium- and rate-based models.

Before the 1950s, distillation column calculations were performed by hand. Although rigorous calculation procedures were available, they were difficult to apply for all but very small columns. Shortcut methods were therefore the primary design tool. Rigorous procedures were seldom used. Inaccuracies and uncertainties in the shortcut procedures were usually accommodated by overdesign.

The introduction of computers has entirely reversed the design procedure. Rigorous calculations can now be performed quickly and efficiently using a computer. In modern distillation practice, rigorous methods are the primary design tool. The role of shortcut calculations is now restricted to eliminating the least desirable design options, providing the designer with an initial estimate for the rigorous step and for troubleshooting the final design (Kister, 1992).

Two different approaches have evolved for the simulation and design of multicomponent distillation columns. The conventional approach is through the use of an *equilibrium stage model* together with methods for estimating the *tray efficiency*. An alternative approach, the *nonequilibrium, rate-based model*, applies rigorous multicomponent mass- and heat-transfer theory to distillation calculations. This nonequilibrium stage model is also applicable, with only minor modifications, to gas absorption, liquid extraction, and to operations in trayed or packed columns (Taylor and Krishna, 1993).

6.9.1 Equilibrium Stage Model

A schematic diagram of a single equilibrium stage is shown in Figure 6.23. The key assumption of this model is that the vapor and liquid streams leaving a stage are in equilibrium with each other. A complete multicomponent distillation or absorption column may be modeled as a sequence of these stages. The equations that model equilibrium stages have been termed the MESH equations after Wang and Henke (1966). MESH is an acronym referring to the different types of equations that form the mathematical model

1. M equations—material balance for each component (C equations for each stage):

$$M_{i,j} = L_{j-1}x_{i,j-1} + V_{j+1}y_{i,j+1} + F_j z_{i,j} - \left(L_j + U_j\right)x_{i,j}$$
$$-\left(V_j + B_j\right)y_{i,j} = 0 \qquad (6\text{-}90)$$

6.9 Rigorous Calculation Procedures for Multicomponent Distillation

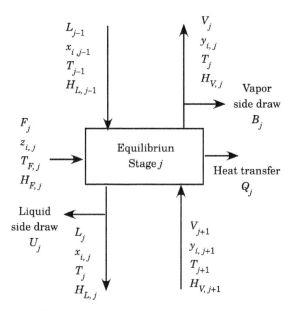

Figure 6.23 Equilibrium stage model.

2. *E* equations—phase equilibrium relation for each component, here modified to include the Murphree efficiency defined by equation 4-60 (*C* equations for each stage):

$$E_{i,j} = \mathbf{E}_{MG,i,j} m_{i,j} x_{i,j} - y_{i,j} - \left(1 - \mathbf{E}_{MG,i,j}\right) y_{i,j+1} = 0 \qquad (6\text{-}91)$$

3. *S* equations—mol fractions summation (2 for each stage):

$$\left(S_y\right)_j = \sum_{i=1}^{C} y_{i,j} - 1 = 0 \qquad \left(S_x\right)_j = \sum_{i=1}^{C} x_{i,j} - 1 = 0 \qquad (6\text{-}92)$$

4. *H* equation—enthalpy balance (one for each stage):

$$H_j = L_{j-1} H_{L,j-1} + V_{j+1} H_{V,j+1} + F_j H_{F,j} - (L_j + U_j) H_{L,j} \\ -(V_j + B_j) H_{V,j} - Q_j = 0 \qquad (6\text{-}93)$$

The MESH equations can be applied to all of the equilibrium stages in the column, including the reboiler and condenser. The result is a set of nonlinear equations that must be solved by iterative techniques.

A wide variety of iterative solution procedures for solving the MESH equations has appeared in the literature. Current practice is based mainly on the bubble-point (BP) method, the simultaneous correction (SC) method, and the inside-out method. The BP method is usually restricted to distillation problems involving narrow-boiling feed mixtures. The SC and inside-out methods are designed to solve any type of column configuration for any type of feed mixture. Because of its computational efficiency, the inside-out method is often the method of choice; however,

370   **Distillation**

it may fail to converge when highly nonideal liquid mixtures are involved, in which case the slower SC method should be tried (Seader and Henley, 1998). Computer implementations of these methods are found in the following widely available programs and simulators: (1) ASPEN PLUS of Aspen Technology, (2) ChemCAD of Chemstations, (3) HYSIM of Hyprotech, and (4) PRO/II of Simulation Sciences.

6.9.2 Nonequilibrium, Rate-Based Model

Although the equilibrium-based model, modified to incorporate stage efficiency, is adequate for binary mixtures and for the major components in nearly ideal multicomponent mixtures, that model has serious deficiencies for more general multicomponent vapor–liquid mixtures. Murphree himself clearly stated the limitations of his development for multicomponent mixtures. He even stated that the theoretical plate should not be the basis of calculation for ternary systems.

When the equilibrium-based model is applied to multicomponent mixtures, a number of problems arise. Values of $\mathbf{E}_{MG}$ differ from component to component and vary from stage to stage. But, at each stage, the number of independent values of $\mathbf{E}_{MG}$ must be determined so as to force the mole fractions in the vapor phase to sum 1. This introduces the possibility that negative values of $\mathbf{E}_{MG}$ can result. This is in contrast to binary mixtures for which the values of $\mathbf{E}_{MG}$ are always positive and are identical for the two components.

Krishna et al. (1977) showed that when the vapor mol-fraction driving force of a component (call it A) is small compared to the other components in the mixture, the transport rate of A is controlled by the other components, with the result that $\mathbf{E}_{MG}$ for A is anywhere in the range from minus infinity to plus infinity. They confirmed this theoretical prediction by conducting experiments with the ethanol/*tert*-butanol/water system and obtained values of $\mathbf{E}_{MG}$ for *tert*-butanol ranging from –2978% to + 527%. In addition, the observed values of $\mathbf{E}_{MG}$ for ethanol and water sometimes differed significantly.

Krishna and Standardt (1979) were the first to show the possibility of applying rigorous multicomponent mass- and heat-transfer theory to calculations of simultaneous transport. The availability of this theory led to the development by Krishnamurthy and Taylor (1985) of the first general rate-based, computer-aided model for application to trayed and packed columns for distillation and other continuous, countercurrent, vapor–liquid separation operations. This model applies the two-resistance theory of mass transfer discussed in Chapter 3, with equilibrium assumed at the interface of the two phases and provides options for vapor and liquid flow configurations in trayed columns, including plug flow and perfectly mixed flow, on each tray. Although the model does not require tray efficiencies, correlations of mass- and heat-transfer coefficients are needed for the particular type of tray or packing employed. The theory was further developed by Taylor and Krishna (1993), and the model was extended by Taylor et al. (1994). The 1994 version, unlike the 1985 model, includes a design mode that estimates column diameter for a specified approach to flooding or pressure drop. Baur et al. (2001) further modified the model to simulate the dynamic, nonequilibrium behavior of reactive distillation tray columns.

6.10 Batch Distillation

In the rate-based models, the mass and energy balances around each equilibrium stage are each replaced by separate balances for each phase around a stage, which can be a tray, a collection of trays, or a segment of a packed section. Rate-based models use the same m-value and enthalpy correlations as the equilibrium-based models. However, the m-values apply only at the equilibrium interphase between the vapor and liquid phases. The accuracy of enthalpies and, particularly, m-values is crucial to equilibrium-based models. For rate-based models, accurate predictions of heat-transfer rates and, particularly, mass-transfer rates are also required. These rates depend upon transport coefficients, interfacial area, and driving forces. It is important that mass-transfer rates account for component-coupling effects through binary pair coefficients.

Rate-based models are implemented in several computer programs, including Aspen Rate-Based Distillation and ChemSep v8.2. Both programs offer considerable flexibility in user specifications. Computing time for these rate-based models is generally about an order of magnitude greater than for an equilibrium-based model.

6.10 BATCH DISTILLATION

Your objectives in studying this section are to be able to:

1. Calculate, by modified McCabe–Thiele methods, residue composition and distillation time for binary batch rectification with constant reflux for a given number of ideal stages, boil-up rate, and specified average distillate composition.
2. Calculate, by modified McCabe–Thiele methods, distillation time for binary batch rectification with constant distillate composition for a given number of ideal stages and specified residue composition.

Continuous distillation is a thermodynamically efficient method for producing large amounts of material of constant composition. However, when small amounts of material of varying product composition are required, *batch distillation* has several advantages. In batch distillation a charge of feed is heated in a reboiler, and after a short startup period, product can be withdrawn from the top of the equipment. When the distillation is finished, the heat is shut off and the material left in the reboiler is removed. Then a new batch can be started. Usually, the distillate is the desired product.

Batch distillation is versatile; a run may last from a few hours to several days. Batch distillation is the choice when the plant does not run continuously and the batch must be completed in one or two shifts (8 to 16 h). It is often used when the same equipment distills several different products at different times. If distillation is required only occasionally, batch distillation would again be the choice.

The simplest form of batch distillation is the differential distillation discussed in Section 6.3, which involves using the simple apparatus shown in Figure 6.4. Practical applications of such an arrangement are limited to wide-boiling mixtures

372 Distillation

such as HCl–H_2O, H_2SO_4–H_2O, and NH_3–H_2O (Seader and Henley, 2006). For other mixtures, to achieve a sharp separation, or to reduce the intermediate cut fraction, a trayed or packed column located above the still and a means of returning reflux to the column are added. Two modes of operation of this improved form of batch distillation are cited most frequently: (1) operation at constant reflux ratio and (2) operation at constant distillate composition.

6.10.1 Binary Batch Distillation with Constant Reflux

If the reflux ratio R or distillate rate D is fixed, instantaneous distillate and bottoms compositions vary with time. For a total condenser, negligible holdup of vapor and liquid in the column and the condenser, equilibrium stages, and constant molar overflow, the Rayleigh equation can now be written as

$$\int_{W}^{F} \frac{dL}{L} = \ln \frac{F}{W} = \int_{x_W}^{x_F} \frac{dx_W}{x_D - x_W} \tag{6-94}$$

The analysis of such a batch rectification for a binary system is facilitated by the McCabe–Thiele diagram. Initially, the composition of the light-key component in the liquid in the reboiler of the column is the charge composition x_F. If there are, for example, two theoretical stages, the initial distillate composition x_{D0} can be found by constructing an operating line of slope $R/(R + 1)$, such that exactly two stages are stepped off from x_F to the line $y = x$ in Figure 6.24. At an arbitrary later time, say time 1, at still pot composition $x_W < x_F$, the instantaneous distillate composition is x_D. A time-dependent series of points for x_D is thus established by trial and error with R and the number of ideal stages held constant.

Equation (6-94) cannot be integrated analytically because the relationship between x_D and x_W depends on the liquid-to-vapor ratio, the number of theoretical stages, and the equilibrium distribution curve. However, it can be integrated graphically with pairs of values for x_D and x_W obtained from the McCabe–Thiele diagram for a series of operating lines with the same slope.

The time t required for a batch distillation at constant reflux and negligible hold up in the column and condenser can be computed by a total material balance based on constant boil-up rate V to give the following equation (Seader and Henley, 2006):

$$t = \frac{R+1}{V}\left(F - W_t\right) = \frac{\left(R+1\right)D_t}{V} \tag{6-95}$$

With a constant-reflux policy, the instantaneous distillate purity is above the specification at the beginning of the run and below specification at the end. By an overall material balance, the average mol fraction of the light-key component in the accumulated distillate at time t is given by

$$x_{D,avg} = \frac{Fx_F - W_t x_{Wt}}{F - W_t} \tag{6-96}$$

6.10 Batch Distillation

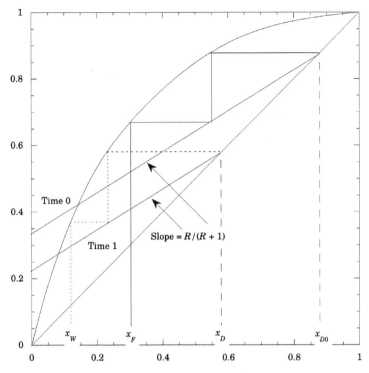

Figure 6.24 Batch binary distillation with fixed R and two theoretical stages.

Example 6.14 Binary Batch Rectification with Constant Reflux

A batch distillation column with three theoretical stages (the first is the still pot) is charged with 100 kmol of a 20 mol% n-hexane in n-octane mixture. At a constant reflux ratio $R = 1.0$, how many mol of the charge must be distilled if an average product composition of 70 mol% n-hexane is required? If the boil-up ratio is 10 kmol/h, calculate the distillation time. The equilibrium distribution curve at the column pressure is given in Figure 6.25.

Solution
A series of operating lines of slope = $R/(R + 1) = 0.5$ are located by the trial-and-error procedure described earlier. Figure 6.25 shows two of those lines: for $x_F = 0.2$ and for $x_W = 0.09$. It is then possible to construct the following table:

x_D	0.85	0.77	0.60	0.50	0.35	0.30
x_W	0.20	0.15	0.09	0.07	0.05	0.035
$(x_D - x_W)^{-1}$	1.54	1.61	1.96	2.33	3.33	3.77

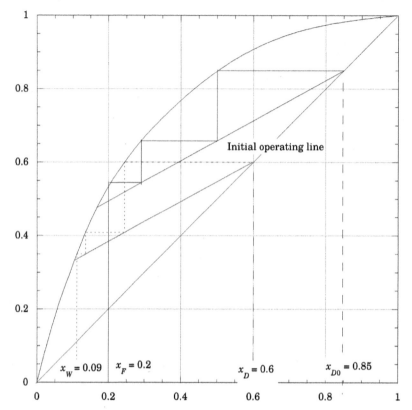

Figure 6.25 McCabe–Thiele diagram for Example 6.14.

Equation (6-94) can be written as

$$W_t(x_{Wt}) = F \exp\left[-\int_{x_{Wt}}^{x_F} \frac{dx_W}{x_D - x_W}\right] = F \exp\left[-\mathbf{I}(x_{Wt})\right] \quad (6\text{-}97)$$

On the other hand, equation (6-96) can be written as

$$x_{Davg}(x_{Wt}) = \frac{Fx_F - W_t(x_{Wt})x_{Wt}}{F - W_t(x_{Wt})} \quad (6\text{-}98)$$

Figure (6.26) shows a Mathcad program to solve equations (6-97) and (6-98) simultaneously for x_{Wt} and $W_t(x_{Wt})$ for the specified value of the average product composition (70 mol% n-hexane). Appendix 6.3 presents the corresponding Python solution. The integral in equation (6-97) is estimated numerically using cubic spline interpolation of the data generated with the help of the McCabe–Thiele diagram.

The answer is x_{Wt} = 0.064 and $W_t(x_{Wt})$ = 78.6 kmol. Therefore, a total of D_t = 100 − 78.6 = 21.4 kmol of the initial charge must be distilled. For the given boilup rate, V = 10 kmol/h, from equation (6-95), t = 4.28 h.

6.10 Batch Distillation

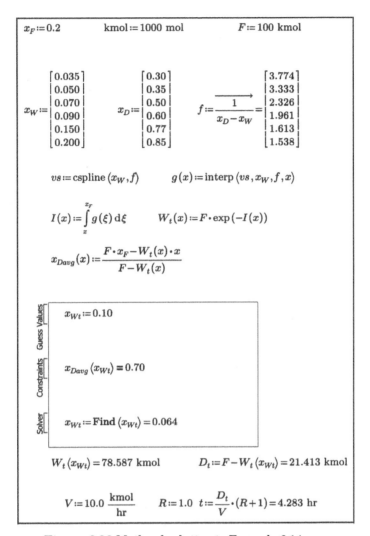

Figure 6.26 Mathcad solution to Example 6.14.

6.10.2 Batch Distillation with Constant Distillate Composition

An alternative to the constant-reflux-ratio policy described above is to maintain a constant-molar-vapor rate, but continuously vary the reflux ratio to achieve a constant distillate composition that meets the specified purity. This policy requires a more complex control system, which may be justified only for large batch distillation systems.

Calculations for the policy of constant distillate composition can also be made with the McCabe–Thiele diagram for binary mixtures. An overall material balance for the light-key component at any time, t, is given by rearrangement of equation (6-96) at constant x_D solving for W as a function of x_W:

$$W = F \left[\frac{x_D - x_F}{x_D - x_W} \right] \tag{6-99}$$

376 **Distillation**

Differentiating equation (6-99) with respect to t for varying W and x_W gives

$$\frac{dW}{dt} = F \frac{x_D - x_F}{\left(x_D - x_W\right)^2} \frac{dx_W}{dt} \qquad (6\text{-}100)$$

For constant molar overflow and negligible liquid holdup, the rate of distillation is given by the rate of loss of the charge, or

$$-\frac{dW}{dt} = \frac{dD}{dt} = V - L \qquad (6\text{-}101)$$

where D now is the amount of distillate, not the distillate rate. Substituting equation (6-101) into (6-100) and integrating yields

$$t = \frac{F\left(x_D - x_F\right)}{V} \int_{x_{W_t}}^{x_F} \frac{\left(R+1\right)}{\left(x_D - x_W\right)^2} dx_W \qquad (6\text{-}102)$$

For fixed values of F, x_F, x_D, V, and number of equilibrium stages, the McCabe–Thiele diagram is used to determine values of R for a series of still composition between x_F and the final value of x_{W_t}. These values are then used with equation (6-102) to determine by numerical integration the time to reach a specified value of still composition. The graphical process can be simplified by recalling Figure 6.7b, where it is shown that

$$y_{\text{int}} = \frac{x_D}{R+1} \qquad (6\text{-}103)$$

where y_{int} is the intersection of the operating line with the y-axis. Substituting equation (6-103) into (6-102) gives us

$$t = \frac{F\left(x_D - x_F\right)x_D}{V} \int_{x_{W_t}}^{x_F} \frac{1}{y_{\text{int}}\left(x_D - x_W\right)^2} dx_W \qquad (6\text{-}104)$$

A series of operating lines are drawn on the McCabe–Thiele diagram from the 45° line at $x = x_D$ to different values of y_{int}. The specified number of equilibrium stages are stepped off, starting at x_D and ending at the corresponding intermediate values of x_W. In this manner, the data required to integrate equation (6-104) numerically are generated. The procedure is illustrated in the following example.

Example 6.15 Binary Batch Rectification with Constant Distillate Composition

A batch distillation column with three theoretical stages (the first stage is the still pot) is charged with 100 kmol of a 34 mol% n-hexane in n-octane mixture. A liquid distillate composition of 95 mol% n-hexane is to be maintained by continuously

6.10 Batch Distillation

377

adjusting the reflux ratio. If the boil-up rate is 20 kmol/h, calculate the distillation time required to reduce the still residue composition to 12 mol% n-hexane. Calculate the fractional recovery of n-hexane in the distillate.

Solution

A series of operating lines are drawn on the McCabe–Thiele diagram from the 45° line at $x = x_D = 0.95$ to different values of y_{int}. The initial reflux ratio is obtained by trial and error until the three theoretical stages can be stepped off from x_D to the initial value of $x_W = x_F = 0.34$. Figure 6.27 shows that for the initial operating line, $y_{int} = 0.5$. Substituting in equation (6-103), the initial reflux ratio is $R = 0.9$. The reflux ratio is then increased gradually (y_{int} reduced) and the corresponding value of x_W is obtained from the McCabe–Thiele diagram until the value of x_W is below the specification of 12 mol% n-hexane. Figure 6.27 shows that for an operating line with $y_{int} = 0.05$ ($R = 18.0$), $x_W = 0.108$. It is then possible to construct the following table:

y_{int}	0.50	0.40	0.30	0.20	0.10	0.05
x_W	0.34	0.28	0.22	0.175	0.13	0.11
R	0.90	1.38	2.17	3.75	8.50	18.00
$[y_{int}(x_D - x_W)]^{-1}$	5.375	5.569	6.255	8.325	14.872	28.345

The integral in equation (6-104) is estimated numerically with a lower limit of $x_{Wt} = 0.12$ using Mathcad (or Python) for cubic spline interpolation of the data generated with the help of the McCabe–Thiele diagram, in a manner very similar to that described in Example (6.14). The value of the integral is found to be 1.64; substituting this value in equation (1-104), $t = 4.75$ h.

At the end of the distillation run, a distillate with 95 mol% hexane and a residue with 12 mol% hexane are obtained from the original charge of 100 kmol of 34 mol% hexane solution. Combining a total material balance with a hexane balance, the products are $D = 26.5$ kmol and $W = 73.5$ kmol. The fractional recovery of hexane in the distillate = $(26.5)(0.95)/34 = 0.74$ (74%).

6.10.3 Multicomponent Batch Distillation

Shortcut methods for handling multicomponent batch distillation have been developed for the two cases of constant reflux and constant distillate composition (Diwekar and Mandhaven, 1991; Sundaram and Evans, 1993). Both methods avoid tedious stage-by-stage calculations of vapor and liquid compositions by employing the Fenske–Underwood–Gilliland (FUG) shortcut procedure for continuous distillation, described in Section 6.8, at successive time steps. In essence, they treat batch distillation as a sequence of continuous, steady-state rectifications. As in the FUG method, no estimation of compositions or temperatures is made for intermediate stages.

For final design studies, complete stage-by-stage temperature, flow, and composition profiles as functions of time are required. Such calculations are tedious but can be carried out conveniently by computer-based simulation programs, such as BatchFrac from Aspen Technology. It uses an algorithm to solve the unsteady-state

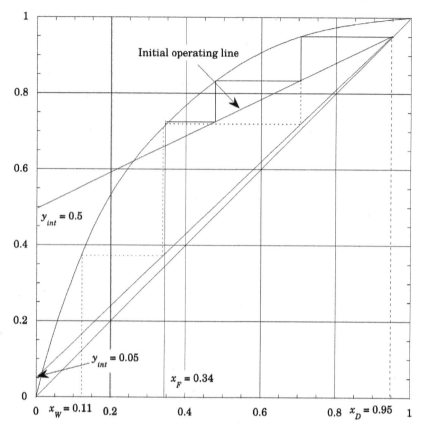

Figure 6.27 McCabe–Thiele diagram for Example 6.15.

heat and material balance equations that describe the behavior of batch distillation processes. A variety of batch distillation problems can be handled, including narrow-boiling system, wide-boiling system, nonideal systems, three-phase systems, and reactive systems. The software can detect the presence of a free-water phase in the condenser, or of any second liquid phase anywhere in the column. The software can even be used to simulate batch distillation columns with chemical reactions.

PROBLEMS

The problems at the end of each chapter have been grouped into four classes (designated by a superscript after the problem number).

Class a: Illustrates direct numerical application of the formulas in the text.
Class b: Requires elementary analysis of physical situations, based on the subject material in the chapter.
Class c: Requires somewhat more mature analysis.
Class d: Requires computer solution.

Problems **379**

6.1[a]. Flash vaporization of a heptane–octane mixture

A liquid mixture containing 50 mol% *n*-heptane (A) and 50 mol% *n*-octane (B), at 303 K, is to be continuously flash-vaporized at a pressure of 1 atm to vaporize 30 mol% of the feed. What will be the composition of the vapor and liquid and the temperature in the separator if it behaves as an ideal stage?

Answer: $x_W = 0.443$

6.2[a]. Flash vaporization of a heptane–octane mixture

A liquid mixture containing 50 mol% *n*-heptane (A) and 50 mol% *n*-octane (B), at 303 K, is to be continuously flash-vaporized at a temperature of 350 K to vaporize 30 mol% of the feed. What will be the composition of the vapor and liquid and the pressure in the separator if it behaves as an ideal stage?

Answer: $y_D = 0.654$

6.3[d]. Flash vaporization of a ternary mixture

Modify the Mathcad program of Figure 6.3 for ternary mixtures. Test your program with the data presented in Example 6.2 for a mixture of benzene (A), toluene (B), and *o*-xylene (C). Critical temperatures and pressures, and the parameters of the Wagner equation for estimating vapor pressure (equation 6-5) are included in the following table (Reid et al., 1987).

Component	T_c, K	P_c, bar	A	B	C	D
Benzene	562.2	48.9	−6.983	1.32	−2.629	−3.334
Toluene	591.8	41.0	−7.286	1.381	−2.834	−2.792
o-Xylene	630.3	37.3	−7.534	1.410	−3.110	−2.860

6.4[b]. Flash vaporization of a ternary mixture

Consider the ternary mixture of Example 6.2 and Problem 6.3. Estimate the temperature, and composition of the liquid and vapor phases when 60% of the mixture has been vaporized at a constant pressure of 1 atm.

Answer: $y_{AD} = 0.630$

6.5[b]. Flash vaporization of a ternary mixture

Consider the ternary mixture of Example 6.2 and Problem 6.3. It is desired to recover in the vapor 75% of the benzene in the feed, and to recover in the liquid 70% of the *o*-xylene in the feed. Calculate the temperature, pressure, fraction of the feed vaporized, and the concentration of the liquid and gas phases.

Answer: $P = 0.756$ atm

6.6[b]. Batch distillation of a heptane–octane mixture

Repeat the calculations of Example 6.3, but for 80 mol% of the liquid distilled.

Answer: $y_{D,av} = 0.573$

380   Distillation

6.7^b. Differential distillation with constant relative volatility

If equation (6-40) describes the equilibrium relation at constant pressure by use of some average relative volatility α over the concentration range involved, show that equation (6-11) becomes

$$\ln\frac{F}{W} = \frac{1}{\alpha-1}\ln\frac{x_F\left(1-x_W\right)}{x_W\left(1-x_F\right)} + \ln\frac{1-x_W}{1-x_F} \qquad (6\text{-}105)$$

6.8^a. Differential distillation with constant relative volatility

Consider the binary batch distillation of Example 6.3. For this system, n-heptane with n-octane at 1 atm, the average relative volatility is α = 2.16 (Treybal, 1980). Using equation (6-105) derived in Problem 6.7, compute the composition of the residue after 60 mol% of the feed is differentially distilled.

6.9^b. Differential distillation with constant relative volatility

Consider the differential distillation of Example (6.3. For this system, n-heptane with n-octane at 1 atm, the average relative volatility is α = 2.16 (Treybal, 1980). The mixture will be differentially distilled until the average concentration of the distillate is 65 mol% heptane. Using equation 6-105 derived in Problem 6.7, compute the composition of the residue, and the fraction of the feed that is distilled.

Answer: x_W = 0.419

6.10^b. Mixtures of light hydrocarbons: m-value correlations

Because of the complex concentration functionality of the m-values, VLE calculations in general require iterative procedures suited only to computer solutions. However, in the case of mixtures of light hydrocarbons, we may assume as a reasonable approximation that both the liquid and the vapor phases are ideal. This allows m-values for light hydrocarbons to be calculated and correlated as functions of T and P. Approximate values can be determined from the monographs prepared by DePriester (1953). The DePriester charts have been fit to the following equation (McWilliams, 1973):

$$\ln m = \frac{a_{T1}}{T^2} + a_{T2} + a_{P1}\ln P + \frac{a_{P2}}{P^2} + \frac{a_{P3}}{P} \qquad (6\text{-}106)$$

where T is in K and P is in kPa. The constants a_{T1}, a_{T2}, a_{P1}, a_{P2}, and a_{P3} are given in Table 6.4. This equation is valid for temperatures from 200 K to 473 K, and pressures from 101.3 kPa to 6000 kPa.

What is the bubble point of a mixture that is 15 mol% isopentane (A), 30 mol% n-pentane (B), and 55 mol% n-hexane (C)? Calculate the composition of the first bubble of vapor. The pressure is 1.0 atm.

Answer: y_A = 0.282

Problems

Table 6.4 Constants in Equation (6-106): T in K, P in kPa

Compound	$-a_{T1}$	a_{T2}	$-a_{P1}$	a_{P2}	a_{P3}
Methane	90,389	9.9730	0.89510	2846.4	0
Ethylene	185,209	9.5411	0.84677	2042.6	0
Ethane	212,114	9.6178	0.88600	2331.8	0
Propylene	285,026	9.4141	0.87871	2267.6	0
Propane	299,595	8.6372	0.76984	0	47.6017
Isobutane	360,138	9.5073	0.92213	0	0
n-Butane	395,234	9.8124	0.96455	0	0
Isopentane	457,279	9.3796	0.93159	0	0
n-Pentane	470,645	9.0527	0.89143	0	0
n-Hexane	549,044	8.6021	0.84634	0	0
n-Heptane	621,544	8.0651	0.79543	0	0

Source: Data from McWilliams (1973).

6.11[b]. Mixtures of light hydrocarbons: m-value correlations

A solution has the following composition, expressed as mol percentage: ethane, 0.25%; propane, 25%; isobutane, 18.5%; n-butane, 56%; isopentane, 0.25%. In the following, the3 pressure is 10 bars. Use equation (6-106) and Table 6.4 to calculate equilibrium distribution coefficients.

(a) Calculate the bubble point.

Answer: 331.2 K

(b) Calculate the dew point.

Answer: 341.9 K

(c) Forty mol% of the feed is flash-vaporized. Calculate the temperature and composition of the products.

Answer: 63.2 mol% n-butane in liquid

6.12[b]. Differential distillation with constant relative volatility

A 30 mol% feed of benzene in toluene is to be differentially distilled. A product having an average composition of 45 mol% benzene is to be produced. Calculate the amount of residue left, assuming that α = 2.5 and F = 100 mol.

Answer: W = 57.4 moles

6.13[b]. Differential distillation of a mixture of isopropanol in water

A mixture of 40 mol% isopropanol in water is to be differentially distilled at 1 atm until 70 mol% of the charge has been vaporized. Calculate the composition of the liquid residue in the still pot, and the average composition of the collected distillate. VLE data for this system, in mol fraction of isopropanol, at 1 atm are (Seader and Henley, 1998):

382   **Distillation**

T, K	366	357	355.1	354.3	353.6	353.2	353.3	354.5
y	0.220	0.462	0.524	0.569	0.593	0.682	0.742	0.916
x	0.012	0.084	0.198	0.350	0.453	0.679	0.769	0.944

Composition of the azeotrope is $x = y = 0.685$; boiling point of the azeotrope = 353.2 K.

Answer: $y_{D,av} = 0.543$

6.14^c. Differential steam distillation

An open kettle contains 50 kmol of a dilute aqueous solution of methanol (2 mol% of methanol), at the bubble point, into which steam is continuously sparged. The entering steam agitates the kettle contents so that they are always of uniform composition, and the vapor produced, always in equilibrium with the liquid, is led away. Operation is adiabatic. For the concentrations encountered, it may be assumed that the enthalpy of the steam and evolved vapor are the same, the enthalpy of the liquid in the kettle is essentially constant, and the relative volatility is constant at 7.6.

a) Show that, under these conditions:

$$V = \frac{F}{\alpha}\left[\ln\left(\frac{x_F}{x_W}\right) + (\alpha - 1)(x_F - x_W)\right] \qquad (6\text{-}107)$$

where V = mol of steam required.

b) Compute the quantity of steam to be introduced in order to reduce the concentration of methanol to 0.1 mol%.

Answer: 20.53 kmol

6.15^a. Continuous distillation of a binary mixture of constant relative volatility

For continuous distillation of a binary mixture of constant relative volatility, Fenske equation (6-58) can be used to estimate the minimum number of equilibrium stages required for the given separation, N_{min}.

Use the Fenske equation to estimate N_{min} for distillation of the benzene–toluene mixture of Example 6.4. Assume that, for this system at 1 atm, the relative volatility is constant at $\alpha = 2.5$.

Answer: 6.43 stages

6.16^a. Continuous distillation of a binary mixture of constant relative volatility

For continuous distillation of a binary mixture of constant relative volatility, the minimum reflux ratio can be determined analytically from the following equation (Treybal, 1980):

$$\frac{R_{min}z_F + qx_D}{R_{min}(1 - z_F) + q(1 - x_D)} = \frac{\alpha\left[x_D(q - 1) + z_F(R_{min} + 1)\right]}{(R_{min} + 1)(1 - z_F) + (q - 1)(1 - x_D)} \qquad (6\text{-}108)$$

Problems

383

Use (6-108) to estimate R_{min} for distillation of the benzene–toluene mixture of Example 6.4. Assume that, for this system at 1 atm, the relative volatility is constant at $\alpha = 2.5$.

Answer: $R_{min} = 1.123$

6.17[b,d]. Algebraic determination of minimum reflux ratio

When the equilibrium curve is always concave downward, the minimum reflux ratio can be calculated algebraically. The required relationship can be developed by solving simultaneously equations (6-18), (6-31), and the equilibrium relationship $y^* = f(x)$. The three unknowns are the coordinates of the point of intersection of the enriching operating line and the q-line (x_{int}, y_{int}) and R_{min}.

Write a Mathcad program to calculate the minimum reflux ratio under these conditions and test it with the data of Example 6.4. In this case, the equilibrium-distribution relationship is a table of VLE values; therefore, an algebraic relationship of the form $y^* = f(x)$ must be developed by cubic spline interpolation.

6.18[c]. Flash calculations: the Rachford–Rice method for ideal mixtures

a) Show that the problem of flash vaporization of a multicomponent ideal mixture can be reformulated as suggested by Rachford and Rice (Doherty and Malone, 2001):

$$f(\phi) = \sum_{i=1}^{C} \frac{z_i(m_i - 1)}{1 + \phi(m_i - 1)} = 0 \quad \text{where} \quad m_i = \frac{P_i}{P} \tag{6-109}$$

$$x_i = \frac{z_i}{1 + \phi(m_i - 1)} \quad i = 1, 2, \cdots, C \tag{6-110}$$

$$y_i = m_i x_i \quad i = 1, 2, \cdots, C \tag{6-111}$$

$$D = \phi F \quad W = (1 - \phi) F \tag{6-112}$$

Equation (6-109) is solved iteratively for ϕ; all other variables are calculated explicitly from equations (6-110) to (6-112).

(b) Solve Example 6.2 using the Rachford–Rice method.

6.19[b]. Continuous rectification of a water–isopropanol mixture

A water–isopropanol mixture at its bubble point containing 10 mol% isopropanol is to be continuously rectified at atmospheric pressure to produce a distillate containing 67.5 mol% isopropanol. Eighty-five percent of the isopropanol in the feed must be recovered. VLE data are given in Problem 6.13. If a reflux ratio of 1.5 times the minimum is used, how many theoretical stages will be required:

(a) If a partial reboiler is used?

(b) If no reboiler is used and saturated steam at 101.3 kPa is introduced below the bottom plate?

384 **Distillation**

6.20[b]. Rectification of a binary mixture: overall efficiency

A solution of carbon tetrachloride and carbon disulfide containing 50 wt% each is to be continuously fractionated at standard atmospheric pressure at the rate of 5000 kg/h. The distillate product is to contain 95 wt% carbon disulfide, the residue 2.5 wt%. The feed will be 30 mol% vaporized before it enters the tower. A total condenser will be used, and reflux will be returned at the bubble point. VLE data at 1 atm (Treybal, 1980), x, y^* = mol fraction of CS_2, are as follows:

T, K	x	y^*	T, K	x	y^*
349.7	0.0	0.0	332.3	0.391	0.634
347.9	0.030	0.082	328.3	0.532	0.747
346.1	0.062	0.156	325.3	0.663	0.829
343.3	0.111	0.266	323.4	0.757	0.878
341.6	0.144	0.333	321.5	0.860	0.932
336.8	0.259	0.495	319.3	1.000	1.000

(a) Determine the product rates, in kg/h.

Answer: Distillate = 2568 kg/h

(b) Determine the minimum number of theoretical stages required.

Answer: 7 stages

(c) Determine the minimum reflux ratio.

Answer: 0.941

(d) Determine the number of theoretical trays required, and the location of the feed tray, at a reflux ratio equal to 1.5 the minimum.

Answer: Feed in stage 6

(e) Estimate the overall tray efficiency of a sieve-tray tower of "conventional design" and the number of real trays. The viscosity of liquid CCl_4 and liquid CS_2 as functions of temperature are given by equation (6-42) with:

Component	A	B	C	$D \times 10^5$
CCl_4	−13.03	2290	0.0234	−2.011
CS_2	−3.442	713.8	0.000	0.000

Source: Data from Reid et al. (1987).

Answer: 20 real trays

6.21[b]. Rectification of an ethanol–water mixture: Murphree efficiency

A distillation column is separating 1000 mol/h of a 32 mol% ethanol, 68 mol% water mixture at atmospheric pressure. The feed enters as a subcooled liquid that will condense 1 mol of vapor in the feed plate for every 4 mol of feed. The column has a total condenser and uses open steam heating. We desire a distillate with an

Problems   **385**

ethanol content of 75 mol%, and a bottoms product with 10 mol% ethanol. The steam used is saturated at 1 atm. Table 6.2 gives VLE data for this system.

(a) Find the minimum reflux ratio.

Answer: 0.63

(b) Find the number of real stages and the optimum feed location for $R = 1.5\,R_{\min}$ if the Murphree vapor efficiency is 0.7 for all stages.

Answer: Feed in stage 10

(c) Find the steam flow rate used.

Answer: 700 mol/h

6.22[d]. Rectification of an acetone–methanol mixture: large number of stages

A distillation column is separating 100 mol/s of a 30 mol% acetone, 70 mol% methanol mixture at atmospheric pressure. The feed enters as a saturated liquid. The column has a total condenser and a partial reboiler. We desire a distillate with an acetone content of 72 mol%, and a bottoms product with 99.9 mol% methanol. A reflux ratio of 1.25 the minimum will be used. Calculate the number of ideal stages required and the optimum feed location. VLE for this system is described by the modified Raoult's law, with the NRTL equation for calculation of liquid-phase activity coefficients, and the Antoine equation for estimation of the vapor pressures.

The Antoine constants for acetone (1) and methanol (2) are (P_i in kPa, T in °C): $A_1 = 14.3916, B_1 = 2795.92, C_1 = 230.0; A_2 = 16.5938, B_2 = 3644.30, C_2 = 239.76$. The NRTL constants are $b_{12} = 184.70$ cal/mol, $b_{21} = 222.64$ cal/mol, $\alpha = 0.3084$ (Smith et al., 1996).

Answer: 28 ideal stages, including the reboiler

6.23[c]. Binary distillation with two feeds

We wish to separate ethanol from water from two different feed solutions in a distillation column with a total condenser and a partial reboiler, at a column pressure of 1 atm. One of the feed solutions, a saturated vapor, flows at the rate of 200 kmol/h and contains 30 mol% ethanol. The other feed, a subcooled liquid, flows at the rate of 300 kmol/h and contains 40 mol% ethanol. One mol of vapor must condense inside the column to heat up 4 mol of the second feed to its boiling point. We desire a bottoms product that is 2 mol% ethanol and a distillate product that is 72 mol% ethanol. The reflux ratio is 1.0; VLE data are available in Table 6.2. The feeds are to be input at their optimum feed locations. Find the optimum feed locations and the total number of equilibrium stages required.

Answer: 7 ideal stages, including the reboiler

6.24[c]. Binary distillation in a random packed column

Redesign the fractionator of Example 6.8 using a random packing. The column is to be packed with 50-mm metal Pall rings. Determine the diameter of the tower, the height of packing in the stripping and rectifying sections, and the total gas-pressure drop. Design for a gas-pressure drop not to exceed 400 Pa/m.

6.25^c. Binary distillation in a structured packed column

Redesign the fractionator of Example 6.8 for a reflux ratio that is twice the minimum. Determine the diameter of the tower, the height of packing in the stripping and rectifying sections, and the total gas-pressure drop. Design for a gas-pressure drop not to exceed 400 Pa/m.

6.26^c. A distillation–membrane hybrid for ethanol dehydration

Many industrially important liquid systems are difficult or impossible to separate by simple continuous distillation because the phase behavior contains an azeotrope, a tangent pinch, or an overall low relative volatility. One solution is to combine distillation with one or more complementary separation technologies to form a hybrid. An example of such a combination is the dehydration of ethanol using a distillation–membrane hybrid, as shown in Figure 6.28.

The membrane performance is specified by defining a *membrane separation factor*, α_m, and *membrane cut*, θ:

$$\alpha_m = \frac{x_P(1-x_R)}{x_R(1-x_P)} \tag{6-113}$$

$$\theta = \frac{P}{D} \tag{6-114}$$

In a given application, 100 mol/s of a saturated liquid containing 37 mol% ethanol and 63 mol% water must be separated to yield a product which is 99 mol% ethanol, and a residue containing 99 mol% water. The solution will be fed to a

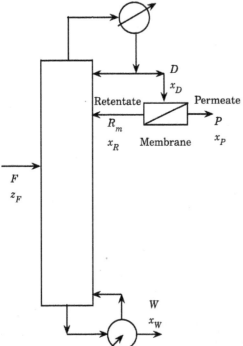

Figure 6.28 Distillation–membrane hybrid.

Problems 387

distillation column operating at atmospheric pressure, with a partial reboiler and a total condenser. The reflux ratio will be 1.5 the minimum. The distillate will enter a membrane with parameters $\alpha_m = 70$ and $\theta = 0.6$. The membrane boosts the concentration so that the permeate stream is the ethanol-rich product ($x_P = 0.99$). The retentate stream is returned as a saturated liquid to the column to the tray at the nearest liquid concentration. Calculate

(a) The molar flow rate of the product and of the residue.

Answer: $P = 36.7$ mol/s

(b) The molar flow rate and composition of the distillate coming out of the column.

Answer: $x_D = 0.828$

(c) The molar flow rate and composition of the retentate returned to the column.

Answer: $x_R = 0.586$

(d) The number of ideal stages required, and the optimal location of the two feeds(the original feed and the retentate recycle) to the distillation column.

6.27[a]. Gilliland correlation applied to a binary system

Use the Gilliland correlation to estimate the number of ideal stages required for the separation of Example 6.4. Assume that, for the system benzene and toluene at atmospheric pressure, the relative volatility is constant at $\alpha = 2.5$. Use the results of Problems 6.15 and 6.16. Use the Kirkbride equation to estimate the optimum feed-stage location.

Answer: 14.4 ideal stages

6.28[b]. Fenske–Underwood–Gilliland method

A distillation column has a feed of 100 kmol/h. The feed is 10 mol% LNK, 55 mol% LK, and 35 mol% HK and is a saturated liquid. The reflux ratio is 1.2 times the minimum. We desire 99.5% recovery of the light key in the distillate. The mol fraction of the light key in the distillate should be 0.75. Use the FUG approach to estimate the number of ideal stages required and the optimal location of the feed stage. Equilibrium data:

$$\alpha_{LNK} = 4.0 \qquad \alpha_{LK} = 1.0 \qquad \alpha_{HK} = 0.75$$

Answer: $N = 48.8$ ideal stages

6.29[b]. Fenske–Underwood–Gilliland method

One hundred kmol/h of a ternary bubble-point mixture to be separated by distillation has the following composition:

Component	Mol fraction	Relative volatility
A	0.4	5
B	0.2	3
C	0.4	1

388   **Distillation**

(a) For a distillate rate of 60 kmol/h, five theoretical stages, and total reflux, calculate the distillate and bottoms composition by the Fenske equation.

Answer: $x_{AD} = 0.662$, $x_{CW} = 0.954$

(b) Using the separation in part (a) for components A and C, determine the minimum reflux ratio by the Underwood equation.

Answer: $R_{min} = 0.367$

(c) For a reflux ratio of 1.2 times the minimum, determine the number of theoretical stages required, and the optimum feed location.

Answer: 14.3 ideal stages

6.31[d]. Binary batch rectification with constant reflux

A batch distillation column with four theoretical stages (first stage is the still pot) is charged with 100 kmol of a 50 mol% benzene in toluene mixture at atmospheric pressure. At a constant reflux ratio $R = 1.5$, how many mol of the charge must be distilled if an average product composition of 84 mol% benzene is required? If the boil-up rate is 15 kmol/h, calculate the distillation time. The equilibrium distribution curve at the column pressure is given in Example 6.4.

6.32[d]. Binary batch rectification with constant distillate composition

A batch distillation column with six theoretical stages (first stage is the still pot) is charged with 100 kmol of a 20 mol% ethanol in water mixture at atmospheric pressure. The boil-up rate is constant at 10 kmol/h. If the distillate composition is to be maintained constant at 80 mol% ethanol by varying the reflux ratio, calculate the distillation time required to reduce the still composition to 5 mol% ethanol. Calculate the fractional recovery of ethanol in the distillate. The equilibrium distribution curve is given in Table 6.2.

6.33[d]. Binary batch rectification with constant distillate composition

A batch distillation column with eight theoretical stages (first stage is the still pot) is charged with 500 kmol of a 48.8 mol% A in B mixture at atmospheric pressure (relative volatility $\alpha_{AB} = 2.0$). The boil-up rate is constant at 213.5 kmol/h. If the distillate composition is to be maintained constant at 95 mol% A by varying the reflux ratio, calculate the distillation time required to reduce the still composition to 19.2 mol% A. Calculate the fractional recovery of A in the distillate.

6.34[d]. Binary batch rectification with constant distillate composition (Seader et al., 2011)

A batch distillation column with three theoretical stages (first stage is the still pot) is charged with 100 kmol of a 50.0 mol% n-hexane in n-octane liquid mixture at atmospheric pressure (the phase-equilibrium curve is given in Figure 6.25). A liquid distillate containing 90 mol% hexane is to be maintained by continuously adjusting the reflux ratio while maintaining a distillate rate of 20.0 kmol/h. What should the reflux ratio be after 1.0 h when the accumulated distillate is 20.0 kmol? Theoretically, when must distillate accumulation be stopped?

REFERENCES

Baur, R. et al., *Chem. Eng. Sci.*, **56**, 1721, 2085 (2001).

Billet, R., and M. Schultes, *Trans IChemE*, **77**, 498–504 (1999).

DePriester, C. L., *Chem. Eng. Progr. Symp. Ser.*, **49**, 42 (1953).

Diwekar, U. M., and K. P. Mandhaven, *Ind. Eng. Chem. Res.*, **30**, 713–721 (1991).

Doherty, M. F., and M. F. Malone, *Conceptual Design of Distillation Systems*, McGraw-Hill, New York (2001).

Fenske, M. R., *Ind. Eng. Chem.*, **24**, 482 (1932).

Gilliland, E. R., *Ind. Eng. Chem.*, **32**, 1220 (1940).

Holman, J. P., *Heat Transfer*, 7th ed., McGraw-Hill, New York (1990).

Hwalek, J. J., Associate Professor, *Chemical and Biological Engineering*, University of Maine, Maine (2001).

King, C. J., *Separation Processes*, 2nd ed., McGraw-Hill, New York (1980).

Kirkbride, C. G., *Petroleum Refiner.*, **23**, 87 (1944).

Kister, H. Z., *Distillation Design*, McGraw-Hill, New York (1992).

Krishna, R. et al., *Trans. Int. Chem. Eng.*, **55**, 178 (1977).

Krishna, R., and G. L. Standardt, *Chem. Eng. Comm.*, **3**, 201 (1979).

Krishnamurthy, R., and R. Taylor, *AIChE J.*, **31**, 449, 456 (1985).

McCabe, W. L., and E. W. Thiele, *Ind. Eng. Chem.*, **17**, 605 (1925).

McWilliams, M. L., *Chem. Eng.*, **80**, 138 (1973).

Molokanov, Y. K. et al., *Int. Chem. Eng.*, **12**, 209 (1972).

Perry, R. H., and C. H. Chilton (eds.), *Chemical Engineers' Handbook*, 5th ed., McGraw-Hill, New York (1973).

Reid, R. C., J. M. Prausnitz, and B. E. Poling, *The Properties of Gases and Liquids*, 4th ed., McGraw-Hill, Boston (1987).

Seader, J. D., and E. J. Henley, *Separation Process Principles*, Wiley, New York (1998).

Seader, J. D., and E. J. Henley, *Separation Process Principles*, 2nd ed., Wiley, New York (2006).

Seader, J. D., E. J. Henley, and D. K. Roper, *Separation Process Principles*, 3rd ed.,Wiley, New York (2011).

Shiras, R. N. et al., *Ind. Eng. Chem.*, **42**, 871 (1950).

Smith, J. M., H. C. Van Ness, and M. M. Abbott, *Introduction to Chemical Engineering Thermodynamics*, 5th ed., McGraw-Hill, New York (1996).

Sundaram, S., and L. B. Evans, *Ind. Eng. Chem. Res.*, **32**, 511–518 (1993).

Taylor, R., and R. Krishna, *Multicomponent Mass Transfer*, Wiley, New York (1993).

Taylor, R. et al., *Comput. Chem. Engng.*, **18**, 205 (1994).

Treybal, R. E., *Mass-Transfer Operations*, 3rd ed., McGraw-Hill, New York (1980).

Wang, J. C., and G. E. Henke, *Hydrocarbon Processing*, **45**, 155 (1966).

Wankat, P. C., *Equilibrium Staged Separations*, Elsevier, New York (1988)

390   **Distillation**

APPENDIX 6.1

Python solution of Example 6.1.

```python
# Flash Vaporization Calculations; Data From Example 6.1
# Parameters of the Wagner Equation for heptane (A)
# and octane (B):
TcA = 540.3; TcB = 568.8; PcA = 27.4; PcB = 24.9
# Temperatures in K and pressures in bar
AA = -7.675; BA = 1.371; CA = -3.536; DA = -3.202
AB = -7.912; BB = 1.380; CB = -3.804; DB = -4.501
import numpy as np
def xA(T):
    return 1 - T/TcA
def xB(T):
    return 1 - T/TcB
def PA(T):
    return PcA*np.exp((AA*xA(T)+BA*(xA(T))**1.5 + \
        CA*(xA(T))**3 + DA*(xA(T))**6)/(1 - xA(T)))
def PB(T):
    return PcB*np.exp((AB*xB(T)+BB*(xB(T))**1.5 + \
        CB*(xB(T))**3 + DB*(xB(T))**6)/(1 - xB(T)))
def mA(T, P):
    return PA(T)/P
def mB(T, P):
    return PB(T)/P
# Operational Data:
D = 0.60; W = 1.0 - D; zF = 0.5; P = 1.013
from scipy.optimize import fsolve
def Flash(p):
    T = p[0]
    yD = p[1]
    xW = p[2]
    return yD - mA(T, P)*xW, 1.0 - yD - mB(T, P)*(1 - xW), \
            (yD - zF)/(xW - zF) + W/D
p = np.zeros(3,dtype = float)
p[0] = 320; p[1] = 0.6; p[2] = 0.4  # Initial estimates
p = fsolve(Flash, p)
print('The vapor concentration, yD = %5.3f' %p[1])
print('The liquid concentration, xW = %5.3f' %p[2])
print ('T (in K) = %6.1f'%p[0])
```

Program results:

The vapor concentration, $y_D = 0.576$

The liquid concentration, $x_W = 0.386$

T (in K) = 385.891

APPENDIX 6.2

Python solution of Example 6.3.

```python
# Differential Distillation Calculations:
# Data are from Example 6.3
# Enter data:

F = 100; D = 60; W = F - D; xF = 0.5
x = [0.317, 0.361, 0.409, 0.460, 0.500, 0.516, 0.577]
f = [5.464, 5.291, 5.236, 5.263, 5.376, 5.435, 5.780]

import numpy as np
from scipy.integrate import quad
from scipy.optimize import fsolve
from scipy.interpolate import CubicSpline
from scipy.integrate import quad

cs = CubicSpline(x, f)
def interp(x):
    return cs(x)
def Int(xW):
    return quad(interp, xW, xF)[0]
def g(xW):
    return np.log(F/W) - Int(xW)
xW = fsolve(g, 0.4)
yDav = (F*xF - W*xW)/D

print('Residue concentration, xW = %5.3f' %xW)
print('Composited distillate, yDav = %5.3f' %yDav)
```

Program results:
Residue concentration, $x_W = 0.327$
Composited distillate, $y_{Dav} = 0.616$

392 Distillation

APPENDIX 6.3

Python solution of Example 6.14.

```python
# Binary Batch Rectification with Constant Reflux;
# Data are from Example 6.14
# Enter data:

F = 100; V = 10; R = 1.0; xF = 0.2
# Units of F are kmol and V are kmol/h

xW = [0.035, 0.050, 0.070, 0.090, 0.150, 0.200]
xD = [0.300, 0.350, 0.500, 0.600, 0.770, 0.850]

import numpy as np
from scipy.integrate import quad
from scipy.optimize import fsolve
from scipy.interpolate import CubicSpline

f = np.zeros(6,dtype=float)
for i in range(6):
    f[i] = 1/(xD[i] - xW[i])
    i+=1
cs = CubicSpline(xW, f)
def interp(x):
    return cs(x)
def Int(x):
    return quad(interp, x, xF)[0]
def Wt(x):
    return F*np.exp(-Int(x))
def xDav(x):
    return (F*xF - Wt(x)*x)/(F - Wt(x))
def g(x):
    return xDav(x) - 0.700
xWt = fsolve(g, 0.10)
print('The bottoms composition, xWt = %5.3f' %xWt)
print('Bottoms amount (in kmol), is Wt = %5.2f'%Wt(xWt))
Dt = F - Wt(xWt)
print('Distilled amount (in kmol),is Dt = %5.2f'%Dt)
t = Dt*(R + 1)/V
print('Time required (in hour), t = %5.3f' %t)
```

Program results:
The bottoms composition, $x_{Wt} = 0.064$
Bottoms amount (in kmol), $W_t = 78.59$
Distilled amount (in kmol), $D_t = 21.41$
Time required (in hour), $t = 4.283$

7

Liquid–Liquid Extraction

7.1 INTRODUCTION

Liquid–liquid extraction, sometimes called *solvent extraction*, is the separation of the constituents of a liquid solution by contact with another insoluble liquid. If the substances constituting the original solution distribute themselves differently between the liquid phases, a certain degree of separation is achieved. This can be enhanced by the use of multiple contacts, or their equivalent, in the manner of gas absorption and distillation.

A simple example will illustrate the scope of the operation and some of its characteristics. Acetic acid is produced by methanol carbonylation or oxidation of acetaldehyde, or as a by-product of cellulose acetate manufacture. In all three cases, a mixture of acetic acid and water must be separated to produce glacial acetic acid (99.8 wt%, minimum). Separation by distillation is expensive because of the need to vaporize large amounts of water with its very high heat of vaporization. Accordingly, an alternative, less expensive liquid–liquid extraction process is often used.

If a solution of acetic acid in water is agitated with a liquid such as ethyl acetate, some of the acid, but relatively little water will enter the ester phase. Since at equilibrium the densities of the aqueous and ester layers are different, they will settle when agitation stops and can be decanted from each other. Since now the ratio of acid to water in the ester layer is different from that in the original solution and also different from that in the residual water solution, a certain degree of separation will have occurred. This is an example of stagewise contact, and it can be carried out either in batch or in continuous fashion. The residual water can be repeatedly extracted with more ester to reduce the acid content still further, or we can arrange a countercurrent cascade of stages. Another possibility is to use some sort of countercurrent continuous-contact device, where discrete stages are not involved. The use of reflux may enhance the ultimate separation.

Principles and Applications of Mass Transfer: The Design of Separation Processes for Chemical and Biochemical Engineering, Fourth Edition. Jaime Benitez.
© 2023 John Wiley & Sons, Inc. Published 2023 by John Wiley & Sons, Inc.

393

394   **Liquid–Liquid Extraction**

There are some other applications of liquid–liquid extraction where it seems uniquely adequate as a separation technique. Many pharmaceutical products (e.g., penicillin) are produced in mixtures so complex that only liquid–liquid extraction is a feasible separation process (Seader and Henley, 2006).

In such separations, the solution which is to be extracted is called the *feed*, and the liquid with which it is contacted is the *solvent*. The solvent-rich product of the operation is called the *extract*, and the residual liquid from which solute has been removed is the *raffinate* (Treybal, 1980).

7.2 LIQUID EQUILIBRIA

Your objectives in studying this section are to be able to:

1. Plot extraction equilibrium data on equilateral- and right-triangular diagrams.
2. Explain the difference between type I and type II extraction equilibrium behavior.
3. Find the saturated extract, saturated raffinate, and conjugate lines on a right-triangular diagram.

Extraction involves the use of systems composed of at least three substances, and although for the most part the insoluble phases are chemically very different, generally all three components appear at least to some extent in both phases. A number of different phase diagrams and computational techniques have been devised to determine the equilibrium compositions in ternary mixtures. In what follows, A and B are pure, substantially insoluble liquids, and C is the distributed solute. Mixtures to be separated by extraction are composed of A and C, and B is the extracting solvent.

Equilateral-triangular diagrams are used extensively in the chemical literature to graphically describe the concentrations in ternary systems. It is the property of an equilateral triangle that the sum of the perpendicular distances from any point within the triangle to the three sides equals the altitude of the triangle. We can, therefore, let the altitude represent 100% composition and the distances to the three sides the percentages or fractions of the three components (refer to Figure 7.1). Each apex of the triangle represents a pure component as marked. The perpendicular distances from any point, such as K to the base AB, represent the percentage of C in the mixture at K, the distance to the base AC the percentage of B, and to the base BC the percentage of A. Thus, the composition at point K in Figure 7.1 is 40% A, 20% B, and 40% C.

Any point on a side of the triangle represents a binary mixture. Point D, for example, is a binary mixture containing 80% A and 20% B. All points on the line DC represent mixtures containing the same ratio of A to B and can be considered as mixtures originally at D to which C has been added.

7.2 Liquid Equilibria

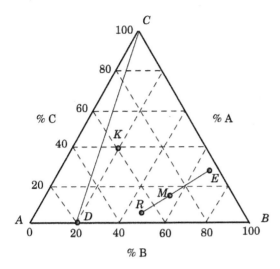

Figure 7.1 Equilateral-triangular diagram.

If R kg of a mixture at point R is added to E kg of a mixture at E, the new mixture is shown on the straight line RE at point M such that (see Problem 7.1)

$$\frac{R}{E} = \frac{\text{line } ME}{\text{line } RM} = \frac{x_E - x_M}{x_M - x_R} \qquad (7\text{-}1)$$

where x is the weight fraction of C in the solvent-lean, or raffinate liquids. This is known as the lever-arm rule. Similarly, if a mixture at M has a mixture of composition E removed from it, the new mixture is on the straight line EM extended in the direction opposite to E and located at R such that equation (7-1) applies.

Extraction systems are noted for the wide variety of equilibrium behavior that can occur in them. The most common type of system behavior is shown in Figure 7.2, often called a type I system, since there is one pair of immiscible binary compounds. The triangular coordinates are used as *isotherms*, or diagrams at constant temperature. Liquid C dissolves completely in A and B, but A and B dissolve only to a limited extent in each other, to give rise to the saturated liquid solutions at L (A-rich) and at K (B-rich). A binary mixture J anywhere between L and K will separate into two insoluble liquid phases of compositions at L and K. The relative amounts of the phases depend upon the position of J, according to the principle of equation (7-1). A typical example of a system exhibiting type I behavior is water (A)–chloroform (B)–acetone (C).

Curve $LRPEK$ is the binodal solubility curve, indicating the change in solubility of the A- and B-rich phases upon addition of C. Any mixture outside this curve will be a homogeneous solution of one liquid phase. Any ternary mixture underneath the curve, such as M, will form two insoluble, saturated liquid phases of equilibrium compositions indicated by R (A-rich) and E (B-rich). The line RE joining these equilibrium compositions is a *tie line*, which must necessarily pass through M representing the mixture as a whole. There are an infinite number of tie lines in the two-phase region. The tie lines converge to point P, the *plait point*, where the two phases become one phase. The plait point is ordinarily not at the maximum value of C on the solubility curve.

The percentage of C in solution E is clearly greater than in solution R, and it is said that in this case, the distribution of C favors the B-rich phase. This is conveniently shown on the distribution diagram xy in Figure 7.2. Here, y is the weight fraction of C in the solvent-rich (B-rich) liquid, or extract. The ratio y^*/x, the *distribution coefficient*, is in this case greater than unity. The distribution curve lies above the diagonal $y = x$. Should the tie lines on Figure 7.2 slope in the opposite direction, the distribution curve will lie below the diagonal.

A less common type of system equilibrium behavior, type II, is illustrated in Figure 7.3, a typical isotherm. In this case, A and C are completely soluble, while the pairs A–B and B–C show

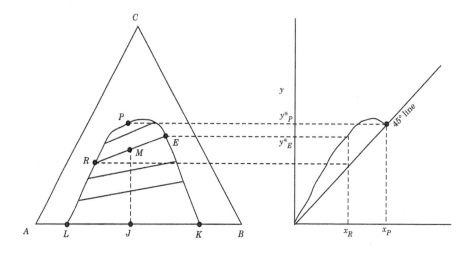

Figure 7.2 Type I equilibrium diagram: A and B partially miscible.

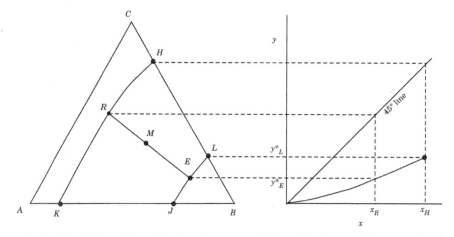

Figure 7.3 Type II equilibrium diagram: A–B and B–C partially miscible.

7.2 Liquid Equilibria

only limited solubility. At the prevailing temperature, points J and K represent the mutual solubilities of A and B, and points H and L those of B and C. Curves KRH (A-rich) and JEL (B-rich) are the ternary solubility curves, and mixtures outside the band between these two curves form homogeneous single-phase liquid solutions. Mixtures such as M, inside the heterogeneous area, form two liquid phases at equilibrium at E and R, joined in the diagram by a tie line. Figure 7.3 also shows the corresponding distribution curve. Notice that type II systems do not have a plait point. This type is exemplified by the system chlorobenzene (A)–water (B)–methyl ethyl ketone (C).

Whether a ternary system is type I or type II often depends on the temperature. For example, data for the system n-hexane (A)–methylcyclopentane (C)–aniline (B) for temperatures of 298 K, 307.5 K, and 318 K are shown in Figure 7.4. At the lowest temperature, we have a type II system. As the temperature increases, the solubility of C in B increases more rapidly than the solubility of A in B until at 34.5 °C the system is at the border between type II and type I behavior. At 45 °C the system is clearly of type I.

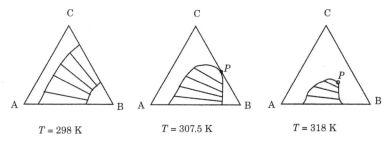

Figure 7.4 Effect of temperature on solubility for the system n-hexane (A)–aniline (B)–methylcyclopentane (C).

The coordinate scales of equilateral triangles are necessarily always the same. In order to be able to expand one concentration scale relative to the other, right-triangular coordinates can be used. Here the concentrations in weight percent of any two of the three components are given; the concentration of the third is obtained by difference from 100 wt%. Figure 7.5 is an example of a right-triangular diagram.

We will use right-triangular diagrams exclusively in the remainder of this chapter because they are easy to read, they do not require special paper, the scales of the axes can be varied, and portions of the diagram can be enlarged as shown in Example 7.1. Although equilateral-triangular diagrams have none of these advantages, they are used extensively in the literature for reporting extraction data; therefore, it is important to be able to read and use this type of extraction diagram (Wankat, 1988). Notice that the lever-arm rule also applies to right-triangular diagrams.

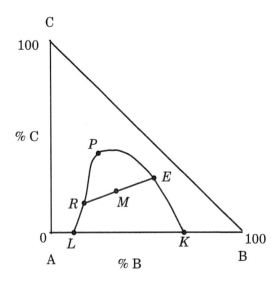

Figure 7.5 Right-triangular diagram.

Example 7.1 Equilibrium: System Water–Chloroform–Acetone

Table 7.1 presents equilibrium extraction data obtained by Alders (1959) for the system water–chloroform–acetone at 298 K and 1 atm. Plot these data on a right-triangular diagram, including in the diagram a conjugate or auxiliary line that will allow graphical construction of tie lines.

Table 7.1 Water–Chloroform–Acetone Equilibrium Data[a]

	Water phase, wt%			Chloroform phase, wt%	
Water	Chloroform	Acetone	Water	Chloroform	Acetone
82.97	1.23	15.80	1.30	70.00	28.70
73.11	1.29	25.00	2.20	55.70	42.10
62.29	1.71	36.00	4.40	42.90	52.70
45.60	5.10	49.30	10.30	28.40	61.30
34.50	9.80	55.70	18.60	20.40	61.00
23.50	16.90	59.60	23.50	16.90	59.60

Source: Data from Alders (1959) at 298 K and 1 atm.
[a] Water and chloroform phases on the same line are in equilibrium with each other.

7.3 Stagewise Liquid–Liquid Extraction

Solution

Figure 7.6 is the right-triangular diagram for this system. We have chosen chloroform as solvent (B), water as diluent (A), and acetone as solute (C). This system exhibits a type I equilibrium behavior. The graphical construction to generate the conjugate line from the tie-line data given (e.g., line RE) is shown, as well as the use of the conjugate line to generate other tie lines not included in the original data. Line LRP is the saturated raffinate line, while line PEK is the saturated extract line.

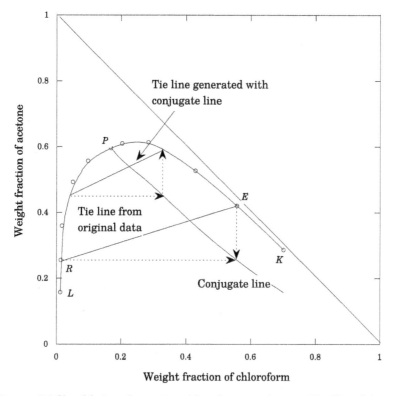

Figure 7.6 Equilibrium for water–chloroform–acetone at 298 K and 1 atm.

7.3 STAGEWISE LIQUID–LIQUID EXTRACTION

Your objectives in studying this section are to be able to:

1. Calculate flows and compositions for single-stage extraction.
2. Calculate flows and compositions for multistage crosscurrent extraction.
3. Analyze countercurrent extraction cascades without reflux.
4. Analyze countercurrent extraction cascades with reflux combining Janecke and McCabe–Thiele diagrams.

400   **Liquid–Liquid Extraction**

Extraction in equipment of the stage type can be carried on according to a variety of flowsheets, depending upon the nature of the system and the extent of separation desired. In the discussion which follows it is to be understood that each stage is a *theoretical or equilibrium stage*,

such that the effluent extract and raffinate solutions are in equilibrium with each other. Each stage must include facilities for contacting the insoluble liquids and separating the product streams. A combination of a mixer and a settler may therefore constitute a stage, and in multistage operations these may be arranged in cascades as desired. In countercurrent multistage operation, it is also possible to use towers of the multistage type as described in previous chapters.

7.3.1 Single-Stage Extraction

This may be a batch or continuous operation (refer to Figure 7.7). Feed of mass F (if batch) of F mass/time (if continuous) contains substances A and C at $x_{C,F}$ weight fraction of C. This is contacted with mass S (or mass/time) of a solvent, principally B, containing $y_{C,S}$ weight fraction of C, to give an equilibrium extract E and raffinate R, each measured in mass or mass/time. Solvent recovery then involves separate removal of solvent B from each product stream (not shown).

The operation can be followed in the right-triangular diagram as shown. If the solvent is pure B ($y_{C,S} = 0$), it will be plotted at the B apex. If the solvent has been recovered from a previous extraction, it may contain small amounts of A and B as shown by the location of S on the diagram. Adding S to F produces in the extraction stage a mixture M which, on settling, separates into the equilibrium phases E and R joined by the tie line through M. A total material balance is

$$F + S = M = E + R \tag{7-2}$$

and point M can be located on the line FS by the lever-arm rule of equation (7-1), but it is usually more satisfactory to locate M by computing its C concentration. Thus, a C balance provides

$$Fx_{CF} + Sy_{CS} = Mx_{CM} \tag{7-3}$$

from which x_{CM} can be computed. Alternatively, the amount of solvent required to provide a given location for M on the line FS can be computed:

$$\frac{S}{F} = \frac{x_{CF} - x_{CM}}{x_{CM} - y_{CS}} \tag{7-4}$$

the quantities of extract and raffinate can be computed from the lever-arm rule, or by the material balance for C:

$$Ey_{CE} + Rx_{CR} = Mx_{CM} \tag{7-5}$$

$$E = M \frac{\left(x_{CM} - x_{CR}\right)}{\left(y_{CE} - x_{CR}\right)} \tag{7-6}$$

7.3 Stagewise Liquid-Liquid Extraction

and R can be determined from equation (7-2).

Since two insoluble phases must form for an extraction operation, point M must lie within the heterogeneous liquid area, as shown. The minimum amount of solvent is thus found by locating M at D, which would then provide an infinitesimal amount of extract at G, and the maximum amount of solvent is found by locating M_1 at K which would then provide an infinitesimal amount of raffinate at L.

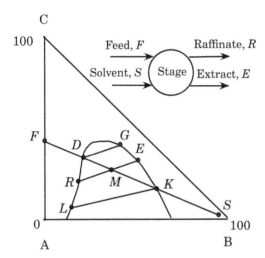

Figure 7.7 Single-stage extraction.

Example 7.2 Single-Stage Extraction

Pure chloroform is used to extract acetone from a feed containing 60 wt% acetone and 40 wt% water. The feed rate is 50 kg/h, and the solvent rate is also 50 kg/h. Operation is at 298 K and 1 atm. Find the extract and raffinate flow rates and compositions when one equilibrium stage is used for the separation.

Solution

This problem can be solved graphically or algebraically. Next, we illustrate both methods of solution.

(a) Graphical solution

The equilibrium data for this system can be obtained from Table 7.1 and Figure 7.6. On Figure 7.6, plot the streams F (x_{CF} = 0.6, x_{BF} = 0.0) and S (y_{CS} = 0.0, y_{BS} = 1.0). After locating points F and S, M is on the line FS; its exact location is found by calculating x_{CM} from

$$x_{CM} = \frac{Fx_{CF} + Sy_{CS}}{F + S} = \frac{50 \times 0.6 + 50 \times 0.0}{50 + 50} = 0.30$$

A tie line is then constructed through M by trial and error, using the conjugate line, and the extract (E) and raffinate (R) locations are obtained (see Figure 7.8). The concentrations are $x_{CR} = 0.189$, $x_{BR} = 0.013$, $y_{CE} = 0.334$, and $y_{BE} = 0.648$. The flow rates of extract and raffinate can be obtained from equations (7.6) and (7.2):

$$E = M \frac{(x_{CM} - x_{CR})}{(y_{CE} - x_{CR})} = (50 + 50)\frac{(0.300 - 0.189)}{(0.334 - 0.189)} = 76.71 \text{ kg/h}$$

$$R = F + S - E = 50 + 50 - 76.71 = 23.29 \text{ kg/h}$$

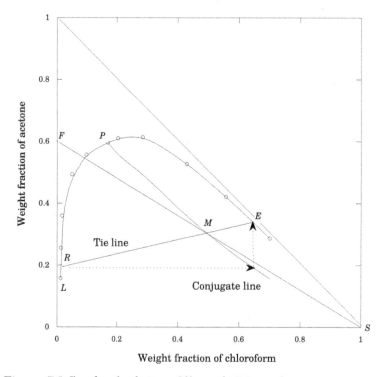

Figure 7.8 Graphical solution of Example 7.2: single-stage extraction.

(b) Algebraic solution

Using the data from Table 7.1, generate a cubic spline interpolation formula for the saturated raffinate curve, one for the saturated extract curve, and one for the tie-line data (see the Mathcad and Python programs in Appendices G-1 and G-2). Equations in the form

$$x_{BR}(x_{CR}) = f_{raf}(x_{CR}) \quad y_{CE}(x_{CR}) = f_{tie}(x_{CR}) \quad y_{BE}(x_{CR}) = f_{ext}(y_{CE}(x_{CR}))$$

are convenient because they reduce the number of unknown equilibrium concentrations from four (x_{BR}, x_{CR}, y_{CE}, and y_{BE}) to only one (x_{CR}). Then, the total material balance represented by equation (7-2) is solved simultaneously with the solute (C) material balance:

$$Fx_{CF} + Sy_{CS} = Rx_{CR} + Ey_{CE}(x_{CR}) \qquad (7\text{-}7)$$

7.3 Stagewise Liquid–Liquid Extraction

and the solvent (B) material balance:

$$Fx_{BF} + Sy_{BS} = Rx_{BR}(x_{CR}) + Ey_{BE}(x_{CR}) \tag{7-8}$$

Equations (7-2), (7-7), and (7-8) constitute a system of three simultaneous algebraic equations in three unknowns ($E, R,$ and x_{CR}) which are easily solved using Mathcad or Python. From Appendices G-1 and G-2, the results are: $x_{CR} = 0.189$, $E = 76.71$ kg/h, and $R = 23.29$ kg/h. These results agree with those obtained graphically.

7.3.2 Multistage Crosscurrent Extraction

This is an extension of single-stage extraction wherein the raffinate is successively contacted with fresh solvent and may be done continuously or in batches. Figure 7.9 is a schematic diagram for a three-stage crosscurrent extraction process. A single final raffinate results, and the extracts can be combined to provide a composited extract, as shown. As many stages as necessary can be used.

Computations are shown in Figure 7.9 on a right-triangular diagram. All the material balances for a single stage now apply, of course, to the first stage. Subsequent stages are dealt with in the same manner, except that the feed to any stage is the raffinate from the previous stage. Unequal amounts of solvent can be used in the various stages. For a given final raffinate concentration, the greater the number of stages the less total solvent will be used.

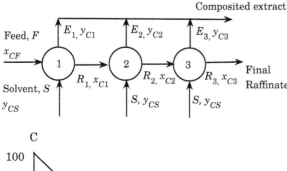

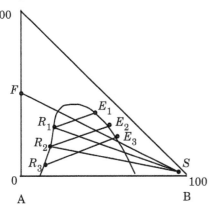

Figure 7.9 Multistage crosscurrent extraction.

Example 7.3 Multistage Crosscurrent Extraction

The feed of Example 7.2 is extracted three times with pure chloroform at 298 K, using 8 kg/h of solvent in each stage. Determine the flow rates and compositions of the various streams.

Solution

This problem can be solved graphically, as shown schematically in Figure 7.9, or algebraically. Appendices G-3 and G-4 present modifications of the Mathcad and Python programs of Appendices G-1 and G-2 that can handle multistage crosscurrent extraction algebraic calculations. The following table summarizes the results, stage by stage.

	Flow rate, kg/h		Raffinate, wt%		Extract, wt%	
Stage #	Raffinate	Extract	Acetone	Chloroform	Acetone	Chloroform
1	36.36	21.64	46.6	4.1	60.4	30.2
2	25.76	18.60	31.1	1.4	48.0	49.1
3	21.42	12.34	18.5	1.3	32.8	65.4

The composited extract flows at the rate of 52.58 kg/h, and its composition is 5.4 wt% water, 45.1 wt% chloroform, and 49.5 wt% acetone. Notice that the acetone content of the final raffinate (18.5 wt%) is slightly lower than the acetone content of the raffinate obtained in Example 7.2 in a single-stage extraction with 50 kg/h of chloroform. Remarkably, this is true even though the total amount of solvent used in the multistage crosscurrent extraction process is only 3 × 8 = 24 kg/h.

7.3.3 Countercurrent Extraction Cascades

A countercurrent cascade allows for more complete removal of the solute, and the solvent is reused so less is needed. Figure 7.10 is a schematic diagram of a countercurrent extraction cascade. Extract and raffinate streams flow from stage to stage in countercurrent fashion and yield two final products, raffinate R_N and extract E_1. For a given degree of separation, this type of operation requires fewer stages for a given amount of solvent, or less solvent for a fixed number of stages, than those required for the crosscurrent extraction method described above.

In the usual design problem, the extraction column temperature and pressure, the flow rates and compositions of streams F and S, and the desired composition (or

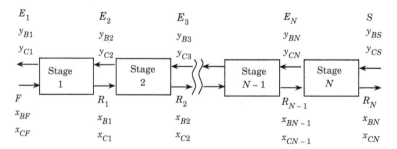

Figure 7.10 Schematic diagram of a countercurrent multistage extraction.

7.3 Stagewise Liquid–Liquid Extraction

percent removal) of solute in the raffinate product are specified. The designer must determine the number of equilibrium stages and the flow rate and composition of the outlet extract stream.

The graphical approach to solution is developed in Figure 7.11 on right-triangular coordinates. A total material balance around the entire cascade is

$$F + S = E_1 + R_N = M \tag{7-9}$$

Point M can be located on the line FS through a material balance for substance C:

$$Fx_{CF} + Sy_{CS} = E_1 y_{C1} + R_N x_{CN} = M x_{CM} \tag{7-10}$$

$$x_{CM} = \frac{Fx_{CF} + Sy_{CS}}{F + S} \tag{7-11}$$

Equation (7-9) indicates that point M lies on line $R_N E_1$ as shown. Rearranging equation (7-9) provides

$$R_N - S = F - E_1 = \Delta_R \tag{7-12}$$

where Δ_R, called a *difference point*, is the net flow outward at the last stage N. According to equation (7-12), the extended lines $E_1 F$ and SR_N must intersect at Δ_R, as shown in Figure 7.11. A material balance for stages n through N is

$$R_{n-1} + S = R_N + E_n \tag{7-13}$$

or

$$R_{n-1} - E_n = R_N - S = \Delta_R \tag{7-14}$$

so that the difference in flow rates at a location between any two adjacent stages is constant, Δ_R. Line $E_n R_{n-1}$ extended must therefore pass through Δ_R as shown on the figure.

The graphical construction is then as follows. After location of points F, S, M, E_1, R_N, and Δ_R, a tie line from E_1 provides R_1 since extract and raffinate from the

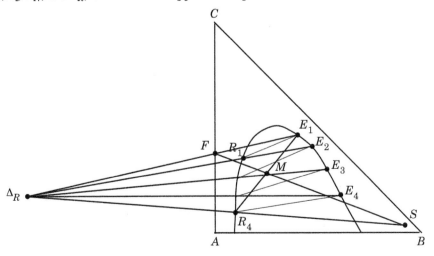

Figure 7.11 Countercurrent multistage extraction on right-triangular diagram.

first ideal stage are in equilibrium. A line from Δ_R through R_1 when extended provides E_2; a tie line from E_2 provides R_2, and so forth. The lowest possible value of x_{CN} is that given by the A-rich end of the tie line which passes through S.

As the amount of solvent is increased, point M representing the overall balance moves toward S on Figure 7.12 and point Δ_R moves farther to the left. At an amount of solvent such that lines E_1F and SR_N are parallel, point Δ_R will be at an infinite distance. Greater amounts of solvent will cause these lines to intersect on the right-hand side of the diagram rather than as shown, with point Δ_R nearer B for increasing solvent quantities. The interpretation of the difference point is, however, still the same: a line from Δ_R intersects the two branches of the solubility curve at points representing extract and raffinate from adjacent stages.

If a line from point Δ_R should coincide with a tie line, an infinite number of stages will be required to reach this condition. The maximum amount of solvent for

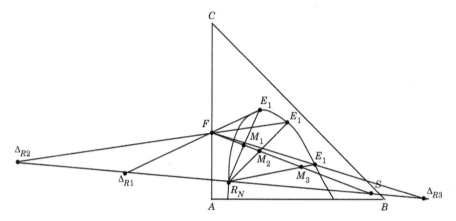

Figure 7.12 Effect of increasing solvent/feed ratio on the location of the difference point $[(S/F)_3 > (S/F)_2 > (S/F)_1]$.

which this occurs corresponds to the minimum solvent/feed ratio which can be used for the specified products. The procedure for determining the minimum amount of solvent is indicated in Figure 7.13. All tie lines below that marked JK are extended to line SR_N to give intersections with line SRN as shown. The intersection farthest from S (if on the left-hand side of the diagram) or nearest S (if on the right) represents the difference point for the minimum solvent, Δ_{Rm}. The actual position of Δ_R must be farther from S (if on the left) or nearer to S (if on the right) for a finite number of stages. Usually, but not in the instance shown, the tie line which when extended passes through F that is, tie line JK, will locate Δ_{Rm} for minimum solvent.

Stepping off a lot of stages on a triangular diagram can be difficult and inaccurate. More accurate calculations can be done with a McCabe–Thiele diagram. Therefore, the construction indicated in Figure 7.14 may be convenient. A few lines are drawn at random from point Δ_R to intersect the two branches of the solubility curve as shown, where the points of intersection do not now necessarily indicate streams between two actual adjacent stages. The concentrations $x_{C,op}$ and $y_{C,op}$ corresponding to these are plotted on x_C, y_C as shown to generate an operating

7.3 Stagewise Liquid–Liquid Extraction

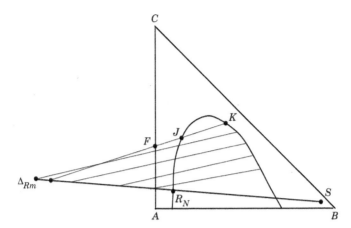

Figure 7.13 Minimum solvent for countercurrent extraction.

curve. Tie-line data provide the equilibrium curve y_C^* versus x_C, and the theoretical stages are stepped off in the resulting McCabe–Thiele diagram in the manner used for distillation and gas absorption.

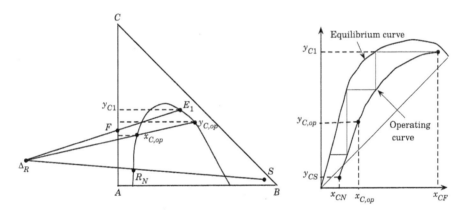

Figure 7.14 From right-triangular coordinates to a McCabe–Thiele diagram.

Example 7.4 Multistage Countercurrent Extraction

A solution of acetic acid (C) in water (A) is to be extracted using isopropyl ether (B) as the solvent. The feed is 1000 kg/h of a solution containing 35 wt% acid and 65 wt% water. The solvent used is essentially pure isopropyl ether. The exiting raffinate stream should contain 2 wt% acetic acid. Operation is at 293 K and 1 atm. Determine:

(a) The minimum amount of solvent which can be used and (b) the number of theoretical stages if the solvent rate used is 60% above the minimum. Table 7.2 gives the equilibrium data.

408   **Liquid–Liquid Extraction**

Table 7.2 Water–Isopropyl Ether–Acetic Acid Equilibrium Data[a]

Water phase, wt%			Isopropyl ether phase, wt%		
Water	Ether	Acetic acid	Water	Ether	Acetic acid
98.10	1.2	0.69	0.5	99.3	0.18
97.10	1.5	1.41	0.7	98.9	0.37
95.50	1.6	2.89	0.8	98.4	0.79
91.70	1.9	6.42	1.0	97.1	1.93
84.40	2.3	13.30	1.9	93.3	4.82
71.10	3.4	25.30	3.9	84.7	11.40
58.90	4.4	36.70	6.9	71.5	21.60
45.10	10.6	44.30	10.8	58.1	31.10
37.10	16.5	46.40	15.1	48.7	36.20

[a]Water and ether phases on the same line are in equilibrium with each other.
Source: Data from Treybal (1980) at 293 K and 1 atm.

Solution

(a) Figure 7.15 is the right-triangular diagram for this system. Locate points F, S, and R_N. Extend the tie line that passes through point F until it meets the extension of line $R_N S$. The intersection of the two lines locates the difference point corresponding to the minimum solvent rate, Δ_{Rm}. Line $F\Delta_{Rm}$ intercepts the extract branch of the solubility curve at point $(E_1)_{max}$, which gives the composition of the final extract if the minimum amount of solvent is used. Draw the line from $(E_1)_{max}$ to R_N. The intersection of this line with the line FS gives $(M)_{min}$. From Figure 7.15, $(x_{CM})_{min} = 0.144$. Equation (7-11) can be written for the minimum solvent rate condition, S_{min}:

$$\left(x_{CM}\right)_{min} = \frac{Fx_{CF} + S_{min}y_{CS}}{F + S_{min}} \qquad (7\text{-}15)$$

For this example, $x_{CF} = 0.35$, $y_{CS} = 0$, $(x_{CM})_{min} = 0.144$, and $F = 1000$ kg/h. substituting in equation (7-15), we obtain $S_{min} = 1431$ kg/h.

(b) For $S = 1.6 S_{min} = 2290$ kg/h, equation (7-11) gives $x_{CM} = 0.106$. Point M is then located on the line FS. Line $R_N E_1$ is located passing through point M and ending on the extract branch of the solubility curve. Extend the line FE_1 until it meets the extension of the line $R_N S$. The intersection of the two lines corresponds to the location of the actual difference point, Δ_R. A few lines are drawn at random from point Δ_R to intersect the two branches of the solubility curve. The concentrations $x_{C,op}$ and $y_{C,op}$ corresponding to these are given in Table 7.3 and plotted on x_C, y_C coordinates as shown in Figure 7.16 to generate an operating curve. Tie-line data provide the equilibrium curve y_C^* versus x_C, and the theoretical stages are stepped

7.3 Stagewise Liquid–Liquid Extraction

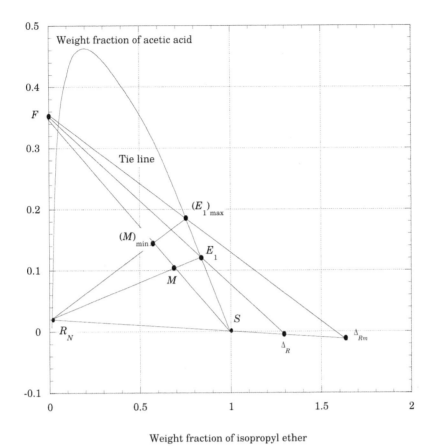

Figure 7.15 Minimum and actual difference points for Example 7.4.

off in the resulting McCabe–Thiele diagram. A total of about 7.4 equilibrium stages are required for the specified separation.

Table 7.3 Operating Line Data for Example 7.4

$x_{C,op}$	0.020	0.050	0.100	0.150	0.200	0.250	0.300	0.350
$y_{C,op}$	0.000	0.009	0.023	0.037	0.054	0.074	0.096	0.121

7.3.4 Insoluble Liquids

When the liquids A and B are insoluble over the range of solute concentrations encountered, as shown in Example 7.5, the stage computation is done more simply in terms of mass flow rates of A (A, mass/time) and B (B, mass/time) and the mass ratios, defined here as:

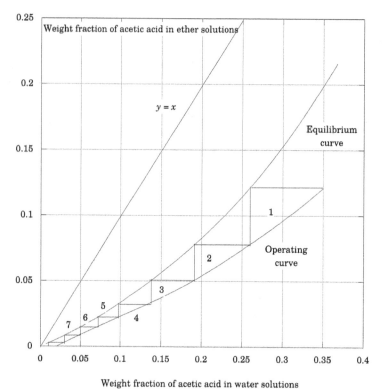

Figure 7.16 McCabe–Thiele diagram for Example 7.4.

x' = mass of C/mass of A in the raffinate liquids
y' = mass of C/mass of B in the extract liquids

An overall plant balance for substance C is

$$By'_S + Ax'_F = Ax'_N + By'_1 \qquad (7\text{-}16)$$

or

$$\frac{A}{B} = \frac{y'_1 - y'_S}{x'_F - x'_N} \qquad (7\text{-}17)$$

which is the equation of a straight line, the operating line, of slope A/B, through points (y'_1, x'_F) and (y'_S, x'_N).

For the special case where the equilibrium curve is of constant slope $K_D = y'_i/x'_i$, a special form of Kremser equations applies:

$$N = \frac{\ln\left[\dfrac{x'_F - y'_S/K_D}{x'_N - y'_S/K_D}\left(1 - \dfrac{1}{EF}\right) + \dfrac{1}{EF}\right]}{\ln(EF)} \qquad (7\text{-}18)$$

where the *extraction factor*, *EF*, is given by

7.3 Stagewise Liquid–Liquid Extraction

$$EF = \frac{K_D B}{A} \tag{7-19}$$

Equation (7-18) is valid for $EF \neq 1$. If the Murphree stage efficiency expressed in terms of extract compositions, $\mathbf{E}_{ME}$, is constant, then the overall efficiency for the cascade is given by

$$\mathbf{E}_O = \frac{\text{equilibrium trays}}{\text{real trays}} = \frac{\ln\left[1 + \mathbf{E}_{ME}\left(EF - 1\right)\right]}{\ln\left(EF\right)} \tag{7-20}$$

Example 7.5 Multistage Extraction: Insoluble Liquids

A solution of nicotine (C) in water (A) is to be extracted at 293 K using pure kerosene (B) as the solvent. Water and kerosene are essentially insoluble at this temperature. The feed is 1000 kg/h of a solution containing 1.0 wt% nicotine. The exiting raffinate stream should contain 0.1 wt% nicotine. Determine

(a) The minimum amount of solvent which can be used.

(b) The number of theoretical stages if the solvent rate used is 20% above the minimum.

(c) The number of real stages if the Murphree stage efficiency, expressed in terms of extract compositions, is $\mathbf{E}_{ME}$ = 0.6. The equilibrium curve in this range of compositions is basically a straight line with slope K_D = 0.926 kg water/kg kerosene (Treybal, 1980).

Solution

(a) Because the solutions are so dilute, the concentrations expressed in terms of mass ratios are virtually equal to the mass fractions. Therefore, $x'_F = 0.01$ kg nicotine/kg water, $x'_N = 0.001$ kg nicotine/kg water, and $y'_S = 0.0$ kg nicotine/kg kerosene. Because, in this case, both the equilibrium and operating lines are straight, if the minimum solvent flow rate ($B_{\min}$) is used, the concentration of the exiting extract, $y'_1(\max)$, will be in equilibrium with x'_F Then,

$$y'_1\left(\max\right) = K_D x'_F = 0.926 \times 0.01 = 0.00926 \text{ kg nicotine/kg kerosene}$$

The feed to the process flows at a rate of 1000 kg/h and is 99% water. Therefore, A = 990 kg water/h. From equation (7-17),

$$B_{\min} = A \frac{x'_F - x'_N}{y'_1\left(\max\right) - y'_S} = 900 \times \frac{0.01 - 0.001}{0.00926 - 0.0} = 962 \text{ kg kerosene/h}$$

(b) For a solvent flow rate that is 20% above the minimum,

$$B = 1.2 B_{\min} = 1154 \text{ kg kerosene/h}$$

Then,

$$EF = K_D B / A = 0.926 \times 1154 / 990 = 1.079$$

Substituting in equation (7-18), $N = 6.66$ equilibrium stages.

(c) For $EF = 1.079$ and $\mathbf{E}_{ME} = 0.6$, equation (7-20) yields $\mathbf{E}_O = 0.609$. Then, the number of real stages required is $6.66/0.609 = 11$ stages.

7.3.5 Continuous Countercurrent Extraction with Reflux

Whereas in ordinary countercurrent operation the richest possible extract product leaving the plant is at best only in equilibrium with the feed solution, the use of reflux at the extract end of the cascade can provide a product even richer, as in the rectifying section of a distillation column. Reflux is not needed at the raffinate end of the cascade since, unlike distillation, where heat must be carried in from the reboiler by a vapor reflux, in extraction the solvent (the analog of heat) can enter without a carrier stream.

An arrangement for this is shown in Figure 7.17. The feed to be separated into its components is introduced at an appropriate place in the cascade, through which extract and raffinate liquids are flowing countercurrently. The concentration of solute C is increased in the extract-enriching section by countercurrent contact with a raffinate liquid rich in C. This is provided by removing the solvent from extract E_1 to produce the solvent-free stream E', part of which is removed as extract product P'_E, and part returned as reflux R_0. The raffinate-stripping section of the cascade is the same as the countercurrent extractor of Figure 7.10, and C is stripped from the raffinate by countercurrent contact with the solvent.

Graphical determination of the number of stages required for such an operation is usually inconvenient to carry out in triangular coordinates because of crowding (Seader and Henley, 2006). Instead, a *Janecke diagram*, used in conjunction with a distribution diagram, is very useful for this purpose. In Janecke diagrams, which use convenient rectangular coordinates, solvent concentration on a solvent-free basis, N_R or N_E, is plotted as the ordinate against solute concentration on a solvent-free basis, X or Y, as the abscissa (Janecke, 1906). These are defined here as:

N_R = mass B/(mass of A + mass of C) in the raffinate liquids
N_E = mass B/(mass of A + mass of C) in the extract liquids
X = mass C/(mass of A + mass of C) in the raffinate liquids
Y = mass C/(mass of A + mass of C) in the extract liquids

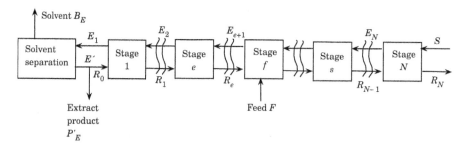

Figure 7.17 Countercurrent multistage extraction with reflux.

7.3 Stagewise Liquid–Liquid Extraction 413

The application of Janecke diagrams to liquid–liquid extraction of a type II system with the use of reflux was considered in detail by Maloney and Schubert (1940), who used an auxiliary distribution diagram of the McCabe–Thiele type, but on a solvent-free basis to facilitate visualization of the stages. This method is also referred to as the *Ponchon–Savarit method for extraction*. Although the Janecke diagram can also be applied to type I systems, it becomes difficult to use when the liquids A and B are highly miscible because the resulting values of the ordinate can become very large.

For the development that follows, primed letters indicate solvent-free quantities. Thus, for example, E' = mass of B-free solution/time. Consider now the extract-enriching section in Figure 7.17. An (A + C) balance around the solvent separator is

$$E_1' = E' = P_E' + R_0' \tag{7-21}$$

Let Δ_E represent the net rate of flow outward from this section. Then, for its (A + C) content,

$$\Delta_E = P_E' \tag{7-22}$$

and for its C content,

$$X_{\Delta E} = X_{PE} \tag{7-23}$$

while for its B content,

$$B_E = \Delta_E N_{\Delta E} \tag{7-24}$$

The point Δ_E is plotted on Figure 7.18, which is drawn for a system of two partly miscible component pairs. For all stages through e, an (A + C) balance is

$$E_{c+1}' = P_E' + R_c' = \Delta_E + R_c' \tag{7-25}$$

or

$$\Delta_E = E_{c+1}' - R_c' \tag{7-26}$$

A balance of component C is

$$\Delta_E X_{\Delta E} = E_{c+1}' Y_{c+1} - R_c' X_c \tag{7-27}$$

and a B balance is

$$\Delta_E N_{\Delta E} = E_{c+1}' N_{E,c+1} - R_c' N_{R,c} \tag{7-28}$$

Since e is any stage in the extract enriching section, lines radiating from point Δ_E cut the solubility curves of Figure 7.18 at points representing extract and raffinate flowing between any two adjacent stages. Δ_E is therefore a difference point, constant for all stages in this section. Then, alternating tie lines and lines from Δ_E establish the equilibrium stages in this section, starting with stage 1 and continuing to the feed stage.

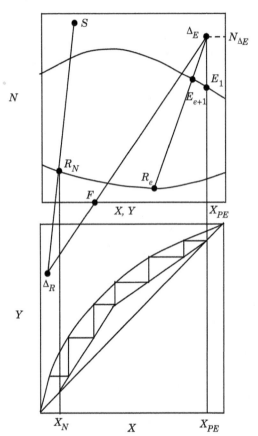

Figure 7.18 Use of Janecke diagram with auxiliary distribution curve for countercurrent extraction with reflux.

Combining equations (7-25), (7-27), and (7-28), we get an expression for the internal reflux ratio at any stage,

$$\frac{R'_c}{E'_{c+1}} = \frac{N_{\Delta E} - N_{E,c+1}}{N_{\Delta E} - N_{R,c}} \quad (7\text{-}29)$$

The external reflux ratio can be calculated from equation (7-29)

$$\frac{R'_0}{P'_E} = \frac{R_0}{P'_E} = \frac{N_{\Delta E} - N_{E,1}}{N_{E,1}} \quad (7\text{-}30)$$

which can be used to locate $_{\Delta E}$ for any specified reflux ratio.

Consider now the raffinate stripping section of Figure 7.17. The balance for stages s through N is

$$R'_N - S' = R'_{s-1} - E'_s = \Delta_R \quad (7\text{-}31)$$

where Δ_R is the difference in solvent-free flow, out minus in, at stage N. It is also the constant difference of solvent-free flows of the streams between any two adjacent stages in this section of the cascade. Δ_R is therefore a difference point, constant for all stages

7.3 Stagewise Liquid–Liquid Extraction 415

in this section. Then, alternating tie lines and lines from Δ_R establish the equilibrium stages in this section, starting with the feed stage and continuing to the last stage N.

Material balances may also be written for the whole cascade. A balance for $(A + C)$ is of the form

$$F' + S' = P'_E + R'_N \qquad (7\text{-}32)$$

Combining equation (7-32) with (7-22) and (7-31) gives us

$$F' = \Delta_E + \Delta_R \qquad (7\text{-}33)$$

where normally $F' = F$. Point F must therefore lie on the line joining the two difference points as shown in Figure 7.18.

The higher the location of Δ_E (and the lower Δ_R), the larger the reflux ratio and the smaller the number of ideal stages. At total reflux, $N_{\Delta E} = \infty$ and the minimum number of stages results. The capacity of the plant falls to zero, feed must be stopped, and solvent B_E is recirculated to become S. The minimum number of stages is easily determined on the XY equilibrium distribution diagram using the 45° diagonal as operating lines for both sections of the cascade.

An infinite number of stages is required if a line radiating from either Δ_E or Δ_R coincides with a tie line, and the greatest reflux ratio for which this occurs is the minimum reflux ratio. Frequently, the tie line that when extended passes through point F, representing the feed condition, will establish the minimum reflux ratio. This will always be the case if the XY equilibrium distribution curve is everywhere concave downward.

Example 7.6 Countercurrent Extraction with Extract Reflux

A solution containing 60% styrene (C) and 40% ethylbenzene (A) is to be separated at 298 K at the rate of 1000 kg/h into products containing 10% and 90% styrene, respectively, with diethylene glycol (B) as solvent. Determine

(a) The minimum number of theoretical stages.

(b) The minimum extract reflux ratio.

(c) The number of theoretical stages and the important flow quantities at an extract reflux ratio of 1.5 times the minimum value.

Solution

Equilibrium data (Boobar et al., 1951) have been converted to a solvent-free basis and are given in Table 7.4 (Treybal, 1980). These are plotted in the form of a Janecke diagram in Figure 7.20. The corresponding equilibrium distribution diagram is Figure 7.19. From the problem statement, $F' = F = 1000$ kg/h, $X_F = 0.6$ wt fraction of styrene, $X_{PE} = 0.9$, and $X_N = 0.1$, all on a solvent-free basis.

(a) Minimum theoretical stages are determined on the XY equilibrium distribution diagram, stepping them off from the diagonal line to the equilibrium curve, beginning at $X_{PE} = 0.9$ and ending at $X_N = 0.1$, as shown in Figure 7.19. The answer is $N_{\min} = 9$ ideal stages.

Table 7.4 Equilibrium Data for Ethylbenzene (A)–Diethylene Glycol (B)–Styrene (C)

Raffinate solutions		Extract solutions	
X	N_R	Y	N_E
kg C/kg (A + C)	kg B/kg (A + C)	kg C/kg (A + C)	kg B/kg (A + C)
0.0000	0.00675	0.0000	8.62
0.0870	0.00617	0.1429	7.71
0.1883	0.00938	0.2730	6.81
0.2880	0.01010	0.3860	6.04
0.3840	0.01101	0.4800	5.44
0.4580	0.01215	0.5570	5.02
0.4640	0.01215	0.5650	4.95
0.5610	0.01410	0.6550	4.46
0.5730	0.01405	0.6740	4.37
0.7810	0.01833	0.8630	3.47
0.9000	0.02300	0.9500	3.10
1.0000	0.02560	1.0000	2.69

Source: Data at 298 K from Boobar et al. (1951) and Treybal (1980).

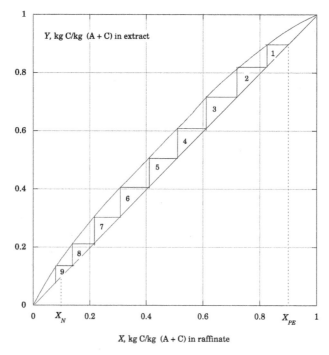

Figure 7.19 Equilibrium-distribution diagram and minimum number of stages for Example 7.6.

7.3 Stagewise Liquid–Liquid Extraction

(b) Since the equilibrium-distribution curve is everywhere concave downward, the tie line which when extended passes through F provides the minimum reflux ratio. From Figure 7.20, $N_{\Delta Em} = 11.04$ and $N_{E1} = 3.1$. From equation (7-30),

$$\left(\frac{R'_0}{P'_E}\right)_{min} = \frac{11.04 - 3.1}{3.1} = 2.561 \text{ kg reflux/kg extract product}$$

(c) For $R_0/P'_E = (1.5)(2.561) = 3.842$ kg reflux/kg extract, equation (7.30) gives $N_{\Delta E} = 15.01$. Point Δ_E is plotted as shown. A straight line from Δ_E through F intersects line $X = X_N = 0.10$ at Δ_R; from the diagram, $N_{\Delta R} = -24.90$. Random lines are drawn from Δ_E for concentrations to the right of F, and from Δ_R for those to the left. Intersections of these with the solubility curves provide the coordinates of the operating lines. Figure 7.21 shows the McCabe–Thiele-type construction to determine the number of equilibrium stages required. A total of 17.5 equilibrium stages is required, and the feed is to be introduced into the fifth stage from the extract-product end of the cascade.

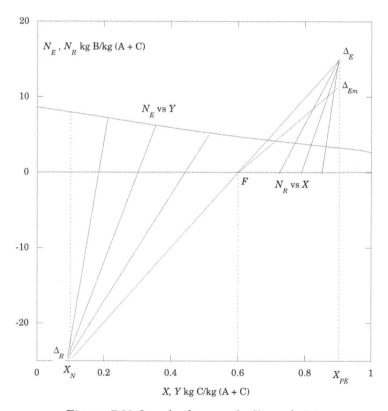

Figure 7.20 Janecke diagram for Example 7.6.

From Figure 7.20, for $X_N = 0.1$, $N_{RN} = 0.0083$. On the basis of 1 h, an overall plant balance is

$$F = 1000 = P'_E + R'_N$$

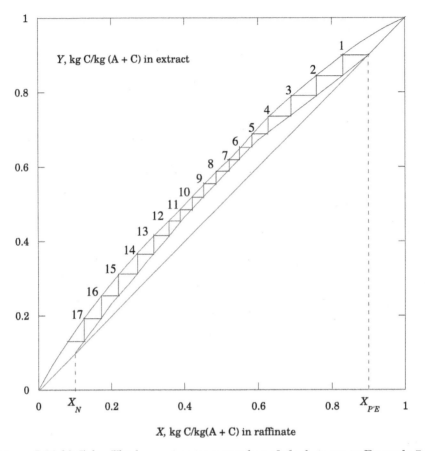

Figure 7.21 McCabe–Thiele construction: number of ideal stages in Example 7.6.

A balance on C is

$$FX_F = 600 = 0.9 \times P'_E + 0.1 \times R'_N$$

Solving simultaneously gives

$$P'_E = 625 \text{ kg/h} \quad R'_N = 375 \text{ kg/h}$$
$$R' = R'_0 = 3.842 \times P'_E = 2401 \text{ kg/h}$$
$$E'_1 = R'_0 + P'_E = 3026 \text{ kg/h}$$

$$B_E = E'_1 N_{E1} = 3026 \times 3.1 = 9381 \text{ kg/h}$$
$$E_1 = B_E + E'_1 = 12,407 \text{ kg/h}$$
$$R_N = R'_N (1 + N_{RN}) = 375(1 + 0.0083) = 378 \text{ kg/h}$$
$$S = B_E + R'_N N_{RN} = 9384 \text{ kg/h}$$

7.4 EQUIPMENT FOR LIQUID–LIQUID EXTRACTION

Your objectives in studying this section are to be able to:

1. Identify the most common types of equipment used for liquid–liquid extraction.
2. Make preliminary estimates of the dimensions of the equipment and of their power requirements.

Given the wide diversity of applications of liquid–liquid extraction, one might suspect a correspondingly large variety of extraction devices. Indeed, such is the case. Equipment such as that used for absorption, stripping, and distillation is sometimes used. However, such devices are inefficient unless liquid viscosities are low and the difference in phase density is high. For that reason, centrifugal and mechanically agitated devices are preferred. Often, the choice of suitable extraction equipment is between a cascade of *mixer-settler* units or a multicompartment column-type extractor with mechanical agitation. Methods for estimating size and power requirements for these two general types of extractors are presented next. Column devices with no mechanical agitation are also considered.

7.4.1 Mixer-Settler Cascades

In mixer-settlers, the two liquids are first mixed and then separated by settling. Any number of mixer-settlers units may be connected to form a multistage, countercurrent cascade. During mixing, one of the liquids is dispersed in the form of small droplets into the other liquid phase. The dispersed phase may be either the lighter or the heavier of the two phases. Ordinarily, the liquid flowing at the smaller volume rate will be dispersed in the other, but some control can be exercised. The mixing step is commonly conducted in an agitated vessel, with sufficient agitation and residence time so that a reasonable approach to equilibrium (e.g., 80 to 90% stage efficiency) is attained. The mixer is usually agitated by impellers. For continuous operation, the vessel should first be filled with the liquid to be continuous and, with agitation in progress, the two liquids are then introduced in the proper ratio. The settling step is by gravity in a second vessel, called a settler or *decanter*.

Sizing of mixer-settler units is done most accurately by scale-up from batch or continuous runs in laboratory or pilot-plant equipment. However, preliminary sizing calculations can be done using available theory and empirical correlations. Experimental data by Flynn and Treybal (1955) show that when liquid viscosities are less than 5 cP and the specific gravity difference between the two liquid phases is greater than about 0.1, the average residence time required of the two liquid phases in the mixing vessel to achieve at least 90% stage efficiency may be as low as 30 s, and is usually not more than 5 min, when an agitator-power input per mixer volume of 0.788 kW/m^3 is used. For a vertical, cylindrical mixer of height H and diameter D_T, the economic ratio of H to D_T is approximately 1. The mixing vessel is closed, with the two liquid phases entering at the bottom, and the effluent, in the form of a two-phase emulsion, leaving at the top.

420   **Liquid–Liquid Extraction**

Based on experiments reported by Ryan et al. (1959), the capacity of a settler vessel can be estimated as 0.2 m³/min of combined extract and raffinate flow per square meter of phase disengaging area. For a horizontal, cylindrical vessel of length L and diameter D_T, the economic ratio of L to D_T is approximately 4. Thus, if the interphase is located at the middle of the vessel, the disengaging area is $D_T L$, or, using the economic L/D_T ratio, $4D_T^2$. Frequently, the settling vessel will be bigger than the mixing vessel, as in the following example.

Example 7.7 Design of a Mixer-Settler Extractor

Benzoic acid is to be continuously extracted from a dilute solution in water using toluene as solvent in a series of mixer-settler vessels operated in countercurrent flow. The flow rates of the feed and solvent are 1.89 and 2.84 m³/min, respectively. The residence time in each mixer is 2 min. Estimate:

(a) The diameter and height of each mixing vessel.

(b) The agitator power for each mixer.

(c) The diameter and length of a settling vessel.

(d) The residence time in the settling vessel, in min.

Solution

(a) Let Q = total flow rate = 1.89 + 2.84 = 4.73 m³/min. For a residence time, t_{res} = 2 min, the vessel volume is $V_T = Q \times t_{res}$ = 4.73 × 2 = 9.46 m³. For a cylindrical vessel with $H = D_T$,

$$V_T = \frac{\pi}{4} D_T^3$$

$$H = D_T = \left(\frac{4V_T}{\pi}\right)^{1/3} = 2.3 \text{ m}$$

(b) Based on a recommendation of Flynn and Treybal (1955),

$$\text{Mixer power} = 0.788 \text{ kW/m}^3 \times 9.46 \text{ m}^3 = 7.45 \text{ kW}$$

(c) Based on the recommendation by Ryan et al. (1959), the disengaging area in the settler is

$$D_T L = 4.73 \text{ m}^3/\text{min} / \left(0.2 \text{ m}^3 / \text{min m}^2\right) = 23.65 \text{ m}^2$$

For L/D_T = 4,

$$D_T = \left(\frac{23.65}{4}\right)^{0.5} = 2.43 \text{ m}$$

$$L = 4 \times 2.43 = 9.72 \text{ m}$$

(d) Total volume of the settler, $V_T = \pi D_T^2 L/4$ = 45.1 m³, $t_{res} = V_T/Q$ = 45.1/4.73 = 9.5 min

7.4 Equipment for Liquid–Liquid Extraction 421

Mixing is accomplished by an appropriate, centrally located impeller, as shown in Figure 7.22. Vertical side baffles are usually installed to prevent vortex formation in open tanks and to minimize swirling and improve circulation patterns in closed tanks. Although no standards exist for vessel and turbine geometry, the values recommended in Figure 7.22 give good dispersion performance in liquid–liquid agitation (Seader and Henley, 2006).

To achieve a high-stage efficiency (between 90 and 100%), it is necessary to provide fairly vigorous agitation. Based on the work of Skelland and Ramsay (1987), a minimum impeller rate of rotation Ω_{min} is required for complete and uniform dispersion of one liquid into another. For a flat-blade impeller in a baffled vessel of the type discussed above, this minimum rotation rate can be estimated from

$$\frac{\Omega_{min}^2 \rho_M D_i}{g \Delta \rho} = 1.03 \phi_D^{0.106} \left(\frac{D_T}{D_i} \right)^{2.76} \left[\frac{\mu_M^2 \sigma}{D_i^5 \rho_M g^2 (\Delta \rho)^2} \right]^{0.084} \tag{7-34}$$

where

$\Delta \rho$ = absolute value of the difference in density between the liquids
σ = interfacial tension between the liquid phases
ϕ_D = fractional holdup of the dispersed liquid phase in the tank
ρ_M = two-phase mixture density, given by

$$\rho_M = \rho_C \phi_C + \rho_D \phi_D \tag{7-35}$$

ϕ_C = fractional holdup in the tank of the continuous liquid phase
ρ_C = density of the continuous liquid phase
ρ_D = density of the dispersed liquid phase
μ_M = two-phase mixture viscosity, given by

$$\mu_M = \frac{\mu_C}{\phi_C} \left[1 + \frac{1.5 \mu_D \phi_D}{\mu_C + \mu_D} \right] \tag{7-36}$$

The viscosity of the mixture given by equation (7-36) can exceed that of either constituent (Treybal, 1980).

The agitator power, P, can be estimated from an empirical correlation in terms of a *power number*, Po, which depends on an *impeller Reynolds number*, Re, where

$$\text{Po} = \frac{P}{\Omega^3 D_i^5 \rho_M} \qquad \text{Re} = \frac{\Omega D_i^2 \rho_M}{\mu_M} \tag{7-37}$$

A correlation of experimental data on Po versus Re for baffled vessels with six-bladed flat-blade impellers was developed by Laity and Treybal (1957). They found that for Re > 10,000, the power number tends to a value of Po = 5.7.

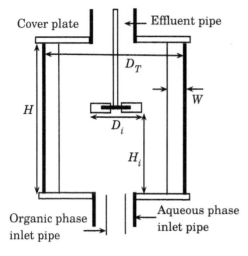

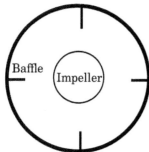

Figure 7.22 Agitated vessel with turbine and vertical baffles (recommended values: $H/D_T = 1$; $D_i/D_T = 1/3$; $W/D_T = 1/12$; $H_i/H = \frac{1}{2}$; number of baffles = 4; number of impeller blades = 6; Seader, J. D., and E. J. Henley, 2006/John Wiley & Sons).

Example 7.8 Power Requirements of a Mixer-Settler Extractor

Furfural is to be continuously extracted from a dilute solution in water using pure toluene as solvent in a series of mixer-settler vessels operated in countercurrent flow. The flow rates of the feed and solvent are 1.61 and 2.24 kg/s, respectively. The residence time in each mixer is 2 min. The density of the feed is 998 kg/m³, and its viscosity is 0.89 cP; the density of the solvent is 868 kg/m³ and its viscosity is 0.59 cP. The interfacial tension is 0.025 N/m (25 dyn/cm). Estimate for the raffinate as the dispersed phase:

(a) The dimensions of the mixing vessel and the diameter of the flat-blade impeller.

(b) The minimum rate of rotation of the impeller for complete and uniform dispersion.

(c) The power requirement of the agitator at 1.20 times the minimum rotation rate.

Solution
Calculate the feed volumetric flow rate, $Q_F = 1.61 \times 60/998 = 0.097$ m³/min, and the solvent volumetric flow rate, $Q_S = 2.24 \times 60/868 = 0.155$ m³/min. Because of

7.4 Equipment for Liquid–Liquid Extraction 423

the dilute concentration of solute in the feed and sufficient agitation to achieve complete and uniform dispersion, assume that the fractional volumetric holdups of raffinate and extract in the vessel are equal to the corresponding volume fractions in the combined feed and solvent entering the mixer:

$$\phi_E = \frac{0.155}{0.155 + 0.097} = 0.615 \qquad \phi_R = 1 - \phi_E = 0.385$$

(a) Let Q = total flow rate = $0.155 + 0.097 = 0.252$ m³/min. For a residence time of 2 min, the vessel volume is $V_T = Qt_{res} = 0.252 \times 2 = 0.504$ m³. For a cylindrical vessel with $H = D_T$,

$$V_T = \frac{\pi}{4} D_T^3$$

$$H = D_T = \left(\frac{4V_T}{\pi}\right)^{1/3} = 0.863 \text{ m}$$

$$D_i = \frac{D_T}{3} = 0.288 \text{ m}$$

(b) For the raffinate phase dispersed,

$$\phi_D = \phi_R = 0.385 \qquad \phi_C = \phi_E = 0.615$$

$$\Delta\rho = 998 - 868 = 130 \text{ kg/m}^3$$

$$\rho_M = 0.385 \times 998 + 0.615 \times 868 = 918 \text{ kg/m}^3$$

$$\mu_M = \frac{0.59}{0.615}\left[1 + \frac{1.5 \times 0.89 \times 0.385}{0.59 + 0.89}\right] = 1.29 \text{ cP}$$

Substituting in equation (7-34), Ω_{min} = 152 rpm.

(c) For $\Omega = 1.2\Omega_{min}$ = 182 rpm, from equation (7-37), Re = 1.78×10^5. Then, according to Laity and Treybal (1957), the power number, Po = 5.7. From (7-37), $P = 0.288$ kW. The power density is $P/V_T = 0.288/0.504 = 0.572$ kW/m³, relatively close to the recommended value of 0.788 kW/m³ (Flynn and Treybal, 1955).

When dispersion is complete and uniform, the contents of the vessel are perfectly mixed with respect to both phases. In that case, the concentration of the solute in each of the two phases in the vessel is uniform and equal to the concentrations in the two-phase emulsion leaving the mixing tank. This called the ideal *CFSTR* (*continuous-flow-stirred-tank-reactor*) *model*, sometimes called the *perfectly mixed model*. Next, we develop an equation to estimate the Murphree stage efficiency for liquid–liquid extraction in a perfectly mixed vessel.

The Murphree dispersed-phase efficiency for liquid–liquid extraction, based on the raffinate as the dispersed phase, can be expressed as the fractional approach to equilibrium. In terms of bulk molar concentrations of the solute,

$$\mathbf{E}_{MD} = \frac{c_{D,in} - c_{D,out}}{c_{D,in} - c^*_D} \tag{7-38}$$

424 **Liquid–Liquid Extraction**

where c^*_D is the solute concentration in the dispersed phase in equilibrium with the solute concentration in the exiting continuous phase, $c_{C,out}$. The rate of mass transfer of solute from dispersed phase to continuous phase, n (kg/s), can be expressed as

$$n = K_{OD}a\left(c_{D,out} - c^*_D\right)V_T \tag{7-39}$$

where the concentration driving force for mass transfer is uniform throughout the well-mixed vessel and is equal to the driving force based on the exit concentrations, a is the interfacial area for mass transfer per unit volume of the liquid phases, V_T is the total volume of the liquid phases in the vessel, and K_{OD} is the overall mass-transfer coefficient based on the dispersed phase. The overall mass-transfer coefficient is given in terms of the separate resistances of the dispersed and continuous phases by

$$\frac{1}{K_{OD}} = \frac{1}{k_D} + \frac{1}{mk_C} \tag{7-40}$$

where equilibrium is assumed at the interphase between the two liquids, and m is the slope of the equilibrium curve for the solute expressed as $c_C = f(c_D)$. For dilute solutions, changes in volumetric flow rate of the raffinate and extract are small, and thus the rate of mass transfer based on the change in solute concentration in the dispersed phase is given by the material balance:

$$n = Q_D\left(c_{D,in} - c_{D,out}\right) \tag{7-41}$$

where Q_D is the volumetric flow rate of the dispersed phase.

To obtain an expression for $\mathbf{E}_{MD}$ in terms of $K_{OD}a$, equations (7-38), (7-39), and (7-41) are combined as follows. From equation (7-38),

$$\frac{\mathbf{E}_{MD}}{1 - \mathbf{E}_{MD}} = \frac{c_{D,in} - c_{D,out}}{c_{D,out} - c^*_D} \tag{7-42}$$

On the other hand, the number of dispersed-phase overall transfer units, N_{tOD}, for this case is given by [compare to equation (5-23) for dilute solutions]

$$N_{tOD} = \int_{c_{D,out}}^{c_{D,in}} \frac{dc_D}{c_D - c^*_D} \tag{7-43}$$

It has been established that the liquids in a well-agitated vessel are thoroughly backmixed, so that everywhere the concentrations are constant at the effluent values (Schindler and Treybal, 1968). Then,

$$N_{tOD} = \int_{c_{D,out}}^{c_{D,in}} \frac{dc_D}{c_D - c^*_D} \simeq \frac{c_{D,in} - c_{D,out}}{c_{D,out} - c^*_D} \tag{7-44}$$

Combining equations (7-39), (7-41), and (7-44) gives us

7.4 Equipment for Liquid–Liquid Extraction 425

$$N_{tOD} \cong \frac{c_{D,in} - c_{D,out}}{c_{D,out} - c^*_D} = \frac{K_{OD} a V_T}{Q_D} \qquad (7\text{-}45)$$

Combining equations (7-42) and (7-45) gives us

$$\mathbf{E}_{MD} = \frac{N_{tOD}}{1 + N_{tOD}} = \frac{K_{OD} a V_T / Q_D}{1 + K_{OD} a V_T / Q_D} \qquad (7\text{-}46)$$

From equations (7-40) and (7-46), it is evident that an estimate of $\mathbf{E}_{MD}$ requires generalized correlations of experimental data for the interfacial area for mass transfer and for the dispersed- and continuous-phase mass-transfer coefficients. The population of dispersed-phase droplets in an agitated vessel will cover a range of sizes and shapes. For each droplet, it is useful to define d_e, the equivalent diameter of a spherical drop, using the method of Lewis et al. (1951):

$$d_e = \left(d_1^2 d_2 \right)^{1/3} \qquad (7\text{-}47)$$

where d_1 and d_2 are the major and minor axis, respectively, of an ellipsoidal drop image. For a spherical drop, d_e is simply the diameter of the drop. For the population of drops, it is useful to define an average or mean diameter. For mass-transfer calculations, the surface-mean diameter d_{vs} (also called the *Sauter mean diameter*) is the most appropriate. It is usually determined from experimental drop-size distribution data. The interfacial area for mass transfer per unit volume of a two-phase mixture in terms of the Sauter mean diameter is given by (Seader and Henley, 2006)

$$a = \frac{6 \phi_D}{d_{vs}} \qquad (7\text{-}48)$$

Early experimental investigations, such as those of Vermeulen et al. (1955), found that d_{vs} is dependent on the Weber number, We, defined as

$$\text{We} = \frac{D_i^2 \Omega^2 \rho_C}{\sigma} \qquad (7\text{-}49)$$

High Weber numbers give small droplets and high interfacial areas. Gnanasundaram et al. (1979) recommended the following correlations:

$$\frac{d_{vs}}{D_i} = 0.052 \text{We}^{-0.6} \exp\left(4 \phi_D\right) \quad \text{for We} \leq 10,000 \qquad (7\text{-}50)$$

$$\frac{d_{vs}}{D_i} = 0.39 \text{We}^{-0.6} \exp\left(4 \phi_D\right) \quad \text{for We} > 10,000 \qquad (7\text{-}51)$$

Typical values of We for industrial extractors are less than 10,000, so equation (7-50) usually applies. Example 7.9 illustrates these calculations.

426   **Liquid–Liquid Extraction**

Example 7.9 Drop Size and Interfacial Area in an Extractor

For the conditions and results of Example 7.8, estimate the Sauter mean drop diameter and the interfacial area.

Solution

From equation (7-49), We = 7612. Equation (7-50) yields d_{vs} = 0.327 mm. Substituting in equation (7-48), a = 7065 m^2/m^3.

Experimental studies conducted since the early 1940s show that mass transfer in mechanically agitated liquid–liquid systems is very complex. The reasons for this complexity are many. For example, the magnitude of k_D depends on drop diameter, solute diffusivity, and fluid motion within the drop. When the drop diameter is less than 1 mm, interfacial tension is higher than 15 dyn/cm, and trace amounts of surface-active ingredients are present, droplets are rigid and behave like solid particles (Davies, 1978). Under those conditions, k_D can be estimated from (Treybal, 1963)

$$\text{Sh}_D = \frac{k_D d_{vs}}{D_D} = 6.6 \tag{7-52}$$

For the continuous-phase mass-transfer coefficient, k_C, Skelland and Moeti (1990) proposed the following:

$$\text{Sh}_C = \frac{k_C d_{vs}}{D_C} = 1.237 \times 10^{-5}\, \text{Re}_C^{2/3}\, \text{Sc}_C^{1/3}\text{Fr}^{5/12}\text{Eo}^{5/4}\phi_D^{-0.5}\left(\frac{D_i}{d_{vs}}\right)^2\left(\frac{d_{vs}}{D_T}\right)^{0.5} \tag{7-53}$$

where

$$\text{Re}_C = D_i^2\Omega\rho_C/\mu_C$$
$$\text{Sc}_C = \mu_C/\rho_C D_C$$
$$\text{Fr} = D_i\Omega^2/g$$
$$\text{Eo} = \rho_D d_{vs}^2 g/\sigma \text{ (known as the Eotvos number)}$$

Example 7.10 illustrates these calculations.

Example 7.10 Mass-Transfer Coefficients in Agitated Extractor

For the conditions and results of Examples 7.8 and 7.9, estimate

(a) The dispersed-phase mass-transfer coefficient.

(b) The continuous-phase mass-transfer coefficient.

(c) The Murphree dispersed-phase efficiency.

(d) The fractional extraction of furfural.

The molecular diffusivities of furfural in water (dispersed) and toluene (continuous) are, respectively, D_D = 1.15 × 10^{-9} m^2/s and D_C = 2.15 × 10^{-9} m^2/s. The equilibrium distribution coefficient m = 10.15 m^3 raffinate/m^3 extract.

7.4 Equipment for Liquid–Liquid Extraction
427

Solution

(a) From equation (7-52), $k_D = 6.6 \times 1.15 \times 10^{-9}/(3.27 \times 10^{-4}) = 2.32 \times 10^{-5}$ m/s.

(b) To apply equation (7-53), compute each of the following dimensionless groups:

$$Sc_C = 316.2 \quad Re_C = 369,400 \quad Fr = 0.270 \qquad Eo = 0.042$$

Substituting in equation (7-53), $Sh_C = 115.5$. Then,

$$k_C = Sh_C D_C / d_{vs} = 7.60 \times 10^{-4} \text{ m/s}$$

(c) From equation (7-40),

$$K_{OD} = \left[\left(2.32 \times 10^{-5} \right)^{-1} + \left(10.15 \times 7.60 \times 10^{-4} \right)^{-1} \right]^{-1} = 2.315 \times 10^{-5} \text{ m/s}$$

From equation (7-45),

$$N_{tOD} = K_{OD} a V_T / Q_D = 2.315 \times 10^{-5} \times 7065 \times 0.504 \times 60 / 0.097 = 51$$

From equation (7-46),

$$\mathbf{E}_{MD} = N_{tOD} / \left(1 + N_{tOD} \right) = 51 / 52 = 0.981$$

(d) For dilute solutions, where the total volumes of the raffinate and extract liquids remain fairly constant, the fractional extraction of furfural, f_{ext}, is defined as

$$f_{ext} = \frac{c_{D,in} - c_{D,out}}{c_{D,in}} = 1 - \frac{c_{D,out}}{c_{D,in}} \tag{7-54}$$

By a material balance on furfural,

$$Q_D \left(c_{D,in} - c_{D,out} \right) = Q_C c_{C,out} \tag{7-55}$$

From the definition of the equilibrium-distribution coefficient,

$$c^*{}_D = \frac{c_{C,out}}{m} \tag{7-56}$$

Combining equations (7-38), (7-54), and (7-56) yields

$$f_{ext} = \frac{\mathbf{E}_{MD}}{1 + \mathbf{E}_{MD} \dfrac{Q_D}{m Q_C}} \tag{7-57}$$

Substituting in equation (7-47), we obtain $f_{ext} = 0.925$.

428   Liquid–Liquid Extraction

7.4.2 Multicompartment Columns

Sizing extraction columns, which may or may not include mechanical agitation, involves the determination of the column diameter and column height. The diameter must be sufficiently large to permit the two phases to flow countercurrently through the column without flooding. The column height must be sufficient to achieve the number of equilibrium stages corresponding to the desired degree of separation.

Sieve-tray towers are very effective, both with respect to liquid-handling capacity and extraction efficiency, particularly for systems of low interfacial tension which do not require mechanical agitation for good dispersion. The general assembly of plates and downspouts is much the same as for gas–liquid contact except that a weir is not required. Towers packed with the same random packing used for gas–liquid contact have also been used for liquid extractors; however, mass-transfer rates are poor. It is recommended instead that sieve-tray towers be used for systems of low interfacial tension and mechanically agitated extractors for those of high interfacial tension (Treybal, 1980).

The most important mechanically agitated columns are those that employ rotating agitators, driven by a shaft that extends axially through the column. The agitators create shear mixing zones, which alternate with settling zones in the column. Differences among the various agitated columns lie primarily in the mixers and settling chambers used.

Perhaps the first mechanically agitated extractor of importance was the *Scheibel column* (Scheibel, 1948). The liquid phases are contacted at fixed intervals by unbaffled, flat-bladed, turbine-type agitators mounted on a vertical shaft. In the unbaffled separation or calming zones, knitted wire-mesh packing is installed to prevent backmixing between mixing zones and to induce coalescence and settling of drops. For more economical designs for larger-diameter installations (more than 1 m), Scheibel added outer and inner horizontal annular baffles to divert the vertical flow of the phases in the mixing zone and to ensure complete mixing. For systems with high interfacial surface tension and viscosities, the wire mesh is removed.

Another type of column with rotating agitators is the *rotating disk contactor* (RDC). On a worldwide basis, it is probably the most extensively used liquid–liquid extraction device (Seader and Henley, 2006). Horizontal disks mounted on a centrally located shaft are the agitation elements. Mounted at the column wall are annular stator rings with an opening larger than the agitator disk diameter. Thus, the agitator shaft assembly is easily removed from the column. Because the rotational speed of the rotor controls the drop size, the rotor speed can be continuously varied over a wide range.

Rather than provide agitation by rotating impellers on a vertical shaft, Karr and Lo (1976) devised a reciprocating perforated-plate extractor, called the *Karr column*, in which the plates move up and down approximately two times per second with a stroke length of about 20 mm. Annular baffle plates are provided periodically in the plate stack to minimize axial mixing. The perforated plates use large holes (typically 14 mm) and a high hole area fraction (typically 58%). The central

7.4 Equipment for Liquid–Liquid Extraction

shaft, which supports both sets of plates, is reciprocated by a drive mechanism located at the top of the column.

A number of industrial centrifugal extractors have been available since 1944 when the *Podbielniak extractor* (POD), with its short residence time, was successfully applied to penicillin extraction (Barson and Beyer, 1953). In the POD, several concentric sieve trays are arranged around a horizontal axis through which the two liquids flow countercurrently. Liquid inlet pressures of 4 to 7 atm are required to overcome pressure drop and centrifugal force. As many as five theoretical stages can be achieved in one unit.

Because of the large number of important variables, an accurate estimation of column diameter for liquid–liquid contacting devices is more complex and less certain than for vapor–liquid contactors. Column diameter is best determined by scale-up from tests run in standard laboratory or pilot-plant test units. The sum of the measured superficial velocities of the two liquid phases ($v_C + v_D$) in the test unit can then be assumed to hold for larger commercial units.

Despite their compartmentalization, mechanically assisted liquid–liquid extraction columns, such as the RDC and Karr columns, operate more nearly like differential contacting devices than like staged contactors. Therefore, it is more common to consider stage efficiency for these columns in terms of HETS (*height equivalent to a theoretical stage*) or as some function of mass-transfer parameters, such as HTU (height of a transfer unit). Although it is not as sound on a theoretical basis as the HTU, the HETS is preferred here because it can be applied directly to determine column height from the number of equilibrium stages.

Because of the great complexity of liquid–liquid systems and the large number of variables that influence contacting efficiency, general correlations for HETS have been difficult to develop. For small-diameter columns, rough estimates of the diameter and height can be made using the results of a study by Stichlmair (1980). Typical results from that study are summarized in Table 7.5, and illustrated in Example 7.11. It is preferred to obtain values of HETS by conducting small-scale laboratory experiments with systems of interest. These values are scaled to commercial-size units by assuming that HETS increases with column diameter raised to an exponent, which varies from 0.2 to 0.4.

Table 7.5 Performance of Several Types of Column Extractors

Extractor type	1/HETS, m^{-1}	($v_D + v_C$), m/h
Sieve-plate column	0.8 – 1.2	27 – 60
Scheibel column	5.0 – 9.0	10 – 14
RDC	2.5 – 3.5	15 – 30
Karr column	3.7 – 7.0	30 – 40
Packed column	1.5 – 2.5	12 – 30

Source: Data from Stichlmair (1980).

430   Liquid–Liquid Extraction

Example 7.11 Preliminary Design of an RDC

Estimate the diameter and height of an RDC to extract acetone from a dilute toluene–acetone solution into water at 293 K. The flow rates for the dispersed organic and continuous aqueous phases are 12,250 and 11,340 kg/h, respectively. The density of the organic phase is 858 kg/m^3 and that of the aqueous phase is 998 kg/m^3. For the desired degree of separation, 12 equilibrium stages are required.

Solution
Calculate the total volume flow rate of the two liquids:

$$Q_D = 12,250 / 858 = 14.28 \text{ m}^3 / \text{h} \quad Q_C = 11,340 / 998 = 11.36 \text{ m}^3 / \text{h}$$

(a) Assume that based on the information in Table 7.5, $(v_D + v_C)$ = 22 m/h. Then, the column cross-sectional area is $A_c = (Q_D + Q_C)/(v_D + v_C)$ = 1.165 m^2. The column diameter is $D_T = (4A_c/\pi)^{0.5}$ = 1.22 m.

(b) Assume that based on the information in Table 7.5, HETS = 0.333 m/theoretical stage. Then, the column height is Z = 12 × 0.333 = 4.0 m.

7.5 LIQUID–LIQUID EXTRACTION OF BIOPRODUCTS

Your objectives in studying this section are to be able to:

1. Identify some advantages of liquid–liquid extraction for separation of bioproducts.
2. Compare organic-solvent and aqueous two-phase extraction (ATPE) for recovery of bioproducts.
3. Calculate the dependence of the partition coefficients on pH for weak organic acids.
4. Calculate partition coefficients for various proteins in ATPE systems.

With the availability of variety of solvents as well as commercial equipment, liquid–liquid extraction finds many applications in the separation of bioproducts. Liquid–liquid extraction process is particularly suitable for biorefinery processes, featuring mild operational conditions and ease of control of process. Small biomolecules such as inhibitory metabolites like ethanol and butanol, and antibiotics such as penicillin, erythromycin, and cephalosporin can be extracted directly from the fermentation broth into an immiscible organic fluid phase (Seader et al., 2011; Bokhary et al., 2021). Penicillin, for example, can be extracted from aqueous fermentation broth using methyl isobutyl ketone (MIBK) as solvent at pH values less than the pK_a of the antibiotic. Larger biopolymers like peptides, proteins, and lipids; cellular particulates such as membrane components; and products from solid feeds, whose activity can be reduced by organic solvents, can be extracted into separate aqueous or supercritical-fluid phases.

7.5 Liquid–Liquid Extraction of Bioproducts 431

Solvent extraction of bioproducts is generally less expensive than membrane or chromatographic operations, allows for continuous operation, and is readily scalable. It can significantly reduce process volume and facilitate further product recovery via evaporation and or crystallization. Currently, various intensification techniques are being applied in the field of liquid–liquid extraction for improving the process efficiency (such as hybrid processes—illustrated in Example 7.12, reactive extraction, and use of ionic liquids) which are gaining importance due to the cost associated with the downstream processing of fermentation products, estimated to be around 20–50% of the total production cost (Antony et al., 2021; Scharzec et al., 2021).

Example 7.12 Hybrid Extraction-Distillation Process for Purification of γ-Valerolactone

γ-Valerolactone (GVL) is an organic compound with the formula $C_5H_8O_2$. This colorless liquid is readily obtained from cellulosic biomass and is a potential "green fuel" as well as "green solvent." Scharzec et al. (2021) studied different processes for purification of diluted GVL aqueous solutions that can be applied to other bio-based production processes, as well as to the recovery of valuable impurities prior to wastewater treatment. They used as a case study an aqueous stream flowing at the rate of 100 mol/s containing 5 mol% of GVL and 95% water. The objective was separating the solution into a product containing 99.99 mol% GVL and a residue containing only 0.01 mol% GVL. The first separation option that they considered was distillation. The preliminary design consisted of a 48-stages column with a partial reboiler and total condenser. They estimated the total annual cost (TAC) of this option, 84% of which was related to the energy costs.

The second separation option considered by these researchers was a hybrid extraction-distillation process. The feed solution was now processed in a 15-stages plate extractor using 34.3 mol/s of toluene as solvent in countercurrent flow. The raffinate from the extractor contained 99.99 mol% water, 0.01% toluene, and only traces of GVL. The extract, flowing at the rate of 39.4 mol/s, had a molar concentration of 12.7% GVL, 86.7% toluene, and 0.6% water. This stream was pre-heated and fed to a 22-stages distillation column. The distillate from the column, almost pure toluene, was recycled to the extractor. The bottoms from the distillation column was the desired product with a concentration of 99.99 mol% GVL and 0.01% toluene. The TAC for the hybrid separation process was around 80% lower than the TAC for the original distillation column. Most of the savings associated with the hybrid were due to the reduction in energy costs. They also considered using butyl acetate as solvent in the extraction column with similar results.

The researchers considered a third separation option consisting of a membrane-assisted hybrid extraction-distillation process that resulted in even more significant savings of money and energy. We will consider that option in more detail in Chapter 9.

Solvents commonly used in biological organic/aqueous extraction are acetone, amyl acetate, butyl acetate, methyl isobutyl ketone (MIBK), methylene chloride, n-butanol, n-octanol, and hexane. The chief high-value bioproducts extracted are antibiotics. They include penicillin, clavulanic acid, erythromycin, gramicidin D, cycloheximide, fusidic acid, antimycin A, chloramphenicol, and virginiamycin.

432   **Liquid–Liquid Extraction**

Another important application deals with yeast fermentation to produce ethanol and butanol from renewable agricultural feedstocks. This process is limited by product inhibition of a catalytic enzyme in the metabolic pathway. This limits broth concentrations of the products to no more than 5 to 10% by volume. *In situ* solvent extraction offers a low-temperature method for recovering biological alcohols from ongoing fermentation and improving productivity. However, identification of a solvent system that provides a suitable partition coefficient, K_D (see equation 3-4), while maintaining cell activity remains a primary challenge since many of the partition coefficients of the most common solvents available are very small (Seader et al., 2011).

Partition coefficients for organic/aqueous extraction are functions of solubility, temperature, pH, ionic strength, and component concentrations. As temperature decreases, partition coefficients increase. The pH influences K_D by changing ionization state. Since un-ionized forms are more soluble in organic solvents, weak biological acids are extracted from fermentation broths at low pH values. Temperature and pH dependencies explain why aqueous solutions of the weak acid penicillin G (pK_a = 2.7) are buffered to pH around 2 to 2.5 and chilled to 0 to 3 °C to optimize extraction into *n*-butyl acetate. Examples 7.13 and 7.14 illustrate other cases of partition coefficients dependence on pH.

Example 7.13 Dependence of Partition Coefficient on pH

Derive an expression that shows the pH dependence of the partition coefficient for a weak organic acid between a fermentation broth and an organic solvent.

Solution

Let us begin by considering the partitioning of an un-ionized weak acid, HA, between an organic solvent (1) and an aqueous fermentation broth (2). The corresponding partition coefficient, $K_{D,U}$, may be written as

$$K_{D,U} = \frac{\left[\mathrm{HA}\right]_{(1)}}{\left[\mathrm{HA}\right]_{(2)}} \tag{7-58}$$

Considering partial ionization of the acid in the aqueous phase, but none in the organic phase, the partition coefficient may be written as

$$K_D = \frac{\left[\mathrm{HA}\right]_{(1)}}{\left[\mathrm{HA}\right]_{(2)} + \left[\mathrm{A}^-\right]_{(2)}} \tag{7-59}$$

From the definition of the ionization constant for the acid (equation 3-10), it can be shown that

$$\left[\mathrm{A}^-\right]_{(2)} = \left[\mathrm{HA}\right]_{(2)} \times 10^{\left(pH - pK_a\right)} \tag{7-60}$$

7.5 Liquid–Liquid Extraction of Bioproducts 433

Substituting equation (7-60) into (7-59)

$$K_D = \frac{[HA]_{(2)}}{[HA]_{(1)} \times \left(1 + 10^{(pH-pK_a)}\right)} = \frac{K_{D,U}}{1 + 10^{(pH-pK_a)}} \quad (7\text{-}61)$$

Equation (7-61) shows that K_D for weak acids in organic solvents decreases significantly as pH increases to values above pK_a.

Example 7.14 Dependence of Penicillin F Partition Coefficient on pH

Penicillin F extracted from fermentation broth into amyl acetate has partition coefficients of 32.0 and 0.423 at pH 4.0 and 6.0, respectively, both at the same temperature. Estimate the value of K_D at pH 7.0.

Solution

Applying equation (7-61) with the two given pairs of K_D and pH, two nonlinear equations in terms of $K_{D,U}$ and pK_a are obtained as follows:

$$32.0 = \frac{K_{D,U}}{1 + 10^{(4.0-pK_a)}} \quad \text{and} \quad 0.423 = \frac{K_{D,U}}{1 + 10^{(6.0-pK_a)}}$$

A nonlinear solver from Mathcad or Python gives pK_a = 3.51 and $K_{D,U}$ = 130.6. Therefore, applying equation (7-61) at pH = 7.0 gives

$$K_D = \frac{130.6}{1 + 10^{(7.0-3.51)}} = 0.042$$

Thus, increasing the pH from 4.0 to 7.0 decreases the ability of amyl acetate to extract penicillin F by a factor of approximately 760.

Water-immiscible organic solvents may denature biopolymers and lyse cells. An alternative separation process to avoid these drawbacks of organic solvents is called aqueous two-phase extraction (ATPE). In this process biopolymers and cells can partition between two aqueous phases, each containing 75 to 90% water, formed by dissolving one or two ionic or nonionic polymers, or a polymer and mineral salt. Polyacrylamide is a common ionic polymer. Nonionic polymers are polyethylene oxide (PEO), polyethylene glycol (PEG), and dextran. Each polymer in a two-phase system is fully soluble in water, but incompatible with the other phase in concentration ranges where two aqueous phases are formed. The most common ATPE systems are PEG–dextran–water and PEG–potassium phosphate–water. For example, an aqueous solution of 5 wt% dextran 500 (average molecular weight of 500,000) and 3.5 wt% PEG 6000 (average molecular weight of 6000) at 20 °C partitions into two aqueous phases: a PEG-rich top phase containing 4.9% PEG, 1.8% dextran, and 93.3% water; and a dextran-rich bottom phase containing 2.6% PEG, 7.3% dextran, and 90.1% water (Seader et al., 2011).

Recently, aqueous polymer two-phase technology has evolved, both experimentally and theoretically, into a separation science with many useful applications in biomolecule purification and bioconversion (Iqbal et al., 2016). Protein processing, for example, demands a high yielding and economical method for purification that cannot be achieved by conventional methods. That is why protein recovery from crude feedstocks at large-scale has been achieved by ATPE and this application has attracted the most interest. Another application of ATPE that is gaining a lot of interest is in the purification of monoclonal antibodies (mAbs). Antibody based therapies have played a key role in the treatment of cancers, autoimmune disorders, and infectious diseases such as the recent COVID-19 pandemic (NIH, 2022). Recent studies suggest that ATPE is an economical and environmentally stable option for the purification of mAbs as compared to currently used alternatives. Other important applications of ATPE being explored include purification of enzymes; isolation of DNA, RNA, and other nucleic acids; and the recovery of virus and virus like particles.

Each of the three components that form an ATPE system—solvent (i.e., water), polymer 1 (e.g., PEG), and polymer 2 (e.g., dextran)—partitions between the top (t) and bottom (b) phases. Aqueous–aqueous phase separation takes place only at compositions in excess of the solubility curve that separates the monophasic region (below the curve) from the biphasic region (above the curve) as shown in Figure 7.23. Tie lines connecting points on the top and bottom phases in equilibrium are characterized using the lever-arm rule. The biopolymer also distributes between the two phases. Biomolecules such as peptides, proteins, nucleic acids, and cells exhibit different solubilities in the two phases and partition accordingly.

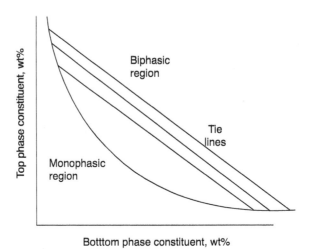

Figure 7.23 Schematic representation of phase diagram for an ATPE system.

For just partitioning of the two polymers between the solvent in the absence of the biopolymer, Diamond and Hsu (1989) developed the following expression:

$$\ln\left(K_{D,i}\right) = \ln\left(\frac{\omega_i^t}{\omega_i^b}\right) = A_i\left(\omega_1^t - \omega_1^b\right) \tag{7-62}$$

where i applies only to the two polymers (i = 1 and 2), the empirical parameter A_i depends on the molecular weights of the two polymers, and ω_i^p is the weight fraction

7.5 Liquid–Liquid Extraction of Bioproducts 435

of the polymer in phase p. Note that the difference term on the RHS of equation (7-62) is always for polymer 1. From measurements of one tie line, the value of A_i can be computed for each polymer. Equation (7-62) can then be used to predict a series of tie lines. However, to do this requires the additional assumption that all tie lines have the same slope on a plot like Figure 7.23. Example 7.15 illustrates these calculations.

Example 7.15 Tie Line Calculations for an ATPE System

Phase-distribution measurements were made for the PEG 8000–dextran T70–water system at 4 °C by Diamond and Hsu (1989). The following table gives the equilibrium concentrations of the two resulting phases in one of the tie lines observed, all in wt%.

Component	Bottom phase	Top phase
PEG (1)	1.56	5.95
Dextran (2)	11.91	2.35
Water	86.53	91.70

Use equation (7-62) to compute the values of A_i for dextran and PEG. Use those values to generate a tie line with one end at 9.71 wt% PEG in the top phase.

Solution

Solving equation (7-62) for A_1 of PEG,

$$A_1 = \frac{\ln\left(\dfrac{\omega_1^t}{\omega_1^b}\right)}{\omega_1^t - \omega_1^b} = \frac{\ln\left(\dfrac{0.0595}{0.0156}\right)}{0.0595 - 0.0156} = 30.494$$

Solving equation (7-62) for A_2 of dextran,

$$A_2 = \frac{\ln\left(\dfrac{\omega_2^t}{\omega_2^b}\right)}{\omega_1^t - \omega_1^b} = \frac{\ln\left(\dfrac{0.0235}{0.1191}\right)}{0.0595 - 0.0156} = -36.970$$

Next, calculate the slope of the observed tie line, S_{TL},

$$S_{TL} = \frac{\omega_1^t - \omega_1^b}{\omega_2^t - \omega_2^b} = \frac{0.0595 - 0.0156}{0.0235 - 0.1191} = -0.459$$

Now, compute a tie line with 9.71 wt% of PEG in the top phase ($\omega_1^t = 0.0971$) using the value of A_1 calculated above

$$A_1 = 30.494 = \frac{\ln\left(\dfrac{0.0971}{\omega_1^b}\right)}{0.0971 - \omega_1^b}$$

436   **Liquid–Liquid Extraction**

Solving, ω_1^b = 0.00604 [compare this prediction to the experimental value of 0.0055 reported by Diamond and Hsu (1989)].

To solve for ω_2^b and ω_2^t of the new tie line with ω_1^t = 0.0971 and ω_1^b = 0.00604, the following two equations must be solved simultaneously (assuming that all tie lines have the same slope):

$$A_2 = \frac{\ln\left(\dfrac{\omega_2^t}{\omega_2^b}\right)}{0.0971 - 0.00604} = -36.970 \quad S_{TL} = \frac{0.0971 - 0.00604}{\omega_2^t - \omega_2^b} = -0.459$$

Solving, ω_2^t = 0.0071 and ω_2^b = 0.2054 [compare these predictions to the experimental values of 0.0055 and 0.2025, respectively, reported by Diamond and Hsu (1989)].

The partition coefficient, $K_{D,3}$, of a biomolecule in an aqueous two-phase system has been found to be a function of many variables. Diamond and Hsu (1990) developed the following relationship for proteins in ATPE which was based on Flory-Huggins solution thermodynamics:

$$\ln\left(K_{D,3}\right) = \ln\left(\frac{\omega_3^t}{\omega_3^b}\right) = A*\left(\omega_1^t - \omega_1^b\right) + b*\left(\omega_1^t - \omega_1^b\right)^2 \qquad (7\text{-}63)$$

In this equation, ω_3^t and ω_3^b are defined as kg of protein/kg of protein-free top phase and kg of protein/kg of protein-free bottom phase, respectively. The parameters $A*$ and $b*$ are function of protein and phase forming polymer molecular weights, pH, and salt type and concentration. Table 7.6 summarizes the value of these parameters for 17 proteins partitioning in the system PEG8000/dextran T500/water, as reported by Diamond and Hsu (1990). Example 7.16 illustrates the use of Table 7.6 for calculations related to thyroglobulin extraction.

Example 7.16 Aqueous Two-Phase Extraction of Thyroglobulin

Thyroglobulin (Tg) is a protein produced by the follicular cells of the thyroid and is used entirely within the thyroid gland. It is the main precursor of the thyroid hormones. Tg levels in the blood are mainly used as a tumor marker for certain kinds of thyroid cancer. Consider the partitioning of Tg in the PEG 8000/dextran T-500/water ATPE system at 4 °C. One tie line for this system—in terms of wt% for PEG, dextran, and water—shows a top-phase composition of 7.1, 0.9, and 92.0, with a bottom-phase composition of 1.1, 13.9, and 85.0. Calculate the partition coefficient for Tg under these conditions using information from Table 7.6.

Solution
From Table 7.6, for thyroglobulin, the parameters for equation (7-63) are $A*$ = 81.2 and $b*$ = – 753.0. From the measured tie-line data, $\left(\omega_1^t - \omega_1^b\right) = 0.071 - 0.011 = 0.06$. Substituting in equation (7-63), we obtain $\ln\left(K_{D,3}\right)$ = 2.1612; therefore, $K_{D,3}$ = 8.681.

Problems

Table 7.6 Parameters of Equation (7-63) for Proteins Partitioning in the PGE 8000/Dextran T-500/Water ATPEs

Protein	Molecular weight	A^*	b^*
Lipase	6669	18.1	−217.0
Cytochrome c	12,400	−17.7	−47.9
Ribonuclease	12,600	−6.7	−25.3
Lysozyme	13,900	−16.1	31.3
Myoglobin	16,900	−7.6	−78.0
Trypsin	23,200	−4.8	−14.8
Rhodanese	37,570	8.0	−178.0
Ovalbumin	44,000	8.7	−74.5
α-Amylase	45,000	−20.8	−10.0
Protease	48,410	5.5	−37.3
Bovine Serum Albumin (BSA)	67,500	2.4	−183.0
Transferrin	77,000	1.1	−186.0
Conalbumin	86,810	−12.0	−122.0
Hexokinase	102,000	11.9	−194.0
Alc. Dehydrogenase	145,000	5.8	−147.0
Invertase	270,000	17.3	−387.0
Thyroglobulin (Tg)	669,000	81.2	−753.0

Source: Diamond and Hsu (1990).

PROBLEMS

The problems at the end of each chapter have been grouped into four classes (designated by a superscript after the problem number).

Class a: Illustrates direct numerical application of the formulas in the text.
Class b: Requires elementary analysis of physical situations, based on the subject material in the chapter.
Class c: Requires somewhat more mature analysis.
Class d: Requires computer solution.

7.1[a]. Lever-arm rule

Derive equation (7-1).

7.2[b]. Single-stage extraction

Repeat Example 7.2 using a solvent rate of 80 kg/h.

Answer: $x_{CR} = 0.136$

7.3[a]. Single-stage extraction: insoluble liquids

A feed of 13,500 kg/h consists of 8 wt% acetic acid in water. Acetic acid will be removed from the solution by extraction with pure methyl isobutyl ketone at 298 K.

438   Liquid–Liquid Extraction

If the raffinate is to contain only 1 wt% of acetic acid, estimate the kilograms/hour
of solvent required if a single equilibrium stage is used. Assume that water and
methyl isobutyl ketone are insoluble. For this system, K_D = 0.657 kg water/kg
methyl isobutyl ketone (Perry and Chilton, 1973).

Answer: 143,800 kg/h

7.4[a]. Single-stage extraction: insoluble liquids

Repeat Problem 7.3 using methyl acetate as solvent, for which K_D = 1.273 kg
water/kg methyl acetate.

Answer: 74,230 kg/h

7.5[b]. Multistage crosscurrent extraction: insoluble liquids

Consider the extraction process of Problem 7.3. Calculate the total amount of
solvent required if the extraction is done in a crosscurrent cascade consisting of 5
ideal stages. Use equation (3-89) from Problem 3.22.

Answer: 50,900 kg/h

7.6[a]. Multistage crosscurrent extraction

Repeat Example 7.3 using 10 kg/h of solvent in each stage.

Answer: 13.6 wt% acetone in final raffinate

7.7[a]. Multistage countercurrent extraction

Determine the number of ideal stages required in Example 7.4 if the solvent
rate used is twice the minimum.

Answer: 5 ideal stages

7.8[a]. Multistage countercurrent extraction: insoluble liquids

A water solution containing 0.005 mole fraction of benzoic acid is to be
extracted using pure benzene as the solvent. If the feed rate is 100 mol/h and the
solvent rate is 10 mol/h, find the number of equilibrium stages required to reduce
the concentration of benzoic acid in the aqueous solution to 0.0001 mol fraction.
Operation is isothermal at 280 K (water and benzene are insoluble at this temperature), where the equilibrium data can be represented as (Wankat, 1988)

mol fraction of benzoic acid in water = 0.0446 × (mol fraction of benzoic acid
in benzene)

Answer: 4.13 ideal stages

Problems 7.9 to 7.12 refer to the system water (A)–chlorobenzene (B)–pyridine (C)
at 298 K. Equilibrium tie-line data taken from Treybal (1980) in weight percent are
given in Table 7.7.

Problems

439

Table 7.7 Water–Chlorobenzene–Pyridine Equilibrium Data[a]

	Chlorobenzene-phase, wt%			Water-phase, wt%	
Water	Chlorobenzene	Pyridine	Water	Chlorobenzene	Pyridine
0.05	99.95	0.00	99.92	0.08	0.00
0.67	88.28	11.05	94.82	0.16	5.02
1.15	79.90	18.95	88.71	0.24	11.05
1.62	74.28	24.10	80.72	0.38	18.90
2.25	69.15	28.60	73.92	0.58	25.50
2.87	65.58	31.55	62.05	1.85	36.10
3.95	61.0	35.05	50.87	4.18	44.95
6.40	53.00	40.60	37.90	8.90	53.20
13.20	37.80	49.00	13.20	37.80	49.00

Source: Data from Treybal (1980) at 298 K and 1 atm.
[a]Water and chlorobenzene phases on the same line are in equilibrium.

7.9[a]. Right-triangular diagrams

Plot the equilibrium data for the system water–chlorobenzene–pyridine in right-triangular coordinates.

7.10[b]. Single-stage extraction

It is desired to reduce the pyridine concentration of 5000 kg/h of an aqueous solution from 50 wt% to 5 wt% in a single batch extraction with pure chlorobenzene. What amount of solvent is required? Solve on right-triangular coordinates. Corroborate your results using the Mathcad or Python programs of Appendix G-1 and G-2.

Answer: 19,050 kg/h

7.11[b]. Multistage crosscurrent extraction

A 5000-kg batch of pyridine–water solution, 50 wt% pyridine, is to be extracted with an equal weight of pure chlorobenzene. The raffinate from the first extraction is to be reextracted with a weight of pure solvent equal to the raffinate weight, and so on ($S_2 = R_1$, $S_3 = R_2$, etc.). How many ideal stages and what total solvent are required to reduce the concentration of pyridine to no more than 2 wt% in the final raffinate?

Answer: 10,580 kg

7.12[b]. Multistage countercurrent extraction

A pyridine–water solution, 50 wt% pyridine, is to be continuously and countercurrently extracted at the rate of 2.25 kg/s with pure chlorobenzene to reduce the pyridine concentration to 2 wt%.

440   **Liquid–Liquid Extraction**

(a) Determine the minimum solvent rate required.

Answer: 1.75 kg/s

(b) If 2.3 kg/s of solvent is used, calculate the number of theoretical stages required, and the flow rates of final extract and raffinate.

7.13[b]. Multistage countercurrent extraction

We wish to remove acetic acid from water using pure isopropyl ether as solvent. The operation is at 293 K and 1 atm (see Table 7.2). The feed is 45 wt% acetic acid and 55 wt% water. The feed flow rate is 2000 kg/h. A multistage countercurrent extraction cascade is used to produce a final extract that is 20 wt% acetic acid and a final raffinate that is also 20 wt% acetic acid. Calculate how much solvent and how many equilibrium stages are required.

Answer: 2500 kg solvent/h

7.14[b]. Penicillin F extraction: insoluble liquids

An aqueous dilute fermentation broth contains 0.5 wt% of penicillin F. It is to be extracted with amyl acetate. At room temperature and pH = 3.2 water and amyl acetate are essentially insoluble, and the distribution coefficient for the penicillin is K_D = 80.

(a) If 100 kg of the fermentation broth is extracted with 6 kg of the pure solvent in a single ideal stage, calculate the fractional recovery of penicillin and the penicillin concentration in the final raffinate and extract.

Answer: 82.8%

(b) What would be the recovery with two-stage crosscurrent extraction if 6 kg of fresh solvent is used in each stage?

Answer: 97%

(c) How many ideal stages would be needed to give the same recovery as in part (b) if a countercurrent cascade were used with 6 kg of solvent/100 kg feed?

Answer: 2.1 stages

7.15[b]. Single-stage and multistage crosscurrent extraction

Water–dioxane solutions form a minimum-boiling-point azeotrope at atmospheric pressure and cannot be separated by ordinary distillation methods. Benzene forms no azeotrope with dioxane and can be used as an extraction solvent. At 298 K, the equilibrium distribution of dioxane between water and benzene is as follows (Treybal, 1980):

| Wt% dioxane in water | 5.1 | 18.9 | 25.2 |
| Wt% dioxane in benzene | 5.2 | 22.5 | 32.0 |

Problems   441

At these concentrations, water and benzene are substantially insoluble. A 1000 kg batch of a 25 wt% dioxane and 75 wt% water solution is to be extracted with pure benzene to remove 95% of the dioxane.

(a) Calculate the solvent requirement for a single batch operation.

Answer: 11,470 kg

(b) If the extraction were done with equal amounts of solvent in five crosscurrent ideal stages, how much solvent would be required?

Answer: 2322 kg

7.16[b]. Multistage countercurrent extraction

A 25 wt% solution of dioxane in water is to be extracted continuously at the rate of 1000 kg/h in countercurrent fashion with pure benzene to remove 95% of the dioxane. Equilibrium data are given in Problem 7.15.

(a) What is the minimum solvent requirement?

Answer: 516 kg/h

(b) If 800 kg/h of solvent is used, how many theoretical stages are required?

(c) What is the number of real stages if the Murphree stage efficiency, expressed in terms of extract compositions, is $\mathbf{E}_{ME} = 0.5$?

Answer: 10 real stages

Problems 7.17 to 7.19 refer to the system cottonseed oil (A)–liquid propane (B)–oleic acid (C) at 372 K and 42.5 atm. Equilibrium tie-line data taken from Treybal (1980) in weight percent are given in Table 7.8. Oleic acid is a monounsaturated fatty acid found naturally in many plant sources and in animal products. It is an omega-nine fatty acid and considered one of the healthier sources of fat in the diet. It is commonly used as a replacement for animal fat sources that are high in saturated fat. As a fat, oleic acid is one of the better ones to consume. As a replacement for other saturated fats, it can lower total cholesterol level and raise levels of high-density lipoproteins (HDLs) while lowering low-density lipoproteins (LDLs), also known as the "bad" cholesterol.

7.17[b]. Janecke diagram for liquid–liquid extraction

Generate the Janecke diagram and the distribution curve for the system cottonseed oil (A)–propane (B)–oleic acid (C) at 372 K and 42.5 atm.

7.18[b]. Crosscurrent extraction on the Janecke diagram

If 100 kg of a cottonseed oil–oleic acid solution containing 25 wt% acid is extracted twice in crosscurrent fashion, each time with 1000 kg of pure propane, determine the compositions (wt%) and the weights of the mixed extract and the

442   **Liquid–Liquid Extraction**

Table 7.8 Equilibrium Tie-Line Data for System Cottonseed Oil (A)–Propane (B)–Oleic Acid (C) at 372 K and 42.5 Atm

Oil	Propane	Acid	Oil	Propane	Acid
63.5	36.5	0.0	2.30	97.7	0.00
57.2	37.3	5.5	1.95	97.3	0.76
52.0	39.0	9.0	1.78	97.0	1.22
42.7	39.5	13.8	1.50	96.6	1.90
39.8	41.5	18.7	1.36	95.9	2.74
31.0	42.7	26.3	1.20	95.0	3.80
26.9	43.7	29.4	1.10	94.5	4.40
21.0	46.6	32.4	1.00	93.9	5.10
14.2	48.4	37.4	0.80	93.1	6.10
8.3	52.2	39.5	0.70	92.1	7.20
4.5	54.4	41.1	0.40	93.5	6.10
0.8	55.5	43.7	0.20	94.3	5.50

Source: Data from Treybal (1980).

final raffinate. Make the computations on a solvent-free basis using the Janecke diagram generated in Problem 7.17.

7.19^c. Countercurrent extraction with extract reflux

If 1000 kg/h of a cottonseed oil–oleic acid solution containing 25 wt% acid is to be continuously separated into two products containing 2 and 90 wt% acid (solvent-free compositions) by countercurrent extraction with propane, calculate:

(a) The minimum number of theoretical stages required.

Answer: N_{min} = 5 stages

(b) The minimum external extract/reflux ratio required.

Answer: $(R_0/P'_E)_{min}$ = 3.75

(c) For an external extract/reflux ratio of 4.5, the number of theoretical stages, the position of the feed stage, and the quantities, in kg/h, of the important streams.

7.20^c. Countercurrent extraction with extract reflux

A feed mixture containing 50 wt% n-heptane and 50 wt% methylcyclohexane (MCH) at 298 K and 1 atm flows at a rate of 100 kg/h and is to be separated by liquid–liquid extraction into one product containing 92.5 wt% MCH and another containing 7.5 wt% MCH (both on a solvent-free basis). Aniline will be used as the solvent. Using the equilibrium data given in Table 7.9 to construct a Janecke diagram, calculate

(a) The minimum number of theoretical stages required.

(b) The minimum external extract/reflux ratio required.

(c) For an external extract/reflux ratio of 7.0, the number of theoretical stages, the position of the feed stage, and the quantities, in kg/h, of the important streams.

Problems

Table 7.9 Equilibrium Data for the System n-Heptane(A)–Aniline (B)–MCH (C) at 298 K (Solvent-Free Basis)

Hydrocarbon layer		Solvent layer	
wt% MCH[a]	kg B/kg (A + C)	wt% MCH[a]	kg B/kg (A + C)
0.0	0.0799	0.0	15.12
9.9	0.0836	11.8	13.72
20.2	0.0870	33.8	11.50
23.9	0.0894	37.0	11.34
36.9	0.0940	50.6	9.98
44.5	0.0952	60.0	9.00
50.5	0.0989	67.3	8.09
66.0	0.1062	76.7	6.83
74.6	0.1111	84.3	6.45
79.7	0.1135	88.8	6.00
82.1	0.1160	90.4	5.90
93.9	0.1272	96.2	5.17
00.0	0.1350	100.0	4.92

Source: Data from Seader and Henley (2006).
[a]Solvent-free basis.

7.21[b]. Design of a mixer-settler extraction unit

Acetic acid is to be extracted from a dilute aqueous solution with isopropyl ether at 298 K in a countercurrent cascade of mixer-settler units. In one of the units, the following conditions apply:

	Raffinate	Raffinate
Flow rate, kg/s	2.646	6.552
Density, kg/m^3	1017	726
Viscosity, cP	1.0	0.5
Interfacial tension = 0.0135 N/m		

If the raffinate is the dispersed phase and the mixer residence time is 2.5 min, estimate for the mixer:

(a) The dimensions of a closed, baffled vessel.

Answer: H = 1.3 m

(b) The diameter of a flat-bladed impeller.

(c) The minimum rate of rotation of the impeller.

Answer: 156 rpm

(d) The power requirement of the agitator at the minimum rate of rotation.

Answer: 1.23 kW

444 Liquid–Liquid Extraction

7.22[c]. Mass transfer in mixer unit for liquid–liquid extraction

For the conditions and results of Example 7.7, involving the extraction of benzoic acid from a dilute aqueous solution using toluene as solvent, determine the following when using a six-flat-blade impeller in a closed vessel with baffles and with the raffinate phase dispersed:

(a) The minimum rate of rotation of the impeller for complete dispersion.

(b) The power requirement of the agitator at the minimum rate of rotation.

Answer: 2.9 kW

(c) The Sauter mean droplet diameter and the interfacial area.

(d) The overall mass-transfer coefficient, K_{OD}.

(e) The fractional extraction of benzoic acid.

Answer: 93.5%

Liquid properties are

	Raffinate	Extract
Density, g/cm^3	0.995	0.860
Viscosity, cP	0.95	0.59
Diffusivity, cm^2/s	2.2×10^{-5}	1.5×10^{-5}

Interfacial tension = 0.022 N/m

Distribution coefficient of benzoic acid = K_D = 21

7.23[a]. Sizing of an RDC extraction column

Estimate the diameter and height of an RDC column to extract acetic acid from water using isopropyl ether for the conditions and data of Problem 7.21. For the desired degree of separation, 15 equilibrium stages are required.

7.24[a]. Sizing of a Karr extraction column

Estimate the diameter and the HETS of a Karr column to extract acetic acid from water using isopropyl ether for the conditions and data of Problem 7.21.

7.25[b]. Effect of pH on the partition coefficient of a bioproduct (Seader et al., 2011)

A monoacidic sugar extracted from water into hexanol has partition coefficients of 4.5 and 2.0 at pH 4.0 and 5.5, respectively, both at the same temperature. Estimate the value of K_D at pH of 7.2.

Answer: K_D = 0.0047

7.26[c] Tie-line calculations for an ATPE system

Phase-distribution measurements were made for the PEG 3400 – dextran T40 – water system at 4 °C by Diamond and Hsu (1989). The following table gives

Problems

445

the equilibrium concentrations of the two resulting phases in one of the tie lines observed, all in wt%.

Component	Bottom phase	Top phase
PEG (1)	3.28	8.82
Dextran (2)	15.83	3.70
Water	80.89	87.48

Use equation (7-62) to compute the values of A_i for dextran and PEG. Use those values to generate a tie line with one end at 11.19 wt% PEG in the top phase.

7.27[a] Aqueous two-phase extraction of alcohol dehydrogenase

Alcohol dehydrogenase (ADH) is an enzyme that occurs in many organisms and facilitates the interconversion between alcohols and aldehydes or ketones with the reduction of nicotinamide adenine dinucleotide (NAD^+) to NADH. In humans and many other animals, it serves to break down alcohols that otherwise are toxic. Consider the partitioning of ADH in the PEG 8000/dextran T-500/water ATPE system at 4 °C. One tie line for this system—in terms of wt% for PEG, dextran, and water—shows a top-phase composition of 7.1, 0.9, and 92.0, with a bottom-phase composition of 1.1, 13.9, and 85.0. Calculate the partition coefficient for ADH under these conditions using information from Table 7.6.

7.28[c] Aqueous two-phase extraction of lysozyme (Seader et al., 2011)

Lysozyme is an antimicrobial enzyme produced by animals that forms part of the innate immune system. Large quantities of recombinant lysozyme can be produced by genetically engineering tobacco, a ubiquitous gene host for high-valued proteins. Consider the following application of ATPE for purifying recombinant lysozyme. A clarified tobacco broth contains 800 kg/h of water, 5 kg/h of lysozyme (L), and 40 kg/h of other proteins (OP). The lysozyme is to be extracted using the PEG3000-water-Na_2SO_4 (salt) system at 20 °C. One tie line for this system—in terms of wt% of PEG, salt, and water—shows a top-phase composition of 40, 1, and 59; with a bottom-phase composition of 0.6, 18, and 81.4. The partition coefficients for lysozyme and the other proteins along this tie line are 20.0 and 2.0, respectively. If 190 kg/h of salt is added to the broth and equilibrium is achieved in a single extraction stage, determine

(a) the flow rate of water in the aqueous PEG solvent if it is to contain 406 kg/h of PEG, and if the overall composition of PEG-salt-water for the system is to lie on the aforementioned tie line;

Answer: 604 kg/h

(b) the composition in wt% on a protein-free basis of the equilibrium top phase and bottom phase;

(c) the percent of lysozyme and other proteins extracted to the top phase;

(d) the % purity of lysozyme in the total extracted protein.

Answer: 15.1%

REFERENCES

Alders, L., *Liquid–Liquid Extraction*, 2nd ed., Elsevier, Amsterdam (1959).

Antony, F. M., D. Pal, and K. Wasewar, *Phys. Sci. Rev.*, **6** (4), 20180065 (2021).

Barson, N., and G. H. Beyer, *Chem. Eng. Prog.*, **49**, 243 (1953).

Bokhary, A., M. Leitch, and B. Q. Liao, *J. Water Process Eng.*, **40**, 101762 (2021).

Boobar, N., et al., *Ind. Eng. Chem.*, **43**, 2922 (1951).

Davies, J. T., *Turbulence Phenomena*, Academic Press, New York (1978).

Diamond, A. D., and J. T. Hsu, *Biotechnol. Bioeng.*, **34**, 1000 (1989).

Diamond, A. D., and J. T. Hsu, *AIChE J.*, **36**, 1017 (1990).

Flynn, A. W., and R. E. Treybal, *AIChE J.*, **1**, 324 (1955).

Gnanasundaram, S., et al., *Can. J. Chem. Eng.*, **57**, 141 (1979).

Iqbal, M., et al., *Biological Procedures Online* (2016).

Janecke, E. Z., *Anorg. Allg. Chem.*, **51**, 132 (1906).

Karr, A. E., and T. C. Lo, *Chem. Eng. Prog.*, **72**, 243 (1976).

Laity, D. S., and R. E. Treybal, *AIChE J.*, **3**, 176 (1957).

Lewis, J. B., et al., *Trans. Inst. Chem. Eng.*, **29**, 126 (1951).

Maloney, J. O., and A. E. Schubert, *Trans. AIChE*, **36**, 741 (1940).

NIH, https://www.covid19treatmentguidelines.nih.gov. (Accessed Jan 7, 2022).

Perry, R. H., and C. H. Chilton (eds.), *Chemical Engineers' Handbook*, 5th ed., McGraw-Hill, New York (1973).

Ryan, A. D., et al., *Chem. Eng. Prog.*, **55**, 70 (1959).

Scharzec, B., K. F. Kruber, and M. Skiborowski, *Chem. Eng. Sci.*, **240, 116650** (2021).

Scheibel, E. G., *Chem. Eng. Prog.*, **44**, 681 (1948).

Schindler, H. D., and R. E. Treybal, *AIChE J.*, **14**, 790 (1968).

Seader, J. D., and E. J. Henley, *Separation Process Principles*, 2nd ed., Wiley, New York (2006).

Seader, J. D., E. J. Henley, and D. K. Roper, *Separation Process Principles*, 3rd ed., Wiley, New York (2011).

Skelland, A. H. P., and L. T. Moeti, *Ind. Eng. Chem. Res.*, **29**, 2258 (1990).

Skelland, A. H. P., and G. G. Ramsey, *Ind. Eng. Chem. Res.*, **26**, 77 (1987).

Stichlmair, J., *Chemie–Ingeniur–Technik*, **52**, 253 (1980).

Treybal, R. E., *Liquid Extraction*, 2nd ed., McGraw-Hill, New York (1963).

Treybal, R. E., *Mass-Transfer Operations*, 3rd ed., McGraw-Hill, New York (1980).

Vermeulen, T., et al., *Chem. Eng. Prog.*, **51**, 85F (1955).

Wankat, P. C., *Equilibrium Staged Separations*, Elsevier, New York (1988).

8

Humidification Operations

8.1 INTRODUCTION

The operations considered in this chapter are concerned with the interphase transfer of mass and energy, which result when a gas is brought into contact with a pure liquid in which it is essentially insoluble. The matter transferred between phases in such cases is the substance constituting the liquid phase, which either vaporizes or condenses. These operations are somewhat simpler—from the point of view of mass transfer—than absorption and stripping, for when the liquid contains only one component, there are no concentration gradients and no resistance to mass transfer in the liquid phase. On the other hand, both heat transfer and gas-phase mass transfer are important and must be considered simultaneously since they influence each other.

While the term *humidification operations* is used to characterize these in a general fashion, the purpose of such operations may include not only humidification of the gas but dehumidification and cooling of the gas, measurement of its vapor content, and cooling of the liquid as well. Water cooling with air is without question the most important of the humidification operations. In this application, water warmed by passage through heat exchangers, condensers, and the like is cooled by contact with atmospheric air for reuse. A small amount of water evaporates driven by a concentration gradient in the gas phase. The latent heat of evaporation required is supplied by sensible heat from the liquid, which cools down in the process. The latent heat of water is so large that a small amount of evaporation produces large cooling effects.

As in all mass-transfer problems, it is necessary for a complete understanding of the process to be familiar with the equilibrium characteristics of the systems. Moreover, since the mass transfer in these cases will invariably be accompanied by simultaneous transfer of energy as well, consideration must also be given to the enthalpy characteristics of the systems.

Principles and Applications of Mass Transfer: The Design of Separation Processes for Chemical and Biochemical Engineering, Fourth Edition. Jaime Benitez.
© 2023 John Wiley & Sons, Inc. Published 2023 by John Wiley & Sons, Inc.

447

8.2 EQUILIBRIUM CONSIDERATIONS

Your objectives in studying this section are to be able to:

1. Calculate the concentration (in terms of absolute humidity and molal absolute humidity) and enthalpy per unit mass of dry gas of saturated and unsaturated gas–vapor mixtures.
2. Define and calculate relative saturation, dew point, adiabatic-saturation temperature, and wet-bulb temperature of unsaturated mixtures.

In humidification and dehumidification operations, the liquid phase is always a single pure component. Then, the equilibrium partial pressure of the solute in the gas phase is equal to the vapor pressure of the liquid at its temperature. By Dalton's law, the equilibrium partial pressure of the solute may be converted to the equilibrium mol fraction $y_{A,i}$ in the gas phase once the total pressure is specified. Since the vapor pressure of any liquid depends only upon the temperature, for a given temperature and total pressure, the equilibrium composition of the gas phase is fixed. Because the liquid is a single pure component, $x_{A,i}$ is always unity. Equilibrium data are often presented as plots of $y_{A,i}$ versus temperature at a given total pressure, as shown for the system air–water at 1 bar in Figure 8.1.

8.2.1 Saturated Gas–Vapor Mixtures

In the discussion that follows, the term *vapor* will be applied to that substance, designated as substance A, in the gaseous state which is relatively near its

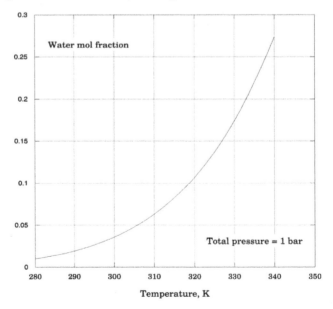

Figure 8.1 Equilibrium concentrations for the system air–water at 1 bar.

8.2 Equilibrium Considerations

condensation temperature at the prevailing pressure and, therefore, could easily condense. The term *gas* will be applied to substance B, which is a highly superheated gas and would never condense under the prevailing conditions.

When operations involve changes in the vapor content of a gas–vapor mixture without changes in the gas content, it is convenient to use concentration units based on the unchanging amount of gas. The ratio mass of vapor to mass of gas is the *absolute humidity*, Y'. If the quantities are expressed in mol, the ratio is the *molal absolute humidity*, Y. Under conditions where the ideal gas law applies,

$$Y = \frac{y_A}{y_B} = \frac{p_A}{p_B} = \frac{p_A}{P - p_A} \frac{\text{mol A}}{\text{mol B}}$$

$$Y' = Y\frac{M_A}{M_B} = \frac{p_A}{P - p_A}\frac{M_A}{M_B}\frac{\text{mass A}}{\text{mass B}} \tag{8-1}$$

If an insoluble gas B is brought into contact with sufficient liquid A, the liquid will evaporate into the gas until ultimately, at equilibrium, the gas will be *saturated* with the vapor and the partial pressure of the vapor in the saturated mixture will be the vapor pressure P_A of the liquid at the prevailing temperature as shown in Example 8.1. Then, at equilibrium

$$Y_s = \frac{y_A}{y_B} = \frac{P_A}{p_B} = \frac{P_A}{P - P_A} \frac{\text{mol A}}{\text{mol B}}$$

$$Y_s' = Y_s\frac{M_A}{M_B} = \frac{P_A}{P - P_A}\frac{M_A}{M_B}\frac{\text{mass A}}{\text{mass B}} \tag{8-2}$$

Example 8.1 Humidity of a Saturated Gas–Vapor Mixture

A gas (B)–benzene (A) mixture is saturated at 1 bar and 320 K. Calculate the absolute humidity if B is (a) nitrogen and (b) carbon dioxide.

Solution

Since the solution is saturated, the partial pressure of benzene equals the equilibrium vapor pressure of liquid benzene at 320 K. From equation (6-5) and using the data for benzene from Example 6.4, at 320 K, $P_A = 0.32$ bar. The molecular weight of benzene is $M_A = 78$ g/mol.

(a) If the gas B is nitrogen, $M_B = 28$ g/mol. From equation (8-2),

$$Y_s = \frac{y_A}{y_B} = \frac{P_A}{p_B} = \frac{P_A}{P - P_A} = \frac{0.32}{1 - 0.32} = 0.471 \frac{\text{mol } C_6H_6}{\text{mol } N_2}$$

$$Y_s' = Y_s\frac{M_A}{M_B} = 0.471 \times \frac{78}{28} = 1.311 \frac{\text{kg } C_6H_6}{\text{kg } N_2}$$

450 Humidification Operations

(b) If the gas B is carbon dioxide, M_B = 44 g/mol. From equation (8-2),

$$Y_s = \frac{y_A}{y_B} = \frac{P_A}{p_B} = \frac{P_A}{P - P_A} = \frac{0.32}{1 - 0.32} = 0.471 \frac{\text{mol C}_6\text{H}_6}{\text{mol CO}_2}$$

$$Y'_s = Y_s \frac{M_A}{M_B} = 0.471 \times \frac{78}{44} = 0.834 \frac{\text{kg C}_6\text{H}_6}{\text{kg CO}_2}$$

Notice that the absolute molal humidity is independent of the identity of the gas, unlike the absolute humidity.

In humidification problems, energy balances are always required, therefore it is necessary to calculate the enthalpy of gas–vapor mixtures. The enthalpy of a gas–vapor mixture is the sum of the enthalpies of the gas and of the vapor content. To calculate the enthalpy of the mixture per unit mass of dry gas, H', two reference states must be chosen, one for the gas and one for the vapor. Let T_0 be the datum temperature chosen for both components and base the enthalpy of component A on liquid at T_0 (T_0 = 273 K for most air–water problems). Let the temperature of the mixture be T and the absolute humidity Y'. The total enthalpy of the mixture is the sum of three items: the sensible heat of the vapor, the latent heat of vaporization of the liquid at T_0, and the sensible heat of the vapor-free gas. Then,

$$H' = C_{p,B}\left(T - T_0\right) + Y'\left[C_{p,A}\left(T - T_0\right) + \lambda_0\right] \tag{8-3}$$

where λ_0 is the latent heat of vaporization per unit mass of liquid A at T_0.

Equation (8-3) shows that the enthalpy of a gas–vapor mixture depends upon the temperature and humidity at a given pressure. If the mixture is saturated, the humidity is determined by the temperature, then the enthalpy of saturated gas–vapor mixtures depends only upon the temperature at a given pressure as shown in Example 8.2.

Example 8.2 Enthalpy of a Saturated Gas–Vapor Mixture

Calculate the enthalpy per unit mass of dry air of a saturated mixture of air (B) and water vapor (A) at 303 K and 1 atm. The reference state is air and liquid water at 273 K. The vapor pressure of water at 303 K is 4.24 kPa; latent heat of vaporization at 273 K is 2502.3 kJ/kg. The average heat capacities between 273 K and 303 K are $C_{p,A}$ = 1.884 kJ/kg·K and $C_{p,B}$ = 1.005 kJ/kg·K.

Solution
For air, M_B = 29 g/mol. From equation (8-2), Y'_s = 0.0272 kg H$_2$O/kg dry air. From equation (8-3), H'_s = 99.75 kJ/kg dry air at 303 K and 1 atm. The following table and Figure 8.2 show the results of similar calculations from 273 K to 333 K.

Temp., K	273	283	293	303	313	323	333
H'_s kJ/kg	9.48	29.36	57.57	99.75	166.79	275.78	461.50

8.2 Equilibrium Considerations

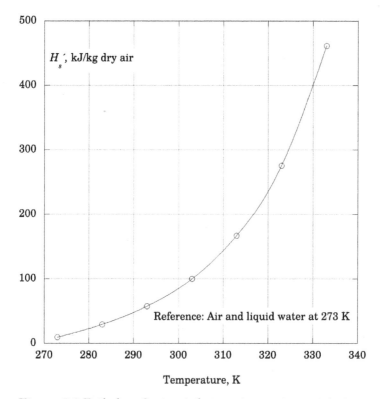

Figure 8.2 Enthalpy of saturated air–water mixtures at 1 atm.

8.2.2 Unsaturated Gas–Vapor Mixtures

If the partial pressure of the vapor in a gas–vapor mixture is less than the equilibrium vapor pressure of the liquid at the same temperature, the mixture is unsaturated. This condition is called *partial saturation*. In this case, to completely define the thermodynamic state of the mixture, the degree of saturation of the mixture must be specified. There are several ways to express the vapor concentration in an unsaturated mixture.

Relative saturation (*relative humidity* for the system air–water), expressed as a percentage is defined as $100\, p_A/P_A$, where P_A is the vapor pressure at the *dry-bulb temperature* of the mixture. Therefore, if the relative saturation, temperature, and pressure of a gas–vapor mixture are specified, all of its thermodynamic properties can be calculated.

Another way of fixing the state of an unsaturated mixture is to specify its *dew point*. This is the temperature at which a gas–vapor mixture becomes saturated when cooled at constant total pressure. In this case, the partial pressure of the vapor in the unsaturated mixture equals the vapor pressure of the liquid at the specified dew-point temperature as shown in Example 8.3.

Example 8.3 Properties of an Unsaturated Gas–Vapor Mixture

An air–water vapor mixture at a dry-bulb temperature of 328 K and a pressure of 1 atm has a relative humidity of 30%. Calculate for this mixture: (a) dew point, (b) absolute humidity, and (c) enthalpy per unit mass of dry air relative to 273 K.

Solution

The vapor pressure of water at 328 K is 15.73 kPa. For a relative humidity of 30%, $p_A = 0.3 \times 15.73 = 4.72$ kPa.

(a) The dew point is that temperature at which $P_A = p_A = 4.72$ kPa. For water, the corresponding temperature is 304.5 K.

(b) From equation (8-1), $Y' = 4.72 \times 18/[(101.3 - 4.73) \times 29] = 0.03$ kg H_2O/kg air.

(c) From equation (8-3), $H' = 1.005 \times (328 - 273) + 0.03 \times [1.884 \times (328 - 273) + 2502.3] = 133.45$ kJ/kg dry air.

8.2.3 Adiabatic-Saturation Curves

Consider the operation indicated schematically in Figure 8.3. Here the entering gas is contacted with liquid, for example, in a spray. As a result of heat and mass transfer between gas and liquid, the gas leaves at conditions of humidity and temperature different from those at the entrance. The operation is adiabatic since no heat is exchanged with the surroundings. A mass balance for substance A gives

$$L' = V_S' \left(Y_2' - Y_1' \right) \qquad (8\text{-}4)$$

An energy balance is

$$V_S' H_1' + L' H_L = V_S' H_2' \qquad (8\text{-}5)$$

Combining equations (8-4) and (8-5) gives us

$$H_1' + \left(Y_2' - Y_1' \right) H_L = H_2' \qquad (8\text{-}6)$$

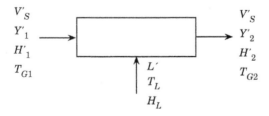

Figure 8.3 Adiabatic gas–liquid contact.

8.2 Equilibrium Considerations 453

This can be expanded by the definition of H' giving

$$
\begin{aligned}
C_{S1}\left(T_{G1} - T_0\right) + Y_1'\lambda_0 &+ \left(Y_2' - Y_1'\right)C_{A,L}\left(T_L - T_0\right) \\
&= C_{S2}\left(T_{G2} - T_0\right) + Y_2'\lambda_0
\end{aligned}
\tag{8-7}
$$

where C_S is known as the *humid heat* and is defined as

$$
C_S = C_{p,B} + Y'C_{p,A}
\tag{8-8}
$$

In the special case where the gas–vapor mixture leaves saturated, and therefore at conditions T_{as}, Y'_{as}, H'_{as}, and the liquid enters at temperature T_{as}, equation (8-7) can be written as (Treybal, 1980)

$$
C_{S1}\left(T_{G1} - T_{as}\right) = \left(Y_{as}' - Y_1'\right)\left[C_{p,A}\left(T_{as} - T_0\right) + \lambda_0 - C_{A,L}\left(T_{as} - T_0\right)\right]
\tag{8-9}
$$

It can be shown (see problem 8.1) that the quantity in brackets in equation (8-9) is the latent heat of evaporation of the liquid A at the temperature T_{as}, λ_{as}. Consequently,

$$
T_{as} = T_{G1} - \left(Y_{as}' - Y_1'\right)\frac{\lambda_{as}}{C_{S1}}
\tag{8-10}
$$

To solve equation (8-10) for the adiabatic-saturation temperature requires a trial-and-error procedure since Y'_{as} and λ_{as} are functions of T_{as}.

Example 8.4 Adiabatic-Saturation Temperature

Air at 356 K, Y' = 0.03 kg water/kg dry air, and 1 atm is contacted with water at the adiabatic-saturation temperature until it becomes saturated. What are the final temperature and humidity of the air?

Solution
To solve equation (8-10), the latent heat of vaporization of water and the absolute humidity of saturated mixtures of air–water vapor must be expressed in terms of the saturation temperature. The method proposed by Watson (Smith et al., 1996) for the latent heat (see Example 2.11) and Antoine equation combined with equation (8-2) for the humidity are used. The trial-and-error procedure required to solve equation (8-10) for T_{as} is easily implemented using the "solve block" feature of Mathcad as shown in Figure 8.4. Appendix 8.1 shows the corresponding Python solution. The results are: T_{as} = 312.7 K, Y'_{as} = 0.048 kg water/kg dry air. Therefore, the air is cooled and humidified in the process.

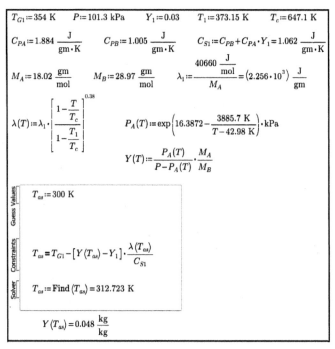

Figure 8.4 Adiabatic saturation calculations with Mathcad.

8.2.4 Wet-Bulb Temperature

The *wet-bulb temperature* is the steady-state temperature reached by a small amount of liquid evaporating into a large amount of unsaturated gas-vapor mixture. Under properly controlled conditions, it can be used to measure the humidity of the mixture. For this purpose, a thermometer whose bulb has been covered with a wick kept wet with the liquid is immersed in a rapidly moving stream of the gas mixture. The temperature indicated by this thermometer will ultimately reach a value lower than the dry-bulb temperature of the gas if the latter is unsaturated. From a knowledge of this value, the humidity is calculated.

Consider a drop of liquid immersed in a rapidly moving stream of unsaturated gas–vapor mixture. If the liquid is initially at a temperature higher than the gas dew point, the vapor pressure of the liquid will be higher at the drop surface than the partial pressure of the vapor in the gas, and the liquid will evaporate and diffuse into the gas. The latent heat required for the evaporation will at first be supplied at the expense of the sensible heat of the liquid drop, which will then cool down. As soon as the liquid temperature drops below the dry-bulb temperature of the gas, heat will flow from the gas to the liquid, at an increasing rate as the temperature difference becomes larger. Eventually, the rate of heat transfer from the gas to the liquid will equal the rate of heat required for the evaporation, and the temperature of the liquid will remain constant at some low value, the wet-bulb temperature T_w. The mechanism of the wet-bulb process is essentially the same as that governing the adiabatic-saturation process, except that in the wet-bulb process the humidity of the rapidly moving gas is assumed to remain constant.

8.2 Equilibrium Considerations
455

Since both heat and mass transfer occur simultaneously during the wet-bulb process, equation (2-77) applies with $q_t = 0$ since no heat passes through the gas–liquid interface. Given that $N_B = 0$, equation (2-76) reduces to

$$q_s = \frac{N_A M_A C_{p,A}}{1 - \exp\left(-\dfrac{N_A M_A C_{p,A}}{h_G}\right)}\left(T_G - T_w\right) \approx h_G\left(T_G - T_w\right) \qquad (8\text{-}11)$$

The approximation of the right-hand side of equation (8-11) is usually satisfactory since ordinarily the rate of mass transfer is small. Further,

$$N_A = F \ln\left[\frac{1 - P_{A,W}/P}{1 - P_A/P}\right] \approx k_G\left(p_A - P_{A,W}\right) \qquad (8\text{-}12)$$

where $P_{A,W}$ is the vapor pressure of A at T_w. Again, the approximation of the right-hand side of equation (8-12) is usually satisfactory since ordinarily the rate of mass transfer is small. Notice that the form of equation (8-12) reflects the fact that N_A is negative if q_s is taken to be positive. Substituting equations (8-11) and (8-12) into equation (2-77) with $N_B = 0$ and $q_t = 0$ gives

$$h_G\left(T_G - T_w\right) + \lambda_w M_A k_G\left(p_A - P_{A,W}\right) = 0 \qquad (8\text{-}13)$$

where λ_w is the latent heat of vaporization of A at the wet-bulb temperature, per unit mass. Equation (8-13) can be written in terms of absolute humidities as

$$T_G - T_w = \frac{-\lambda_w M_A k_G\left(p_A - P_{A,W}\right)}{h_G} = \frac{\lambda_w M_B p_{B,M} k_G\left(Y_w' - Y'\right)}{h_G} \qquad (8\text{-}14)$$

where $p_{B,M}$ is the average partial pressure of the gas and Y_w' is the humidity of a gas–vapor mixture saturated at T_w. Define a new mass-transfer coefficient, k_Y, for diffusion of A through stagnant B in dilute solutions and with the driving force expressed in terms of humidity: $k_Y = M_B p_{B,M} k_G$ (units of kg A/m^2·s). Then, equation (8-14) can be written as

$$T_G - T_w = \frac{\lambda_w\left(Y_w' - Y'\right)}{h_G / k_Y} \qquad (8\text{-}15)$$

which is the form of the relationship commonly used. The quantity $T_G - T_w$ is known as the *wet-bulb depression*.

In order to use equation (8-15) for determination of Y', it is necessary to know the value of h_G/k_Y, known as the *psychrometric ratio*. Values of h_G and k_Y can be estimated independently for the particular shape of the wetted surface by correlations like those presented in Chapter2, using the heat-transfer mass-transfer analogy if necessary. Alternatively, experimental values of the psychrometric ratio can be used. Henry and Epstein (1970) have examined the data and methods of

456   **Humidification Operations**

measurement and have produced some measurements of their own. For turbulent flow of gases past cylinders, such as wet-bulb thermometers, and past single spheres, the results of 18 gas–vapor systems are well correlated by

$$\frac{h_G}{k_Y C_S} = \left(\frac{Sc}{Pr}\right)^{0.567} = Le^{0.567} \tag{8-16}$$

where the dimensionless numbers are evaluated at the bulk-gas conditions. For the system air–water, a thorough analysis by Parker and Treybal (1960) of the abundant experimental data available led to the value $h_G/k_Y = 0.950$ kJ/kg·K, which is recommended for this system. It agrees closely with equation (8-16).

Notice the similarities between equations (8-10) and (8-15). The adiabatic saturation temperature, T_{as}, would be identical to the wet-bulb temperature, T_w, if $C_{S1} = h_G/k_Y$. For the system air–water, these are nearly equal at moderate humidities, that is, $h_G/k_Y C_S \approx 1$. This is called the *Lewis relation* (Lewis, 1922). *It must be emphasized that this is not the case for most other systems.* This is shown in Examples 8.5 and 8.6. Not only does the Lewis relation lead to the near equality of the wet-bulb and adiabatic-saturation temperatures in the case of air–water mixtures, but also to other important simplifications to be developed later.

Example 8.5 Wet-Bulb Temperature of Air–Water Mixture

For an air–water mixture of dry-bulb temperature 340 K and 1 atm, a wet-bulb temperature of 320 K was measured. Calculate the absolute humidity of the air.

Solution
At $T_w = 320$ K, $\lambda_w = 2413$ kJ/kg, and $Y'_w = 0.073$ kg water/kg dry air (saturated at 320 K and 1 atm). Substituting in equation (8-15),

$$Y' = 0.073 - (340 - 320)(0.95/2413) = 0.065 \text{ kg water/kg dry air.}$$

Example 8.6 Wet-Bulb and Adiabatic-Saturation Temperatures of Air–Toluene Mixture

Estimate the adiabatic-saturation and wet-bulb temperatures for a toluene–air mixture of 333 K dry-bulb temperature, $Y' = 0.050$ kg vapor/kg dry air, 1 atm pressure.

Solution

(a) Use the Mathcad program of Figure 8.4 or the Python version of Appendix 8.1 to calculate the adiabatic-saturation temperature. The following data, required by the program, are available for toluene vapor (Smith et al., 1996):

8.3 Adiabatic Gas–Liquid Contact Operations

Molecular weight: M_A = 92 g/mol
Critical temperature: T_c = 591.8 K
Normal boiling point: T_n = 383.8 K
Latent heat of vaporization at T_n: λ_n = 33.18 kJ/mol
Heat capacity at 333 K: $C_{p,A}$ = 1.256 kJ/kg·K
Humid heat: C_{S1} = 1.005 + 0.05 × 1.256 = 1.068 kJ/kg·K
Constants in Antoine equation (P_A kPa): A = 13.9320, B = 3057 K, C = –55.52 K

The answer is T_{as} = 300 K, Y'_{as} = 0.136 kg vapor/kg dry air.

(b) Estimate the psychrometric ratio for a mixture of air–toluene at 333 K, 1 atm, and 0.05 kg vapor/kg dry air. The thermodynamic properties of such a dilute mixture are essentially those for dry air at the given temperature and pressure, namely: ρ = 1.06 kg/m^3, μ = 195 μP, and Pr = 0.7. The diffusivity is from the Wilke–Lee equation, D_{AB} = 0.1 cm^2/s. Calculate Sc = 1.95 × 10^{-5}/(1.06 × 1.0 × 10^{-5}) = 1.84; calculate h_G/k_Y = 1.068 × (1.84/0.7)$^{0.567}$ = 1.847 kJ/kg·K. Modify the Mathcad program of Figure 8.4, or the Python version of Appendix 8.1, to calculate the dew-point temperature by simply replacing the humid heat by the psychrometric ratio. The answer is T_w = 305 K, Y'_w = 0.177 kg vapor/kg dry air.

8.3 ADIABATIC GAS–LIQUID CONTACT OPERATIONS

Your objectives in studying this section are to be able to:

1. Combine material and energy balances to develop the equation for the operating line for a countercurrent adiabatic gas–liquid contact operation.
2. Develop the design equation for water-cooling towers and dehumidifiers using air.

Adiabatic direct contact of a gas with a pure liquid may have any of several purposes:

1. *Cooling a liquid*. The cooling occurs by transfer of sensible heat and also by evaporation. The principal application is cooling of water by contact with atmospheric air.

2. *Cooling a hot gas*. Direct contact provides a nonfouling heat exchanger which is very effective, provided the presence in the gas of some of the vapor of the liquid is not objectionable.

3. *Humidifying or dehumidifying a gas*. This can be used for controlling the moisture content of air for drying, for example.

8.3.1 Fundamental Relationships

Adiabatic gas–liquid contact is usually carried out in some sort of packed tower, frequently with countercurrent flow of gas and liquid. Figure 8.5 is a schematic diagram of such a tower of cross-sectional area S. A mass balance for substance A over the lower part of the tower is

$$L' - L'_1 = V'_s \left(Y' - Y'_1 \right) \tag{8-17}$$

or

$$dL' = V'_S dY' \tag{8-18}$$

Similarly, an energy balance is

$$L'H_L + V'_S H'_1 = L'_1 H_{L1} + V'_S H' \tag{8-19}$$

These can be applied to the entire column by putting subscript 2 on the unnumbered terms.

Since the change in liquid flow rate in the tower is usually only 1 to 2%, it is assumed to be constant. Then, equation (8-19) can be written as

Although operations of this sort are simple in the sense that mass transfer is confined to the gas phase, they are nevertheless complex owing to the large heat effects which accompany evaporation or condensation.

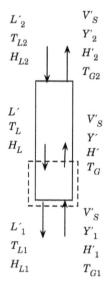

Figure 8.5 Continuous countercurrent adiabatic gas–liquid contact.

8.3 Adiabatic Gas–Liquid Contact Operations

$$V_S' dH' = L' dH_L = L' C_{A,L} dT_L \tag{8-20}$$

$$V_S' \left(H_2' - H_1' \right) = L' C_{A,L} \left(T_{L2} - T_{L1} \right) \tag{8-21}$$

Equation (8-21) is the operating line for the tower and is a straight line on a diagram of gas enthalpy versus liquid temperature, of slope $L'C_{A,L}/V_S'$, as shown on Figure 8.6. The equilibrium line gives the enthalpy of the gas saturated with the liquid as a function of liquid temperature.

The rate of sensible-heat transfer from the liquid to the interface in a differential height of the packed tower, dz, is

$$L' C_{A,L} dT_L = h_L a_H \left(T_L - T_i \right) S dz \tag{8-22}$$

or

$$G_x C_{A,L} dT_L = h_L a_H \left(T_L - T_i \right) dz \tag{8-23}$$

where G_x is the superficial liquid velocity and $h_L a_H$ is the volumetric heat-transfer coefficient for the liquid. The rate of heat transfer from the interface to the gas is

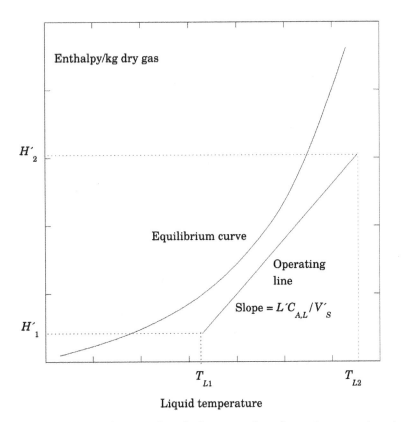

Figure 8.6 Operating diagram for adiabatic gas–liquid countercurrent contact.

460   **Humidification Operations**

$$G'_S C_S dT_G = \frac{N_A M_A C_{p,A}}{1 - \exp\left(-N_A M_A C_{p,A} / h_G\right)} a_H \left(T_i - T_G\right) dz$$
$$= h'_G a_H \left(T_i - T_G\right) dz \qquad (8\text{-}24)$$

where G'_S is the mass velocity of the vapor-free gas and $h'_G a_H$ is the effective volumetric heat-transfer coefficient for the gas. If the rate of mass transfer is small, as it usually is, $h'_G a_H = h_G a_H$, and

$$G'_S dY' = k_Y a_M \left(Y'_i - Y'\right) dz \qquad (8\text{-}25)$$

where $k_Y a_M$ is the volumetric mass-transfer coefficient. If the packing is incompletely wetted by the liquid, the surface for mass transfer a_M —which is the liquid–gas interfacial area—will be smaller than that for heat transfer a_H since heat transfer may also occur between the packing and the fluids. For a thoroughly irrigated tower packing, we will assume that $a_H = a_M = a_h$ (as defined in Chapter 4).

Differentiating the definition of the total enthalpy of the mixture, equation (8-3), and substituting the definition of the humid heat, equation (8-8):

$$dH' = C_S dT_G + \lambda_0 dY' \qquad (8\text{-}26)$$

Multiplying both sides of equation (8-26) by G'_S and substituting in the resulting expression equations (8-24) and (8-25) yields

$$G'_S dH' = h_G a_H \left(T_i - T_G\right) dz + \lambda_0 k_Y a_M \left(Y'_i - Y'\right) dz \qquad (8\text{-}27)$$

Equation (8-20) can be written in terms of the mass velocities dividing both sides of the equation by the cross-sectional area, S, and the result, combined with equations (8-23) and (8-27), yields

$$G'_S dH' = h_G a_H \left(T_i - T_G\right) dz + \lambda_0 k_Y a_M \left(Y'_i - Y'\right) dz$$
$$= h_L a_H \left(T_L - T_i\right) dz \qquad (8\text{-}28)$$

Equation (8-28) is the fundamental relationship for the design of adiabatic, countercurrent, gas–liquid contact processes.

8.3.2 Water Cooling with Air

When cooling water with air in an adiabatic process, the Lewis relation applies reasonably well and equation (8-28) can be simplified. Assuming a thoroughly irrigated tower packing, $a_H = a_M = a_h$, and application of the Lewis relation, the left-hand side of equation (8-28) can be written as

$$G'_S dH' = k_Y a_h \left[\frac{h_G}{k_Y} \left(T_i - T_G\right) + \lambda_0 \left(Y'_i - Y'\right)\right] dz$$
$$= k_Y a_h \left[C_S \left(T_i - T_G\right) + \lambda_0 \left(Y'_i - Y'\right)\right] dz \qquad (8\text{-}29)$$
$$= k_Y a_h \left[H'_i - H'\right] dz$$

8.3 Adiabatic Gas–Liquid Contact Operations 461

Equation (8-29) is remarkable in that a mass-transfer coefficient is used with an enthalpy driving force! Substituting equation (8-29) into (8-28)

$$G'_S dH' = k_Y a_h \left[H'_i - H' \right] dz = h_L a_h \left(T_L - T_i \right) dz \qquad (8\text{-}30)$$

The right-hand-side equality of equation (8-30) can be simplified as

$$\frac{H'_i - H'}{T_i - T_L} = -\frac{h_L}{k_Y} \qquad (8\text{-}31)$$

This is the equation of a straight line that goes from the equilibrium curve (at point T on Figure equation (8.7) to the operating line (at point U) with slope $= -h_L/k_Y$. The distance TR is the enthalpy driving force within the gas, $H'_i - H'$. By making constructions like the triangle RTU at several places along the operating line, corresponding values of $(H'_i - H')$ can be obtained. Separating variables on the left-hand side of equation (8-30), assuming that the volumetric mass-transfer coefficient is constant, and integrating, we have

$$Z = \frac{G'_S}{k_Y a_h} \int_{H'_1}^{H'_2} \frac{dH'}{H'_i - H'} = H_{tG} N_{tG} \qquad (8\text{-}32)$$

Equation (8-32) is the design equation for water cooling towers. The integral corresponding to the number of transfer units, N_{tG}, is evaluated numerically following the procedure illustrated in Chapter 5 for absorbers and strippers if the individual coefficients h_L and k_Y are known. The equilibrium curve for cooling water with air at 1 atm is that shown on Figure 8.2.

It frequently happens that, for commercial cooling-tower packings, only overall coefficients of the form $K_Y a$, and not the individual phase coefficients, are available. In that case, an overall driving force representing the enthalpy difference for the bulk phases but expressed in terms of H' must be used, such as the vertical distance SU on Figure 8.7. Then,

$$Z = \frac{G'_S}{K_Y a} \int_{H'_1}^{H'_2} \frac{dH'}{H^* - H'} = H_{tOG} N_{tOG} \qquad (8\text{-}33)$$

Actually, the use of overall coefficients is correct (see Chapter 3) only if the equilibrium enthalpy curve of Figure 8.7 is straight, which is not strictly so, or if $h_L a_h$ is infinite, so that the interface temperature equals the bulk-liquid temperature. Therefore, some error is always introduced when overall coefficients are used for this type of calculations; however, frequently the available data leave the designer no other choice.

Just as with concentration (see Chapter 3), an operating line on the enthalpy coordinates of Figure 8.7 which anywhere touches the equilibrium curve results in a zero driving force and consequently an infinite height Z to accomplish a given temperature change in the liquid. This condition would then represent the limiting ratio $L'/V'_S(\text{min})$ permissible. Because of the curvature of the equilibrium line, the minimum air rate is frequently determined by a line tangent to the curve. A numerical method using Mathcad or Python, such as described in Problem 3.16, gives more accurate results than a graphical method. The air rate is generally chosen to be 1.2 to 2.0 times the minimum value (McCabe et al., 2005).

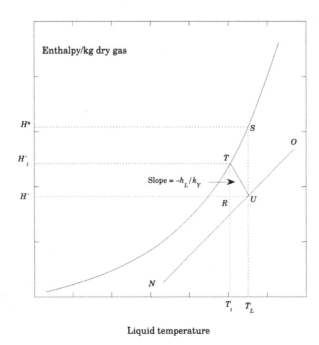

Figure 8.7 Operating diagram for water cooler with air.

It is clear from Figure 8.7 that point N, for example, will be below the equilibrium curve as long as the entering-air enthalpy H'_1 is less than the saturation enthalpy H^*_1 for air at T_{L1}. Since the enthalpy of the mixture is for most practical purposes only a function of the adiabatic-saturation temperature (or, for air–water, the wet-bulb temperature), the entering-air wet-bulb temperature must be below T_{L1}, *but its dry-bulb temperature need not be.* For this reason, it is perfectly possible to cool water to a value of T_{L1} less than the entering-air dry-bulb temperature T_{G1}. The difference between the exit-liquid temperature and the entering-air wet-bulb temperature, $T_{L1} - T_{w1}$, called the *wet-bulb temperature approach*, is then a measure of the driving force available for diffusion at the lower end of the equipment. In the design of cooling towers, this is ordinarily specified to be from 2.5 K to 5 K (Treybal, 1980).

Example 8.7 Water Cooling Using Air: Graphical Solution

A plant requires 15 kg/s of cooling water to flow through its distillation-equipment condensers. The water will leave the condensers at 318 K. It is planned to cool the water for reuse by contact with air in a countercurrent adiabatic cooling tower. The dry-bulb temperature of the entering air is 303 K and its wet-bulb temperature is 297 K. The wet-bulb temperature approach must be 5 K. The air rate to be used is 1.5 times the minimum. A specialized packing will be used for which $K_Y a$ is expected to be 0.9 kg/m³·s if the liquid mass velocity is at least 2.7 kg/m²·s and the gas mass velocity is at least 2.0 kg/m²·s. Compute the dimensions of the packed section and the make-up water requirement due to evaporation in the tower.

8.3 Adiabatic Gas–Liquid Contact Operations 463

Solution

The entering-air humidity and enthalpy are those of air saturated at the wet-bulb temperature of 297 K. The vapor pressure of water at 297 K is 2.982 kPa. From equation (8-2), $Y'_1 = 0.019$ kg water/kg dry air. From equation (8-3), for air saturated at 297 K, $H'_1 = 72.5$ kJ/kg of dry air. For the specified wet-bulb temperature approach of 5 K, the leaving-water temperature is $T_{L1} = 297 + 5 = 302$ K. The operating diagram, Figure 8.8, contains the saturated-air-enthalpy curve and point N representing conditions at the bottom of the tower ($T_{L1} = 302$ K, $H'_1 = 72.5$ kJ/kg of dry air). The operating line will start at N and end at $T_{L2} = 318$ K. For the minimum value of V'_S the operating line is tangent to the equilibrium curve and pass through point O' where $H' = 210$ kJ/kg of dry air. The slope of the line $O'N$ is therefore

$$\frac{L'C_{A,L}}{V'_S\left(\min\right)} = \frac{15 \times 4.187}{V'_S\left(\min\right)} = \frac{210 - 72.5}{318 - 302} = 8.594$$

Therefore, $V'_S(\min) = 7.31$ kg of dry air/s. For a gas rate of 1.5 times the minimum, $V'_S = 10.97$ kg of dry air/s. Therefore,

$$\frac{H'_2 - 72.5}{318 - 302} = \frac{15 \times 4.187}{10.97} = 5.725$$

from which $H'_2 = 164.2$ kJ/kg of dry air, plotted at point O on Figure 8.8. The operating line is therefore ON.

For a liquid mass velocity $G_x = 2.7$ kg/m²·s, the tower cross-sectional area would be $15/2.7 = 5.56$ m². For a gas mass velocity $G'_S = 2.0$ kg/m²·s, the tower cross-sectional area would be $10.97/2.0 = 5.50$ m². Therefore, choosing a cross-sectional area of 5.50 m² would ensure that $G_x = 2.73 > 2.7$ kg/m²·s, and $K_Y a = 0.9$ kg/m³·s.

The height of the packed section, Z, is from equation (8-33). Calculate $H_{tOG} = G'_S / K_Y a = 2.0/0.9 = 2.22$ m. To calculate N_{tOG}, the overall driving force $H^* - H'$ must be evaluated from Figure 8.8 at frequent intervals of T_L as follows:

T_L, K	H^*, kJ/kg	H', kJ/kg	$(H^* - H')^{-1}$ kg/kJ
302.0	100.0	72.5	0.03636
305.5	114.0	92.0	0.04545
308.0	129.8	106.5	0.04292
310.5	147.0	121.0	0.03846
313.0	166.8	135.5	0.03195
315.5	191.0	149.5	0.02410
318.0	216.0	164.2	0.01931

The last two columns are plotted against each other, with H' as the abscissa, and the area under the curve is estimated numerically as $N_{tOG} = 3.28$. Therefore, $Z = 3.28 \times 2.22 = 7.28$ m of packed height.

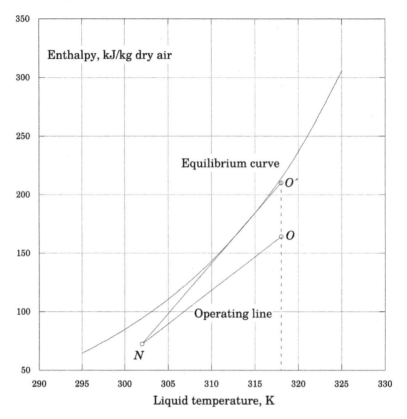

Figure 8.8 Operating diagram for Example 8.7.

To estimate the make-up water requirement due to evaporation in the tower, it is assumed that the outlet air ($H_2' = 164.2$ kJ/kg) is essentially saturated. This corresponds to an exit-air temperature of 312.8 K and humidity $Y_2' = 0.048$ kg water/kg of dry air. The approximate rate of evaporation is then $V_S' \times (Y_2' - Y_1') = 0.318$ kg/s. This is about 2.1% of the total water circulation rate. (See Figure 8.9 for a picture of an industrial application.)

Example 8.8 Water Cooling Using Air: Numerical Solution

A countercurrent cooling tower operates with inlet- and exit-water temperatures of 314 K and 303 K when the air dry-bulb and wet-bulb temperatures are, respectively, 306 K and 298 K. The tower is packed with a specialized plastic fill to a depth of 3.0 m, and the mass velocities are $G_x = 3.0$ kg/m²·s and $G_S = 2.7$ kg/m²·s.

(a) Estimate the overall volumetric mass-transfer coefficient, $K_Y a$.
(b) If the cooling load, and the water and air flow rates remain the same, but the air temperature drops to 294 K with a wet-bulb temperature of 288 K, predict the outlet-water temperature and wet-bulb temperature approach.

8.3 Adiabatic Gas–Liquid Contact Operations

Figure 8.9 Natural-draft cooling towers in a nuclear power plant. The packing occupies only the bottom section; the rest of the tower acts as a chimney to create the flow of air (Courtesy of Michael Rennhack).

Solution

(a) The entering-air humidity and enthalpy are those of air saturated at the wet-bulb temperature of 298 K. They are given by the Mathcad program of Figure 8.10 as $Y_1' = 0.02$ kg water/kg of dry air and $H_1' = 76.17$ kJ/kg dry air. From an overall energy balance, the enthalpy of the exit air is $H_2' = 127.35$ kJ/kg dry air. The program calculates the number of overall mass transfer units as $N_{tOG} = 1.642$. Then, $H_{tOG} = Z/N_{tOG} = 3/1.642 = 1.827$ m. From the definition of H_{tOG}, $K_Y a = G_S'/H_{tOG} = 2.7/1.827 = 1.478$ kg/m^3·s. Appendix 8.2 presents the corresponding Python solution.

(b) At the new entering-air wet-bulb temperature of 288 K, $H_1' = 41.88$ kJ/kg dry air. Since the cooling load and water flow rate are to remain constant, the change in water temperature through the tower must remain the same as in part (a), namely, $T_{L2} - T_{L1} = 11$ K. However, since the water recirculates (from the cooling tower to the heat-transfer equipment, and back to the cooling tower) the new inlet-water temperature will depend on the corresponding outlet-water temperature, so both are unknown. On the other hand, the packed height and mass velocities of gas and liquid are fixed; therefore, the number of transfer units is fixed at $N_{tOG} = 1.642$. Since N_{tOG} is a function of both water temperatures, this provides the second relation needed to calculate the new values of T_{L2} and T_{L1}. The two equations are solved simultaneously by trial-and-error using the "solve block" feature of Mathcad in Figure 8.10. The answer is $T_{L2} = 308$ K and $T_{L1} = 297$ K. The wet-bulb temperature approach is $297 - 288 = 9$ K. Appendix 8.2 presents the corresponding Python solution.

466 **Humidification Operations**

8.3.3 Dehumidification of Air–Water Vapor

If a warm gas–vapor mixture is contacted with cold liquid so that the humidity of the gas is greater than that at the gas–liquid interface, vapor will diffuse toward the liquid and the gas will be dehumidified. In addition, sensible heat can be

$$T_0 := 273 \text{ K} \qquad C_{AL} := 4.187 \frac{\text{J}}{\text{gm} \cdot \text{K}} \qquad n := 5 \qquad P := 101.3 \text{ kPa}$$

$$C_{PA} := 1.884 \frac{\text{J}}{\text{gm} \cdot \text{K}} \qquad C_{PB} := 1.005 \frac{\text{J}}{\text{gm} \cdot \text{K}} \qquad \lambda_0 := 2502.3 \frac{\text{J}}{\text{gm}}$$

$$M_A := 18.02 \frac{\text{gm}}{\text{mol}} \qquad M_B := 28.97 \frac{\text{gm}}{\text{mol}} \qquad P_A(T) := \exp\left(16.3872 - \frac{3885.7 \text{ K}}{T - 42.98 \text{ K}}\right) \cdot \text{kPa}$$

$$Heq(T_L) := \left\| DATAT_L \leftarrow \begin{bmatrix} 273 \\ 283 \\ 293 \\ 303 \\ 313 \\ 323 \\ 333 \end{bmatrix} \cdot \text{K} \right.$$

Equilibrium gas enthalpy as a function of liquid temperature

$$\left\| DATAHG \leftarrow \begin{bmatrix} 9.48 \\ 29.36 \\ 57.57 \\ 99.75 \\ 166.79 \\ 275.58 \\ 461.5 \end{bmatrix} \cdot \frac{\text{J}}{\text{gm}} \right.$$

$$\left\| vs \leftarrow \text{cspline}(DATAT_L, DATAHG) \right.$$
$$\left\| \text{interp}(vs, DATAT_L, DATAHG, T_L) \right.$$

$$Y_m(T) := \frac{P_A(T)}{P - P_A(T)} \cdot \frac{M_A}{M_B} \qquad \text{Humidity of saturated mixture, kg water/kg dry air}$$

$$H_G(T) := C_{PB} \cdot (T - T_0) + Y_m(T) \cdot (C_{PA} \cdot (T - T_0) + \lambda_0) \qquad \text{Enthalpy of saturated mixture, J/gm of dry air}$$

a) Operational data:

$$T_{L1} := 303 \text{ K} \qquad T_{L2} := 314 \text{ K} \qquad G_x := 3.0 \frac{\text{kg}}{\text{m}^2 \cdot \text{s}} \qquad G_S := 2.7 \frac{\text{kg}}{\text{m}^2 \cdot \text{s}}$$

$$T_{W1} := 298 \text{ K} \qquad H_1 := H_G(T_{W1}) = 76.174 \frac{\text{J}}{\text{gm}} \qquad Z := 3.0 \text{ m}$$

$$Y_{m1} := Y_m(T_{W1}) = 0.02 \qquad C_{AL} := 4.1 \frac{\text{J}}{\text{gm} \cdot \text{K}}$$

$$H_2 := H_1 + \frac{G_x \cdot C_{AL} \cdot (T_{L2} - T_{L1})}{G_S} = 126.285 \frac{\text{J}}{\text{gm}} \qquad \text{Energy balance}$$

Figure 8.10 Mathcad numerical solution of Example 8.8.

8.3 Adiabatic Gas–Liquid Contact Operations

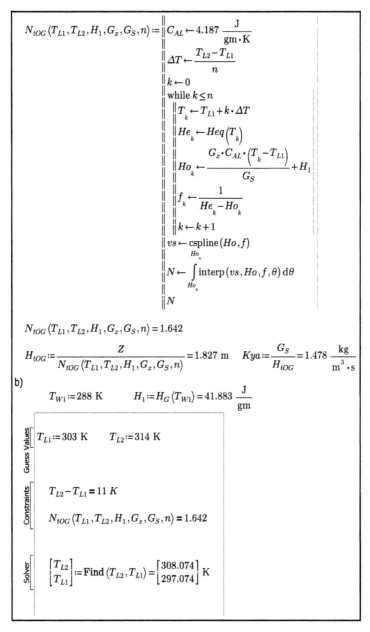

Figure 8.10 Mathcad numerical solution of Example 8.8 (continuation).

transferred as a result of temperature differences within the system. For air–water mixtures (Le = 1) contacted with cold water, the methods of analyzing water cooling presented in the previous section apply with only obvious modifications. The operating line on the H' versus T_L diagram will be above the equilibrium curve. Likewise, the driving force is now $(H' - H^*)$, and equation (8-33) can be used with this driving force.

468   **Humidification Operations**

PROBLEMS

**The problems at the end of each chapter have been grouped into four
classes (designated by a superscript after the problem number).**

Class a: Illustrates direct numerical application of the formulas in the text.
Class b: Requires elementary analysis of physical situations, based on
the subject material in the chapter.
Class c: Requires somewhat more mature analysis.
Class d: Requires computer solution.

8.1^a. Latent heat of vaporization

The latent heat of vaporization per unit mass of a pure substance at a given
temperature, λ, is defined as the difference in enthalpy between the saturated
vapor and saturated liquid at the given temperature, T. Since enthalpy is a thermodynamic function of state, show that λ can be evaluated from a known value of λ_0
at a reference temperature T_0 from the equation

$$\lambda = \lambda_0 + C_{p,A}\left(T - T_0\right) - C_{A,L}\left(T - T_0\right) \qquad (8\text{-}34)$$

where $C_{p,A}$ and $C_{A,L}$ are, respectively, the average heat capacities of gas and liquid
between T_0 and T.

8.2^a. Unsaturated air–water mixture

Air at 345 K has a dew-point temperature of 312 K. Calculate the absolute
and relative humidity, and the enthalpy of the mixture relative to liquid water and
air at 273 K, in kJ/kg of dry air.

Answer: Relative humidity = 20.3%

8.3^b. Unsaturated benzene–nitrogen mixture

Calculate the absolute humidity, relative saturation, and dew-point temperature of a mixture of benzene–nitrogen at a pressure of 1.5 bar and temperature of
320 K if the partial pressure of benzene in the mixture is 0.2 bar.

Answer: $Y' = 0.429$ kg benzene/kg nitrogen

8.4^b. Unsaturated acetone–nitrogen mixture

In a plant for the recovery of acetone which has been used as a solvent, the acetone is evaporated into a stream of nitrogen gas. A mixture of acetone vapor and
nitrogen flows through a duct, 0.3 by 0.3 m cross section. The pressure and temperature at one point in the duct are 106.6 kPa and 300 K. At this point, the average
velocity of the gas is 3.0 m/s. A wet-bulb temperature (wick wet with acetone) indicates a temperature at this point of 270 K. Calculate the kilograms of acetone per

Problems **469**

second carried by the duct. For N_2 at 300 K and 106.6 kPa, Pr = 0.716, $C_{p,B}$ = 1.041 kJ/kg·K, μ = 178 μP (Incropera et al., 2007). For acetone vapor, $C_{p,A}$ = 1.30 kJ/kg·K.

Answer: 0.02 kg/s

8.5[c]. Properties of air–water mixtures

A drier requires 1.5 m^3/s of air at 338 K, 1 atm, and 20% relative humidity. This is to be prepared from air at 300 K dry-bulb, 291 K wet-bulb temperatures by direct injection of steam into the airstream followed by passage of the air over steam-heated finned tubes. The available steam is saturated at 383 K. Calculate the kilograms of steam per second required for (a) direct injection and (b) the heat exchanger. Assume that in the heat exchanger the steam condenses at constant temperature.

Answer: (a) 0.034 kg/s

8.6[d]. Minimum air flow for water cooling

Modify the Mathcad program developed in Problem 3.16 to estimate the minimum gas flow rate in strippers so that it can be used to estimate the minimum air flow required for water cooling. Test your program with the data from Example 8.7.

Answer: V'_S(min) = 7.331 kg/s

8.7[c, d]. Effect of entering-air conditions on cooling tower performance

In the cooling tower of Example 8.7, to what temperature would the water be cooled if, after the tower was built and operated at the design water and air rates, the air entered at a dry-bulb temperature of 305 K and a wet-bulb temperature of 301 K. Assume that the cooling load of the tower remains the same as in Example 8.7.

Answer: 304.5 K

8.8[c, d]. Guaranteed performance of a cooling tower

A recently installed induced-draft cooling tower (with the fan located at the top) was guaranteed by the manufacturer to cool 450 m^3/h of water from 316 K to 303 K when the available air has a wet-bulb temperature of 297 K. A test on the tower, when operated at full fan capacity, provided the following results:

Inlet water, 450 m^3/h, 319 K

Outlet water, 298.6 K

Inlet air, 297 K dry-bulb, 288.6 K wet-bulb temperature

Outlet air, 310.6 K essentially saturated

(a) What is the fan capacity, m^3/s?

Answer: 76 m^3/s

470   **Humidification Operations**

(b) Can the tower be expected to meet the guaranteed conditions? Note that, to do so, N_{tOG} in the test must be at least equal to the guaranteed value if H_{tOG} is unchanged.

Answer: Yes

8.9[b, d]. Minimum air flow and number of transfer units

Water is to be cooled at the rate of 10 kg/s from 317 K to 300 K in a countercurrent cooling tower under conditions such that H_{tOG} = 2.5 m. Air enters at the bottom of the tower at 297 K and a wet-bulb temperature of 294 K. The air rate to use is 1.33 times the minimum.

(a) Using the results of Problem (8.6), calculate the minimum air rate.

(b) For the specified air flow rate [1.33 V'_S(min)], calculate the number of overall transfer units and tower packed height.

Answer: Z = 11.4 m

8.10[c, d]. Effect of air flow rate on cooling tower performance

A cooling tower cools 227 m³/h of water from 314 K to 306 K using a countercurrent forced draft of air entering at 317 K and wet-bulb temperature of 299 K. Measurements indicate that the air leaves at 309 K with a wet-bulb temperature of 308 K. The plant manager wishes to cool the water to as cold a temperature as possible. One possibility is to increase the air flow rate, and it is found that the fan speed can be increased without overloading of the motor so that the airflow is 1.5 times the previously used. Tower flooding will not occur at this higher gas rate. Previous experience with the type of packing use leads the plant engineer to predict that $K_Y a$ is proportional to $G'_S{}^{0.8}$. Assuming that the cooling load remains the same, what will be the outlet temperature attained with the higher air rate?

Answer: 303.6 K

8.11[b, d]. Cooling towers in a nuclear power plant

A nuclear power plant produces 1000 megawatts of electricity with a power cycle thermodynamic efficiency of 30%. The heat rejected is removed by cooling water that enters the condenser at 293 K and is heated to 313 K. The hot water flows to two identical natural-draft cooling towers where it is recooled to 293 K, and make-up water is added as necessary. The available air is at 298 K with a wet-bulb temperature of 285 K, and it will flow at a rate 1.2 times the minimum. A specialized packing will be used for which $K_Y a$ is expected to be 1.0 kg/m³·s if the liquid mass velocity is at least 3.4 kg/m²·s, and the gas mass velocity is at least 2.75 kg/m²·s. Compute the dimensions of the packed sections of the cooling towers, and the make-up water requirement due to evaporation.

Answer: Z = 17.9 m

8.12[c, d]. Design of a dehumidifier using random packing

It is desired to dehumidify 1.2 m³/s of air, available at 311 K with a wet-bulb temperature of 303 K, to a wet-bulb temperature of 288 K in a countercurrent tower using water chilled to 283 K. The packing will be 25-mm ceramic Raschig rings, and the tower will be designed for a maximum gas-pressure drop of 300 Pa/m. A liquid rate of 1.5 times the minimum will be used. Assuming a well-irrigated packing, estimate the cross-sectional area and height of the packed portion of the tower. The heat-transfer coefficient for the liquid phase can be estimated by the following correlation for Raschig rings and Berl saddles (Treybal, 1980)

$$\text{Nu}_L = \frac{h_l d_s}{k} = 25.1 \left(\frac{d_s G_x}{\mu_L} \right)^{0.45} \text{Pr}_L^{0.5} \qquad (8\text{-}35)$$

where d_s is the diameter of a sphere of the same surface area of a single packing particle (35.6 mm for 25-mm Raschig rings) and k is the thermal conductivity of the liquid.

8.13[c]. Recirculating liquid, gas humidification, and cooling

This is a special case of countercurrent gas–liquid contact in which the liquid enters the equipment at the adiabatic-saturation temperature of the entering gas. This can be achieved by continuously reintroducing the exit liquid to the contactor without removal or addition of heat, as shown in Figure equation (8.11. For this case, 8-25) becomes

$$G'_S dY' = k_Y a_M \left(Y'_{as} - Y' \right) dz \qquad (8\text{-}36)$$

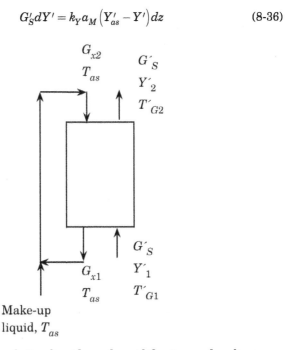

Figure 8.11 Recirculating liquid–gas humidification and cooling.

472   **Humidification Operations**

(a) For Y'_{as} constant, show that, integrating equation (8-36)

$$N_{tG} = \ln\left[\frac{Y'_{as} - Y'_1}{Y'_{as} - Y'_2}\right]$$ (8-37)

(b) A horizontal spray chamber with recirculated water is used for adiabatic humid-ification and cooling of air. The active part of the chamber is 2 m long and has a cross section of 2 m². With an air flow rate of 3.5 m³/s at a dry-bulb temperature of 338 K and Y' = 0.017 kg water/kg dry air, the air is cooled and humidified to a dry-bulb temperature of 315 K. If a duplicate spray chamber operated in the same manner were to be added in series with the existing chamber, what outlet condi-tions would be expected for the air?

Answer: T_G = 308 K

(c) Calculate the make-up water flow rate.

Answer: 139 kg/h

REFERENCES

Henry, H. C., and N. Epstein, *Can. J. Chem. Eng.*, **48**, 595, 602, 609 (1970).
Incropera, F. P., et al., *Introduction to Heat Transfer*, 5th ed., John Wiley and Sons, Hoboken, NJ (2007).
Lewis, W. K., *Trans. AIME*, **44**, 325 (1922).
McCabe, W. L., J. C. Smith, and P. Harriott, *Unit Operations of Chemical Engineering*, 7th ed., McGraw-Hill, New York (2005).
Parker, R. O., and R. E. Treybal, *Chem. Eng. Prog. Symp. Ser.*, **57** (32), 138 (1960).
Smith, J. M., H. C. Van Ness, and M. M. Abbott, *Introduction to Chemical Engineering Thermodynamics*, 5th ed., McGraw-Hill, New York (1996).
Treybal, R. E., *Mass-Transfer Operations*, 3rd ed., McGraw-Hill, New York (1980).

Appendix 8.1   **473**

APPENDIX 8.1

Python solution of Example 8.4.

```python
# Adiabatic Saturation Calculations; Example 8.4
import numpy as np
from scipy.optimize import fsolve

# Enter data:

# Temperatures in K; pressure in kPa
TG1 = 354; T1 = 373.15; Tc = 647.1; P = 101.3
# Heat capacities are in J/gm-K
CpA = 1.884; CpB = 1.005; Y1 = 0.03
# Molecular weights in gm/mol
MA = 18.02; MB = 28.97
# Heat of vaporization at T1 = 373.15 K, J/gm
Hvap1 = 40660/MA

# Calculations:

CS1 = CpB + Y1*CpA
def Hvap(T):
    return Hvap1*((Tc - T)/(Tc - T1))**0.38
def PA(T):
    return np.exp(16.3872 - 3885.7/(T - 42.98))
def Y(T):
    return PA(T)*MA/((P - PA(T))*MB)
def g(Tas):
    return Tas - TG1 + (Y(Tas) - Y1)*Hvap(Tas)/CS1
Tas = fsolve(g, 300)

# Print results:

print('Adiabatic saturation temperature (K), Tas = %5.2f' %Tas)
print('Humidity at Tas (in kg/kg) is Y(Tas) = %5.3f' % Y(Tas))
```

Program results:
Adiabatic saturation temperature (K), T_{as} = 312.7
Humidity at T_{as} (in kg/kg) is $Y(T_{as})$ = 0.048

474   **Humidification Operations**

APPENDIX 8.2

Python solution of Example 8.8.

```python
# Water Cooling Using Air; Data are from Example 8.8

import numpy as np
from scipy.optimize import fsolve
from scipy.interpolate import CubicSpline
from scipy.integrate import quad
# Enter data:
#Temperatures in K; Heat capacities in J/gm-K
T0 = 273.0; CAL = 4.187; CpA = 1.884; CpB = 1.005
# Molecular weights in gm/mol
MA = 18.02; MB = 28.97
#Heat of vaporization at T0 in J/gm; pressure in kPa
Hvap0 = 2502.3; P = 101.3
# Calculations:
def PA(T):
    return np.exp(16.3872 - 3885.7/(T - 42.98))
# Equilibrium gas enthalpy as a function of liquid temperature:
# Liquid temperature, K
DATATL = [273, 283, 293, 303, 313, 323, 333]
# Gas enthalpy, J/gm
DATAHG = [9.48, 29.36, 57.57, 99.75, 166.79, 275.58, 461.50]
cs = CubicSpline(DATATL, DATAHG)
def Heq(TL):
    return cs(TL)
#Calculate the humidity and enthalpy of saturated mixtures:
#Humidity of saturated mixture, kg/kg
def Ym(T):
    return PA(T)*MA/((P - PA(T))*MB)
#Enthalpy of saturated mixture, J/gm of dry air
def HG(T):
    return CpB*(T - T0) + Ym(T)*(CpA*(T - T0) + Hvap0)


# a) Operational Data:
#Water temperatures in K; mass velocities in kg/m^2-s
TL1 = 303; TL2 = 314; Gx = 3.0; GS = 2.7
#Entering air wet-bulb temperature, in K; packing depth, in m
TW1 = 298; Z = 3.0
H1 = HG(TW1); Ym1 = Ym(TW1)
#Energy balance
```

Appendix 8.2

475

Python solution of Example 8.8 (continuation).

```python
H2 = H1 + Gx*CAL*(TL2 - TL1)/GS
print('Part a) Results')
print('The enthalpy of the inlet air (kJ/kg dry air), H1 = %5.2f' %H1)
print('The enthalpy of the exit air (in kJ/kg dry air), H2 = %5.2f' %H2)
def NtOG(TL1,TL2,H1,Gx,GS,n):
    delT = (TL2 - TL1)/n
    T = np.zeros(n+1,dtype=float)
    He = np.zeros(n+1,dtype=float)
    Ho = np.zeros(n+1,dtype=float)
    f = np.zeros(n+1,dtype=float)
    for i in range(n+1):
        T[i] = TL1 + i*delT
        He[i] = Heq(T[i])
        Ho[i] = H1 + Gx*CAL*(T[i] - TL1)/GS
        f[i] = 1/(He[i] - Ho[i])
        i+=1
    cs = CubicSpline(Ho,f)
    def interp(x):
        return cs(x)
    return quad(interp, Ho[0], Ho[n])[0]
HtOG = Z/NtOG(TL1,TL2,H1,Gx,GS,5)
KYa = GS/HtOG
NTOG = NtOG(TL1,TL2,H1,Gx,GS,5)
print('Number of overall transfer units, NtOG = %5.3f' %NTOG)
print('Height of an overall transfer unit (m) is HtOG = %5.3f' %HtOG)
print('Overall  mass-transfer coefficient (kg/m^3-s), KYa = %5.3f' %KYa)

# b) Air temperature drops to 294 K with wet-bulb temperature of 288 K

TW1 = 288
H1 = HG(TW1)
def g(p):
    TL1 = p[0]
    TL2 = p[1]
    return TL2 - TL1 -11, NtOG(TL1,TL2, H1, Gx,GS,5) - 1.642
p = np.zeros(2,dtype=float)
p[0] = 303; p[1] = 314  #Initial estimates, K
p = fsolve(g, p)
print('')
print('Part b) Results')
print('The inlet water temperature (in K), TL2 = %5.3f' %p[1])
print('The outlet water temperature (in K), TL1 = %5.3f' %p[0])
print('The wet-bulb temperature approach (in K) = %3.1f' %(p[0]- TW1) )
```

476 **Humidification Operations**

Program results:

Part a)
The enthalpy of the inlet air (kJ/kg dry air), H_1 = 76.17
The enthalpy of the exit air (kJ/kg dry air), H_2 = 127.35
Number of overall transfer units, N_{tOG} = 1.642
Height of an overall transfer unit (m) is H_{tOG} = 1.827
Overall mass-transfer coefficient (kg/m^3-s), $K_Y a$ = 1.478

Part b)
The inlet water temperature (in K), T_{L2} = 308.07
The outlet water temperature (in K), T_{L1} = 297.07
The wet-bulb temperature approach (in K) = 9.1

9

Membranes and Other Solid Sorption Agents

9.1 INTRODUCTION

Membrane separations, adsorption, ion exchange, and chromatography are sorption operations in which certain components of a fluid phase are selectively transferred to a solid interface, such as that of a membrane, or to insoluble, rigid particles suspended in a vessel or packed in a column. In recent years, the number of industrial applications of sorption processes have greatly increased because of progress in producing selective membranes and adsorbents (Seader and Henley, 2006 Scharzec et al., 2021). This chapter presents a discussion of mass-transfer considerations in membranes and *porous adsorbents*. It also presents calculation methods for the design of the more widely used continuous membrane separation processes (including *dialysis, electrodialysis, reverse osmosis, microfiltration, and ultrafiltration*). Other sorption separation processes (including *fixed-bed adsorption, pressure-swing adsorption, ion exchange, electrophoresis, and chromatographic separations*) are also discussed.

In membrane-separation processes, a feed consisting of a mixture of two or more components is partially separated by means of a semipermeable membrane through which one species move faster than the others. That part of the feed that passes through the membrane is called the *permeate*, while the portion that does not pass is called the *retentate*. The membranes may be thin layers of a rigid material such as porous glass or sintered metal, but more often they are flexible films of synthetic polymers prepared to have a high permeability for certain types of molecules.

Many industrial separation processes that have been traditionally implemented using the classical mass-transfer operations—such as distillation or stripping—can be replaced by membrane separations that can be more environmentally friendly (such as the boiler feedwater deaerator of Example 2.14) or less energy intensive (such as the distillation-membrane hybrid of Problem 6.26). This

---

Principles and Applications of Mass Transfer: The Design of Separation Processes for Chemical and Biochemical Engineering, Fourth Edition. Jaime Benitez.
© 2023 John Wiley & Sons, Inc. Published 2023 by John Wiley & Sons, Inc.

478   **Membranes and Other Solid Sorption Agents**

replacement can be very challenging since it requires the production of high-mass-transfer-flux, long-life, defect-free membranes on a large scale, and the fabrication of the membranes into compact, economical modules of high surface area per unit volume (Seader et al., 2011).

In the adsorption process, molecules, or atoms in a fluid phase, diffuse to the surface of a highly porous solid where they bond with the solid surface or are held there by weak intermolecular forces. In an ion-exchange process, positive ions (cations) or negative ions (anions) in a liquid solution replace dissimilar and displaceable ions of the same charge contained in a solid ion exchanger. During adsorption and ion exchange, the sorbent becomes saturated or nearly so with the substance transferred from the fluid phase. To recover the sorbed substance and reuse the sorbent, it must be regenerated by desorption. Accordingly, these two separation processes are carried out in a cyclic manner. Chromatography is a process similar to adsorption in that fluid mixtures are passed through a bed of solid particles, but in this case the bed is continuously regenerated by passage of a carrier fluid and can be operated for long periods of time. The individual components move through the bed at different rates because of repeated sorption, desorption cycles, and are, therefore, separated. Electrophoresis separates charged molecules, such as proteins, by application of an electric field. Electrophoretic methods are similar to chromatographic methods in that fixed barrier phases (ranging from paper to polymer gels) are employed to facilitate separation.

For all these sorption processes, the performance depends on fluid–solid equilibria and on mass-transfer rates, which are discussed in the following sections.

9.2 MASS TRANSFER IN MEMBRANES

Your objectives in studying this section are to be able to:

1. Define and apply the concepts permeability, permeance, permselectivity, separation factor, and cut.
2. Understand the solution-diffusion model for separation of liquid and gas mixtures using dense, nonporous membranes.
3. Explain the four idealized flow patterns in membrane modules.
4. Apply the perfect-mixing and cross-flow models to solve gas permeation problems in membrane modules.

To be effective for separating mixtures of chemical components, a membrane must possess high *permeance* and a high permeance ratio, or *permselectivity*, for the two species being separated. The permeance for a given species diffusing through a membrane of a given thickness is analogous to a mass-transfer coefficient. The molar transmembrane flux of species i can then be expressed as

$$N_i = \frac{q_i}{l_M} \times \text{driving force} = Q_i \times \text{driving force} \qquad (9\text{-}1)$$

9.2 Mass Transfer in Membranes 479

where q_i is the *permeability* of the membrane to species i, l_M is the membrane thickness, and Q_i is the permeance.

Membranes can be macroporous, microporous, or dense (nonporous). A microporous membrane contains interconnected pores that are small (on the order of 10–100,000 Å), but large in comparison to the size of the molecules to be transferred. Only microporous and dense membranes are permselective. However, macroporous membranes are widely used to support thin microporous or dense membranes when significant pressure differences across the membrane are necessary to achieve a reasonable flux.

The theory of steady-state diffusion through porous solids was discussed in detail in Section 1.5. The transport of components through nonporous membranes is usually described by the *solution-diffusion model* of Lonsdale et al. (1965). This model assumes that the solute dissolves in the membrane at the feed side, diffuses through the solid, and desorbs at the permeate side. Fick's law for diffusion through the solid is assumed to apply based on the concentration driving force $(c_{i0} - c_{iL})$ shown in Figure 9.1, where the concentrations are those for the solute dissolved in the membrane. Thermodynamic equilibrium is assumed at the fluid–membrane interfaces, validated experimentally by Montanedian et al. (1990) for permeation of light gases through dense cellulose acetate membranes at up to 90 atm.

Included in this diagram is the drop in concentration across the membrane, and also possible drops due to resistances in the fluid boundary layers or films on either side of the membrane.

9.2.1 Solution-Diffusion for Liquid Mixtures

As shown in Figure 9.1 for a nonporous membrane, there is a solute concentration discontinuity at both liquid–membrane interfaces. Solute concentration c_{i0}' is that in the feed liquid just adjacent to the upstream membrane surface, whereas c_{i0} is that in the membrane just adjacent to the upstream surface. The two are related by a thermodynamic equilibrium *partition coefficient* K_i defined by

$$K_i = \frac{c_{i0}}{c_{i0}'} = \frac{c_{iL}}{c_{iL}'} \tag{9-2}$$

where it is assumed that the partition coefficient is constant, independent of concentration. Fick's law applied to diffusion across the membrane of Figure 9.1 is

$$N_i = \frac{D_i}{l_M}\left(c_{i0} - c_{iL}\right) = \frac{K_i D_i}{l_M}\left(c_{i0}' - c_{iL}'\right) \tag{9-3}$$

If the mass-transfer resistances in the two fluid films are negligible,

$$N_i = \frac{K_i D_i}{l_M}\left(c_{iF} - c_{iP}\right) = \frac{q_i}{l_M}\left(c_{iF} - c_{iP}\right) \tag{9-4}$$

where $K_i D_i$ is the permeability, q_i, for the solution-diffusion model. Both K_i and D_i depend on the solute and the membrane. Because D_i is generally very small, it is important that the membrane material offers a large value for K_i and/or a small membrane thickness.

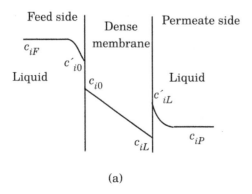

(a)

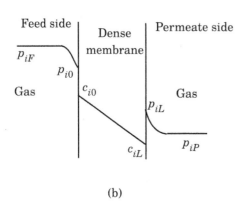

(b)

Figure 9.1 Concentration and partial pressure profiles for transport through dense membranes: (a) liquid mixture and (b) gas mixture.

When mass-transfer resistances external to the membrane are not negligible, concentration gradients exist in the films adjacent to the membrane surfaces, as is illustrated in Figure 9.1. In that case, the total mass-transfer resistance is the sum of three individual resistances, and the transmembrane flux of solute i is given by

$$N_i = \frac{c_{iF} - c_{iP}}{\dfrac{1}{k_{iF}} + \dfrac{l_M}{q_i} + \dfrac{1}{k_{iP}}} = \frac{c_{iF} - c_{iP}}{\dfrac{1}{k_{iF}} + \dfrac{1}{Q_i} + \dfrac{1}{k_{iP}}} \qquad (9\text{-}5)$$

where k_{iF} and k_{iP} are the mass-transfer coefficients for the feed-side and permeate-side films. In general, these coefficients depend on fluid properties, flow-channel geometry, and flow regime (laminar or turbulent).

Example 9.1 Liquid Flux in Tubular Membrane

A dilute solution of solute A in solvent B flows through a tubular-membrane separator where the feed flows through the tubes. At a certain location, the solute

9.2 Mass Transfer in Membranes

concentrations on the feed and permeate sides are 50 mol/m^3 and 15 mol/m^3, respectively. The membrane thickness is 2 µm, and its permeability for solute A is 176 barrer. The tube inside diameter is 0.4 cm, the tube-side Reynolds number is 20,000, and the feed-side solute Schmidt number is 450. The diffusivity of the solute is 5 × 10^{-5} cm^2/s. The mass-transfer coefficient on the permeate side of the membrane is 0.12 cm/s. Estimate the transmembrane flux of solute A.

Solution

The permeability of the membrane is expressed in terms of *barrer*. The barrer unit is named for R. M. Barrer who published an early article on the nature of diffusion in a membrane, followed by a widely referenced monograph on diffusion in and through solids (Barrer, 1951). A barrer is defined as

$$1 \text{ barrer} = \frac{10^{-10} \text{ cm}^3 \text{ gas (STP)} \times \text{ cm thickness}}{\text{cm}^2 \text{ membrane area} \times \text{ cm Hg pressure} \times \text{ s}} = 8.3 \times 10^{-9} \text{ cm}^2/\text{s} \quad (9\text{-}6)$$

Therefore, the permeability of the membrane for solute A is q_A = 176 × 8.3 × 10^{-9} = 1.46 × 10^{-6} cm^2/s. The permeance is Q_A = 1.46 × 10^{-6}/(2 × 10^{-4}) = 7.3 × 10^{-3} cm/s. The mass-transfer coefficient on the feed-side is from equation (2-75) for turbulent flow of a liquid inside a circular pipe:

$$\text{Sh} = \frac{k_{AF}D}{D_A} = 0.023 \times (20,000)^{0.83} \times (450)^{0.33} = 641$$

from which k_{AF} = 641 × 5 × 10^{-5}/0.4 = 0.08 cm/s. The total resistance to mass transfer is, then, equal to (1/0.08 + 1/0.0073 + 1/0.12) = (12.5 + 137.0 + 8.3) = 157.8 s/cm. The transmembrane flux of solute A is

$$N_A = \frac{(50-15) \text{ mol}}{\text{m}^3} \left| \frac{\text{cm}}{157.8 \text{ s}} \right| \frac{1 \text{ m}}{100 \text{ cm}} = 2.22 \times 10^{-3} \frac{\text{mol}}{\text{m}^2 \cdot \text{s}}$$

Notice that, in this case, the membrane resistance is (137/157.8) × 100 = 87% of the total.

9.2.2 Solution-Diffusion for Gas Mixtures

As shown in Figure 9.1 for a nonporous membrane, there is a solute concentration discontinuity at both gas–membrane interfaces. Solute partial pressure p_{i0} is that in the feed gas just adjacent to the upstream membrane surface, whereas c_{i0} is the solute concentration in the membrane just adjacent to the upstream surface. The two are related by a thermodynamic equilibrium described by Henry's law that is most conveniently written for membrane applications as

$$H_i = \frac{c_{i0}}{p_{i0}} = \frac{c_{iL}}{p_{iL}} \quad (9\text{-}7)$$

482   Membranes and Other Solid Sorption Agents

where it is assumed that H_i is independent of total pressure and that the temperature is the same on both sides of the membrane. Assuming again that Fick's law applies for diffusion across the membrane,

$$N_i = \frac{D_i}{l_M}\left(c_{i0} - c_{iL}\right) = \frac{H_i D_i}{l_M}\left(p_{i0} - p_{iL}\right) \tag{9-8}$$

If the mass-transfer resistances in the two fluid films are negligible,

$$N_i = \frac{H_i D_i}{l_M}\left(p_{iF} - p_{iP}\right) = \frac{q_i}{l_M}\left(p_{iF} - p_{iP}\right) \tag{9-9}$$

where $H_i D_i$ is the permeability, q_i for the solution-diffusion model. Thus, the permeability depends on both the solubility of the gas component in the membrane and the diffusivity of that component in the membrane material.

The ideal dense membrane has a high permeance for the penetrant molecules and a high selectivity between the components to be separated. The selectivity is expressed in terms of the *separation factor*, α_{AB}, defined similarly to the relative volatility in distillation:

$$\alpha_{AB} = \frac{y_A / x_A}{y_B / x_B} \tag{9-10}$$

where y_i is the mol fraction in the permeate leaving the membrane corresponding to the partial pressure p_{iP} in Figure 9.1b, while x_i is the mol fraction in the retentate on the feed side of the membrane corresponding to the partial pressure p_{iF} in Figure 9.1b. Unlike the case of distillation, y_i and x_i are not in equilibrium with each other.

For the separation of a binary mixture of species A and B, in the absence of external film mass-transfer resistances, the transport fluxes are given by equation (9-9) as

$$N_A = \frac{H_A D_A}{l_M}\left(p_{AF} - p_{AP}\right) = \frac{H_A D_A}{l_M}\left(x_A P_F - y_A P_P\right) \tag{9-11}$$

$$N_B = \frac{H_B D_B}{l_M}\left(p_{BF} - p_{BP}\right) = \frac{H_B D_B}{l_M}\left(x_B P_F - y_B P_P\right) \tag{9-12}$$

When no sweep gas is used, the ratio of N_A to N_B fixes the composition of the permeate so that

$$\frac{N_A}{N_B} = \frac{y_A}{y_B} = \frac{H_A D_A \left(x_A P_F - y_A P_P\right)}{H_B D_B \left(x_B P_F - y_B P_P\right)} \tag{9-13}$$

If the permeate pressure, P_P, is negligible compared to the feed pressure, P_F, equation (9-13) can be simplified, rearranged, and combined with equation (9-10) to give an *ideal separation factor*

$$\alpha_{id} = \frac{H_A D_A}{H_B D_B} = \frac{q_A}{q_B} \tag{9-14}$$

9.2 Mass Transfer in Membranes 483

When the permeate pressure is not negligible, equation (9-13) can be rearranged to obtain an expression for α_{AB} in terms of the pressure ratio, $r = P_P / P_F$, the ideal separation factor, and the mol fraction of A on the feed side of the membrane (Seader and Henley, 2006)

$$\alpha_{AB} = \alpha_{id} \left[\frac{x_A \left(\alpha_{AB} - 1 \right) - r\alpha_{AB}}{x_A \left(\alpha_{AB} - 1 \right) + 1 - r} \right] \qquad (9\text{-}15)$$

Equation (9-15) is an implicit equation for α_{AB} that is readily solved once the other parameters in the expression are known.

Another measure of membrane performance is the *cut*, θ, which refers to the ratio of molar flow rate of permeate to molar flow rate of feed. If all the components of the feed have finite permeabilities, the cut can vary from 0 to 1. For a cut of 1, all of the feed becomes permeate and no separation is accomplished.

Example 9.2 Oxygen-Enriched Air by Gas Permeation

Air can be separated into oxygen-enriched and nitrogen-enriched streams by gas permeation through dense membranes. A polyethylmethacrylate (PEMA) membrane in the form of a thin film (0.2 μm thick) is being considered for this purpose. The permeabilities of this membrane for oxygen (A) and nitrogen (B) at 298 K are 3.97×10^{-13} and 0.76×10^{-13} mol/(m·s·kPa), respectively (Gülmüs and Yilmaz, 2007). An air flow rate of 1.0 m^3/s (at STP) is sent to the membrane separator at 298 K and a pressure of 1.0 MPa. The pressure on the permeate side will be 0.1 MPa. Calculate the concentration of the permeate and retentate, and the membrane area required for the separation as functions of the cut, θ, for $0.1 \le \theta \le 0.9$. Assume perfect mixing on both sides of the membrane.

Solution
From the given permeabilities and membrane thickness, the ideal separation factor, $\alpha_{id} = 3.97/0.76 = 5.224$; $Q_A = 3.97 \times 10^{-13}/(0.2 \times 10^{-6}) = 1.985 \times 10^{-6}$ mol/$(m^2 \cdot s \cdot kPa)$. From the given air volumetric flow rate, the feed molar flow rate, $L_F = 1.0/0.0224 = 44.64$ mol/s. The pressure ratio, $r = P_P / P_F = 0.1$. Express the molar flow rates of permeate (V_P) and retentate (L_R) in terms of the cut: $V_P = \theta L_F$, $L_R = (1 - \theta)L_F$. A material balance on oxygen gives

$$x_{F,A} L_F = y_A \theta L_F + x_A \left(1 - \theta \right) L_F \text{ or simplifying,}$$

$$x_A = \frac{x_{F,A} - y_A \theta}{1 - \theta} \qquad (9\text{-}16)$$

From the definition of the separation factor,

$$\alpha_{AB} = \frac{y_A / x_A}{\left(1 - y_A \right) / \left(1 - x_A \right)} \qquad (9\text{-}17)$$

484   **Membranes and Other Solid Sorption Agents**

For a given value of θ, equations (9-17), (9-16), and (9-15) can be solved simultaneously by trial-and-error for the value of the separation factor and the compositions of retentate and permeate. Then, equation (9-11) can be rewritten as

$$N_A A_M = y_A \theta L_F = Q_A \left(x_A P_F - y_A P_P \right) A_M \tag{9-18}$$

Solving for the membrane area, A_M,

$$A_M = \frac{y_A \theta L_F}{Q_A \left(x_A P_F - y_A P_P \right)} \tag{9-19}$$

Figure 9.2 shows a Mathcad computer program to implement the computational scheme outlined for a set of values of θ. Appendix 9.1 presents the corresponding Python solution. Notice that the separation factor remains relatively constant, varying by about 6% with a value about 75% of the ideal value. The maximum oxygen content of the permeate (47.5%) occurs with the smallest cut ($\theta = 0.1$). The maximum nitrogen content of the retentate (93%) occurs at the largest cut ($\theta = 0.9$). The membrane area requirements are very large (for example, $A_M = 60{,}100$ m^2 for $\theta = 0.6$) even though the volumetric flow rate of air is relatively small.

9.2.3 Module Flow Patterns

In Example 9.2, perfect mixing was assumed on both sides of the membrane. Three other idealized flow patterns, common to other mass-transfer processes, have been studied: countercurrent flow, cocurrent flow, and cross flow. For a given cut, the flow pattern can significantly affect the degree of separation achieved and the membrane area required. For a given membrane module geometry, it is not always obvious which idealized flow pattern to assume. Hollow-fiber modules are the most versatile since they may be designed to approximate any of the three flow patterns mentioned above.

Consider separation of a binary mixture in a membrane module with the cross-flow pattern shown in Figure 9.3. The feed passes across the upstream membrane surface in plug flow with no longitudinal mixing. The pressure ratio and the ideal separation factor are assumed to remain constant. Film mass-transfer resistances external to the membrane are assumed to be negligible. A total mass balance around the differential-volume element gives

$$dV = dL \tag{9-20}$$

A material balance for A, combined with equation (9-20), gives

$$y\,dV = y\,dL = d\left(xL \right) = x\,dL + L\,dx \quad \text{simplifying:}$$

$$\frac{dL}{L} = \frac{dx}{y - x} \tag{9-21}$$

which is identical to the Rayleigh equation for differential distillation. Integrating equation (9-21) gives the retentate flow change with composition along the module

9.2 Mass Transfer in Membranes

$V := 1 \dfrac{m^3}{s}$ $P_F := 1000 \text{ kPa}$ $P_P := 100 \text{ kPa}$ $\text{ORIGIN} := 1$

$r := \dfrac{P_P}{P_F} = 0.1$ $T := 300 \text{ K}$ $x_{FA} := 0.21$ $x_{FB} := 0.79$ $j := 1..9$

$L_F := \dfrac{V}{22.4 \dfrac{L}{mol}} = 44.643 \dfrac{mol}{s}$ $Q_A := 1.985 \cdot 10^{-6} \cdot \dfrac{mol}{m^2 \cdot s \cdot kPa}$

$Q_B := 0.380 \cdot 10^{-6} \cdot \dfrac{mol}{m^2 \cdot s \cdot kPa}$ $\alpha_{id} := \dfrac{Q_A}{Q_B} = 5.224$ $\theta_j := 0.1 \cdot j$

Guess Values: $x_A := 0.2$ $y_A := 0.2$ $\alpha := \alpha_{id}$

Constraints:

$\alpha = \dfrac{y_A \cdot (1 - x_A)}{x_A \cdot (1 - y_A)}$ $x_A = \dfrac{x_{FA} - y_A \cdot \theta}{1 - \theta}$

$\alpha = \alpha_{id} \cdot \left(\dfrac{(\alpha - 1) \cdot x_A + 1 - r \cdot \alpha}{(\alpha - 1) \cdot x_A + 1 - r} \right)$

Solver: $f(\theta) := \text{Find}(\alpha, x_A, y_A)$

$\alpha_j := f(\theta_j)_1$ $x_{A_j} := f(\theta_j)_2$ $y_{A_j} := f(\theta_j)_3$ $A_M := \dfrac{\overrightarrow{y_A \cdot \theta \cdot L_F}}{(x_A \cdot P_F - y_A \cdot P_P) \cdot Q_A}$

$\text{Solution} := \text{augment}(\theta, \alpha, x_A, y_A) = \begin{bmatrix} 0.1 & 4.112 & 0.181 & 0.475 \\ 0.2 & 4.062 & 0.156 & 0.428 \\ 0.3 & 4.018 & 0.135 & 0.385 \\ 0.4 & 3.98 & 0.118 & 0.348 \\ 0.5 & 3.949 & 0.105 & 0.315 \\ 0.6 & 3.922 & 0.093 & 0.288 \\ 0.7 & 3.9 & 0.084 & 0.264 \\ 0.8 & 3.881 & 0.077 & 0.243 \\ 0.9 & 3.864 & 0.07 & 0.226 \end{bmatrix}$ $A_M = \begin{bmatrix} 8037.1 \\ 17074.1 \\ 26963.3 \\ 37530.9 \\ 48618.4 \\ 60098.7 \\ 71875.7 \\ 83878.8 \\ 96056.3 \end{bmatrix} m^2$

Figure 9.2 Mathcad solution of Example 9.2.

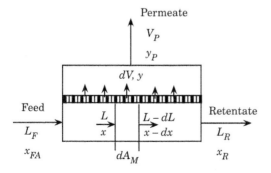

Figure 9.3 Cross-flow model for membrane module.

486   **Membranes and Other Solid Sorption Agents**

$$L = L_F \exp\left[\int_{x_{FA}}^{x} \frac{dx}{y-x}\right] \tag{9-22}$$

The definition of the separation factor, equation (9-9), can be used to eliminate y from equation (9-22). After some algebraic manipulation,

$$L = L_F \exp\left[\int_{x_{FA}}^{x} \frac{1+(\alpha_{AB}-1)x}{x(\alpha_{AB}-1)(1-x)} dx\right] \tag{9-23}$$

where

$$\alpha_{AB} = \alpha_{id} \left[\frac{x(\alpha_{AB}-1)-r\alpha_{AB}}{x(\alpha_{AB}-1)+1-r}\right] \tag{9-24}$$

Equation (9-23) can be integrated numerically using equation (9-24) to evaluate α_{AB} at each value of x for which the integrand is evaluated. When the upper limit of integration in equation (9-23) is $x = x_R$, $L = L_R = L_F(1-\theta)$. Then,

$$\theta = 1 - \exp\left[-\int_{x_R}^{x_{FA}} \frac{1+(\alpha_{AB}-1)x}{x(\alpha_{AB}-1)(1-x)} dx\right] \tag{9-25}$$

Equation (9-25) establishes the relation between the cut and the concentration of the retentate leaving the module. Furthermore, the relation between the local permeate and retentate concentrations can be easily derived (Seader and Henley, 2006):

$$\frac{y}{1-y} = \alpha_{id} \frac{x-ry}{(1-x)-r(1-y)} \tag{9-26}$$

The composited permeate concentration, y_P, can be evaluated from an overall material balance on component A:

$$y_P = \frac{x_{FA} - x_R(1-\theta)}{\theta} \tag{9-27}$$

The local flux of A across the membrane in the differential element is

$$y\frac{dV}{dA_M} = y\frac{dL}{dA_M} = Q_A(xP_F - yP_P) = Q_A P_F(x-ry) \tag{9-28}$$

from which the total membrane surface area can be obtained by integration:

$$A_M = \frac{1}{Q_A P_F} \int_{L_R}^{L_F} \frac{y\,dL}{x-ry} = \frac{1}{Q_A P_F} \int_{x_R}^{x_{FA}} \frac{yL\,dx}{(x-ry)(y-x)} \tag{9-29}$$

9.2 Mass Transfer in Membranes 487

Equation (9-21) was used in the integral of the right-hand side of equation (9-29) to change the integration variable from L to x.

The integrals in equations (9-23), (9-25), and (9-29) must be evaluated numerically. Given values of L_F, x_{FA}, r, Q_A, P_F, and α_{id}, apply the following computational scheme to calculate θ, A_M, and y_P for different values of x_R:

1. Choose a value of $x_R < x_{FA}$.
2. Divide the interval $(x_{FA} - x_R)$ into n equally spaced values of x.
3. For each value of x, calculate α_{AB} from equation (9-24), y from equation (9-26), and L from equation (9-22).
4. Use the values generated in (3) to evaluate θ from equation (9-25) and A_M from equation (9-29).
5. Calculate y_P from equation (9-27).

Figure 9.4 presents a Mathcad computer program to implement this scheme. Appendix 9.2 presents the corresponding Python version.

Example 9.3 Cross-flow Gas Permeation

Consider the air separation problem of Example 9.2. In this case, the separation will be performed in a hollow-fiber module using the same PEMA membrane and operating in a cross-flow mode. For the same operating pressures considered in Example 9.2, calculate the cut, composited permeate concentration, and membrane area required to produce a retentate oxygen concentration of 7%. Compare these results to those obtained for perfect mixing in Example 9.2.

Solution

Figure 9.4 shows that, for an oxygen retentate concentration of $x_R = 0.07$ in cross-flow mode, the cut must be $\theta = 0.442$, the permeate composition is $y_P = 0.387$, and the membrane surface area required is $A_M = 39,695$ m^2. The corresponding values for perfect mixing mode obtained in Example 9.2 are $\theta = 0.9$, $y_P = 0.226$, and $A_M = 96,056$ m^2. Comparing these results, we see that cross-flow operation is definitely more efficient than perfect mixing, as might be expected.

Also included in Figure 9.4 is the calculated degree of separation of the stage, α_S, defined on the basis of the concentrations in the permeate and retentate leaving the stage as

$$\alpha_S = \frac{y_P / x_R}{(1 - y_P)/(1 - x_R)} = 8.371 \tag{9-30}$$

Recall that the ideal separation factor for this example is $\alpha_{id} = 5.224$. For the perfect mixing case of Example 9.2, it was found that the stage separation factor was always smaller that the ideal separation factor. Such is not the case for cross flow, where α_S increases monotonically with θ reaching values that can be much higher than α_{id}.

Calculation of the degree of separation of a binary mixture in a membrane module for cocurrent or countercurrent flow patterns involve the numerical solution of a system of two nonlinear, coupled, ordinary differential equations (Walander

488   **Membranes and Other Solid Sorption Agents**

$$x_{FA} := 0.21 \qquad P_F := 1000 \text{ kPa} \qquad P_P := 100 \text{ kPa} \qquad n := 100 \qquad \textbf{\textit{ORIGIN}} := 1$$

$$r := \frac{P_P}{P_F} = 0.1 \qquad Q_A := 1.985 \cdot 10^{-6} \cdot \frac{\text{mol}}{\text{m}^2 \cdot \text{s} \cdot \text{kPa}} \qquad x_{FB} := 0.79 \qquad i := 1 .. n$$

$$L_F := 44.643 \, \frac{\text{mol}}{\text{s}} \quad Q_B := 0.380 \cdot 10^{-6} \cdot \frac{\text{mol}}{\text{m}^2 \cdot \text{s} \cdot \text{kPa}} \qquad \alpha_{id} := \frac{Q_A}{Q_B} = 5.224$$

$$x_R := 0.07 \qquad \Delta x := \frac{x_{FA} - x_R}{n} \qquad x_i := x_R + i \cdot \Delta x$$

Solver | Constraints | Guess Values

$$\alpha := \alpha_{id}$$

$$\alpha = \alpha_{id} \cdot \left(\frac{(\alpha - 1) \cdot x + 1 - r \cdot \alpha}{(\alpha - 1) \cdot x + 1 - r} \right)$$

$$a(x) := \text{Find}(\alpha)$$

Solver | Constraints | Guess Values

$$y := x_{FA}$$

$$\frac{y}{1-y} = \alpha_{id} \cdot \left[\frac{x - r \cdot y}{1 - x - r \cdot (1 - y)} \right]$$

$$b(x) := \text{Find}(y)$$

$$\alpha_i := a\left(x_i\right) \qquad\qquad\qquad\qquad\qquad y_i := b\left(x_i\right)$$

$$h(x) := \overrightarrow{\frac{1}{y - x}} \qquad g(x) := \overrightarrow{\frac{1 + (\alpha - 1) \cdot x}{(\alpha - 1) \cdot x \cdot (1 - x)}}$$

$$vs := \text{cspline}(x, g(x)) \qquad vs1 := \text{cspline}(x, h(x))$$

$$I1 := \int_{x_R}^{x_{FA}} \text{interp}(vs, x, g(x), \varepsilon) \, d\varepsilon = 0.584 \qquad I2 := \overrightarrow{\int_x^{x_{FA}} \text{interp}(vs1, x, h(x), \varepsilon) \, d\varepsilon}$$

$$\theta := 1 - \exp(-I1) = 0.442 \qquad\qquad L := \overrightarrow{L_F \cdot \exp(-I2)}$$

$$y_P := \frac{x_{FA} - x_R \cdot (1 - \theta)}{\theta} = 0.387 \qquad \phi(x) := \overrightarrow{\frac{y \cdot L}{(y - x) \cdot (x - y \cdot r)}} \qquad vs2 := \text{cspline}(x, \phi(x))$$

$$A_M := \frac{\displaystyle\int_{x_R}^{x_{FA}} \text{interp}(vs2, x, \phi(x), \varepsilon) \, d\varepsilon}{Q_A \cdot P_F} = \left(3.969 \cdot 10^4\right) \text{ m}^2 \qquad \alpha_S := \frac{y_P}{x_R} \cdot \frac{(1 - x_R)}{(1 - y_P)} = 8.37$$

Figure 9.4 Mathcad solution of Example 9.3.

and Stern, 1972). For a given cut, the best separation is achieved with countercurrent flow, followed by cross-flow, cocurrent flow, and perfect mixing, in that order. The cross-flow case is considered to be a good, conservative estimate of module membrane performance (Seader and Henley, 2006).

9.3 EQUILIBRIUM CONSIDERATIONS IN POROUS SORBENTS

Your objectives in studying this section are to be able to:

1. Define the concept of an adsorption isotherm and describe the characteristics of the four general types of isotherms.
2. Fit experimental data to the Langmuir and Freundlich adsorption isotherm models.
3. Explain the ion-exchange process, and how it differs from adsorption and chromatography; define the molar selectivity coefficient.
4. Generate ion-exchange distribution curves from published data of relative selectivity coefficients.

To be suitable for commercial applications, a sorbent should have a high selectivity to enable sharp separations, high capacity to minimize the amount of sorbent needed, and the capability of being regenerated for reuse. These properties depend upon the dynamic equilibrium distribution of the solute between the fluid and the solid surface. Unlike vapor–liquid and liquid–liquid equilibria where theory is often applied to estimate phase distribution of the solute, no acceptable theory has been developed to predict solid-sorbent equilibria. Thus, it is necessary to obtain experimental equilibrium data for a particular solute–sorbent combination. The equilibria behavior for adsorption and chromatography are very similar. Ion-exchange equilibria is significantly different and is considered separately.

9.3.1 Adsorption and Chromatography Equilibria

The experimental equilibrium data in this case are usually expressed in terms of (1) the solute loading on the adsorbent and (2) the concentration of the solute in the fluid phase. If the data are taken over a range of fluid concentrations at a constant temperature, the resulting plot of solid loading versus fluid concentration is called an *adsorption isotherm*. For gases, the concentration is usually given as a partial pressure, p; for liquids, it is often expressed in mass or molar units, such as g/L or mol/L. The solid loading is given as mass of solute adsorbed per unit mass of the original adsorbent.

Figure 9.5 shows schematically some typical adsorption isotherm shapes. The *linear isotherm* goes through the origin, and the amount adsorbed is proportional to the concentration in the fluid. Isotherms that are concave downward are called *favorable* because a relatively high solid loading can be obtained at low concentration in the fluid. The limiting case of a very favorable isotherm is *irreversible adsorption*, where the amount adsorbed is independent of the fluid concentration down to very low values. All systems show a decrease in the amount adsorbed with an increase in temperature, and adsorbate can be desorbed by raising the temperature even for the cases labeled "irreversible." However, desorption requires a much higher temperature when the adsorption is strongly favorable or irreversible. An

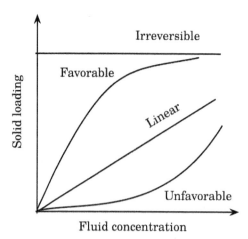

Figure 9.5 Adsorption isotherms.

isotherm that is concave upward is called *unfavorable* because relatively low solid loadings are obtained. Isotherms of this shape are rare, but they are worth studying to help understand the regeneration process.

For gases, most adsorption isotherms follow an almost linear behavior at low partial pressures of the solute and the following form of Henry's law is obeyed:

$$q = kp \tag{9-31}$$

where k is a temperature-dependent constant. Since adsorption is an exothermic process, somewhat analogous to condensation, as the temperature increases the value of k decreases because of Le Chatelier's principle.

At higher partial pressures, the behavior becomes nonlinear and more complex models are required to describe the observed equilibrium data. A frequently used model for monomolecular layer adsorption is the *Langmuir isotherm equation*. This equation is derived from simple mass-action kinetics. It assumes that the surface of the pores of the adsorbent is homogeneous and that the forces of interaction between the adsorbed molecules are negligible. Let f be the fraction of the surface covered by adsorbed molecules. Therefore, $(1-f)$ is the fraction of the bare surface. Then, the net rate of adsorption is the difference between the rate of adsorption on the bare surface and desorption from the covered surface:

$$\frac{dq}{dt} = k_a p(1-f) - k_d f \tag{9-32}$$

At equilibrium, $dq/dt = 0$ and equation (9-32) reduces to

$$f = \frac{Kp}{1+Kp} \tag{9-33}$$

where K is the adsorption-equilibrium constant, $K = k_a/k_d$. Also, $f = q/q_m$ where q_m is the maximum loading corresponding to complete coverage of the adsorbent surface by the solute. Then, equation (9-33) can be written as

9.3 Equilibrium Considerations in Porous Sorbents 491

$$q = \frac{Kq_m p}{1 + Kp} \qquad (9\text{-}34)$$

Equation (9-34) is the Langmuir isotherm equation. At low partial pressures, if $Kp \ll 1$, it reduces to the linear Henry's law form of equation (9-31), while at high partial pressures, if $Kp \gg 1$, $q = q_m$.

The parameters K and q_m are treated as empirical constants, obtained by fitting the nonlinear equation directly to experimental adsorption data or by first linearizing the Langmuir equation to the form

$$\frac{p}{q} = \frac{p}{q_m} + \frac{1}{q_m K} \qquad (9\text{-}35)$$

Equation (9-35) represents a straight line when p/q is plotted against p; the slope of the straight line is $1/q_m$ while the intercept is $1/q_m K$.

Another adsorption equilibrium model, attributed to Freundlich, is the empirical relation

$$q = kp^{1/n} \qquad (9\text{-}36)$$

where k and n are temperature-dependent constants. Generally, n lies in the range 1 to 5. Experimental q-p data can be fitted to equation (9-36) by first linearizing it to the form

$$\ln q = \frac{1}{n} \ln p + \ln k \qquad (9\text{-}37)$$

While the Langmuir equation has a theoretical foundation, the Freundlich equation is purely empirical.

Example 9.4 Freundlich and Langmuir Adsorption Isotherms

The following experimental data for equilibrium adsorption of pure methane on activated carbon (PCB from Calgon Corp.) at 296 K were obtained by Ritter and Yang (1987):

cm^3 (STP) CH_4/g carbon	45.5	91.5	113	121	125	126	126
pressure, MPa	0.276	1.138	2.413	3.758	5.240	6.274	6.688

Fit the data to the Langmuir and Freundlich isotherms and compare your results.

Solution

Figure 9.6a shows a Mathcad computer program to convert the raw data to the proper units of solid loading (mg CH_4/g of carbon), linearize the data for both the Langmuir and Freundlich isotherms, and obtain the best estimate of the parameters using the intrinsic Mathcad functions $slope(x, y)$ and $intercept(x, y)$. Appendix 9.3 presents the corresponding Python solution. For the Langmuir isotherm, the

492 Membranes and Other Solid Sorption Agents

$$M_A := 16 \ \frac{\text{gm}}{\text{mol}} \qquad\qquad VSTP := 22.4 \ \frac{\text{L}}{\text{mol}}$$

Experimental Data:

$$p_{exp} := \begin{bmatrix} 0.276 \\ 1.138 \\ 2.413 \\ 3.758 \\ 5.240 \\ 6.274 \\ 6.688 \end{bmatrix} \cdot \text{MPa} \qquad V := \begin{bmatrix} 45.5 \\ 91.5 \\ 113.0 \\ 121.0 \\ 125.0 \\ 126.0 \\ 126.0 \end{bmatrix} \cdot \frac{\text{cm}^3}{\text{gm}} \qquad q := \frac{\overrightarrow{V}}{VSTP} \cdot M_A$$

$$q = \begin{bmatrix} 32.5 \\ 65.4 \\ 80.7 \\ 86.4 \\ 89.3 \\ 90 \\ 90 \end{bmatrix} \frac{\text{mg}}{\text{gm}}$$

Linearize data for Langmuir isotherm:

$$y := \frac{\overrightarrow{p_{exp}}}{q} \qquad q_m := \frac{1}{\text{slope}\,(p_{exp}, y)} = 97.834 \ \frac{\text{mg}}{\text{gm}}$$

$$K := \frac{1}{q_m \cdot \text{intercept}\,(p_{exp}, y)} = 1.863 \ \frac{1}{\text{MPa}}$$

Linearize data for Freundlich isotherm:
$$y := \ln\overrightarrow{\left(\dfrac{q}{\frac{\text{mg}}{\text{gm}}}\right)} \qquad x := \ln\overrightarrow{\left(\dfrac{p_{exp}}{\text{MPa}}\right)}$$

$$n := \text{slope}\,(x, y)^{-1} = 3.224 \qquad k := \exp\,(\text{intercept}\,(x, y)) = 54.638$$

Figure 9.6a Mathcad solution of Example 9.4.

results are q_m = 97.83 mg CH$_4$/g of carbon, K = 1.863 (MPa)$^{-1}$. For the Freundlich isotherm, the results are n = 3.224, k = 54.64 (mg CH$_4$/g of carbon·MPa$^{-1/n}$).

Figure 9.6b shows a q-p plot of the experimental data and the corresponding predictions of the Langmuir and Freundlich isotherms. It is evident from the plot that, in this case, the Langmuir isotherm fits the data significantly better than the Freundlich isotherm.

Frequently, adsorption applications involve mixtures rather than pure gases. If the adsorption of two or more components of the mixture is significant, the situation can become quite complicated, depending upon the interactions of the adsorbed molecules. Assuming no interactions, the Langmuir isotherm can be extended to a multicomponent mixture of j components:

$$q_i = \frac{q_{i,m} K_i p_i}{1 + \sum_{i=1}^{j} K_i p_i} \tag{9-38}$$

As with multicomponent vapor–liquid and liquid–liquid equilibria, experimental data for multicomponent gas–solid adsorbent equilibria are scarce and less accurate than the corresponding data for pure gases.

9.3 Equilibrium Considerations in Porous Sorbents

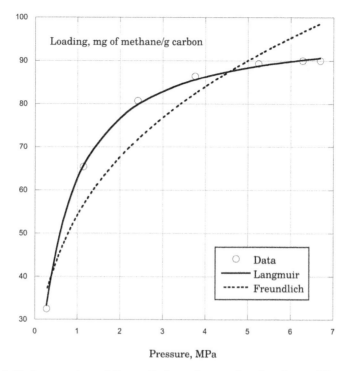

Figure 9.6b Langmuir and Freundlich isotherms fitted to data of Example 9.4.

Liquid adsorption equilibrium data are usually more complex to obtain and analyze than described above for gas adsorption. When porous adsorbent particles are immersed in a gas, the pores fill with the gas and the amount of adsorbed gas is determined by the decrease in pressure. With a liquid, the pressure does not change, and no simple experimental procedure has been devised for determining the extent of adsorption of a liquid. If the liquid is a homogeneous binary mixture, it is customary to designate one component the solute (A) and the other the solvent (B). It is assumed that only the solute adsorbs. If the liquid mixture is dilute in the solute, the consequences of this assumption are not serious and isotherms of shapes similar to those for gases are obtained. If, however, experimental data are obtained over the entire concentration range, the resulting isotherms can exhibit curious shapes that are unlike those obtained for gases (Kipling, 1965).

For dilute liquid binary solutions, it is common to fit the adsorption data with concentration forms of the Langmuir or Freundlich equations, such as

$$q = \frac{K q_m c}{1 + K c} \tag{9-39}$$

$$q = k c^{1/n} \tag{9-40}$$

For liquid mixtures that are dilute in two or more solutes, the multicomponent adsorption may be estimated from a concentration form of the extended Langmuir equation (9-38).

494   **Membranes and Other Solid Sorption Agents**

9.3.2 Ion-Exchange Equilibria

Ion-exchange capability exists in some natural materials such as clays and zeolites, but most industrial ion-exchange processes use synthetic resins (McCabe et al., 2005). These are prepared from organic polymers such as cross-linked polystyrene to which ionizable groups are added. Cation (positive charge) exchangers include strong-acid resins with sulfonic acid groups ($-SO_3^-$), weak-acid resins with carboxylic acid groups ($-CO_2^-$), and other types with intermediate acid strength. Anion (negative charge) exchangers can have strong-base quaternary ammonium groups [$-N^+(CH_3)_3$] or weak-base amine groups ($-N^+H_3$).

In both cation and anion exchangers, the acid or base groups are chemically bonded to the resin matrix, and the resins have a high concentration of fixed negative or positive ions. These are balanced by mobile counterions such as Na^+, H^+, or Ca^{2+} for cation resins and Cl^-, OH^-, or NO_3^- for anion resins, so that electrical neutrality is always maintained in the resin particles. Ion exchange takes place when the activity of ions in the external solution differs from that of the mobile ions in the solid phase. For example, exposing a hydrogen-form resin HR to a solution with Na^+ and H^+ will result in diffusion of some Na^+ ions into the resin and diffusion of some H^+ ions into the solution.

Ion-exchange resins are insoluble in water, but they swell in aqueous solution to an extent determined by the degree of cross-linking, the concentration of fixed charges, and the concentration of electrolytes in solution. Some swelling is desirable to increase diffusion rates inside the particles but swelling also decreases the capacity of the resin per unit volume of the bed. Resins are available as spherical beads with sizes between 0.3 and 1.2 mm and are generally used in packed beds similar to those used for adsorption from liquids (McCabe et al., 2005).

Ion exchange differs from adsorption and chromatography in that one sorbate (a counterion) is exchanged for a solute ion, and the exchange is governed by a reversible, stoichiometric, chemical-reaction equation. For ion exchange, the law of mass action is applied to obtain an equilibrium ratio rather than fit data to a sorption isotherm such as the Langmuir or Freundlich equation.

In describing ion-exchange equilibria, two cases are important. In the first case, the counterion initially in the resin is exchanged with a counterion from an acid or base. For example,

$$Na^+\left(aq\right)+OH^-\left(aq\right)+HR\left(s\right)\rightleftarrows NaR\left(s\right)+H_2O\left(l\right) \qquad (9\text{-}41)$$

Notice that the hydrogen ion leaving the resin immediately reacts with the hydroxyl ion in the aqueous solution to form water leaving no counterion on the right-hand side of the reaction. Accordingly, the ion exchange will continue until the aqueous solution is depleted of sodium ions or the resin is depleted of hydrogen ions.

In the second case, which is more common than the first, the counterion being transferred from the resin to the fluid phase remains as an ion. For example, the exchange of counterions A and B is expressed as

$$A^{n\pm}\left(l\right)+nBR\left(s\right)\rightleftarrows AR_n\left(s\right)+nB^{\pm}\left(l\right) \qquad (9\text{-}42)$$

9.3 Equilibrium Considerations in Porous Sorbents 495

where A and B must both be cations or anions. The equilibrium constant for this reaction is expressed in terms of molar concentrations (c_i for the liquid and q_i for the solid) and activity coefficients (γ_i) as

$$K_{eq} = \frac{q_{AR_n} c_{B^\pm}^n}{q_{BR}^n c_{A^{n\pm}}} \frac{\gamma_{AR_n} \gamma_{B^\pm}^n}{\gamma_{BR}^n \gamma_{A^{n\pm}}} \tag{9-43}$$

For dilute solutions, the activity coefficients do not change much with concentrations, and a simple concentration-based equilibrium constant called the *molar selectivity coefficient*, K_{AB}, can be defined as

$$K_{AB} = \frac{q_{AR_n} c_{B^\pm}^n}{q_{BR}^n c_{A^{n\pm}}} \tag{9-44}$$

When exchange is between two counterions of equal charge, equation (9-44) can be reduced to a simple equation in terms of the equilibrium concentrations of A in the liquid solution and in the ion-exchange resin. Because of equation (9-42), the total concentrations C and Q in equivalents of counterions in the liquid and resin, respectively, remain constant during the exchange process. Therefore,

$$c_i = \frac{Cx_i}{z_i} \qquad q_i = \frac{Qy_i}{z_i} \tag{9-45}$$

where z_i is the valence of counterion i, and x_i and y_i are equivalent fractions, rather than mole fractions, of A and B. Combining equations (9-44) and (9-45) for $n = 1$,

$$K_{AB} = \frac{y_A(1-x_A)}{x_A(1-y_A)} \tag{9-46}$$

Thus, at equilibrium x_A and y_A are independent of the total equivalent concentrations C and Q. Such is not the case when the two counterions have unequal charge. For the general case where the charge ratio is n,

$$K_{AB} = \left(\frac{C}{Q}\right)^{n-1} \frac{y_A(1-x_A)^n}{x_A(1-y_A)^n} \tag{9-47}$$

Then, K_{AB} depends in general on the ratio C/Q and on the charge ratio n.

When experimental data for K_{AB} for a particular binary system of counterions with a particular resin is not available, an approximate value of the selectivity coefficient can be estimated from

$$K_{ij} = \frac{K_i}{K_j} \tag{9-48}$$

where values for relative molar selectivities K_i and K_j are tabulated for the particular resin. Table 9.1 gives relative molar selectivities for some anions and

496   **Membranes and Other Solid Sorption Agents**

Table 9.1 Relative Molar Selectivities for Cation and Anion with 8% DVB[a] Cross-Linked Polystyrene Resins

Strong-acid resin		Strong-base resin	
Counterion	K	Counterion	K
Li^+	1.00	Iodide	8.70
H^+	1.27	Nitrate	3.80
Na^+	1.98	Bromide	2.80
NH^{4+}	2.55	Cyanide	1.60
K^+	2.90	Chloride	1.00
Mg^{2+}	3.29	Bicarbonate	0.30
Cu^{2+}	3.47	Acetate	0.20
Ni^{2+}	3.93	Sulfate	0.16
Ca^{2+}	5.16	Fluoride	0.09
Ba^{2+}	11.50	Hydroxide	0.05–0.07

[a]Divinylbenzene.
Source: Data from Perry and Green (1997).

cations with 8% cross-linked polystyrene resins. For values of K in this table, the units of C and Q are, respectively, eq/L of solution and eq/L of bulk bed volume of water-swelled resin.

A typical cation-exchange resin of the sulfonated styrene-divinylbenzene (DVB) type, such as Dowex 50, has an exchangeable ion capacity of 5.0 meq/g of dry resin. As shipped, the water-wet resin might contain 41.4 wt% water. Thus, the wet capacity is 2.9 meq/g of wet resin. If the bulk density of a drained bed of wet resin is 0.83 g/cm^3, the bed capacity is 2.4 eq/L of resin bed (Seader and Henley, 2006).

Example 9.5 Ion-Exchange Equilibrium

For the Cu^{2+}/Na^+ exchange with a strong-acid resin, show how the fraction CuR_2 in the resin (y_A) varies with the fraction Cu^{2+} in solution (x_A) for a total solution normality of 0.05. Assume a bed capacity of 2.4 eq/L of resin bed.

Solution
The equilibrium ion-exchange reaction is of the divalent-monovalent type

$$Cu^{2+}(l) + 2NaR(s) \rightleftarrows CuR_2(s) + 2Na^+(l)$$

From equation (9-48), using the data from Table 9.1,

$$K_{AB} = \frac{K_{Cu^{2+}}}{K_{Na^+}} = \frac{3.47}{1.98} = 1.75$$

9.4 Mass Transfer in Fixed Beds of Porous Sorbents

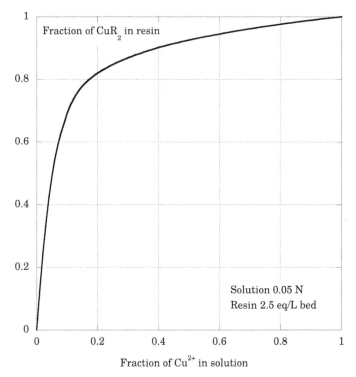

Figure 9.7 Predicted distribution curve for Cu^{2+}/Na^+ ion exchange in Example 9.5.

In this case, the total concentrations are C = 0.05 eq/L solution, Q = 2.4 eq/L resin bed; the charge ratio is n = 2. Substituting in equation (9-47) gives us

$$\frac{y_A(1-x_A)^2}{x_A(1-y_A)^2} = \frac{1.75 \times 2.4}{0.05} = 84$$

Substituting values of x_A in the range $0 \leq x_A \leq 1.0$, we generate the distribution curve shown in Figure 9.7. The curve is similar in shape to an adsorption isotherm of the very favorable type.

9.4 MASS TRANSFER IN FIXED BEDS OF POROUS SORBENTS

Your objectives in studying this section are to be able to:

1. Define the concepts of mass-transfer zone (MTZ) and breakthrough curve in fixed-bed sorption.
2. Generate breakthrough curves for fixed-bed adsorption with linear and Langmuir isotherms.

3. Explain the concept length of unused bed (LUB) and use it for fixed-bed scale-up.
4. Explain the concept of shock-wave front theory and use it to estimate breakthrough time in ion exchangers.
5. Generate chromatograms for linear sorption isotherms.
6. Explain the different modes and applications of electrophoresis.

In fixed-bed sorption, the concentration of the fluid phase and of the solid phase change with time as well as with position in the bed. At first, most of the mass transfer takes place near the inlet of the bed, where the fluid contacts the sorbent first. If the solid contains no solute at the start, the concentration of the fluid drops exponentially with distance essentially to zero before the end of the bed is reached. This concentration profile is shown by curve t_1 in Figure 9.8a, where c/c_F is the concentration in the fluid relative to the feed concentration for a total bed depth of Z. After some time, the solid near the inlet becomes nearly saturated, and most of the mass transfer takes place farther from the inlet. The concentration gradient becomes S-shaped as shown by curve t_2. The region where most of the change in concentration occurs is called the *mass-transfer zone*, MTZ, and its limits are arbitrarily chosen, often taken as the region in the bed where $0.05 \leq c/c_F \leq 0.95$.

Figure 9.8b is a typical plot of the outlet-to-inlet solute concentration in the fluid as a function of time from the start of flow, measured at the column outlet

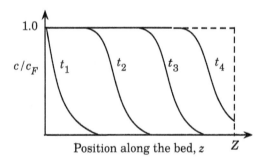

Figure 9.8a Concentration profiles for sorption in a fixed bed.

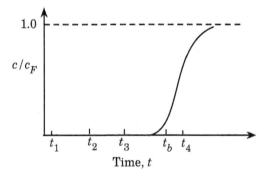

Figure 9.8b Breakthrough curve for sorption in a fixed bed (at $z = Z$).

9.4 Mass Transfer in Fixed Beds of Porous Sorbents

$z = Z$. The resulting S-shaped curve is called the *breakthrough curve*. At times t_1, t_2, and t_3, the exit concentration is practically zero. At time t_b when the concentration reaches some limiting permissible value, or *break point*, the flow is stopped or diverted to a fresh sorbent bed.

9.4.1 Basic Equations for Adsorption

Although adsorbers are generally designed from laboratory data, the approximate performance can sometimes be predicted from equilibrium data and mass-transfer considerations. A mass balance on the solute for the flow of fluid through a differential adsorption-bed length, dz, over a differential-time duration, dt, gives

$$\varepsilon \frac{\partial c}{\partial t} + v_0 \frac{\partial c}{\partial z} + (1-\varepsilon)\rho_p \frac{\partial q}{\partial t} = 0 \qquad (9\text{-}49)$$

where ε is the external bed porosity, v_0 is the superficial fluid velocity, and ρ_p is the density of the adsorbent particles. For adsorption from a gas or dilute solution, the first term in equation (9-49)—corresponding to accumulation in the fluid—is usually negligible compared to accumulation in the solid.

The mechanism of solute transfer to the porous solid includes diffusion through the fluid film around the particle, and diffusion inside the pores to internal adsorption sites. The actual process of adsorption is practically instantaneous and equilibrium is assumed to exist between the surface and the fluid at each point inside the particle. The transfer process is approximated using an overall volumetric mass-transfer coefficient ($K_c a$) and an overall driving force:

$$(1-\varepsilon)\rho_p \frac{\partial q}{\partial t} = K_c a(q^* - q) \qquad (9\text{-}50)$$

where q^* is the adsorbate loading in equilibrium with the solute concentration c in the bulk fluid. The mass-transfer area per unit bed volume, a, is taken as the external surface of the particles, which is $6(1 - \varepsilon)/d_p$ for spherical particles.

The overall coefficient K_c depends on the external film coefficient, $k_{c,\text{ext}}$, and on an effective internal coefficient, $k_{c,\text{int}}$. The external film coefficient can be estimated from the correlations for packed beds given in Section 2.6.5. The internal coefficient may be estimated by (Vermeulen et al., 1973)

$$k_{c,\text{int}} = \frac{10 \mathrm{D}_{eff}}{d_p} \qquad (9\text{-}51)$$

For gases, the effective diffusivity, D_{eff}, can be estimated from

$$\mathrm{D}_{eff} = \frac{\varepsilon_p}{\tau} \left[\frac{1}{(1/\mathrm{D}_{AB}) + (1/\mathrm{D}_K)} \right] \qquad (9\text{-}52)$$

where the Knudsen diffusivity, D_K, is from equation (1-103), and ε_p is the particle internal porosity, not to be confused with the bed porosity, ε. For liquids, the effective diffusivity is from equation (1-110), using ε_p instead of ε.

500 **Membranes and Other Solid Sorption Agents**

9.4.2 Linear Isotherm

For the case of a linear isotherm of the form $q = \kappa c$, solutions for equations (9-49) and (9-50) are readily available. The equations in this case are of the same form as those for passage of a temperature wave through a packed bed. The solution to the heat-transfer problem was first obtained by Anzelius (1926). Modified for adsorption in a fixed-bed with a linear isotherm, the solution is

$$X\left(t,z\right) = \frac{c\left(t,z\right)}{c_F} = J\left(N,NT\right) \tag{9-53}$$

$$Y\left(t,z\right) = \frac{q\left(t,z\right)}{q_F^*} = 1 - J\left(NT,N\right) \tag{9-54}$$

where

$$N = N\left(z\right) = \frac{K_c az}{v_0} \qquad K_c = \left[\frac{1}{k_c} + \frac{c_F}{q_F^* k_{c,\mathrm{int}} \rho_p \left(1-\varepsilon\right)}\right]^{-1} \tag{9-55}$$

$$T = T(t,z) = \frac{v_0 c_F \left(t - z\varepsilon/v_0\right)}{\rho_p \left(1-\varepsilon\right) z q_F^*} \quad \text{where } q_F^* = \kappa c_F \text{ (linear)} \tag{9-56}$$

$$J\left(\alpha,\beta\right) = 1 - e^{-\beta} \int_0^\alpha e^{-x} I_0 \left(2\sqrt{\beta x}\right) dx \tag{9-57}$$

and $I_0(x)$ is the *modified Bessel function of the first kind and zero order*. The integral function defined in equation (9-57) is easily evaluated using Mathcad or Python.

Example 9.6 Fixed-Bed Adsorption with Linear Isotherm

Air at 294 K and 1 atm enters a fixed-bed adsorber at a flow rate of 0.146 m³/s with a benzene vapor concentration of 29 g/m³. The cylindrical adsorber is 0.61 m inside diameter and is packed to a height of 1.83 m with 331 kg of silica gel particles having an effective diameter of 2.6 mm and an external porosity of 50%. The adsorption isotherm for benzene has been determined experimentally and found to be linear over the concentration range of interest, given by $q = \kappa c$, where q is in kg benzene/kg gel, c is in kg benzene/m³ of gas, and $\kappa = 4.127$ m³ of gas/kg of gel. It has been estimated that the overall volumetric mass-transfer coefficient for the conditions prevailing in the bed is $K_c a = 8.79$ s⁻¹. Assuming isothermal and isobaric operation, calculate

(a) The breakthrough curve.

(b) The breakthrough time, t_b, when the concentration in the exiting air is 5% of the inlet concentration.

9.4 Mass Transfer in Fixed Beds of Porous Sorbents

(c) The fraction of the bed adsorption capacity that has been used at the breakthrough time.

(d) The width of the MTZ (defined here for $0.05 \le c/c_F \le 0.95$).

Solution

(a) From the given volumetric flow rate of the feed and the bed dimensions, calculate the superficial fluid velocity, $v_0 = 0.5$ m/s. From the given mass of silica gel and the bed dimensions, calculate the bed bulk density, $\rho_b = \rho_p(1 - \varepsilon) = 619.2$ kg/m^3. The solid loading in equilibrium with the feed concentration, $q^*_F = \kappa c_F = 4.127 \times 0.029 = 0.12$ kg benzene/kg silica gel. The Mathcad program of Figure 9.9 defines the special function J (α,β) of equation (9-57) using the intrinsic I0(x) Mathcad function. (The corresponding solution using Python is presented in Appendix 9.4). The breakthrough curve is generated by plotting $X(t, Z)$ where $Z = 1.83$ m.

(b) The breakthrough time is determined using the "solve block" feature of Mathcad by specifying that $X(t_b, Z) = 0.05$. The answer is $t_b = 96.6$ min.

(c) The fraction of the bed adsorption capacity that has been used at the breakthrough time, Y_{av}, is calculated by

$$Y_{av} = \frac{\int\limits_0^Z Y(t_b, z) dz}{Z} \tag{9-58}$$

The answer is $Y_{av} = 61.4\%$. From this result, we conclude that the bed adsorption capacity is being utilized inefficiently. This is typical of a linear isotherm, an equilibrium condition that is not favorable for adsorption.

(d) To determine the width of the MTZ, we specify in a "solve block" that $X(t_b, Z - MTZ) = 0.95$. The answer is MTZ = 1.19 m. Therefore, in this case, the MTZ covers 1.19/1.83 = 65% of the bed, again the result of an unfavorable adsorption isotherm.

9.4.3 Langmuir Isotherm

For the case of a Langmuir isotherm, equation (9-34), solutions for equations (9-49) and (9-50) are also available. The solutions are (Thomas, 1944)

$$X = \frac{J(RN_R, N_R T)}{J(RN_R, N_R T) + \left[1 - J(N_R, RN_R T)\right] \exp\left[(R-1)N_R(T-1)\right]} \tag{9-59}$$

$$Y = \frac{1 - J(N_R T, RN_R)}{J(RN_R, N_R T) + \left[1 - J(N_R, RN_R T)\right] \exp\left[(R-1)N_R(T-1)\right]} \tag{9-60}$$

Membranes and Other Solid Sorption Agents

where

$$R = \frac{1}{1 + Kc_F} \tag{9-61}$$

$$N_R(z) = \frac{2}{R+1}\frac{K_c az}{v_0} \tag{9-62}$$

$$T = T(t,z) = \frac{v_0 c_F\left(t - z\varepsilon/v_0\right)}{\rho_p\left(1-\varepsilon\right)zq_F^*} \quad \text{where } q_F^* = \frac{Kq_m c_F}{1 + Kc_F} \tag{9-63}$$

$Q := 0.146\ \dfrac{\text{m}^3}{\text{s}} \qquad D := 0.61\ \text{m} \qquad Z := 1.829\ \text{m} \qquad A_c := \dfrac{\pi \cdot D^2}{4} = 0.292\ \text{m}^2$

$v_0 := \dfrac{Q}{A_c} = 0.5\ \dfrac{\text{m}}{\text{s}} \qquad c_F := 0.029\ \dfrac{\text{kg}}{\text{m}^3} \qquad m_s := 331\ \text{kg} \qquad \varepsilon := 0.5$

$V_b := A_c \cdot Z = 0.535\ \text{m}^3 \qquad \rho_b := \dfrac{m_s}{V_b} = 619.248\ \dfrac{\text{kg}}{\text{m}^3} \qquad \kappa := 4.134\ \dfrac{\text{m}^3}{\text{kg}}$

$Kca := 8.789\ \text{s}^{-1} \qquad qast := \kappa \cdot c_F = 0.12 \qquad J(\alpha,\beta) := 1 - e^{-\beta} \cdot \displaystyle\int_0^\alpha e^{-x} \cdot \text{I0}\left(2 \cdot \sqrt{\beta \cdot x}\right)\,dx$

$N(z) := \dfrac{Kca \cdot z}{v_0} \qquad T(z,t) := \dfrac{v_0 \cdot \left(t - \dfrac{\varepsilon \cdot z}{v_0}\right)}{\rho_b \cdot z \cdot \kappa}$

$Y(t,z) = \dfrac{q(t,z)}{q_F} \qquad X(t,z) = \dfrac{c(t,z)}{c_F} \qquad\qquad$ Dimensionless solutions

$X(t,z) := J(N(z), N(z) \cdot T(z,t)) \qquad Y(t,z) := 1 - J(N(z) \cdot T(z,t), N(z))$

a. Breakthrough curve

$t := 0\ \text{s}, 60\ \text{s} .. 15000\ \text{s}$

X (t, Z)

t (min)

Figure 9.9 Mathcad solution of Example 9.6.

9.4 Mass Transfer in Fixed Beds of Porous Sorbents

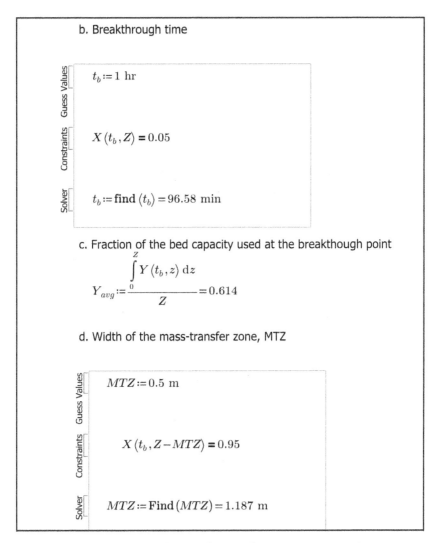

Figure 9.9 Mathcad solution of Example 9.6 (continuation).

Example 9.7 Fixed-Bed Adsorption with Langmuir Isotherm

Repeat the calculations of Example 9.6, but for a Langmuir isotherm with parameters q_m = 0.2 kg benzene/kg gel, and K = 51.72 m³ gas/kg gel. Notice that this isotherm gives the same value of q^*_F = 0.12 kg benzene/kg gel as the linear isotherm of Example 9.6. Assume that all of the other values given in the previous example apply.

Solution
(a) The Mathcad program of Figure 9.10 generates the breakthrough curve. It is much steeper than the corresponding curve for the linear isotherm of Example 9.6. (The corresponding solution using Python is presented in Appendix 9.5.)

(b) The breakthrough time is 139.7 min, compared to the previous value of 96.6 min.

(c) The fraction of the bed capacity used at breakthrough is 89.1% compared to 61.4% for the linear isotherm.

(d) The width of the MTZ is 0.39 m compared to 1.19 m for Example 9.6.

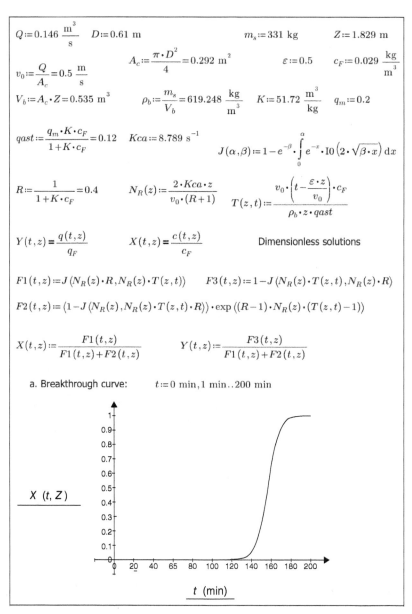

Figure 9.10 Mathcad solution of Example 9.7.

9.4 Mass Transfer in Fixed Beds of Porous Sorbents

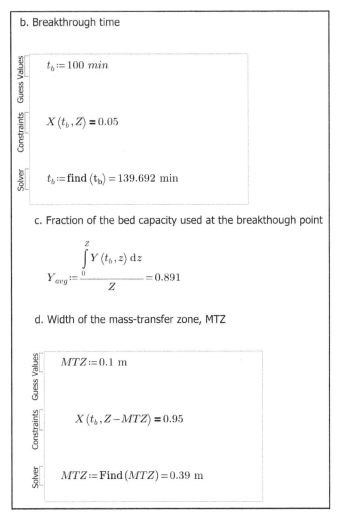

Figure 9.10 Mathcad solution of Example 9.7 (continuation).

9.4.4 Length of Unused Bed

For systems with a favorable isotherm, the concentration profile in the MTZ soon acquires a characteristic shape and width that do not change as the zone moves down the bed. Thus, tests with different bed lengths give breakthrough curves of the same shape, but with longer beds the MTZ is a smaller fraction of the bed length and a greater fraction of the bed is utilized at the breakthrough point. At the break point, the solid between the bed inlet and the start of the MTZ is completely saturated. The solid in the MTZ goes from nearly saturated to nearly free of adsorbate. For a rough average, this solid could be assumed to be about one-half saturated. This is equivalent to having one-half of the solid in the MTZ fully saturated and one-half unused. The scale-up principle is that the amount of unused

506   **Membranes and Other Solid Sorption Agents**

solid—or, equivalently, the *length of unused bed* (LUB)—does not change with the total bed length.

To calculate the LUB from the breakthrough curve, the fraction of the bed capacity used at the break point, Y_{av}, is calculated, as shown in Examples 9.6 and 9.7. Then, $(1 - Y_{av})$ is the fraction of unused bed, and LUB $= (1 - Y_{av}) Z$, which is assumed to be constant if the bed length changes under the same operating conditions. Thus, if the breakthrough time, t_{b1} is known for a bed of length Z_1 and known LUB, increasing the bed length to Z_2 would result in a new breakthrough time t_{b2} given by (as illustrated in Example 9.8)

$$t_{b2} = t_{b1} \frac{Z_2}{Z_1} \left[\frac{1 - \text{LUB} / Z_2}{1 - \text{LUB} / Z_1} \right] \tag{9-64}$$

Example 9.8 Fixed-Bed Scale-up Using LUB

Consider the adsorber of Example 9.7. If the bed length is increased to 3.0 m, estimate the new time of breakthrough assuming constant LUB.

Solution

From the results of Example 9.7, Y_{av} = 0.891. For a bed length of Z_1 = 1.829 m, LUB = $(1 - 0.891) \times 1.829$ = 0.2 m. For this bed length, t_{b1} = 139.7 min. If the bed length is increased to Z_2 = 3.0 m, the new time of breakthrough is from equation (9-64):

$$t_{b\,2} = 139.7 \times (3.0 / 1.829) \times \left[(1 - 0.2 / 3.0) / (1 - 0.2 / 1.829) \right] = 240 \text{ min.}$$

9.4.5 Mass-Transfer Rates in Ion Exchangers

The rates of ion-exchange reactions are limited by the rates of diffusion of the counterions in the external solution and in the pores of the resin particles. Equilibrium can be assumed to exist at any point inside each particle. With favorable equilibrium curves, such as shown in Figure 9.7, the MTZ soon acquires a constant length and shape. The steepness of this curve depends on the number of mass-transfer units and the nature of the controlling resistance.

The exact shape of the breakthrough curve and the value of t_b are hard to predict. Therefore, the design of large units is usually based on experience or tests with a small column. If the particle size and liquid velocity are kept constant, and care is taken to ensure uniform distribution of the feed, the LUB should not change in going to a larger column.

With a very favorable equilibrium curve, the LUB will be only a very small fraction of the normal bed length (McCabe et al., 2005). In that case, which is very common in ion-exchange applications, the process is well approximated using simple stoichiometric or *shock-wave front theory*. In this simplified analysis of fixed-bed sorption phenomena, it is assumed that equilibrium between the fluid and the sorbent is achieved instantaneously resulting in a shock-like wave, called a

9.4 Mass Transfer in Fixed Beds of Porous Sorbents

stoichiometric front, that moves as a sharp concentration front through the bed. In other words, the breakthrough curve becomes a straight vertical line, and the width of the MTZ is negligible. The location of the concentration wave front, z, as a function of time is obtained solely by material balances and equilibrium considerations. Assuming that all of the counterion A fed to the bed up to time t is transferred to the resin, and that its concentration in the solid has increased from zero to the equilibrium value q^*_F:

$$v_0 c_F t = z q^*_F \rho_p \left(1 - \varepsilon\right) \quad \text{for } z \leq Z \tag{9-65}$$

The ideal breakthrough time, t^*, is

$$t^* = \frac{Z q^*_F \rho_p \left(1 - \varepsilon\right)}{v_0 c_F} \tag{9-66}$$

The use of these equations is illustrated in Example 9.9.

Example 9.9 Ion-Exchanger Ideal Break Time

Water containing 7.0 meq/L of Ca^{2+} and 0.86 meq/L of Na^+ will flow at the rate of 13 L/s through a fixed bed of gel resin with a cation capacity of 2.9 eq/kg of wetted resin. The cylindrical bed is 2.6 m in diameter and 3.0 m long. The resin beads have a density of 1.34 kg/L and are packed with a bed porosity of 0.38. Estimate the ideal breakthrough time for the ion exchanger.

Solution

The superficial velocity, v_0 is calculated dividing the feed flow rate by the bed cross-sectional area, $v_0 = 0.013$ m/s. The feed concentration, c_F, refers to the concentration in the feed solution of the cations to be removed, in this case, Ca^{2+}. Therefore, $c_F = 7.0 \times 10^{-3}$ eq/L, and $q^*_F = 2.9$ eq/kg. Other properties of the resin are $\rho_p = 1.34$ kg/L and $\varepsilon = 0.38$. Substituting in equation (9-66), $t^* = 22$ h.

9.4.6 Mass-Transfer Rates in Chromatographic Separations

The separation of multicomponent mixtures into more than two products usually requires more than one separation device. Chromatography is one of the few separation techniques that can separate a multicomponent mixture into nearly pure components in a single device, generally a column packed with a suitable sorbent. The degree of separation depends upon the column length and the differences in component affinities for the sorbent used.

A pulse of the feed mixture, insufficient to saturate the sorbent, is introduced into the feed end of the chromatographic column. During the pulse, the components of the feed mixture are adsorbed by a relatively small fraction of the sorbent in the column. An *elutant*, such as a carrier gas or solvent that has little or no affinity for the sorbent, is now introduced continuously into the feed end of the column,

508   **Membranes and Other Solid Sorption Agents**

causing the components to desorb. The components with less affinity for the sorbent desorb more readily. However, as the desorbed components are carried down the bed by the elutant into cleaner regions of the bed, the components are successively readsorbed and then redesorbed to produce concentration waves. Because of the difference in affinities for the sorbent, the waves, which initially overlap considerably, gradually overlap less. If the column is long enough, the concentration waves become completely separated and the components elute from the column one at a time.

For each of the n solutes in the feed, neglecting axial dispersion and assuming a constant interstitial velocity, equations (9-49) and (9-50) apply. The boundary conditions, however, are different from those of adsorption in a fixed bed. In this case, periodic injections of rectangular feed pulses of duration t_F are followed by elution periods of duration t_E. Then, for the first feed pulse the boundary conditions are

Initial condition:

$$c_i(0,z) = 0 \qquad i = 1, 2, \cdots, n \tag{9-67}$$

Feed pulse:

$$c_i(t,0) = c_{iF} \qquad 0 \le t \le t_F \qquad i = 1, 2, \cdots, n \tag{9-68}$$

Elution period:

$$c_i(t,0) = 0 \qquad t_F \le t \le t_E \qquad i = 1, 2, \cdots, n \tag{9-69}$$

For the case of linear adsorption isotherm, the system of equations (9-49) and (9-50) with boundary conditions (9-67) to (9-69) have been solved assuming no interaction between the solutes (Vermeulen et al., 1973):

$$
\begin{aligned}
X(t,z) = \frac{c(t,z)}{c_F} &= 0 && \text{if } t < \left(\varepsilon z / v_0\right) \\
&= J(N, NT) && \text{if } \left(\varepsilon z / v_0\right) \le t \le \left(\varepsilon z / v_0\right) + t_F \quad (9\text{-}70) \\
&= J(N, NT) - J(N, NT') && \text{if } t > \left(\varepsilon z / v_0\right) + t_F
\end{aligned}
$$

where $T' = T(t - t_F, z)$.

Example 9.10 Chromatographic Separation of Sugars

An aqueous solution containing 2 g/cm^3 each of glucose and fructose is to be separated in a chromatographic column packed with a special resin. In the range of expected solute concentrations, the sorption isotherms are lineal and independent with $q_i = \kappa_i c_i$. The superficial velocity is 0.05 cm/s, the bed porosity is 39%, and the bulk density of the bed is 0.5 g/cm^3. The length of the bed is 2.0 m; the duration of the feed pulse is 500 s followed by elution with pure water. Generate the

9.4 Mass Transfer in Fixed Beds of Porous Sorbents 509

chromatogram (plot of solute exit concentrations in the elutant as a function of time) for these conditions. Does a significant overlap of the peaks result? The following data apply:

Property	Glucose	Fructose
κ, cm^3/g	0.6	1.4
$K_c a$, (1/s)[a]	0.0263	0.0665

[a]Resistance to diffusion in the pores dominates.

Solution

Figure 9.11 shows the Mathcad program that implements equation (9-70) for both glucose and fructose. It also shows the resulting chromatograph. (The corresponding solution using Python is presented in Appendix 9.6.) The glucose peak starts to elute

$$v_0 := 0.05 \; \frac{\text{cm}}{\text{s}} \qquad Z := 2.0 \text{ m} \qquad \varepsilon := 0.39 \qquad \rho_b := 500 \; \frac{\text{kg}}{\text{m}^3} \qquad t_F := 500 \text{ s}$$

$$J(\alpha,\beta) := 1 - e^{-\beta} \cdot \int_0^\alpha e^{-x} \cdot \text{I0}\left(2 \cdot \sqrt{\beta \cdot x}\right) dx \qquad t_1 := \frac{\varepsilon \cdot Z}{v_0} = \left(1.56 \cdot 10^3\right) \text{ s}$$

Glucose: $\quad Kca_g := 0.0263 \text{ s}^{-1} \qquad \kappa_g := 0.0006 \; \frac{\text{m}^3}{\text{kg}} \qquad TG(z,t) := \dfrac{v_0 \cdot \left(t - \dfrac{\varepsilon \cdot z}{v_0}\right)}{\rho_b \cdot z \cdot \kappa_g}$

$$NG(z) := \frac{Kca_g \cdot z}{v_0}$$

$$XG(t,t_1,t_F,Z) := \left\| \begin{array}{l} \text{if } t \leq t_1 \\ \quad \left\| \; 0 \right. \\ \text{else if } t_1 < t \leq t_1 + t_F \\ \quad \left\| \; J(NG(Z),NG(Z) \cdot TG(Z,t)) \right. \\ \text{else} \\ \quad \left\| \; J(NG(Z),NG(Z) \cdot TG(Z,t)) - J(NG(Z),NG(Z) \cdot TG(Z,t-t_F)) \right. \end{array} \right.$$

Fructose: $\quad Kca_f := 0.0665 \text{ s}^{-1} \qquad \kappa_f := 0.0014 \; \frac{\text{m}^3}{\text{kg}} \qquad TF(z,t) := \dfrac{v_0 \cdot \left(t - \dfrac{\varepsilon \cdot z}{v_0}\right)}{\rho_b \cdot z \cdot \kappa_f}$

$$NF(z) := \frac{Kca_f \cdot z}{v_0}$$

$$XF(t,t_1,t_F,Z) := \left\| \begin{array}{l} \text{if } t \leq t_1 \\ \quad \left\| \; 0 \right. \\ \text{else if } t_1 < t \leq t_1 + t_F \\ \quad \left\| \; J(NF(Z),NF(Z) \cdot TF(Z,t)) \right. \\ \text{else} \\ \quad \left\| \; J(NF(Z),NF(Z) \cdot TF(Z,t)) - J(NF(Z),NF(Z) \cdot TF(Z,t-t_F)) \right. \end{array} \right.$$

Figure 9.11 Mathcad solution of Example 9.10.

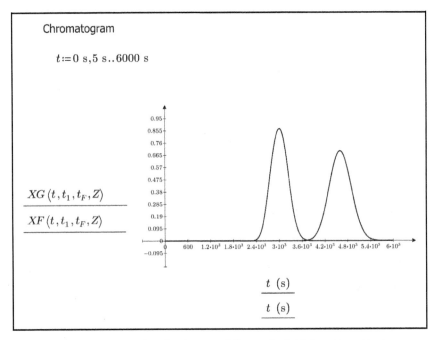

Figure 9.11 Mathcad solution of Example 9.10 (continuation).

at 2400 s, reaches its maximum value at 3000 s, and vanishes at 3600 s. The fructose peak, on the other hand, starts to elute at 3800 s, reaches its maximum value at 4620 s, and vanishes at 5600 s. Therefore, there is no significant overlap of the peaks and the separation process will produce virtually pure glucose and fructose.

9.4.7 Electrophoresis

Electrophoresis separates charged molecules, such as nucleic acids and proteins, according to their size, shape, and charge in an electric field. Electrophoretic methods are similar to chromatographic methods in that fixed barrier phases (ranging from paper to polymer gels) are employed to facilitate separation. Unlike in other bioseparation processes, however, the electrophoretic driving force for creating motion—the electric field—can create unwanted heating due to the Joule effect. Many electrophoresis devices are designed to minimize this type of heating (García et al., 1999).

The application of an external electric field is often used in the analytical separation of charged molecules. Scale-up and commercial applications of this technology, however, are hampered by equipment design limitations for processing large solution volumes. Despite these potential problems, new devices and applications using electrophoretic methods for bioseparations continue to be developed. Traditionally, electrophoresis is conducted in several operating modes to be described next.

Gel electrophoresis is known as a zonal electrophoresis method because the apparatus separates the sample into well-defined zones. The applied sample

9.4 Mass Transfer in Fixed Beds of Porous Sorbents 511

migrates trough a polymer matrix usually composed of polyacrylamide, starch, agar, or agarose. The gel is sandwiched between two slabs. Samples are applied in distinct locations known as lanes along which the components migrate under the influence of the applied electric field. The resulting systems can be automated to perform quantitative analysis and to handle large-volume samples. The most popular laboratory-scale gel electrophoresis method is polyacrylamide gel electrophoresis (PAGE).

Pulsed-field gel electrophoresis (PFGE) is useful in separations involving high-molecular weight DNA. In this application, the choice of gel is agarose. For low-molecular weight DNA, the mobility in a constant field is simply related to size because of the rod-like structure of these molecules. As the size of the DNA increases, a limit is reached where all DNA molecules are coiled and project nearly the same surface area exhibiting approximately the same mobility. To overcome this limitation, in PFGE the electric field is applied for duration t_{pulse} in a direction relative to a primary x-axis given by an angle between 90° and 270°. Changing the angle intermittently causes long, linear DNA strands to kink and become entangled in the agarose. Longer strands move in slow, zigzag motion relative to shorter, rod-like strands that quickly reorient to the new direction and move more readily through the agarose maze.

Capillary electrophoresis (CE) does not require the use of a supporting medium and can take place in free aqueous solution. The most commonly used capillary material is fused silica. Capillary tubes must be small, ranging in size from 25 μm to 75 μm, as larger capillaries show poor performance due to convection from Joule heating. Corporate and academic laboratories face economic and social pressures to minimize solvent wastes. CE, which uses aqueous salts as the buffers, is therefore a very popular and powerful tool for separating charged biomolecules.

Isoelectric focusing (IEF) capitalizes on the maintenance of a pH gradient between two electrodes. Proteins applied at the anode will migrate toward the cathode since they have a net positive charge at low pH. They travel through the electrophoresis media until they reach a pH equal to their isoelectric point (pI). At that point, the protein has no net charge and hence no net force acts on it because of the imposed electric field. Therefore, the protein stops migrating. IEF offers high resolution, being able to separate proteins with a pI difference of as little as 0.02 and can concentrate the sample by restricting the protein to a small volume.

In *isotachophoresis* (ITP) bands in a protein sample are sandwiched between a leading electrolyte with high mobility and a trailing electrolyte with lower mobility in a two-buffer system. Components are separated on the basis of mobility per unit electric field. Two ions possessing like charges, but different mobilities, segregate until all faster ions lead the slower ones. Counterions traveling in the opposite direction must be balanced at each location which prevents formation of a void between adjacent ions, maintains identical velocities of each component, and forms a sharp boundary between adjacent ions. Field strength self-adjusts becoming lowest for the high-mobility leading band, highest for the trailing band, and constant across each band. The development of sharp zones makes this method very desirable for large-scale continuous electrophoretic separation of proteins.

9.5 APPLICATIONS OF MEMBRANE-SEPARATION PROCESSES

Your objectives in studying this section are to be able to:

1. Describe the different types of membrane materials and shapes.
2. Describe four types of membrane modules.
3. Define and enumerate applications of dialysis, reverse osmosis, gas permeation, ultrafiltration, and microfiltration.
4. Explain and quantify the effect of concentration polarization in reverse osmosis and ultrafiltration.
5. Calculate mass-transfer rates for dialysis, reverse osmosis, gas permeation, and ultrafiltration.

Almost all membrane materials are made from natural or synthetic polymers (macromolecules). Natural polymers include wool, rubber, and cellulose. A wide variety of synthetic polymers has been developed and commercialized since 1930. In order to achieve high permeabilities and high separation factors for small molecules, asymmetric membranes have been developed in recent years. They consist of a thin dense skin about 0.1 to 1.0 μm thick formed over a much thicker microporous layer that provides support for the skin.

The application of polymer membranes is generally limited to temperatures below 475 K and to the separation of mixtures that are chemically inert. Otherwise, membranes made of inorganic materials can be used. These include mainly microporous ceramics, metals, and carbon, and dense metals, such as palladium, that allow the selective diffusion of very small molecules such as hydrogen and helium.

Membrane materials are available in various shapes, such as flat sheets, tubular, hollow fiber, and monolithic. Flat sheets have typical dimensions of 1 m by 1 m by 200 μm thickness. Tubular membranes are typically 0.5 to 5.0 cm in diameter and up to 6 m in length. The thin, dense layer is on either the inside or the outside of the tube. Very small-diameter hollow fibers are typically 42 μm i.d. by 85 μm o.d. by 1.2 m long. They provide a very large surface area per unit volume. Honeycomb, monolithic elements of inorganic oxide membranes are available in hexagonal or circular cross section. The circular flow channels are typically 0.3 to 0.6 cm in diameter (Seader and Henley, 2006).

The membrane shapes described are usually incorporated into compact commercial modules and cartridges. The four more common types of modules are (1) plate-and-frame, (2) spiral-wound, (3) tubular, and (4) hollow-fiber. Table 9.2 is a comparison of the characteristics of these four types of modules. The packing density refers to the surface area per unit volume of module, for which the hollow-fiber modules are clearly superior. However, hollow-fiber modules are highly susceptible to fouling and very difficult to clean. The spiral-wound module is very popular for most applications because of its low cost and reasonable resistance to fouling.

9.5 Applications of Membrane-Separation Processes 513

Table 9.2 Characteristics of Membrane Modules

Characteristic	Plate-frame	Spiral-wound	Tubular	Hollow-fiber
Packing density, m^{-1}	30 to 500	200 to 800	30 to 200	500 to 9000
Resistance to fouling	Good	Moderate	Very good	Poor
Ease of cleaning	Good	Fair	Excellent	Poor
Relative cost	High	Low	High	Low
Main applications[a]	D, RO, UF, MF, PV	D, RO, GP, UF, MF	RO, UF	D, RO, UF, GP

[a]D, dialysis; RO, reverse osmosis; GP, gas permeation; PV, pervaporation; UF, ultrafiltration; MF, microfiltration.
Source: Data from Seader et al. (2011).

9.5.1 Dialysis

In a dialysis membrane-separation process, the feed is a liquid containing solvent, solutes of type A, solutes of type B, and/or insoluble but dispersed colloidal matter. A sweep liquid of the same solvent is fed to the other side of the membrane. The membrane is thin with micropores of a size such that solutes of type A can pass through by a concentration driving force. Solutes of type B are larger in molecular size and pass through the membrane with difficulty or not at all. Colloids do not pass through the membrane. By elevating the pressure on the feed side slightly above that on the permeate side, the transport of the solvent in the opposite direction driven by a concentration gradient (*osmosis*) can be reduced or eliminated. In dialysis, the permeate is called *diffusate*, while the retentate is called *dialysate*.

Dialysis is attractive when the concentration differences for the main diffusing solutes are large and the permeability differences between those solutes and the other solutes and/or colloids is large. Although commercial applications of dialysis do not rival reverse osmosis and gas permeation, it has been applied to a number of important separation processes, such as purification of pharmaceuticals, production of a reduced-alcohol beer, and hemodialysis (described in detail in Example 1.3).

Typical microporous-membrane materials used in dialysis are hydrophilic, including cellulose, cellulose acetate, and various acid-resistant polyvinyl copolymers, typically less than 50 μm thick and with pore diameters of 15 to 100 Å. Dialysis membranes can be thin because pressures on either side of the membrane are essentially equal. The most common membrane modules are plate-and-frame and hollow-fiber. Compact hollow-fiber hemodialyzers, which are widely used, typically contain several thousand 200-μm-diameter fibers with a wall thickness of 20 to 30 μm and a length of 10 to 30 cm (Seader et al., 2011).

In a plate-and-frame dialyzer, the flow pattern is nearly countercurrent. Because total flow rates change little and solute concentrations are typically small,

514   **Membranes and Other Solid Sorption Agents**

it is common to estimate the solute transport rate by assuming a constant overall mass-transfer coefficient and using a log-mean concentration driving force, $(\Delta c_i)_M$. Thus,

$$n_i = K_i \left(\Delta c_i \right)_M \tag{9-71}$$

Although the transport of solvents, such as water, which usually occurs in a direction opposite the solute, could be formulated in similar terms, it is more common to report the so-called *water-transport number*, which is the ratio of the water flux to the solute flux. A negative value of the water-transport number indicates transport of solvent in the same direction as the solute. Ideally, the absolute value of the water transport number should be less than 1.0.

Example 9.11 Dialysis for Sulfuric Acid Purification

A countercurrent plate-and-frame dialyzer is to be sized to process 1.0 m³/h of an aqueous solution containing 25 wt% H_2SO_4 and smaller amounts of copper and nickel sulfates. A wash water rate of 1000 kg/h is to be used, and it is desired to recover 60% of the acid at 298 K. From laboratory experiments with an acid-resistant vinyl membrane, a permeance of 0.03 cm/min for the acid and a water-transport number of +0.8 were reported. Transmembrane transport of copper and nickel sulfates is negligible. For these operating conditions, it has been estimated that the combined external mass-transfer coefficients will be 0.02 cm/min. Estimate the membrane area required.

Solution

A schematic diagram of the process is shown in Figure 9.12. Estimate the overall mass-transfer coefficient for the sulfuric acid: $K = (1/0.03 + 1/0.02)^{-1} = 0.012$ cm/min. The density of a 25 wt% aqueous sulfuric acid solution at 298 K is 1175 kg/m³ (Perry and Green, 1997). Therefore, the flow of sulfuric acid in the feed = 1.0 × 1175 × 0.25 = 294 kg/h. For 60% recovery, the transmembrane flow of the acid = 294 × 0.6 = 176 kg/h. From the given water-transport number, the transmembrane counterflow of water = 176 × 0.8 = 141 kg/h. The next step is to calculate all of the inlet and outlet concentrations from material balances. The flow of acid in the dialysate = 294 × 0.4 = 118 kg/h. The total dialysate flow = 1175 − 176 + 141 = 1140 kg/h. The mass fraction of acid in the dialysate = 118/1140 = 0.103. The density of a 10.3 wt% aqueous solution of sulfuric acid at 298 K is 1064 kg/m³. Then, c_R = 0.103 × 1064 = 110 kg/m³. The flow of acid in the diffusate = 294 × 0.6 = 176 kg/h. The total diffusate flow = 1000 − 141 + 176 = 1035 kg/h. The mass fraction of acid in the diffusate = 176/1035 = 0.17. The density of a 17 wt% aqueous solution of sulfuric acid at 298 K is 1114 kg/m³. Then, c_P = 0.17 × 1114 = 189 kg/m³. At the feed end of the dialyzer, Δc = 294 − 189 = 105 kg/m³. At the dialysate end, Δc = 110 − 0 = 110 kg/m³. The log-mean driving force = 107.5 kg/m³. Then,

$$A_M = \frac{176 \times 100}{0.012 \times 107.5 \times 60} = 227 \text{ m}^2$$

9.5 Applications of Membrane-Separation Processes

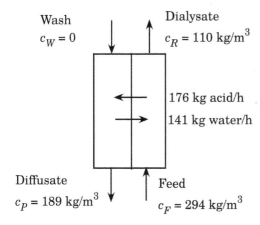

Figure 9.12 Schematic diagram of dialysis process of Example 9.11.

9.5.2 Reverse Osmosis

When miscible solutions of different concentrations are separated by a membrane that is permeable to the solvent but nearly impermeable to the solute, diffusion of solvent occurs from the less concentrated solution to the more concentrated solution, where the solvent activity is lower. The diffusion of the solvent is called osmosis, and osmotic transfer of water occurs in many plant and animal cells. The transfer of solvent can be stopped by increasing the pressure on the concentrated solution until the activity of the solvent is the same on both sides of the membrane. If pure solvent is on one side of the membrane, the pressure required to equalize the solvent activities is the osmotic pressure of the solution π. If a pressure higher than the osmotic pressure is applied, solvent will diffuse from the concentrated solution through the membrane into the dilute solution. This phenomenon is called *reverse osmosis* because the solvent flow is opposite to the normal osmotic flow.

In a reverse osmosis (RO) membrane-separation process, the feed is a liquid at high pressure P_1. No sweep liquid is used, but the other side of the membrane is kept at a much lower pressure, P_2. A dense membrane such as an acetate or aromatic polyamide is used that is permselective for the solvent. The membrane must be thick to withstand the large pressure differential; therefore, asymmetric membranes are commonly used. The products of RO are a permeate of almost pure solvent and a solvent-depleted retentate. Only a fraction of the solvent in the feed is transferred to the permeate.

The most important application of RO is the desalinization of seawater. Since 1990, RO has become the dominant process for seawater desalinization. Seawater contains about 3.5 wt% dissolved salts and has an osmotic pressure of 24.1 bar. The preferred RO membrane for desalinization is a spiral-wound module of polyamide membrane operating at a feed pressure of 55 to 70 bar. With a transmembrane water flux of 365 kg/m^2-day, this module can recover 45% of the water at a purity of 99.95 wt%. A typical cylindrical module is 20 cm in diameter by 1.0 m long, containing 34 m^2 of surface area. Such modules resist fouling by colloidal and particulate matter, but the seawater must be treated with sodium bisulfate to remove oxygen and chlorine.

516   **Membranes and Other Solid Sorption Agents**

Although the driving force for transport of water through the dense membrane is the concentration or activity difference across it, common practice is to use a driving force based on osmotic pressure. At thermodynamic equilibrium, the solvent fugacity on the feed side of the membrane (1) must be equal to the solvent fugacity on the permeate side (2). Thus,

$$f_A^{(1)} = f_A^{(2)} \tag{9-72}$$

If the permeate is pure water, equation (9-72) can be written as

$$x_A^{(1)} \gamma_A^{(1)} f_A^0 \left\{T, P_1\right\} = f_A^0 \left\{T, P_2\right\} \tag{9-73}$$

where f_A^0 is the standard-state, pure component fugacity. The effect of pressure on f_A^0 is given by the Poynting correction factor, which for an incompressible liquid of specific volume v_{AL} is given by (Smith et al., 1996)

$$f_A^0 \left\{T, P_2\right\} = f_A^0 \left\{T, P_1\right\} \exp \left[\frac{v_{AL}\left(P_2 - P_1\right)}{RT}\right] \tag{9-74}$$

Substitution of equation (9-74) into (9-73) gives

$$\pi = P_1 - P_2 = -\frac{RT}{v_{AL}} \ln\left[x_A^{(1)} \gamma_A^{(1)}\right] \tag{9-75}$$

For a mixture on the feed or retentate side, which is dilute in the solid, $\gamma_A^{(1)} \approx 1.0$. Also,

$$x_A^{(1)} = 1 - x_B^{(1)} \qquad \ln\left[1 - x_B^{(1)}\right] \approx -x_B^{(1)}$$

Substituting in (9-75) gives the celebrated van't Hoff equation (see Problem 9.1):

$$\pi \approx \frac{RTx_B^{(1)}}{v_{AL}} \approx \frac{RTc_B}{M_B} \tag{9-76}$$

In the general case, when there is solute on both sides of the membrane, the driving force for solvent transport is $\Delta P - \Delta\pi$, and the solvent flux is given by

$$n_{H_2O} = \frac{q_{H_2O}}{l_M}\left(\Delta P - \Delta\pi\right) \tag{9-77}$$

The flux of solute is given by equation (9-3) in terms of membrane concentrations, and thus is independent of the pressure differential across the membrane. Accordingly, the higher the ΔP, the purer the permeate water. Alternatively, the flux of solute can be expressed in terms of *salt rejection, SR*, defined as

$$SR = 1 - \frac{\left(c_{salt}\right)_{permeate}}{\left(c_{salt}\right)_{feed}} \tag{9-78}$$

9.5 Applications of Membrane-Separation Processes 517

Values of SR are typically in the range of 97 to 99.5%.

A phenomenon that can reduce the water flux through the membrane in RO is *concentration polarization*. The flux of water to the membrane carries with it salt by bulk flow. However, because the salt cannot readily penetrate the membrane, its concentration in the liquid adjacent to the membrane surface, c_{si}, is higher than in the bulk of the feed, c_{sF}. This difference causes diffusion of salt from the membrane surface back to the bulk feed until a steady-state salt-concentration profile develops. The value of c_{si} is very important because it fixes the osmotic pressure, therefore influencing the driving force for water transport according to equation (9-77).

Consider steady-state transport of water with back-diffusion of salt. A salt balance at the upstream membrane surface gives

$$n_{H_2O} c_{sF} \times SR = k_s \left(c_{si} - c_{sF} \right) \tag{9-79}$$

where k_s is the salt mass-transfer coefficient. Solving for c_{si} gives

$$c_{si} = c_{sF} \left[1 + \frac{n_{H_2O} \times SR}{k_s} \right] \tag{9-80}$$

A quantitative estimate of the importance of concentration polarization can be derived by defining the *concentration polarization factor*, Γ, as

$$\Gamma = \frac{c_{si} - c_{sF}}{c_{sF}} = \frac{n_{H_2O} \times SR}{k_s} \tag{9-81}$$

If the polarization factor is less than 0.1, its effect can be neglected (McCabe et al., 2005). If Γ is large, the change in salt rejection and water flux can be estimated by solving equations (9-76), (9-77), and (9-79) simultaneously. However, a large value of Γ suggests that design changes to reduce the effect of concentration polarization should be considered, as illustrated by Example 9.12.

Example 9.12 Water Desalinization by Reverse Osmosis

A hollow-fiber membrane module similar to the one described in Example 2.14 is used for water desalinization. The feed water flows on the shell side at a superficial velocity of 5 cm/s, 298 K, 70 bar, and 2 wt% NaCl. The permeate flows in the fibers lumen at a pressure of 3 bar and a salt content of 0.05 wt%. For this particular membrane, a water permeance of 1.1×10^{-5} g/cm^2·s·bar, and a salt rejection of 97% have been measured.

(a) Calculate the transmembrane flux of water, in m^3/m^2·day.

(b) Estimate the concentration polarization factor.

Solution

(a) Calculate the driving force for solvent flux through the membrane. The applied pressure differential is $\Delta P = 70 - 3 = 67$ bar. Calculate the osmotic pressure on both

518   **Membranes and Other Solid Sorption Agents**

sides of the membrane using equation (9-76). Assume that the density of both the feed and the permeate is 1 g/cm^3. Notice that dissolved NaCl gives 2 ions per molecule. The bulk-feed salt concentration is $c_{sF} = 0.02 \times 2 \times 1000/58.5 = 0.684$ kmol/m^3. Then, $\pi_F = 8.314 \times 298 \times 0.684/100 = 16.94$ bar. A similar calculation for the permeate gives $\pi_P = 0.42$ bar. Then, $\Delta\pi = 16.94 - 0.42 = 16.52$ bar and $\Delta P - \Delta\pi = 50.48$ bar. The transmembrane flux of water is given by $n_{H2O} = 1.1 \times 10^{-5} \times 50.48 = 5.55 \times 10^{-4}$ $g/cm^2{\cdot}s = 480$ kg/$m^2{\cdot}$day $= 0.48$ $m^3/m^2{\cdot}$day.

(b) The concentration polarization factor is from equation (9-81). To estimate the salt mass-transfer coefficient, k_s, use equation (2-97). The properties of water are $\rho = 1.0$ g/cm^3, $\mu = 0.9$ cP. The diffusivity of NaCl in water at 298 K is 1.6×10^{-5} cm^2/s (McCabe et al., equation (2005). The outside diameter of the fibers is 290 μm, and the packing factor $\phi = 0.4$. For a superficial velocity of 5.0 cm/s, Re = 16.1, Sc = 563, and Sh = 8.63. Then, $k_s = 8.63 \times 1.6 \times 10^{-5}/0.029 = 4.76 \times 10^{-3}$ cm/s. From equation (9-81), $\Gamma = 5.55 \times 10^{-4} \times 0.97/(4.76 \times 10^{-3}) = 0.113$. Therefore, in this case, the concentration polarization phenomenon could become significant if good flow distribution is not maintained; sections receiving little flow might exhibit significant polarization.

9.5.3 Gas Permeation

Gas permeation (GP) was described in detail in Examples 9.2 and 9.3. Since the early 1980s, applications of GP with dense polymeric membranes have increased dramatically. Applications include (1) separation of hydrogen from methane; (2) adjustment of the H_2/CO ratio in synthesis gas; (3) oxygen enrichment of air; (4) nitrogen enrichment of air; (5) separation of uranium isotopes; (6) drying of natural gas and air; (7) removal of helium; and (8) recovery of methane from biogas (Seader et al., 2011).

Most applications of GP use dense membranes of cellulose acetates and polysulfones. For high-temperature applications where polymers cannot be used, membranes of glass, carbon, and inorganic oxides are available, but they are limited in their selectivity. Almost all large-scale applications of GP use spiral-wound or hollow-fiber modules, because of their high packing density.

9.5.4 Ultrafiltration and Microfiltration

As with reverse osmosis, ultrafiltration (UF) and microfiltration (MF) are pressure-driven membrane separation processes, with the membrane permselective for the solvent, usually water. MF and UF separate mainly by size exclusion of the solutes. MF retains particles of micrometer size; UF retains particles of submicrometer size by ultramicroporous membranes. Typically, UF retains solutes in the 300 to 500,000 molecular weight range, including biomolecules, polymers, sugars, and colloidal particles.

The ideal membrane for cross-flow filtration would have a high porosity and a narrow pore size distribution, with the largest pores slightly smaller than the particles or molecules to be retained. An asymmetric membrane is usually

9.5 Applications of Membrane-Separation Processes 519

preferred. The filtration effectiveness of a membrane is defined in terms of the *rejection*, σ, defined in terms of solute concentration in the retentate (c_R) and in the permeate (c_P) as

$$\sigma_i = 1 - \frac{c_{iP}}{c_{iR}} \tag{9-82}$$

UF membranes are characterized by a nominal *molecular weight cutoff* (MWCO), defined as the smallest solute molecular weight for which the membrane has at least a 90% rejection. Most commercial UF membranes operate in the $1000 \leq MCO \leq 10,000$ range.

For industrial applications, large surface areas are obtained using modules with many tubular or hollow-fiber membranes or by using large sheets in a spiral-wound arrangement. Tubular membranes are 5 to 25 mm inside diameter and up to 3 m long. Hollow-fiber UF membranes have diameters of 0.2 to 2 mm, and thousands of fibers are sealed in each cylindrical module. Spiral-wound modules of the type used for RO are widely used for UF.

The performance of a UF membrane can be characterized by the permeate flux, the rejection, and the concentration of solute in the retentate. The permeate flux often decreases with time because of membrane fouling, but fouling may increase the rejection. Fouling in UF occurs mainly by severe concentration polarization. Under these conditions, the concentration of the solute next to the membrane surface may attain its solubility limit. A precipitate of gel may then form, the result being fouling of the membrane surface or within the pores. This is known as *gel-layer consolidation*. Usually, the gel layer will form at high fluxes.

Consider first a clean membrane processing solutions with significant osmotic pressure. The solvent is assumed to pass by laminar flow through the small pores of the permselective layer, and the driving force is the applied pressure differential minus the difference of osmotic pressure across the membrane. The *volume flux, v,* which is the superficial permeate velocity normal to the membrane surface, can be written for a clean membrane as

$$v = Q_m \left(\Delta P - \Delta \pi \right) = \frac{\Delta P - \Delta \pi}{R_m} \tag{9-83}$$

where R_m, the membrane resistance, is the reciprocal of Q_m. If gel-layer consolidation occurs, equation (9-83) is modified to include the resistance of the layer, R_{gel}:

$$v = \frac{\Delta P - \Delta \pi}{R_m + R_{gel}} \tag{9-84}$$

The osmotic pressure for solutions of polymers, proteins, and other large molecules increases strongly with concentration. This can make $\Delta \pi$ a significant portion of ΔP even when the osmotic pressure of the feed solution is negligible.

When concentration polarization in RO was discussed, the solute balance at the upstream membrane surface was treated using the simple mass-transfer equation (9-79). This approach is satisfactory where the surface concentration is only moderately higher than the bulk concentration. For UF, the large change in

520 Membranes and Other Solid Sorption Agents

concentration near the surface requires integration to get the concentration profile. The basic equation states that the solute flux due to convection plus diffusion is constant in the boundary layer and equal to the solute flux in the permeate:

$$vc + D_s \frac{dc}{dx} = vc_F \qquad (9\text{-}85)$$

Assuming constant D_s, equation (9-85) is integrated with the boundary conditions $c = c_s$ at $x = 0$ and $c = c_R$ at $x = \delta$, the thickness of the concentration boundary layer:

$$\ln\left[\frac{c_s - c_P}{c_R - c_P}\right] = \frac{v\delta}{D_s} = \frac{v}{k_s} \qquad (9\text{-}86)$$

For the simple case of complete solute rejection, $\sigma = 1$, $c_P = 0$, and equation (9-86) becomes

$$c_s = c_R \exp\left[\frac{v}{k_s}\right] \qquad (9\text{-}87)$$

Because of concentration polarization, the permeate flux is a nonlinear function of ΔP, and a trial-and-error solution is needed to calculate v for a given ΔP. However, when there is complete rejection of the solute and no gel resistance, for a specified volume flux of the solvent, equation (9-87) can be used to determine c_s, osmotic pressure data to get $\Delta\pi$, and equation (9-83) to calculate the required ΔP.

Example 9.13 Ultrafiltration of Cheese Whey Proteins

The coagulation process to produce cheese out of milk generates two phases: a solid portion known as the curd, and a liquid portion known as whey. Only the curd is used to make cheese; the whey is essentially a waste product. The whey produced at a cheese plant is to be further processed by UF at 293 K to recover the valuable proteins that it contains. Tubular membranes will be used with a diameter of 2 cm, a clean-membrane water permeance of 250 $L/m^2 \cdot h \cdot atm$, and a rejection of 100% for the whey proteins. The average diffusivity of the whey proteins at this temperature is 4×10^{-7} cm^2/s. The protein concentration of the whey at a certain point in the equipment is 40 g/L, and the whey flows through the tube at a velocity of 1.5 m/s. The osmotic pressure of whey can be related to its protein concentration through the equation:

$$\Delta\pi\,(\text{atm}) = 3.1646\,c + 8.749\,c^2 + 57.915\,c^3 \quad \text{for } c \leq 0.35 \text{ kg/L}$$

(a) Calculate the required pressure differential to produce a water transmembrane volume flux of 25 $L/m^2 \cdot h$ when the membrane is clean.

(b) If the membrane permeance is reduced fivefold by fouling, calculate the water flux if the applied pressure differential remains the same as calculated in part (a).

9.5 Applications of Membrane-Separation Processes 521

Solution

(a) Assume that the bulk whey solution have the same density and viscosity of pure water at 293 K. For flow inside the 2-cm-diameter tube, Re = 2 × 150 × 1/0.01 = 30,000. Similarly, Sc = 0.01/(1 × 4 × 10^{-7}) = 25,000. To estimate k_s, we need a correlation for turbulent flow inside circular pipes, but at extremely high Sc numbers. The following was developed by Harriott and Hamilton (1965), modified as suggested by McCabe et al. (2005):

$$\text{Sh} = 0.0048\,\text{Re}^{0.913}\,\text{Sc}^{0.346}$$
$$\text{Re} > 10,000 \tag{9-88}$$
$$430 < \text{Sc} < 100,000$$

Substituting in (9-88) gives Sh = 1950. Then, k_s = 1950 × 4 × 10^{-7}/2 = 3.9 × 10^{-4} cm/s. By units conversion, v = 25 L/m²·h = 6.94 × 10^{-4} cm/s. For c_R = 40 g/L, equation (9-87) yields c_s = 237 g/L. Therefore, the concentration at the membrane surface is almost six times higher than the concentration of the bulk solution. From the given equation, for c_s = 0.237 kg/L the corresponding osmotic pressure is 2.0 atm. For 100% rejection, $\Delta\pi = \pi$ = 2.0 atm. From equation (9-83), $\Delta P - \Delta\pi = v/Q_m$ = 0.1 atm. Then, ΔP = 2.1 atm. Notice that most of the driving force is needed to overcome the osmotic pressure difference caused by concentration polarization.

(b) The membrane permeance is reduced to Q_m = 250/5 = 50 L/m²·h·atm. If the applied ΔP remains constant at 2.1 atm, v, c_s, and $\Delta\pi$ will change. Equations (9-83) and (9-87), and the osmotic pressure versus concentration equation must be solved simultaneously by trial and error to calculate new values for these three variables. The results are c_s = 213 g/L, $\Delta\pi$ = 1.63 atm, and v = 6.53 × 10^{-4} cm/s = 23.5 L/m²h. This represents only a 6% reduction in the transmembrane flux of water. Therefore, in this case, the membrane permeance has little effect on the resulting water flux.

There is no sharp dividing line between MF, which treats suspensions of small particles (more commonly from 0.1 μm to 10 μm), and UF, which usually deals with solutions of large molecules. For very small particles, such as 0.1-μm spheres of polymer in a latex paint, either term could be applied. MF is used to separate large colloids, blood cells, yeast, bacteria, and other microbial cells, and very large and soluble macromolecules.

Membrane structures for MF include *screen filters*, which collect retained matter on the surface, and *depth filters*, which trap particles at constrictions within the membrane. Depth filters have a much sharper cutoff, resulting in enhanced separation factors. For example, a Nuclepore membrane of type 2 can separate a male-determining sperm from a female-determining sperm (Seader and Henley, 2006). Nuclepore MF membranes come in pore sizes from 0.03 μm to 8.0 μm with water permeate flux rates, at 294 K and a transmembrane pressure difference of 70 kPa, ranging from 15 L/m²·h to 350,000 L/m²·h.

Two different modes of microfiltration are employed. In *dead-end filtration* (DEF), the entire solvent is forced under pressure through the membrane. With

522   **Membranes and Other Solid Sorption Agents**

time, as retained material accumulates on or within the membrane, the pressure required to maintain a desired flow rate must increase or permeate flux will decrease. In *tangential-flow filtration* (TFF), the feed flows along the surface with only a fraction of the solvent passing through the membrane. Ideally, the retained matter is carried out of the filter with the retentate fluid. The TFF mode is usually accompanied by a large retentate recycle. DEF is most suitable for dilute solutions, while TFF is preferred for concentrated solutions. Membrane modules for MF are mainly DEF plate-and-frame and TFF pleated cartridges. Compared to TFF, DEF has a lower capital cost, higher operating cost, and a simpler operation.

9.5.5 Bioseparations

Semipermeable membranes are widely used to selectively retain and/or permeate biological species. Operations for bio-products include reverse osmosis, electrodialysis, pervaporation, microfiltration, ultrafiltration, nanofiltration (NF), and virus filtration (VF). MF, NF, UF, and VF of bio-products may be conducted in either the DEF or TFF modes, depending on the solution concentration.

Conventional (DEF) filtration is typically used when a product has been secreted from cells and the cells must be removed to obtain the product that is dissolved in liquid. Antibiotics and steroids are often processed by DEF to remove the cells. This mode of operation is also commonly used for sterile filtration in biopharmaceutical production. In DEF, a batch of feed solution is forced under pressure through the membrane causing retained material to accumulate on and within the membrane. The pressure required to maintain a desired flow rate must increase or permeate flux will decrease. A combined mode in which constant-flow operation is employed to a limiting pressure, followed by constant-pressure operation until a minimum flux is reached is superior to either constant-pressure or constant-flux operation.

Cross-flow (TFF) filtration has been used in a wide variety of applications including the removal of cell debris from cells that have been lysed, the concentration of protein solutions, the exchange or removal of salts in a protein solution, and the removal of viruses from protein solutions. This mode of operation is more suitable for large-scale continuous filtration of bio-products. Feed flows along the surface of the membrane with only a fraction of the solvent passing through the membrane, while retained matter is carried out with the retentate fluid. Retentate is usually recycled to the filter at tangential-flow velocities parallel to the membrane surface in the range of about 1 to 8 m/s. TFF gives up to 10-fold-higher flux values than DEF.

High-performance TFF (HPTFF) uses optimal values of buffer pH, ionic strength, and membrane charge to enhance mass throughput and selectivity as a function of local pressure-dependent flux. Covalent surface modification with quaternary amine or sulfonic-acid groups improves membrane selectivity for HPTFF applications. HPTFF can separate equally sized proteins based on charge differences, monomers from dimers, and single-amino-acid variants in real, dilute feeds significantly improving yield and purification factors.

Nanofiltration (NF) is an energy-saving and environmentally friendly separation technology that plays an ever more important role in purification and recovery

9.5 Applications of Membrane-Separation Processes

523

of high-value bio-products (Cao et al., 2021). NF membranes with properties including pore size smaller than 2 nm, molecular weight cut-off (MWCO) of 100–2000 Da and generally charged surface in aqueous environments, have drawn particular attention due to their predominance in selectively rejecting biomolecules. As a result of these features, NF membranes are suitable for various applications such as decolorization, desalination, concentration, and biomolecules separation. Compared to conventional purification strategies, the greatest advantage of NF is the ability to recycle valuable substances (e.g. pigments, inorganic salts and water) without generating new contaminants. In addition, compared to evaporation concentration processes, NF can achieve the effective removal of impurities with less than 20% of the energy consumption and avoid massive chemical reagents consumption. Especially for the biomolecules with similar molecular weight, the pore structure and surface charge pattern of NF membrane can be designed according to the difference of the solutes in size, shape and charge, so as to improve the separation selectivity, as shown in Example 9.14.

Example 9.14 Membrane-Assisted Hybrid Extraction-Distillation Process for Purification of γ-Valerolactone

γ-Valerolactone (GVL) is an organic compound with the formula $C_5H_8O_2$. This colorless liquid is readily obtained from cellulosic biomass and is a potential "green fuel" as well as "green solvent." Scharzec et al. (2021) studied different processes for purification of diluted GVL aqueous solutions that can be applied to other bio-based production processes, as well as to the recovery of valuable impurities prior to wastewater treatment. They used as a case study an aqueous stream flowing at the rate of 100 mol/s containing 5 mol% of GVL and 95% water. The objective was separating the solution into a product containing 99.99 mol% GVL and a residue containing only 0.01 mol% GVL. Example 7.12 presented in detail the first two options considered by the researchers: (1) distillation and (2) a hybrid extraction-distillation process.

The third separation option considered was a membrane-assisted hybrid extraction-distillation process. The feed solution was now preconcentrated in a module consisting of 443 m^2 of NanoPro S-3011 NF membrane operating at a pressure of 60 bar. The retentate from the membrane, flowing at the rate of 46 mol/s with a GVL concentration of 10.9%, was processed in a 15-stages plate extractor using 15.3 mol/s of toluene as solvent in countercurrent flow. The raffinate from the extractor contained 99.99 mol% water, 0.01% toluene, and only traces of GVL. The extract, flowing at the rate of 20.6 mol/s, had a molar concentration of 24.3% GVL, 74.4% toluene, and 1.3% water. This stream was preheated and fed to a 29-stages distillation column. The distillate from the column, almost pure toluene, was recycled to the extractor. The bottoms from the distillation column was the desired product with a concentration of 99.99 mol% GVL and 0.01% toluene. The TAC for the membrane-assisted hybrid separation process was around 86% lower than the TAC for the original distillation column. Most of the savings associated with the membrane-assisted hybrid were due to a 91% reduction in energy costs.

524   **Membranes and Other Solid Sorption Agents**

9.6 APPLICATIONS OF SORPTION-SEPARATION PROCESSES

Your objectives in studying this section are to be able to:

1. Enumerate the most common materials used for adsorption, for ion exchange, and for chromatography.
2. Define monoclonal antibodies, affinity adsorption, steric structure, and volatile organic compounds.
3. Use experimental breakthrough data to estimate the volumetric overall mass-transfer coefficient for fixed-bed adsorption.
4. Calculate volumetric overall mass-transfer coefficients in fixed-bed adsorption from correlations for the individual coefficients and use them for breakthrough predictions.

Most solids are able to adsorb species from gases and liquids. However, only a few have a sufficient selectivity and capacity to make them serious candidates for commercial sorbents. A good sorbent must have a large specific area (area per unit mass). Typical commercial sorbents, such as *activated carbon* or *silica gel*, have specific surface areas from 300 m^2/g to 1200 m^2/g. Thus, a few grams of activated carbon have a surface area equal to that of a football field! Such a large area is possible by a particle porosity from 30% to 85% with average pore diameters from 10 Å to 200 Å. They may be granules, spheres, cylindrical pellets, flakes, and/or powders ranging in size from 50 µm to 1.2 cm.

Commercial ion exchangers in hydrogen, sodium, and chloride form are available under the trade names Amberlite, Duolite, Dowex, Ionac, and Purolite. Typically, they are in the form of spherical beads from about 40 µm to 1.2 mm in diameter. The two most common adsorbents used in chromatography are porous alumina and porous silica gel. Of lesser importance are carbon, magnesium oxide, and various carbonates.

The following examples illustrate some applications of sorption-separation processes to the fields of biotechnology (Example 9.15) and air pollution control (Example 9.16).

Example 9.15 Monoclonal Antibody Purification by Affinity Adsorption

Monoclonal antibodies (MAbs) are antibodies produced by *hybridoma cells*. These cells are clones resulting from cell fusion of *lymphocytes* (antibody-secreting cells) and *myeloma cells* (malignant tumor cells) that have the capability of secreting single species of antibodies and can be cultivated. *Antibodies* are immunoglobins normally produced by the immune system of living organisms to combat the invasion of foreign substances, or *antigens*.

Over the last few years, commercial interest in MAbs has rapidly increased. They have great potential as therapeutic agents for hematologic malignancies and

9.6 Applications of Sorption-Separation Processes 525

other diseases. They are also important as affinity ligands for purifying other pharmaceutical substances of importance. Therefore, cost-efficient methods for purifying large quantities of MAbs are important (Montgomery et al., 1998).

The biological activity of molecules, such as proteins, cells, and viruses, can easily be destroyed by processing conditions that do not conform to their natural environment. Therefore, traditional separation processes, such as distillation or solvent extraction, are seldom used to isolate them. *Affinity adsorption* is one of the most effective methods for the direct isolation and purification of biomolecules from complex mixtures (Camperi et al., 2003). It is based on recognition between a pair of molecules determined by the *steric structure* (three-dimensional arrangement of its atoms) of the molecules. When molecules have complementary steric structures, they can interact to maximize the hydrogen bonds and electrostatic interactions. Affinity adsorption allows a separation process with high specificity and purity.

Camperi et al. (2003) employed affinity adsorption to isolate and purify a monoclonal antibody against GM-CSF (granulocyte macrophage–colony stimulating factor) using a peptide affinity ligand attached to *agarose* (a polysaccharide extracted from agar) in a fixed bed. The adsorbed MAb was recovered quantitatively by elution of the bed with 5 M LiCl. They performed preliminary adsorption equilibrium experiments and found a Langmuir-type isotherm of the form

$$q' = \rho_p \left(1 - \varepsilon\right) q = \frac{K q'_m c}{1 + Kc} \tag{9-89}$$

with q in mg/mL of bed and c in mg/mL of solution. They reported a value of $q'_m = 9.1 \, \text{mg} / \text{mL}.$.

These researchers generated experimental breakthrough curves using a 1.0-mL column packed with the affinity adsorption matrix. The solution containing different feed concentrations of the MAb (0.1, 0.2, and 1.0 mg/mL) was pumped through the bed at a superficial velocity $v_0 = 0.5$ cm/min. The breakthrough results were expressed in terms of the relative MAb concentration in the effluent (X_{exp}) versus cumulative mass of the MAb (m_{MAb}, mg) applied to the column. Besides the total volume, no further details were given about the column. For computational purposes, we will assume a column length $Z = 10$ cm, cross-sectional area $A_c = 0.1$ cm^2, and porosity $\varepsilon = 40\%$. The breakthrough experimental results for $c_F = 1.0$ mg MAb/mL solution are given in Table 9.3. The cumulative mass of MAb applied to the column can be converted to elapsed time from the relation $t_{exp} = m_{MAb}/v_0 A_c c_F$.

According to the adsorption model with Langmuir isotherm presented in Section equation (9.4.3), the value of the dimensionless time parameter T [see equation (9-63)] must be close to 1.0 for $X = 50\%$. From Table 9.3, $t_{exp} = 73.6$ min for $X_{exp} = 50\%$. Substituting in equation (9-63) and letting $T = 1.0$, $\rho_p(1 - \varepsilon)q^* = 3.28$ mg MAb/mL bed. Assuming that the value of $q'_m = 9.1 \, \text{mg} / \text{mL}$ is accurate, from equation (9-89), $K = 0.563$ mL/mg.

The dynamics of the fixed-bed affinity adsorption process were simulated using equations (9-59) to (9-63). The value of the overall volumetric mass-transfer coefficient $K_c a$ was selected by trial and error until a good fit to the experimental

526 Membranes and Other Solid Sorption Agents

Table 9.3 Breakthrough MAb Experimental Data

Cumulative mass, m_{Mab}, mg[a]	Elapsed time, t_{exp}, min	Relative concentration, X_{exp}, %[a]
2.0	40.0	0.0
3.0	60.0	4.0
3.2	64.0	6.6
3.38	67.6	11.6
3.60	72.0	36.3
3.68	73.6	50.0
3.80	76.0	70.3
4.00	80.0	81.7
4.40	88.0	91.9
4.60	92.0	100.0

[a] c_F = 1.0 mg MAb/mL.
Source: Data from Camperi et al. (2003).

data was achieved. Figure 9.13 shows that excellent agreement between the experimental results of Camperi et al. (2003) and the model predictions was achieved for $K_c a = 0.024$ s^{-1} using Mathcad. Appendix 9.7 shows the calculations using Python.

Example 9.16 Control of VOC Emissions by Adsorption on Activated Carbon (Benítez, 1993)

Volatile organic compounds (VOCs) comprise a very important category of air pollutants. They include all organic compounds with appreciable vapor pressure at ambient temperatures. Some are hydrocarbons, but others may be aldehydes, ketones, chlorinated solvents, refrigerants, and so on. The major anthropogenic sources of VOCs are industrial processes and automobiles. Compounds in this category have a wide range of reactivity in the atmosphere. Particularly reactive VOCs combine with the oxides of nitrogen in the presence of sunlight to form *photochemical oxidants*, including *ozone* (O_3) and *peroxyacetyl nitrate* (PAN). These oxidants are severe eye, nose, and throat irritants. They attack synthetic rubber, textiles, paints, and other materials, and they extensively damage plant life.

Adsorption on activated carbon is a very attractive technique to reduce emissions of VOCs from industrial facilities. Activated carbon has a very high affinity for most VOCs, achieving virtually 100% removal efficiency up to the break point for adsorption on a fixed bed. Not only emissions of the VOCs are avoided, but the adsorbed substances can be recovered and recycled by thermal desorption of the bed.

Consider the following application of fixed-bed, activated carbon adsorption for the control of VOC emissions. An industrial waste gas consists of 0.5 vol% acetone in air at 300 K and 1 atm. It flows at the rate of 2.3 kg/s through a fixed bed packed with activated carbon. The bed has a cross-sectional area of 5.0 m^2 and is

9.6 Applications of Sorption-Separation Processes

$$v_0 := 30 \; \frac{cm}{hr} \qquad Z := 10 \; cm \qquad Kca := 0.024 \; s^{-1} \qquad qast := 3.28 \; \frac{mg}{mL}$$

$$\varepsilon := 0.4 \qquad A_c := 0.1 \; cm^2 \quad c_F := 1 \; \frac{mg}{mL} \qquad K := 0.563 \; \frac{mL}{mg}$$

$$R := \frac{1}{1 + K \cdot c_F} = 0.64 \qquad N_R(z) := \frac{2 \cdot Kca \cdot z}{v_0 \cdot (R+1)} \qquad T(z,t) := \frac{v_0 \cdot \left(t - \dfrac{\varepsilon \cdot z}{v_0}\right) \cdot c_F}{z \cdot qast}$$

$$J(\alpha, \beta) := 1 - e^{-\beta} \cdot \int_0^{\alpha} e^{-x} \cdot \mathrm{I0}\left(2 \cdot \sqrt{\beta \cdot x}\right) dx$$

$$F1(t,z) := J\left(N_R(z) \cdot R, N_R(z) \cdot T(z,t)\right)$$

$$F2(t,z) := \left(1 - J\left(N_R(z), N_R(z) \cdot T(z,t) \cdot R\right)\right) \cdot \exp\left((R-1) \cdot N_R(z) \cdot (T(z,t)-1)\right)$$

$$F3(t,z) := 1 - J\left(N_R(z) \cdot T(z,t), N_R(z) \cdot R\right) \qquad X(t,z) := \frac{F1(t,z)}{F1(t,z) + F2(t,z)}$$

Experimental Results:

$$m_{Mab} := \begin{bmatrix} 2 \\ 3 \\ 3.2 \\ 3.38 \\ 3.6 \\ 3.68 \\ 3.8 \\ 4.0 \\ 4.4 \\ 4.6 \end{bmatrix} \cdot mg \qquad X_{exp} := \begin{bmatrix} 0 \\ 0.04 \\ 0.066 \\ 0.116 \\ 0.363 \\ 0.500 \\ 0.703 \\ 0.817 \\ 0.919 \\ 1.0 \end{bmatrix} \qquad t_{exp} := \frac{\overrightarrow{m_{Mab}}}{A_c \cdot v_0 \cdot c_F} = \begin{bmatrix} 40 \\ 60 \\ 64 \\ 67.6 \\ 72 \\ 73.6 \\ 76 \\ 80 \\ 88 \\ 92 \end{bmatrix} \; min$$

Figure 9.13 Mathcad solution of Example 9.15.

packed to a depth of 0.3 m. The external porosity of the bed is 40%, its bulk density is 630 kg/m^3, and the average particle size is 6 mm. The average pore size of the activated carbon particles is 20 Å, the internal porosity is 60%, and the tortuosity factor is 4.0. A Langmuir-type adsorption isotherm applies with q_m = 0.378 kg VOC/kg of carbon, K = 0.867 kPa^{-1}. At the break point, the effluent concentration will be 5% of the feed concentration. Calculate

(a) The breakthrough time.

(b) The fraction of the bed adsorption capacity that has been used at the breakthrough time.

(c) The width of the MTZ (defined here for $0.05 \leq c/c_F \leq 0.95$).

Solution

(a) The average molecular weight of a 0.5 mol% acetone (58 g/mol) in air mixture is 29.14 kg/kmol. At 300 K and 1 atm, the density of the mixture is ρ = 1.184 kg/m^3. The volumetric flow rate is, then, 2.3/1.184 = 1.943 m^3/s. Given a cross-sectional area of 5.0 m^2, the superficial fluid velocity v_0 = 1.943/5 = 0.389 m/s. The feed concentration is c_F = 0.005 × 58 × 101.3/(8.314 × 300) = 0.0118 kg VOC/m^3 air, corresponding to a

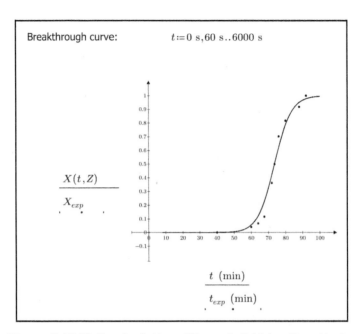

Figure 9.13 Mathcad solution of Example 9.15 (continuation).

partial pressure of acetone p_F = 0.005 × 101.3 = 0.5065 kPa. From equation (9-34), q^* = 0.116 kg VOC/kg carbon. From equation (9-61), R = 0.695.

To calculate k_c, use equation (2-90) for gases in packed beds. Assume that the viscosity of the dilute mixture is that of air at 300 K, μ = 1.85 × 10^{-5} Pa·s. Then, the Reynolds number is Re = 1.184 × 0.389/1.85 × 10^{-5} = 150, which is within the appropriate range (10 to 2500). The diffusivity is from the Wilke–Lee equation, D_{AB} = 0.12 cm²/s. The corresponding Schmidt number is Sc = 1.3. Then, j_D = 1.17/$150^{0.415}$ = 0.146; Sh = 0.146 × 150 × $1.3^{0.33}$ = 23.9 ≈ $k_c d_p/D_{AB}$. Therefore, k_c = 0.048 m/s.

To estimate diffusion inside the particle, calculate first the molecular mean free path from equation (1-102). From Appendix B and equation (1-45) for acetone and air, σ_{AB} = 4.11 Å. Substituting in (1-102), λ = 544 Å. The Knudsen number, Kn = 544/20 = 27.2; therefore, Knudsen diffusion is important. The Knudsen diffusivity for acetone is from equation (1-103), D_K = 2.2 × 10^{-7} m²/s. The effective diffusivity is from equation (9-52), D_{eff} = 3.25 × 10^{-8} m²/s. From equation (9-51), $k_{c,\text{int}}$ = 5.42 × 10^{-5} m/s.

The interfacial area for mass transfer is from equation (2-92), then, a = 600 m^{-1}. The overall mass-transfer coefficient is from (9-55), K_c = 0.042 m/s; the volumetric coefficient is $K_c a$ = 25.0 s^{-1}. Changing the corresponding numerical values in the Mathcad program of Figure 9.10, we obtain the following results:

(a) The break point is reached at 50.8 min.

(b) The fraction of the bed capacity used at the break point is 66.9%.

(c) The width of the MTZ is 17.8 cm.

PROBLEMS

The problems at the end of each chapter have been grouped into four classes (designated by a superscript after the problem number).

Class a: Illustrates direct numerical application of the formulas in the text.
Class b: Requires elementary analysis of physical situations, based on the subject material in the chapter.
Class c: Requires somewhat more mature analysis.
Class d: Requires computer solution.

9.1[a]. van't Hoff equation for osmotic pressure

Demonstrate equation (9-76). Hint: For dilute solutions of solute B in solvent A, $x_B \approx n_B/n_A$.

9.2[a]. Liquid flux in tubular membrane

The membrane in Example 9.1 is replaced with another of the same thickness, but with a permeability of 300 barrer for solute A. Calculate the transmembrane flux of solute A if the film coefficients remain unchanged. What fraction of the total resistance corresponds to the membrane?

Answer: 3.46×10^{-3} mol/m^2·s

9.3[d]. Effect of feed pressure on gas permeation

Repeat Example 9.2, but with a feed pressure of 5.0 MPa. Compare your results to those obtained at a feed pressure of 1.0 MPa.

9.4[d]. Effect of feed pressure on cross-flow gas permeation

Repeat Example 9.3, but with a feed pressure of 5.0 MPa. Compare your results to those obtained at a feed pressure of 1.0 MPa.

9.5[b]. Membrane separation factor

Laboratory tests of a membrane for H_2/CH_4 separation gave a permeate composition of 80% H_2 and a 20% cut when the feed was 50% H_2 and the feed and permeate absolute pressures were 700 and 105 kPa, respectively.

(a) Calculate the composition of the retentate.

Answer: 42.5% H_2

(b) What is the membrane separation factor?

530   Membranes and Other Solid Sorption Agents

9.6[b]. Adsorption equilibrium isotherms

The following data have been reported for equilibrium adsorption of benzene on certain activated carbon at 306 K.

(a) Fit a Langmuir-type isotherm to these data.

Answer: $K = 0.039$ Pa^{-1}

(b) Fit a Freundlich-type isotherm to these data.

(c) Comment on which model best describes the data.

Benzene partial pressure, Pa	Benzene adsorbed; kg/kg solid
15.3	0.124
33.5	0.206
133.3	0.278
374.5	0.313
1042.3	0.348

9.7[b]. Adsorption equilibrium isotherms

The equilibrium adsorption of methane on a given activated carbon was studied by Grant et al. (1962). They proposed a Langmuir-type adsorption isotherm with parameters $q_m = 48$ g CH$_4$/kg carbon, and

$$K = 0.346 \left[2200 \left(\frac{1}{T,K} - \frac{1}{298.1} \right) \right] \text{ atm}^{-1} \qquad (9\text{-}90)$$

If the equilibrium partial pressure of CH$_4$ in a mixture is 0.2 atm, at what temperature will the solid loading be 4.0 g/kg?

9.8[b]. Ion-exchange equilibrium

Repeat the calculations of Example 9.5, but for a total solution normality of 0.5.

9.9[d]. Fixed-bed adsorption with linear isotherm

In the benzene adsorber of Example 9.6, the flow rate is increased to 0.25 m^3/s. Calculate the breakthrough time and the fraction of the bed adsorption capacity that has been used at breakthrough. At the increased flow rate, the new value of $K_c a = 10.5$ s^{-1}.

Answer: $t_b = 50.2$ min

9.10[d]. Fixed-bed adsorption with Langmuir isotherm

In the benzene adsorber of Example 9.7, the flow rate is increased to 0.25 m^3/s. Calculate the breakthrough time and the fraction of the bed adsorption capacity

Problems **531**

that has been used at breakthrough. At the increased flow rate, the new value of $K_c a = 10.5$ s^{-1}.

Answer: t_b = 77.4 min

9.11^b. Mass-transfer coefficients in an ion exchanger

A bed of ion-exchange beads 1.5 mm in diameter is used to deionize water at 293 K with a superficial velocity of 0.6 cm/s. The feed concentration is 0.02 M NaCl, and the resin has an equilibrium capacity of 2.4 eq/L of resin bed. The porosity of the 2-m-depth bed is 0.35. For monovalent ions, the effective diffusivity in the pores of the resin is about one-tenth of the normal liquid diffusivity (McCabe et al., 2005). Calculate the volumetric overall mass-transfer coefficient, $K_c a$, and the number of mass-transfer units, N. Use equation (1-141) to calculate the diffusivity of NaCl in water.

Answer: N = 23.1

9.12^b. Ion-exchanger ideal break time

Calculate the ideal break time for the ion exchanger of Problem 9.11.

Answer: 11.1 h

9.13^d. Chromatographic separation of sugars

If the aqueous solution of Example 9.10 also contained 2 g/cm^3 of sucrose ($K_c a$ = 0.0429 s^{-1}, κ = 0.9 cm^3/g), generate the corresponding chromatogram, including the three peaks.

9.14^b. Membrane area required for dialysis

An aqueous process stream flows at the rate of 1.0 L/s at 293 K and contains 8.0 wt% Na_2SO_4 and 6.0 wt% of a high-molecular-weight substance, A. This stream is processed in a continuous countercurrent dialyzer using a pure water sweep of the same flow rate. The membrane is microporous cellophane with properties as described in Problem 1.31. The molecules to be separated have the following properties:

Property	Na_2SO_4	A
Molecular weight	142	1000
Molecular diameter, Å	5.5	15.0
Diffusivity, cm^2/s × 10^5	0.77	0.25

Calculate the membrane area for a 10% transfer of A through the membrane, assuming no water transfer. What is the percent recovery of Na_2SO_4 in the diffusate? Use log-mean concentration differences and assume that the mass-transfer resistances on each side of the membrane are each 25% of the total, both for Na_2SO_4

532   **Membranes and Other Solid Sorption Agents**

and for component A. Assume that the density of the solution is approximately equal to the density of pure water.

9.15[b]. Fruit juice concentration by osmosis (Cussler, 1997)

Eastern European farmers produce a variety of fruit juices, which they wish to dehydrate to prolong shelf-life and facilitate transportation. One very simple dehydration method is to put the juice in a plastic bag and drop the bag into brine at 283 K. If the bag is permeable to water but not to salt or juice components, then osmotic flow will concentrate the juice. Is the osmotic pressure generated in this way significant? Assume that the juice contains a solids concentration equivalent to 1.0 wt% sucrose (342 g/mol molecular weight) and that the brine contains 35 g of NaCl per 100 g of water. For simplicity, assume that the juice and the brine are ideal solutions.

9.16[b]. Water desalinization by reverse osmosis

RO is used to treat 1.31 m^3/s of seawater at 293 K containing 3.5 wt% dissolved solids to produce 0.44 m^3/s of potable water with 500 ppm of dissolved solids. The feed-side pressure is 138 bar, while the permeate pressure is 3.4 bar. A single stage of spiral-wound membrane is used that approximates cross flow. If the total membrane area is 0.10 km^2, estimate the permeance for water and the salt rejection. Assume that the densities of the seawater and of the brine are approximately equal to the density of pure water.

9.17[c]. Pervaporation for ethanol purification

Pervaporation (PV) is a membrane-separation process in which one or more components of a liquid mixture diffuse through a selective membrane, evaporate under low pressure on the downstream side, and are removed by a vacuum pump or a chilled condenser (McCabe et al., 2005). Composite membranes are used with the dense layer in contact with the liquid and the porous supporting layer exposed to the vapor. A frequently used PV model is that of Wijmans and Baker (1993). They express the driving force for diffusion across the membrane as a partial-pressure difference $(\gamma_A x_A P_A - y_A P_2)$, where x and y are mol fractions in the liquid and gas, P_A is the vapor pressure of component A at the temperature of the liquid, and P_2 is the total pressure on the permeate side of the membrane.

The first commercial application of PV was for ethanol purification. The dilute ethanol solution produced by fermentation is distilled to produce an overhead product with 90 to 95 wt% alcohol (close to the azeotrope). This solution is fed to the PV unit where water is selectively removed to give nearly pure ethanol (99.9 wt%). The permeate stream, with about 20 to 40 wt% alcohol is recycled to the distillation column.

Laboratory tests of a PV membrane exposed to a liquid with 77.9 mol% ethanol in water at 333 K showed a flux of 0.20 kg/m^2·h and a permeate composition of 2.9 mol% ethanol when the downstream pressure was 2.0 kPa. Calculate the permeance of the membrane to ethanol and to water at the test conditions.

Problems 533

Liquid-phase activity coefficients at 333 K for the ethanol (1)–water (2) system are given by the Van Laar equations (Seader and Henley, 2006):

$$\ln \gamma_1 = 1.6276 \left[\frac{0.9232x_2}{1.6276x_1 + 0.9232x_2} \right]^2$$

$$\ln \gamma_2 = 0.9232 \left[\frac{1.6276x_2}{1.6276x_1 + 0.9232x_2} \right]^2 \tag{9-91}$$

9.18[b]. Ultrafiltration of cheese whey proteins

Repeat the calculations of Example 9.13, but for an initial protein concentration of the whey of 30 g/L.

9.19[b]. Diafiltration of proteins (Harrison, 2014)

Ultrafiltration of a protein solution at constant volume is achieved by the addition of water or buffer to the feed in an operation called *diafiltration*. Consider a protein diafiltration system where the flow channels are tubes 0.2 cm in diameter and 1.0 m long. The protein, which has a molecular weight of 490 kDa, has a diffusion coefficient, $D_s = 1.2 \times 10^{-7}$ cm^2/s. The solution has a viscosity of 1.2 cP and a density of 1100 kg/m^3. The system operates at a bulk stream velocity of 6.0 m/s. Under these operating conditions, it has been observed that the concentration polarization modulus, $c_s/c_R = 3.45$. Assuming complete solute rejection by the membrane, estimate the transmembrane flux, in units of L/m^2-h.

9.20[c, d]. Control of VOCs emissions by adsorption on activated carbon

Redesign the VOCs adsorber of Example 9.15 for a breakthrough time of 4.0 h. The pressure drop through the bed [calculated using the Ergun equation (2-95)] should not exceed 1.0 kPa. Calculate the new dimensions of the bed and the total amount of activated carbon required.

9.21[c, d]. Thermal regeneration of a fixed-bed adsorber

Thermal regeneration of a fixed-bed adsorber is based on the fact that the adsorption process is exothermic; therefore, heating the saturated adsorbent will result in desorption. The dynamics of the desorption process is very similar to that of adsorption, and can be modeled, for a Langmuir-type isotherm, by the Thomas solution (Vermeulen et al., 1973):

$$X = \frac{c}{c_0^*} = \frac{1 - J(RN_R, N_R T)}{1 - J(RN_R, N_R T) + [J(N_R, RN_R T)] \exp[(R-1)N_R(T-1)]} \tag{9-92}$$

534   **Membranes and Other Solid Sorption Agents**

where $c_0{}^*$ is the fluid concentration in equilibrium with the initial adsorbent loading, q_0, and $R = 1/(1 + Kp_0{}^*)$, where $p_0{}^*$ is the partial pressure corresponding to $c_0{}^*$, and

$$K_c = \left[\frac{1}{k_c} + \frac{c_0{}^*}{q_0 k_{c,int}\rho_p (1-\varepsilon)}\right]^{-1} \tag{9-93}$$

$$T = T(t,z) = \frac{v_0 c_0{}^*(t - z\varepsilon/v_0)}{\rho_p(1-\varepsilon)zq_0} \tag{9-94}$$

The fixed-bed adsorber of Example 9.15 will be regenerated using saturated steam at 373 K at the rate of 0.25 m^3/s. The bed will be considered completely regenerated when the acetone content in the exit stream is 1% of the gas-phase concentration in equilibrium with the initial solid loading [i.e., $X(t, Z) = 0.01$]. At 373 K, the adsorption equilibrium is described by a Langmuir-type isotherm with $q_m = 0.378$ kg acetone/kg carbon, $K = 0.074$ kPa^{-1}. Calculate the time required for regeneration of the bed.

Hint: During thermal regeneration, the gaseous phase is not necessarily a dilute solution of adsorbate in the inert gas. On a mass basis, the adsorbate can even be the most abundant species. The fluid properties required for estimation of the mass-transfer coefficients must be calculated carefully. They should be estimated at a fluid composition that is the arithmetic average of the composition of the fluid at the solid–fluid interface (in equilibrium) and that of the bulk fluid (basically, the inert gas).

9.22[c, d]. The Bohart–Adams model for fixed-bed adsorber

The Bohart–Adams model is an exact analytic solution for fixed-bed adsorbers when the adsorption isotherm is irreversible, $q = q_m$ (see Figure 9.5). The sorbate concentration in the solution is given by

$$\frac{c(t,z)}{c_F} = \frac{\exp(\alpha)}{\exp(\alpha) + \exp(\beta) - 1} \tag{9-95}$$

where

$$\alpha = K_c a(t - z\varepsilon/v_0) \qquad \beta = \frac{K_c a\rho_p q_m(1-\varepsilon)}{c_F v_0} \tag{9-96}$$

The Bohart–Adams model is commonly, but mistakenly, referred to as the Thomas model in the environmental sorption literature (Chu, 2010). Consider a fixed-bed adsorber with a total depth of 10 cm, a porosity of 40%, and a bulk density of 1.5 g/cm^3. A dilute solution flows through the bed at a superficial velocity of 0.8 cm/s with a feed sorbate concentration of 0.05 mg/cm^3. For these conditions, $K_c a = 5 \times 10^{-4}$ s^{-1}. The adsorption isotherm is of the Langmuir type with $q_m = 100$ mg/g and $K = 50$ cm^3/mg. Approximate the breakthrough curve using the Bohart–Adams model and compare it to the actual curve generated using the Thomas equation.

REFERENCES

Anzelius, A. Z., *Angew: Math Mech.*, **6**, 291–294 (1926).

Barrer, R. M., *Diffusion in and through Solids*, Cambridge University Press, London (1951).

Benítez, J., *Process Engineering and Design for Air Pollution Control*, Prentice Hall, Upper Saddle River, NJ (1993).

Camperi, S. A., et al., *Biotechnol. Lett.*, **25**, 1545–1548 (2003).

Cao, Y., et al., *Eng. Life Sci.*, **21**, 405–416 (2021).

Chu, K. H., *J. Hazard. Mater.*, **177**, 1006–1012 (2010).

Cussler, E. L., *Diffusion Mass Transfer in Fluid Systems*, 2nd ed., Cambridge University Press, Cambridge UK (1997).

Garcia, A. A., et al., *Bioseparation Process Science*, Blackwell Science, Malden, MA (1999).

Grant, R. J., M. Manes, and S. B. Smith, *AIChE J.*, **8**, 403 (1962).

Gülmüs, S. A., and L. Gilmaz, *J. Polym. Sci. Part B: Polym. Phys.*, **45**, 3025–3033 (2007).

Harriott, P., and R. M. Hamilton, *Chem. Eng. Sci.*, **20**, 1073 (1965).

Harrison, R. G., *Chem. Eng. Progress*, Vol. 110, 36–42 (October 2014).

Kipling, J. J., *Adsorption from Solutions of Nonelectrolytes*, Academic Press, London (1965).

Lonsdale, H. K., U. Merten, and R. L. Riley, *J. Applied Polym. Sci.*, **9**, 1341–1362 (1965).

McCabe, W. L., J. C. Smith, and P. Harriott, *Unit Operations of Chemical Engineering*, 7th ed., McGraw-Hill, New York (2005).

Montamedian, S., et al., *Proceedings of the 1990 International Congress on Membranes and Membrane Processes*, Chicago, Vol. **II**, pp. 841–843 (1990).

Montgomery, S., et al., *Chemical Engineering Fundamentals in Biological Systems*, University of Michigan, distributed by CACHE Corp., Austin, TX (1998).

Perry, R. H., and D. W. Green (eds.), *Perry's Chemical Engineers' Handbook*, 7th ed., McGraw-Hill, New York (1997).

Ritter, J. A., and R. T. Yang, *Ind. Eng. Chem. Res.*, **26**, 1679–1686 (1987).

Scharzec, B., et al., *Chem. Eng. Sci.*, **240** (August 2021).

Seader, J. D., and E. J. Henley, *Separation Process Principles*, 2nd ed., Wiley, Hoboken, NJ (2006).

Seader, J. D., et al., *Separation Process Principles*, 3rd ed., Wiley, Hoboken, NJ (2011).

Smith, J. M., H. C. Van Ness, and M. M. Abbott, *Introduction to Chemical Engineering Thermodynamics*, 5th ed., McGraw-Hill, New York (1996).

Thomas, H., *J. Amer. Chem. Soc.*, **66**, 1664 (1944).

Vermeulen, T., G. Klein, and N. K. Hiester, in J. H. Perry (ed.), *Perry's Chemical Engineers' Handbook*, 5th ed., McGraw-Hill, New York (1973).

Walawender, W. P., and S. A. Stern, *Separation Sci.*, **7**, 553–584 (1972).

Wijmans, J. G., and R. W. Baker, *J. Membrane Sci.*, **79**, 101–113 (1993).

APPENDIX 9.1

Python solution of Example 9.2.

```python
import numpy as np
from scipy.optimize import fsolve
# Enter data:
#Volumetric flow rate in L/s; pressures in kPa
V = 1.0e03; PF = 1000.0; PP = 100.0; r = PP/PF
#Temperature in K
T = 300; xFA = 0.21; xFB = 0.79
#Permeances are in mol/(m^2-s-kPa)
QA = 1.985e-06; QB = 0.380e-06
#Ideal separation factor; Feed molar flow rate in mol/s
al_id = QA/QB; LF = V/22.4

Solution = np.zeros((9,4))
for i in range (1, 10):
    Solution[i-1, 0] = 0.1*i
    th = 0.1*i
    def f(p):
        al = p[0]
        xA = p[1]
        yA = p[2]
        return al - yA*(1-xA)/(xA*(1-yA)), xA-(xFA-yA*th)/(1-th),\
        al-al_id*((al-1)*xA+1-r*al)/((al-1)*xA+1-r)
    p = np.zeros(3,dtype=float)
    p[0] = al_id
    p[1] = 0.2
    p[2] = 0.2
    p = fsolve(f,p)
    Solution[i-1, 1]= p[0]
    Solution[i-1, 2]= p[1]
    Solution[i-1, 3]= p[2]
    i+=1
print('Solution ')
print(' theta  alpha    xA      yA')
print(np.array2string(Solution,formatter={'float_kind': \
                                  '{0:.3f}'.format}))
theta = np.zeros(9,dtype=float)
alpha = np.zeros(9,dtype=float)
yA = np.zeros(9,dtype=float)
xA = np.zeros(9,dtype=float)
```

Appendix 9.1 537

Python solution of Example 9.2 (continuation).

```
AM = np.zeros((2,9),dtype=float)
ST = Solution.conj().transpose()
theta[:] = ST[0,:]
alpha[:] = ST[1,:]
xA[:] = ST[2,:]
yA[:] = ST[3,:]
for i in range (0, 9):
    AM[0,i]=theta[i]
    AM[1,i] = yA[i]*theta[i]*LF/((xA[i]*PF - yA[i]*PP)*QA)
    i+=1
print('')
print ('AM in m^2')
print(' theta  AM')
print(np.array2string(AM.transpose(),formatter={'float_kind':\
                               '{0:.1f}'.format}))
```

Program results:

Solution

Theta	Alpha	x_A	y_A	AM, m^2
0.1	4.112	0.181	0.475	8037
0.2	4.062	0.156	0.428	17,074
0.3	4.018	0.135	0.385	29,963
0.4	3.980	0.118	0.348	37,531
0.5	3.949	0.105	0.315	48,618
0.6	3.922	0.093	0.288	60,099
0.7	3.900	0.084	0.264	71,876
0.8	3.881	0.077	0.243	83,879
0.9	3.864	0.070	0.226	96,056

538 **Membranes and Other Solid Sorption Agents**

APPENDIX 9.2

Appendix 9.2 Python solution of Example 9.3.

```python
import numpy as np
from scipy.integrate import quad
from scipy.optimize import fsolve
from scipy.interpolate import CubicSpline

#Enter data:
#Volumetric flow rate in L(STP)/s; pressures in kPa
V = 1.0e03; PF = 1000.0; PP = 100.0; r = PP/PF
xFA = 0.21; xFB = 0.79
#Permeances are in mol/(m^2-s-kPa)
QA = 1.985e-06; QB = 0.380e-06
#Ideal separation factor; Feed molar flow rate in mol/s
al_id = QA/QB; LF = V/22.4
xR = 0.07; n = 100; delx = (xFA - xR)/n

x = np.zeros(101)
y = np.zeros(101)
h = np.zeros(101)
g = np.zeros(101)
phi = np.zeros(101)
L = np.zeros(101)
for i in range(101):
    x[i] = xR + i*delx
    def f(al):
        return al-al_id*((al-1)*x[i]+1-r*al)/((al-1)*x[i]+1-r)
    al = fsolve(f,al_id)
    g[i] = (1 + (al - 1)*x[i])/((al - 1)*x[i]*(1 - x[i]))
    i+=1
cs = CubicSpline(x, g)
def interp(x):
    return cs(x)
def I1(x):
    return quad(interp, x, xFA)[0]
theta = 1 - np.exp(-I1(xR))
yP = (xFA - xR*(1 - theta))/theta
for i in range(101):
    def c(y):
        return y/(1-y) - al_id*(x[i]-r*y)/(1-x[i]-r*(1-y))
    y[i] = fsolve(c, xFA)
```

Appendix 9.2   **539**

Python solution of Example 9.3 (continuation).

```
    h[i] = 1/(y[i] - x[i])
    cs1 = CubicSpline(x,h)
    def interp1(x):
        return cs1(x)
    def I2(x):
        return quad(interp1, x, xFA)[0]
    L[i] = LF*np.exp(-I2(x[i]))
    phi[i] = y[i]*L[i]/((y[i] - x[i])*(x[i] - r*y[i]))
    i+=1
cs2 = CubicSpline(x,phi)
def interp2(x):
    return cs2(x)
AM = (quad(interp2, xR, xFA)[0])/(QA*PF)
alphaS = yP*(1 - xR)/(xR*(1 - yP))
print('The cut is = %5.3f' %theta)
print('The permeate composition is = %5.3f' %yP)
print('The membrane surface area required (in m^2)is = %6.1f' %AM)
print('The degree of separation of the stage is = %5.3f' %alphaS)
```

Program results:

The cut is = 0.442

The permeate composition is = 0.387

The membrane surface area (in m^2) is = 39.695

The degree of separation of the stage is = 8.370

APPENDIX 9.3

Appendix 9.3 Python solution of Example 9.4.

```python
import numpy as np
from scipy import stats
import matplotlib.pyplot as plt

# Enter data:
#Molecular mass of methane; molar volume, ideal gas at STP, L/mol
MA = 16; VSTP = 22.4
#Partial pressure in MPa
pexp = [0.276, 1.138, 2.413, 3.758, 5.240, 6.274, 6.688]
#Adsorbed volume at STP, in cm^3/gm
V = [45.5, 91.5, 113.0, 121.0, 125.0, 126.0, 126.0]
q = np.zeros(7,dtype=float)
for i in range(7):
    q[i] = V[i]*MA/VSTP
    i+=1
# Linearize data for Langmuir Isotherm
y = np.zeros(7,dtype=float)
for i in range(7):
    y[i] = pexp[i]/q[i]
    i+=1
reg = stats.linregress(pexp, y)
qm = 1/reg[0]
K = 1/(qm*reg[1])
print('The maximum loading (in mg/gm), qm = %6.3f' %qm)
print('The equilibrium constant (in 1/MPa), K = %5.3f' %K)

# Linearize data for Freundlich Isotherm
y1 = np.zeros(7,dtype=float)
x = np.zeros(7,dtype=float)
for i in range(7):
    y1[i] = np.log(q[i])
    x[i] = np.log([pexp[i]])
    i+=1
reg1 = stats.linregress(x, y1)
n = 1/reg1[0]
k = np.exp(reg1[1])
print('The Freundlich parameters are:')
```

Appendix 9.3

Appendix 9.3 Python solution of Example 9.4 (continuation).

```
print('n = %5.3f' %n)
print('k = %5.3f' %k)

# Plot the results:
p = np.zeros(101,dtype=float)
qL = np.zeros(101,dtype=float)
qF = np.zeros(101,dtype=float)
for i in range(101):
    p[i]= 0.07*i
    qL[i] = K*qm*p[i]/(1 + K*p[i]) #Langmuir isotherm
    qF[i] = k*p[i]**(1/n)   #Freundlich isotherm
    i+=1
plt.plot(p, qL,'-k',label='Langmuir')
plt.plot(p, qF, '--k', label='Freundlich')
plt.plot(pexp, q, '*k', label='Experimental')
plt.legend(loc='best')
plt.xlabel('Pressure, MPa')
plt.ylabel('Loading, mg of methane/g carbon')
plt.show()
```

Program results:

Langmuir isotherm:

    The maximum loading (in mg/gm), q_m = 97.834

    The equilibrium constant (in 1/MPa), K = 1.863

Freundlich isotherm parameters: n = 3.224; k = 54.638

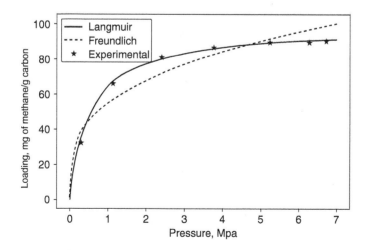

APPENDIX 9.4

Appendix 9.4 Python solution of Example 9.6.

```python
import numpy as np
import matplotlib.pyplot as plt
from scipy.integrate import quad
from scipy.optimize import fsolve
import sys
import warnings
if not sys.warnoptions:
    warnings.simplefilter("ignore")

# Enter data
#Volumetric flow rate in m^3/s; dimensions in m.
Q = 0.146; D = 0.61; Z = 1.829
#Vapor concentration in kg/m^3; mass of silica gel in kg, bed porosity.
cF = 0.029; ms = 331; eps = 0.5
#Linear isotherm constant in m^3/kg; volumetric mass-transfer coefficient, 1/s.
kappa = 4.134; Kca = 8.789
Ac = 3.1416*D**2/4; Vb = Ac*Z; rhob = ms/Vb
v0 = Q/Ac

# Define the special function J(alpha, beta):
def J(al,be):
    def f(x):
        return np.exp(-x)*np.i0(2*(be*x)**0.5)
    return 1.0 - np.exp(-be)*quad(f, 0, al)[0]

# Define dimensionless variables:
def N(z):
    return Kca*z/v0
def T(z,t):
    return v0*(t - eps*z/v0)/(rhob*z*kappa)
def X(t,z):
    return J(N(z), N(z)*T(z,t))
def Y(t,z):
    return 1 - J(N(z)*T(z,t), N(z))

# a) Generate the breakthrough curve:
t = np.zeros(251,dtype=float)
x = np.zeros(251,dtype=float)
for i in range(251):
```

Appendix 9.4

Python solution of Example 9.6 (continuation).

```
    t[i] = 60*i
    x[i] = X(t[i], Z)
    i+=1
plt.plot(t/60,x)
plt.xlabel('Time in min')
plt.ylabel('X(t, Z)')
plt.title('Breakthrough Curve')
plt.grid(True)
plt.show()

# b) Breakthrough time:
def tb(t):
    return(X(t,Z) - 0.05)
solution = fsolve(tb, 1000)
breaktime = solution/60   #Conversion to minutes
print('b) The breakthrough time (in min) = %5.2f' %breaktime)

# c) Fraction of the bed capacity used at breakthrough:
def Ytb(z):
    return Y(solution,z)
Yavg = quad(Ytb, 0, Z)[0]/Z
print('c) The fraction of the bed capacity used at breakthrough = %4.3f' %Yavg)

# d) Width of the mass-transfer zone, MTZ:
def g(z):
    return X(solution, z) - 0.95
MTZ = Z -fsolve(g, 0.5)
print('d) The width of the mass-transfer zone (in m), MTZ = %4.3f' %MTZ)
```

Program results:

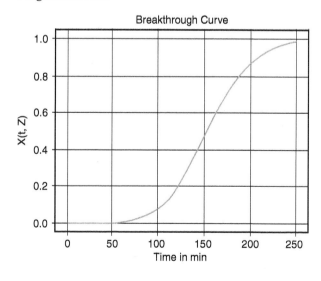

(a) Breakthrough time = 139.69 min

(b) Fraction of the bed used at breakthrough = 0.891

(c) Width of the mass-transfer zone, MTZ = 0.390 m

APPENDIX 9.5

Appendix 9.5 Python solution of Example 9.7.

```python
import numpy as np
import matplotlib.pyplot as plt
from scipy.integrate import quad
from scipy.optimize import fsolve
import sys
import warnings
if not sys.warnoptions:
    warnings.simplefilter("ignore")
# Enter data:
#Volumetric flow rate in m^3/s; dimensions in m.
Q = 0.146; D = 0.61; Z = 1.829
#Vapor concentration, kg/m^3; mass of silica gel, kg, bed porosity.
cF = 0.029; ms = 331; eps = 0.5
qm = 0.2; K = 51.72 #Langmuir isotherm parameters.
Kca = 8.789 # Volumetric mass-transfer coefficient, 1/s.
Ac = 3.1416*D**2/4; Vb = Ac*Z; rhob = ms/Vb
v0 = Q/Ac; qast = qm*K*cF/(1 + K*cF)
# Define the special function J(alpha, beta):
def J(al,be):
    def f(x):
        return np.exp(-x)*np.i0(2*(be*x)**0.5)
    return 1.0 - np.exp(-be)*quad(f, 0, al)[0]
# Define dimensionless variables:
R = 1/(1 + K*cF)
def NR(z):
    return 2*Kca*z/(v0*(R + 1))
def T(z,t):
    return v0*(t - eps*z/v0)*cF/(rhob*z*qast)
# Define dimensionless solutions:
def F1(t,z):
    return J(NR(z)*R, NR(z)*T(z,t))
def F2(t,z):
    return (1-J(NR(z),NR(z)*T(z,t)*R))*np.exp((R-1)*NR(z)*(T(z,t)-1))
def F3(t,z):
    return 1 - J(NR(z)*T(z,t),NR(z)*R)
def X(t,z):
    return F1(t,z)/(F1(t,z)+F2(t,z))
def Y(t,z):
    return F3(t,z)/(F1(t,z)+F2(t,z))
```

Appendix 9.5

Appendix 9.5 Python solution of Example 9.7 (continuation).

```
# a) Generate the breakthrough cirve:
t = np.zeros(201,dtype=float)
x = np.zeros(201,dtype=float)
for i in range(201):
    t[i] = 60*i
    x[i] = X(t[i], Z)
    i+=1
plt.plot(t/60,x,'-k')
plt.xlabel('Time in min')
plt.ylabel('X(t, Z)')
plt.title('Breakthrough Curve')
plt.grid(True)
plt.show()

#b) Breakthrough time
def tb(t):
    return(X(t,Z) - 0.05)
solution = fsolve(tb, 1000)
breaktime = solution/60  #Conversion to minutes
print('b) Breakthrough time (in min) = %5.2f' %breaktime)

# c) Fraction of the bed capacity used at breakthrough:
def Ytb(z):
    return Y(solution,z)
Yavg = quad(Ytb, 0, Z)[0]/Z
print('c) Fraction of the bed capacity used at breakthrough = %4.3f' %Yavg)

# d) Width of the mass-transfer zone, MTZ:
def g(z):
    return X(solution, z) - 0.95
MTZ = Z -fsolve(g, 1)
print('d) Width of the mass-transfer zone (in m), MTZ = %4.3f' %MTZ)
```

Program results:

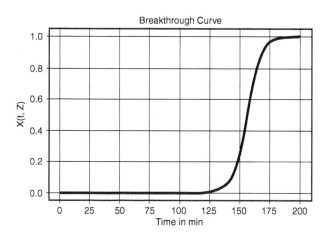

(a) Breakthrough time (in min) = 139.69

(b) Fraction of the bed capacity used at breakthrough = 0.891

(c) Width of the mass-transfer zone (in m), MTZ = 0.390

APPENDIX 9.6

Appendix 9.6 Python solution of Example 9.10.

```python
import numpy as np
import matplotlib.pyplot as plt
from scipy.integrate import quad
from scipy.optimize import fsolve
import sys
import warnings
if not sys.warnoptions:
    warnings.simplefilter("ignore")
#Enter data:
#Superficial velocity (m/s); bed length (m); bed porosity
v0 = 5e-04; Z = 2.0; eps = 0.39
#Duration of feed pulse (s); bulk density of the bed (kg/m^3)
tF = 500; rhob = 500
t1 = eps*Z/v0
# Define the special function J(alpha, beta):
def J(al,be):
    def f(x):
        return np.exp(-x)*np.i0(2*(be*x)**0.5)
    return 1.0 - np.exp(-be)*quad(f, 0, al)[0]


# GLUCOSE


Kcag = 0.0263 # Volumetric mass-transfer coefficient, 1/s
kappag = 6e-04 #Linear isotherm constant in m^3/kg;
# Define dimensionless variables for glucose:
def NG(z):
    return Kcag*z/v0
def TG(z,t):
    return v0*(t - eps*z/v0)/(rhob*z*kappag)
def XG(t,t1,tF,Z):
    if t <= t1:
        return 0
    elif t<=(t1 + tF):
        return J(NG(Z),NG(Z)*TG(Z,t))
    else:
        return J(NG(Z),NG(Z)*TG(Z,t)) - J(NG(Z),NG(Z)*TG(Z,t - tF))


#FRUCTOSE
```

Appendix 9.6

Python solution of Example 9.10 (continuation).

```
Kcaf = 0.0665 # Volumetric mass-transfer coefficient, 1/s
kappaf = 1.4e-03 #Linear isotherm constant in m^3/kg;
# Define dimensionless variables for fructose:
def NF(z):
    return Kcaf*z/v0
def TF(z,t):
    return v0*(t - eps*z/v0)/(rhob*z*kappaf)
def XF(t,t1,tF,Z):
    if t <= t1:
        return 0
    elif t<=(t1 + tF):
        return J(NF(Z),NF(Z)*TF(Z,t))
    else:
        return J(NF(Z),NF(Z)*TF(Z,t)) - J(NF(Z),NF(Z)*TF(Z,t - tF))
#CHROMATOGRAM
t = np.zeros(1201,dtype=float)
xF = np.zeros(1201,dtype=float)
xG = np.zeros(1201,dtype=float)
for i in range(1201):
    t[i] = 5*i
    xF[i] = XF(t[i],t1,tF, Z)
    xG[i] = XG(t[i],t1,tF, Z)
    i+=1
plt.plot(t,xF,'--k',label='Fructose')
plt.plot(t,xG, '-k', label='Glucose')
plt.xlabel('Time in s')
plt.ylabel('XG and XF, dimensionless')
plt.title('Chromatogram')
plt.legend()
plt.show()
```

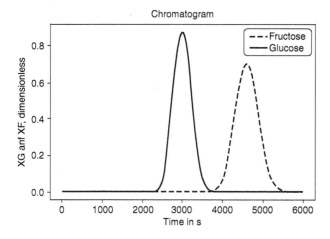

548   **Membranes and Other Solid Sorption Agents**

APPENDIX 9.7

Python solution of Example 9.15.

```python
import numpy as np
import matplotlib.pyplot as plt
from scipy.integrate import quad
from scipy.optimize import fsolve
from scipy.optimize import curve_fit
import sys
import warnings
if not sys.warnoptions:
    warnings.simplefilter("ignore")
# Enter data:
#Superficial velocity(cm/min);bed depth(cm); cross-sectional area(cm^2)
v0 = 0.5; Z = 10.0; Ac = 0.1
# Feed concentration (mg/mL); Langmuir parameter (mL/mg)
cF = 1.0; K = 0.563
# Langmuir parameter (mg/mL); volumetric mass-transfer coefficient (1/s)
qast = 3.28; Kca = 0.024
eps = 0.4 # Bed porosity
# Experimental results:
#Cumulative MAb mass (mg)
mMAb = [2.0, 3.0, 3.2, 3.38, 3.60, 3.68, 3.80, 4.0, 4.40, 4.60]
#Relative concentration, (%)
Xexp = [0, 4, 6.6, 11.6, 36.3, 50, 70.3, 81.7, 91.9, 100]
#Calculate elapsed time from the experimental results
texp = np.zeros(10,dtype=float)
for i in range(10):
    texp[i] = mMAb[i]/(Ac*v0*cF)
    i+=1
# Define the special function J(alpha,beta)
def J(al,be):
    def f(x):
        return np.exp(-x)*np.i0(2*(be*x)**0.5)
    return 1.0 - np.exp(-be)*quad(f, 0, al)[0]
# Define dimensionless variables at z = Z in terms of time
R = 1/(1 + K*cF)
Kcam = 60*Kca
NR = 2*Kcam*Z/(v0*(R + 1))
def T(t):
    return v0*(t - eps*Z/v0)*cF/(Z*qast)
# Define dimensionless breakthrough solutions
```

Appendix 9.7

Python solution of Example 9.15 (continuation).

```
def X(t):
    def F1(t):
        return J(NR*R, NR*T(t))
    def F2(t):
        return (1-J(NR,NR*T(t)*R))*np.exp((R-1)*NR*(T(t)-1))
    return F1(t)/(F1(t)+F2(t))
t = np.zeros(51,dtype=float)
Xfitted = np.zeros(51,dtype=float)
for i in range(51):
    t[i] = 2*i
    Xfitted[i] = X(t[i])*100
    i+=1
plt.plot(texp,Xexp,'ok', label='Experimental data')
plt.plot(t,Xfitted,'-k', label='Fit with Kca = 0.024 1/s')
plt.legend()
plt.title('Breakthrough Curve')
plt.show()
```

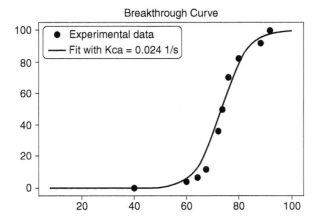

Appendix A
Binary Diffusion Coefficients

Table A.1 Mass Diffusivities in Gases

System	T, K	$D_{AB}P$, m^2·Pa/s	System	T, K	$D_{AB}P$, m^2·Pa/s
Air			Carbon dioxide		
Ammonia	273	2.006	Ethanol	273	0.702
Aniline	298	0.735	Ethyl ether	273	0.548
Benzene	298	0.974	Hydrogen	273	5.572
Bromine	293	0.923	Methane	273	1.550
Carbon dioxide	273	1.378	Methanol	298	1.064
Carbon disulfide	273	0.894	Nitrogen	298	1.672
Chlorine	273	1.256	Nitrous oxide	298	1.185
Diphenyl	491	1.621	Propane	298	0.874
Ethyl acetate	273	0.718	Water	298	1.661
Ethanol	298	1.337	Hydrogen		
Ethyl ether	293	0.908	Ammonia	293	8.600
Iodine	298	0.845	Argon	293	7.800
Methanol	298	1.641	Benzene	273	3.211
Mercury	614	4.791	Ethane	273	4.447
Naphthalene	303	0.870	Methane	273	6.331
Nitrobenzene	298	0.879	Oxygen	273	7.061
n-Octane	298	0.610	Pyridine	318	4.427
Oxygen	273	1.773	Nitrogen		
Propyl acetate	315	0.932	Ammonia	293	2.441
Sulfur dioxide	273	1.236	Ethylene	298	1.651
Toluene	298	0.855	Hydrogen	288	7.527
Water	298	2.634	Iodine	273	0.709
Ammonia			Oxygen	273	1.834
Ethylene	293	1.793	Oxygen		
Carbon dioxide			Ammonia	293	2.563
Benzene	318	0.724	Benzene	296	0.951
Carbon disulfide	318	0.724	Ethylene	293	1.844
Ethyl acetate	319	0.675	Water	308	2.857

Source: Welty, J. R., et al. 1984/John Wiley & Sons.

Principles and Applications of Mass Transfer: The Design of Separation Processes for Chemical and Biochemical Engineering, Fourth Edition. Jaime Benitez.
© 2023 John Wiley & Sons, Inc. Published 2023 by John Wiley & Sons, Inc.

552   **Appendix A Binary Diffusion Coefficients**

Table A 2. Mass Diffusivities in Liquids at Infinite Dilution

System	T, K	$D^0_{AB} \times 10^5$, cm^2/s	System	T, K	$D^0_{AB} \times 10^5$, cm^2/s
Chloroform (solvent B)			Ethyl acetate (solvent B)		
Acetone	298	2.35	Acetic acid	293	2.18
Benzene	288	2.51	Acetone	293	3.18
Ethanol	288	2.20	Ethyl benzoate	293	1.85
Ethyl ether	298	2.13	MEK	303	2.93
Ethyl acetate	298	2.02	Nitrobenzene	293	2.25
MEK	298	2.13	Water	298	3.20
Benzene (solvent B)			Water (solvent B)		
Acetic acid	298	2.09	Methane	298	1.49
Aniline	298	1.96	Air	298	2.00
Benzoic acid	298	1.38	Carbon dioxide	298	1.92
Bromobenzene	281	1.45	Chlorine	298	1.25
Cyclohexane	298	2.09	Argon	298	2.00
Ethanol	288	2.25	Benzene	298	1.02
Formic acid	298	2.28	Ethanol	288	1.00
n-Heptane	353	4.25	Ethane	298	1.20
MEK	303	2.09	Oxygen	298	2.10
Naphthalene	281	1.19	Pyridine	288	0.58
Toluene	298	1.85	Aniline	293	0.92
Vinyl chloride	281	1.77	Ammonia	298	1.64
Acetone (solvent B)			Ethylene	298	1.87
Acetic acid	288	2.92	Allyl alcohol	288	0.90
Acetic acid	313	4.04	Acetic acid	293	1.19
Benzoic acid	298	2.62	Benzoic acid	298	1.00
Formic acid	298	3.77	Propionic acid	298	1.06
Water	298	4.56	Vinyl chloride	298	1.34
Ethanol (solvent B)			Ethylbenzene	293	0.81
Benzene	298	1.81	Sulfuric acid	298	1.73
Water	298	1.24	Nitric acid	298	2.60

Sources: Reid, R. C., J. M. Prausnitz, and B. E. Poling, *The Properties of Gases and Liquids*, 4th ed., McGraw-Hill, New York (1987); Cussler, E. L., *Diffusion, Mass Transfer in Fluid Systems*, 2nd ed., Cambridge University Press, Cambridge, UK (1997).

Appendix A Binary Diffusion Coefficients

Table A.3 Mass Diffusivities in the Solid State

System	T, K	Diffusivity, cm^2/s
Hydrogen in iron	283	1.66×10^{-9}
	323	11.4×10^{-9}
	373	12.4×10^{-8}
Hydrogen in nickel	358	11.6×10^{-9}
	438	10.5×10^{-8}
Carbon monoxide in nickel	1223	4.00×10^{-8}
	1323	14.0×10^{-8}
Aluminum in copper	293	1.30×10^{-30}
	1123	2.2×10^{-9}
Uranium in tungsten	2000	1.3×10^{-11}
Cerium in tungsten	2000	95×10^{-11}
Yttrium in tungsten	2000	1820×10^{-11}
Tin in lead	558	1.60×10^{-10}
Gold in lead	558	4.60×10^{-10}
Gold in silver	1033	3.60×10^{-10}
Antimony in silver	293	3.60×10^{-10}
Zinc in aluminum	773	2.00×10^{-9}
Silver in aluminum	323	1.20×10^{-9}
Bismuth in lead	293	1.10×10^{-16}
Cadmium in copper	293	2.70×10^{-15}
Carbon in iron	1073	1.50×10^{-8}
	1373	45.0×10^{-8}
Helium in silica	293	4.00×10^{-10}
	773	7.80×10^{-8}
Hydrogen in silica	473	6.50×10^{-10}
	773	1.30×10^{-8}
Helium in Pyrex	293	4.50×10^{-11}
	773	2.00×10^{-8}

Sources: Barrer, R. M., *Diffusion in and through Solids*, Macmillan, New York (1941); American Society for Metals, *Diffusion*, ASM (1973); Cussler, E. L., *Diffusion, Mass Transfer in Fluid Systems*, 2nd ed., Cambridge University Press, Cambridge, UK (1997).

Appendix B
Lennard–Jones Constants

Determined from Viscosity Data

Compound		σ, Å	ϵ/κ, K
Ar	Argon	3.542	93.3
He	Helium	2.551	10.22
Kr	Krypton	3.655	178.9
Ne	Neon	2.820	32.8
Xe	Xenon	4.047	231.0
Air	Air	3.620	97.0
AsH_3	Arsine	4.145	259.8
BCl_3	Boron chloride	5.127	337.7
BF_3	Boron fluoride	4.198	186.3
$B(OCH_3)_3$	Methyl borate	5.503	396.7
Br_2	Bromine	4.296	507.9
CCl_4	Carbon tetrachloride	5.947	322.7
CF_4	Carbon tetrafluoride	4.662	134.0
$CHCl_3$	Chloroform	5.389	340.2
CH_2Cl_2	Methylene chloride	4.898	356.3
CH_3Br	Methyl bromide	4.118	449.2
CH_3Cl	Methyl chloride	4.182	350.0
CH_3OH	Methanol	3.626	481.8
CH_4	Methane	3.758	148.6
CO	Carbon monoxide	3.690	91.7
COS	Carbonyl sulfide	4.130	336.0
CO_2	Carbon dioxide	3.941	195.2
CS_2	Carbon disulfide	4.483	467.0
C2H2	Acetylene	4.033	231.8
C2H4	Ethylene	4.163	224.7
C2H6	Ethane	4.443	215.7
C_2H_5Cl	Ethyl chloride	4.898	300.0
C_2H_5OH	Ethanol	4.530	362.6
C2N2	Cyanogen	4.361	348.6
CH_3OCH_3	Methyl ether	4.307	395.0
CH_2CHCH_3	Propylene	4.678	298.9
CH_3CHH	Methylacetylene	4.761	251.8
C_3H_6	Cyclopropane	4.807	248.9
C_3H_8	Propane	5.118	237.1

Principles and Applications of Mass Transfer: The Design of Separation Processes for Chemical and Biochemical Engineering, Fourth Edition. Jaime Benitez.
© 2023 John Wiley & Sons, Inc. Published 2023 by John Wiley & Sons, Inc.

Lennard–Jones Constants Determined from Viscosity Data

Compound		σ, Å	ϵ/κ, K
$n\text{-}C_3H_7OH$	n-Propyl alcohol	4.549	576.7
CH_3COCH_3	Acetone	4.600	560.2
CH_3COOCH_3	Methyl acetate	4.936	469.8
$n\text{-}C_4H_{10}$	n-Butane	4.687	531.4
iso-C_4H_{10}	Isobutane	5.278	330.1
$C_2H_5OC_2H_5$	Ethyl ether	5.678	313.8
$CH_3COOC_2H_5$	Ethyl acetate	5.205	521.3
$n\text{-}C_5H_{12}$	n-Pentane	5.784	341.1
$C(CH_3)_4$	2,2-Dimethylpropanone	6.464	193.4
C_6H_6	Benzene	5.349	412.3
C_6H_{12}	Cyclohexane	6.182	297.1
$n\text{-}C_6H_{14}$	n-Hexane	5.949	399.3
Cl_2	Chlorine	4.217	316.0
F_2	Fluorine	3.357	112.6
HBr	Hydrogen bromide	3.353	449.0
HCN	Hydrogen cyanide	3.630	569.1
HCl	Hydrogen chloride	3.339	344.7
HF	Hydrogen fluoride	3.148	330.0
HI	Hydrogen iodide	4.211	288.7
H_2	Hydrogen	2.827	59.7
H_2O	Water	2.641	809.1
H_2O_2	Hydrogen peroxide	4.196	289.3
H_2S	Hydrogen sulfide	3.623	301.1
Hg	Mercury	2.969	750.0
$HgBr_2$	Mercuric bromide	5.080	686.2
$HgCl_2$	Mercuric chloride	4.550	750.0
HgI_2	Mercuric iodide	5.625	695.6
I_2	Iodine	5.160	474.2
NH_3	Ammonia	2.900	558.3
NO	Nitric oxide	3.492	116.7
N_2	Nitrogen	3.798	71.4
N_2O	Nitrous oxide	3.828	232.4
O_2	Oxygen	3.467	106.7
SF_6	Sulfur hexafluoride	5.128	222.1
SO_2	Sulfur dioxide	4.112	335.4
UF_6	Uranium hexafluoride	5.967	236.8

Source: Reid, R. C., J. M. Praunitz, and B. E. Poling, *The Properties of Gases and Liquids*, 4th ed., McGraw-Hill, New York (1987).

Appendix C-1
Maxwell–Stefan Equations (Mathcad)

Enter data:

$\text{ORIGIN} := 1 \qquad P := 1.013 \text{ bar} \qquad T := 548.0 \text{ K} \qquad \delta := 1.0 \text{ mm} \qquad num := 100$

$R := 8.314 \dfrac{\text{J}}{\text{mol} \cdot \text{K}} \qquad D12 := 72 \dfrac{\text{mm}^2}{\text{s}} \qquad D13 := 230 \dfrac{\text{mm}^2}{\text{s}} \qquad D23 := 230 \dfrac{\text{mm}^2}{\text{s}}$

Preliminary calculations:

$c := \dfrac{P}{R \cdot T} = 22.234 \dfrac{\text{mol}}{\text{m}^3} \qquad F12 := \dfrac{c \cdot D12}{\delta} = 1.601 \dfrac{\text{mol}}{\text{m}^2 \cdot \text{s}}$

$F13 := \dfrac{c \cdot D13}{\delta} = 5.114 \dfrac{\text{mol}}{\text{m}^2 \cdot \text{s}} \qquad r_1 := \dfrac{F13}{F12} = 3.194$

$k_r := 10 \dfrac{\text{mol}}{\text{m}^2 \cdot \text{s}} \qquad \kappa := \dfrac{k_r}{F12} = 6.247$

$F23 := \dfrac{c \cdot D23}{\delta} = 5.114 \dfrac{\text{mol}}{\text{m}^2 \cdot \text{s}} \qquad r_2 := \dfrac{F23}{F12} = 3.194$

Define dimensionless dependent variables:

$u_1 = y_1 \qquad u_2 = y_2 \qquad u_3 = \dfrac{N1}{F12} \qquad u_4 = \dfrac{N2}{F12} \qquad u_5 = \dfrac{N3}{F12}$

Fourth order Runge-Kutta:

$$D(\eta, u) := \begin{bmatrix} \dfrac{u_1 \cdot u_4 - u_2 \cdot u_3}{1} + \dfrac{u_1 \cdot u_5 - \left(1 - u_1 - u_2\right) \cdot u_3}{r_1} \\ \dfrac{u_2 \cdot u_3 - u_1 \cdot u_4}{1} + \dfrac{u_2 \cdot u_5 - \left(1 - u_1 - u_2\right) \cdot u_4}{r_2} \\ 0 \\ 0 \\ 0 \end{bmatrix}$$

Matrix of first derivatives

$$S1(u30, u40, u50) := \text{rkfixed}\left(\begin{bmatrix} 0.60 \\ 0.20 \\ u30 \\ u40 \\ u50 \end{bmatrix}, 0, 1, num, D \right)$$

Function in terms of unknown initial values and Runge-Kutta solution

$u1(u30, u40, u50) := S1(u30, u40, u50)^{(2)}$

$u2(u30, u40, u50) := S1(u30, u40, u50)^{(3)}$

The solutions u1 and u2 are the second and third columns of S1 and depend on the guess values of u30, u40, and u50.

Principles and Applications of Mass Transfer: The Design of Separation Processes for Chemical and Biochemical Engineering, Fourth Edition. Jaime Benitez.
© 2023 John Wiley & Sons, Inc. Published 2023 by John Wiley & Sons, Inc.

Appendix C-1 Maxwell–Stefan Equations (Mathcad)

Shooting method:

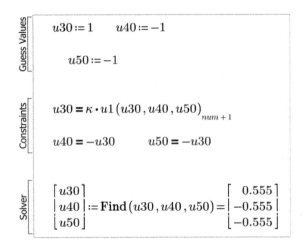

$$N_1 := u30 \cdot F12 = 0.888 \, \frac{mol}{m^2 \cdot s}$$

$$y_1 := u1(u30, u40, u50) \qquad y_2 := u2(u30, u40, u50)$$

$$y_3 := 1 - y_1 - y_2 \qquad z := \delta \cdot S1(u30, u40, u50)^{\langle 1 \rangle}$$

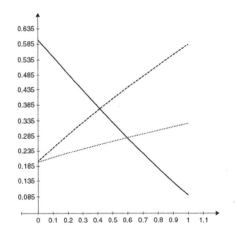

Interface composition:

$$y1_{num+1} = 0.089$$

$$y2_{num+1} = 0.584$$

$$y3_{num+1} = 0.328$$

Appendix C-2
Maxwell–Stefan Equations
(Python)

```python
import numpy as np
import scipy as sp
from scipy.optimize import fsolve
import matplotlib.pyplot as plt
from scipy.integrate import odeint

# Data are from Example 1-18
# Enter data:
# Pressure in bar; temperature in K; length diffusion path in m
P = 1.013; T = 548.0; delta = 0.001
#Ideal gas constant in bar*m^3/mol-K; diffusivities in m^2/s
R = 8.314e-05; D12 = 72.0e-06; D13 = 230.0e-06; D23 = 230.0e-06
kr = 10.0 #Reaction rate constant in mol/m^2-s

# Preliminary calculations:
c = P/(R*T); F12 = c*D12/delta; F13 = c*D13/delta; F23 = c*D23/delta
r1 = F13/F12; r2 = F23/F12; kappa = kr/F12

# Define dimensionless dependent variables:
# u0 = y1; u1 = y2; u2 = N1/F12; u3 = N2/F12; n4 = N3/F12
# Fourth-order Runge-Kutta method; Shooting method
def D(u,t):
    z = np.zeros(5,dtype=float)
    z[0] = u[0]*u[3] - u[1]*u[2] + (u[0]*u[4] - (1 - u[0] - u[1])*u[2])/r1
    z[1] = u[1]*u[2] - u[0]*u[3] + (u[1]*u[4] - (1 - u[0] - u[1])*u[3])/r2
    z[2] = 0
    z[3] = 0
    z[4] = 0
    return[z[0], z[1], z[2], z[3], z[4]]
def r(u0):
    a = u0[0]
    b = u0[1]
    c = u0[2]
    distance = np.arange(0,1.01,0.01)
    z = odeint(D, [0.600, 0.200, a, b, c], distance)
    return a - kappa*z[100,0], a + b, a + c #Bootstrap conditions
u0 = np.zeros(3,dtype=float)
u0[0] = 1
u0[1] = -1
```

Principles and Applications of Mass Transfer: The Design of Separation Processes for Chemical and Biochemical Engineering, Fourth Edition. Jaime Benitez.
© 2023 John Wiley & Sons, Inc. Published 2023 by John Wiley & Sons, Inc.

```python
u0[2] = -1
u0 = fsolve(r,u0)
N1 = u0[0]*F12; N2 = u0[1]*F12;N3 = u0[2]*F12


# Plot the concentration profiles:
distance = np.arange(0,1.01,0.01)
u = odeint(D, [0.600, 0.200, u0[0], u0[1], u0[2]], distance)
y1 = np.zeros(101,dtype=float)
y2 = np.zeros(101,dtype=float)
y3 = np.zeros(101,dtype=float)
for i in range(101):
    y1[i] = u[i,0]
    y2[i] = u[i,1]
    y3[i] = 1 - y1[i] - y2[i]
    i+=1
#Calculate distance along diffusion path in mm.
x = distance*delta*1000
plt.plot(x, y1,'-.k', label = 'Ethanol')
plt.plot(x, y2, '--k',label='Acetaldehyde')
plt.plot(x, y3, '-k',label='Hydrogen')
plt.legend()
plt.xlabel('Distance along diffusion path, mm')
plt.ylabel('Molar fraction')
plt.title('Concentration Profiles')
plt.grid(True)
plt.show()
print('Overall rate of reaction (mol/m^2-s) = %5.3f' %N1)


# Interface Composition
print(' '); print('Interface composition:')
print('Ethanol mole fraction = %5.3f' %y1[100])
print('Acetaldehyde mole fraction = %5.3f' %y2[100])
print('Hydrogen mole fraction = %5.3f' %y3[100])
```

Appendix C-2 Maxwell–Stefan Equations (Python)

Program results:
Overall rate of reaction = 0.888 mol/m^2-s
Interface composition:
Ethanol mol fraction = 0.089
Acetaldehyde mol fraction = 0.584
Hydrogen mol fraction = 0.328

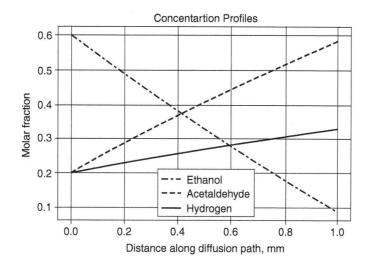

Appendix D-1
Packed–Column Design
(Mathcad)

This program calculates the diameter of a packed column to satisfy a given pressure drop criterion and also estimates the volumetric mass-transfer coefficients. Data presented are from Example 4.4.

Enter data related to the gas and liquid streams:

Enter liquid flow rate

$$mL := 0.804 \ \frac{kg}{s}$$

Enter gas flow rate

$$mG := 2.202 \ \frac{kg}{s}$$

Enter liquid density

$$\rho L := 986 \ \frac{kg}{m^3}$$

Enter gas density

$$\rho G := 1.923 \ \frac{kg}{m^3}$$

Enter liquid viscosity

$$\mu L := 0.000631 \ Pa \cdot s$$

Enter gas viscosity

$$\mu G := 1.45 \cdot 10^{-5} \ Pa \cdot s$$

Enter temperature

$$T := 303 \ K$$

Enter total pressure

$$P := 1.1 \cdot 10^{5} \cdot Pa$$

Enter the gas-phase and liquid-phase diffusion coefficients

$$DL := 1.91 \cdot 10^{-9} \cdot \frac{m^2}{s}$$

$$DG := 8.5 \cdot 10^{-6} \cdot \frac{m^2}{s}$$

Enter the value of the gas constant:

$$R := 8.314 \ \frac{J}{mol \cdot K}$$

Allowed pressure drop, Pa/m

$$\Delta P = 100 \ \frac{Pa}{m}$$

Enter data related to the packing:

Enter loading contant, CST

$$C_{ST} := 2.702$$

Enter specific area

$$a := 92.3 \ m^{-1}$$

Principles and Applications of Mass Transfer: The Design of Separation Processes for Chemical and Biochemical Engineering, Fourth Edition. Jaime Benitez.
© 2023 John Wiley & Sons, Inc. Published 2023 by John Wiley & Sons, Inc.

563

564   Appendix D-1 Packed–Column Design (Mathcad)

Enter flooding constant, CFT

$$C_{FT} := 1.626 \qquad\qquad\qquad\qquad g := 9.8 \, \frac{m}{s^2}$$

Enter porosity fraction   Enter loading constant

$$\varepsilon := 0.977 \qquad\qquad\qquad\qquad C_h := 0.876$$

Enter pressure drop constant

$$C_p := 0.421$$

Enter packing coefficients for mass-transfer correlations:

$$C_L := 1.168 \qquad\qquad\qquad\qquad C_V := 0.408$$

Calculate the flow parameter, X $X := \dfrac{mL}{mG} \cdot \sqrt{\dfrac{\rho G}{\rho L}} = 0.016$

$$C_S := \begin{Vmatrix} \text{if } X < 0.4 \\ \quad \| C_{ST} \\ \text{else} \\ \quad \| \; 0.695 \cdot C_{ST} \cdot \left(\dfrac{\mu L}{\mu G}\right)^{0.1588} \end{Vmatrix} \qquad = 2.702$$

$$n_s := \begin{Vmatrix} \text{if } X < 0.4 \\ \quad \| -0.326 \\ \text{else} \\ \quad \| -0.723 \end{Vmatrix} = -0.326 \qquad C_F := \begin{Vmatrix} \text{if } X < 0.4 \\ \quad \| C_{FT} \\ \text{else} \\ \quad \| \; 0.6244 \cdot C_{FT} \cdot \left(\dfrac{\mu L}{\mu G}\right)^{0.1028} \end{Vmatrix} \qquad = 1.626$$

$$n_F := \begin{Vmatrix} \text{if } X < 0.4 \\ \quad \| -0.194 \\ \text{else} \\ \quad \| -0.708 \end{Vmatrix} = -0.194$$

Appendix D-1 Packed–Column Design (Mathcad) 565

Guess Values

$$v_{LS} := .002 \, \frac{m}{s} \quad v_{GS} := 3 \, \frac{m}{s} \qquad \psi_s := 1$$

Constraints

$$v_{GS} = {}^{2}\!\sqrt{\frac{g}{\psi_s}} \cdot \left(\frac{\varepsilon}{a^{\frac{1}{6}}} - a^{\frac{1}{2}} \cdot \left(\frac{12}{g} \cdot \frac{\mu L}{\rho L} \cdot v_{LS} \right)^{\frac{1}{3}} \right) \cdot \left(\frac{12}{g} \cdot \frac{\mu L}{\rho L} \cdot v_{LS} \right)^{\frac{1}{6}} \cdot {}^{2}\!\sqrt{\frac{\rho L}{\rho G}}$$

$$\psi_s = \frac{9.8}{C_S^{\,2}} \cdot \left(X \cdot \left(\frac{\mu L}{\mu G} \right)^{0.4} \right)^{-2 \cdot n_s} \qquad\qquad v_{LS} = v_{GS} \cdot \frac{mL}{mG} \cdot \frac{\rho G}{\rho L}$$

Solver

$$\begin{bmatrix} v_{LS} \\ v_{GS} \\ \psi_s \end{bmatrix} := \mathrm{Find}\,(v_{LS}, v_{GS}, \psi_s) = \begin{bmatrix} 0.001 \, \frac{m}{s} \\ 2.096 \, \frac{m}{s} \\ 0.244 \end{bmatrix}$$

Guess Values

$$v_{LF} := .005 \, \frac{m}{s} \qquad v_{GF} := 3.0 \, \frac{m}{s} \quad \psi_F := 1$$

$$h_{LF} := 0.5$$

Constraints

$$\psi_F = \frac{9.8}{C_F^{\,2}} \cdot \left(X \cdot \left(\frac{\mu L}{\mu G} \right)^{0.2} \right)^{-2 \cdot n_F}$$

$$v_{LF} = v_{GF} \cdot \frac{mL}{mG} \cdot \frac{\rho G}{\rho L}$$

$$v_{GF} = \sqrt{\frac{2 \cdot g \cdot (\varepsilon - h_{LF})^3 \cdot h_{LF} \cdot \rho L}{\psi_F \cdot \varepsilon \cdot a \cdot \rho G}}$$

$$h_{LF}^{\,3} \cdot (3 \cdot h_{LF} - \varepsilon) = \frac{6 \cdot a^2 \cdot \varepsilon \cdot \mu L \cdot mL \cdot \rho G \cdot v_{GF}}{g \cdot \rho L^{\,2} \cdot mG}$$

Solver

$$\begin{bmatrix} v_{LF} \\ v_{GF} \\ \psi_F \\ h_{LF} \end{bmatrix} := \mathrm{Find}\,(v_{LF}, v_{GF}, \psi_F, h_{LF}) = \begin{bmatrix} 0.002 \, \frac{m}{s} \\ 3.164 \, \frac{m}{s} \\ 1.002 \\ 0.326 \end{bmatrix}$$

566   Appendix D-1 Packed–Column Design (Mathcad)

$$QG := \frac{mG}{\rho G} = 1.145 \ \frac{\text{m}^3}{\text{s}} \qquad\qquad QL := \frac{mL}{\rho L} = \left(8.154 \cdot 10^{-4}\right) \frac{\text{m}^3}{\text{s}}$$

$$ReLs := \frac{v_{LS} \cdot \rho L}{a \cdot \mu L} = 25.265 \qquad\qquad FrLs := \frac{v_{LS}^{\ 2} \cdot a}{g} = 2.098 \cdot 10^{-5}$$

$$ahs := \begin{Vmatrix} \text{if } ReLs \geq 5 \\ \quad \begin{Vmatrix} 0.85 \cdot a \cdot C_h \cdot ReLs^{0.25} \cdot FrLs^{0.1} \end{Vmatrix} \\ \text{else} \\ \quad \begin{Vmatrix} a \cdot C_h \cdot ReLs^{0.5} \cdot FrLs^{0.1} \end{Vmatrix} \end{Vmatrix} = 52.472 \ \frac{1}{\text{m}}$$

$$hLs := \left(\frac{12 \cdot FrLs}{ReLs}\right)^{0.333} \cdot \left(\frac{ahs}{a}\right)^{0.667} = 0.015$$

Calculate effective particle size

$$dp := \frac{6 \cdot (1 - \varepsilon)}{a} = \left(1.495 \cdot 10^{-3}\right) \text{m}$$

Correlations in terms of the unknown diameter:

$$vL(D) := \frac{4 \cdot QL}{\pi \cdot D^2} \qquad ReL(D) := \frac{vL(D) \cdot \rho L}{a \cdot \mu L}$$

$$FrL(D) := \frac{vL(D)^2 \cdot a}{g} \qquad vG(D) := \frac{4 \cdot QG}{\pi \cdot D^2} \qquad KW(D) := \left[1 + \frac{2 \cdot dp}{3 \cdot D \cdot (1 - \varepsilon)}\right]^{-1}$$

$$ReG(D) := \frac{vG(D) \cdot dp \cdot \rho G \cdot KW(D)}{(1 - \varepsilon) \cdot \mu G} \qquad \Psi 0(D) := C_p \cdot \left(\frac{64}{ReG(D)} + \frac{1.8}{ReG(D)^{0.08}}\right)$$

$$\Delta P0(D) := \frac{\Psi 0(D) \cdot a \cdot \rho G \cdot vG(D)^2}{2 \cdot \varepsilon^3 \cdot KW(D)}$$

$$ah(D) := \begin{Vmatrix} \text{if } ReL(D) \geq 5 \\ \quad \begin{Vmatrix} 0.85 \cdot a \cdot C_h \cdot ReL(D)^{0.25} \cdot FrL(D)^{0.1} \end{Vmatrix} \\ \text{else} \\ \quad \begin{Vmatrix} a \cdot C_h \cdot ReL(D)^{0.5} \cdot FrL(D)^{0.1} \end{Vmatrix} \end{Vmatrix}$$

Appendix D-1 Packed–Column Design (Mathcad)   567

$$hL(D) := \left(\frac{12 \cdot FrL(D)}{ReL(D)}\right)^{0.333} \cdot \left(\frac{ah(D)}{a}\right)^{0.667}$$

Iterate to find the tower diameter for the given pressure drop

$$\Delta P \equiv 100 \ \frac{\text{Pa}}{\text{m}}$$

Guess Values

$$D := 0.8 \ \text{m}$$

Constraints

$$\Delta P = \Delta P0(D) \cdot \left(\frac{\varepsilon}{\varepsilon - hL(D)}\right)^{1.5} \cdot \left(\left(\frac{hL(D)}{hLs}\right)^{0.3} \cdot exp\left(13300 \ \sqrt{\frac{FrL(D)}{(a \cdot m)^3}}\right)\right)$$

$$D > 0$$

Solver

$$D := \text{Find}(D) = 0.938 \ \text{m} \qquad \textbf{Column diameter}$$

$$\Delta P0(D) = 97.116 \ \frac{\text{Pa}}{\text{m}}$$

Calculate the fractional approach to flooding:

$$vG(D) = 1.656 \ \frac{\text{m}}{\text{s}} \qquad\qquad f := \frac{vG(D)}{v_{GF}} = 0.523 \qquad\qquad v_{GS} = 2.096 \ \frac{\text{m}}{\text{s}}$$

Calculate the specific interfacial area: $\qquad ah(D) = 47.194 \ \dfrac{1}{\text{m}}$

Calculate the liquid fractional holdup: $\qquad hL(D) = 0.013$

Calculate the liquid superficial velocity: $\qquad vL(D) = 1.179 \ \dfrac{\text{mm}}{\text{s}}$

Calculate the liquid-phase volumetric mass-transfer coefficient:

$$kL(D) := 0.757 \cdot C_L \cdot \sqrt{\frac{DL \cdot a \cdot vL(D)}{\varepsilon \cdot hL(D)}} \qquad\qquad kL(D) = \left(1.141 \cdot 10^{-4}\right) \frac{\text{m}}{\text{s}}$$

$$kL(D) \cdot ah(D) = \left(5.387 \cdot 10^{-3}\right) \frac{1}{\text{s}} \qquad\qquad ReG(D) = \left[1.365 \cdot 10^4\right]$$

Appendix D-1 Packed–Column Design (Mathcad)

Calculate the gas-phase volumetric mass-transfer coefficient:

$$ScG := \frac{\mu G}{\rho G \cdot DG} = 0.887 \qquad\qquad ReL(D) = 19.962$$

$$ky(D) := 0.1304 \cdot C_V \cdot DG \cdot \frac{P \cdot a \cdot \left(\dfrac{ReG(D)}{KW(D)}\right)^{0.75} \cdot ScG^{0.333}}{R \cdot T \cdot \left[\varepsilon \cdot (\varepsilon - hL(D))\right]^{0.5}}$$

$$ky(D) = [2.357]\,\frac{\text{mol}}{\text{m}^2 \cdot \text{s}} \qquad\qquad ky(D) \cdot ah(D) = [111.214]\,\frac{\text{mol}}{\text{m}^3 \cdot \text{s}}$$

Appendix D-2
Packed–Column Design
(Python)

```python
import numpy as np
import scipy as sp
from scipy.optimize import fsolve
import matplotlib.pyplot as plt
# Data are from Example 4.4
# Enter data related to the gas and liquid streams:
#Liquid and gas flow rates, in kg/s
mL = 0.804; mG = 2.202
#Liquid and gas densities, in kg/m^3
rhoL = 986; rhoG = 1.923
#Liquid and gas viscosities, in Pa-s
muL = 6.31e-04; muG = 1.45e-05
#Temperature, in K; total pressure, in Pa
T = 303; P = 1.1e05
#Diffusion coefficients, in m^2/s
DL = 1.91e-09; DG = 8.5e-06
# Ideal gas constant, in J/mol-K; allowed pressure drop, Pa/m
R = 8.314; DelP_allowed = 100
#Enter data related to the packing:
#Loading constant, dimensionless; specific area, 1/m
CST = 2.702; a = 92.3
# Flooding constant, dimensionless; acceleation of gravity, m/s^2
CFT = 1.626; g = 9.8
# Porosity and loading constant, dimensionless
eps = 0.977; Ch = 0.876
Cp = 0.421; CL = 1.168; CV = 0.408
# Calculate effective particle size (m):
dp = 6*(1-eps)/a

# Calculate the flow parameter, X, and related parameters:
X = (mL/mG)*np.sqrt(rhoG/rhoL)
if X < 0.4:
    CS = CST; CF = CFT; ns = -0.326; nF = -0.194
```

Principles and Applications of Mass Transfer: The Design of Separation Processes for Chemical and Biochemical Engineering, Fourth Edition. Jaime Benitez.
© 2023 John Wiley & Sons, Inc. Published 2023 by John Wiley & Sons, Inc.

570 Appendix D-2 Packed–Column Design (Python)

```python
else:
    CS = 0.695*CST*(muL/muG)**0.1588
    CF = 0.6244*CFT*(muL/muG)**0.1028
    ns = -0.723; nF = -0.708
PsiS = (9.8/CS**2)*(X*(muL/muG)**0.4)**(-2*ns)
PsiF = (9.8/CF**2)*(X*(muL/muG)**0.2)**(-2*nF)
KS1 = np.sqrt(g*rhoL/(PsiS*rhoG))
KS2 = (12*muL/(g*rhoL))**(1/3)
KF1 = np.sqrt(2*g*rhoL/(PsiF*eps*a*rhoG))
KF2 = 6*a**2*eps*muL*mL*rhoG/(g*rhoL**2*mG)

# Calculate loading and flooding velocities:
def loading(p):
    vLS = p[0]; vGS = p[1]
    return vLS - vGS*(mL/mG)*(rhoG/rhoL), vGS-KS1*(eps/a**(1/6)-\
                np.sqrt(a)*KS2*vLS**(1/3))*np.sqrt(KS2)*vLS**(1/6)
p = np.zeros(2,dtype=float)
p[0] = 0.002; p[1] = 3 # Initial estimates
p = fsolve(loading, p)
vLS = p[0]
vGS = p[1]
print('The loading liquid velocity, in m/s, vLS = %5.4f' %vLS)
print('The loading gas velocity, in m/s, vGS = %5.3f' %vGS)
def flooding(m):
    vLF = m[0]; vGF = m[1]; hLF = m[2]
    return vLF - vGF*(mL/mG)*(rhoG/rhoL), vGF-KF1*np.sqrt(hLF*\
                (eps-hLF)**3), hLF**3*(3*hLF-eps)-KF2*vGF
m = np.zeros(3,dtype=float)
m[0] = 0.002; m[1] = 3; m[2] = 0.5 # Initial estimates
m = fsolve(flooding, m)
vLF = m[0]
vGF = m[1]
hLF = m[2]
print('The flooding liquid velocity, in m/s, vLF = %5.4f' %vLF)
print('The flooding gas velocity, in m/s, vGF = %5.3f' %vGF)
print('The flooding specific liquid holdup, hLF = %5.3f' %hLF)

# Iterate to find the tower diameter for the allowed pressure drop
# Calculate the volumetric flow rates, in m^3/s
QG = mG/rhoG; QL = mL/rhoL
ReLS = vLS*rhoL/(a*muL)
FrLS = vLS**2*a/g
if ReLS>=5:
    ahS = 0.85*a*Ch*ReLS**0.25*FrLS**0.1
else:
    ahS = a*Ch*ReLS**0.5*FrLS**0.1
hLS = (12*FrLS/ReLS)**0.333*(ahS/a)**0.667
def vL(D):
    return 4*QL/(np.pi*D**2)
def ReL(D):
    return vL(D)*rhoL/(a*muL)
```

Appendix D-2 Packed–Column Design (Python)   571

```python
def FrL(D):
    return vL(D)**2*a/g
def vG(D):
    return 4*QG/(np.pi*D**2)
def KW(D):
    return 1/(1 + 2*dp/(3*D*(1 - eps)))
def ReG(D):
    return vG(D)*dp*rhoG*KW(D)/((1 - eps)*muG)
def psi0(D):
    return Cp*(64/ReG(D) + 1.8/ReG(D)**0.08)
def DelP0(D):
    return psi0(D)*a*rhoG*vG(D)**2/(2*eps**3*KW(D))
def ah(D):
    if ReL(D) >= 5:
        return 0.85*a*Ch*ReL(D)**0.25*FrL(D)**0.1
    else:
        return a*Ch*ReL(D)**0.5*FrL(D)**0.1
def hL(D):
    return (12*FrL(D)/ReL(D))**0.333*(ah(D)/a)**0.667
def pressure(D):
    return DelP_allowed - DelP0(D)*(eps/(eps-hL(D)))**1.5*\
    (hL(D)/hLS)**0.3*np.exp(13300*np.sqrt(FrL(D)/a**3))

D = fsolve(pressure, 2.0)
print('The column diameter, in m, D = %5.3f' %D)

# Calculate various characteristics of the final design:
# Fractional approach to flooding, f:
f = vG(D)/vGF
print('Fractional approach to flooding, f = %4.3f' %f)
# Specific interfacial area:
print('Specific interfacial area,in 1/m, a = %5.3f' %ah(D))
# Liquid superficial velocity:
print('Liquid superficial velocity, in m/s, = %6.5f' %vL(D))

# Calculate the liquid-phase volumetric mass-transfer coefficient:
kL = 0.757*CL*np.sqrt(DL*a*vL(D)/(eps*hL(D)))
kLah = kL*ah(D)
print('Liquid volumetric coefficient,1/s, kLah = %5.3e' %kLah)

# Calculate the gas-phase volumetric mass-transfer coefficient:
ScG = muG/(rhoG*DG)
ky = 0.1304*CV*DG*P*a*(ReG(D)/KW(D))**0.75*ScG**0.333/\
    (R*T*(eps*(eps - hL(D)))**0.5)
kyah = ky*ah(D)
print('Gas volumetric coefficient, mol/m^3*s, kyah = %7.3f' %kyah)
```

572   **Appendix D-2 Packed–Column Design (Python)**

Program results:

Loading liquid velocity, $v_{LS} = 0.0015$ m/s   Loading gas velocity, $v_{GS} = 2.096$ m/s

Flooding liquid velocity, $v_{LF} = 0.0023$ m/s   Flooding gas velocity, $v_{GF} = 3.164$ m/s

Flooding specific liquid holdup, $h_{LF} = 0.326$ Column diameter, $D = 0.938$ m

Fractional approach to flooding, $f = 0.523$ Specific interfacial area, $a_h = 47.194$ m^{-1}

Liquid superficial velocity, $v_L = 0.00118$ m/s

Liquid volumetric mass-transfer coefficient, $k_L a_h = 0.005387$ s^{-1}

Gas volumetric mass-transfer coefficient, $k_y a_h = 111.214$ mol/m^3-s

Appendix E-1
Sieve-Tray Design (Mathcad)

This program calculates the diameter of sieve-tray tower to satisfy an approach to flooding criterion, and estimates the tray efficiency. The data presented are from Example 4.10.

Enter data related to the gas and liquid streams:

Enter liquid flow rate

$$mL := 0.347 \frac{kg}{s}$$

Enter liquid density

$$\rho L := 791 \frac{kg}{m^3}$$

Enter liquid surface tension

$$\sigma := 21 \frac{dyne}{cm}$$

Enter temperature

$$T := 353 \text{ K}$$

Enter gas flow rate

$$mG := 0.494 \frac{kg}{s}$$

Enter gas density

$$\rho G := 1.18 \frac{kg}{m^3}$$

Enter gas viscosity

$$\mu G := 1.05 \cdot 10^{-5} \text{ Pa} \cdot s$$

Enter total pressure

$$P := 1.013 \cdot 10^5 \cdot \text{Pa}$$

Enter the gas-phase and liquid-phase diffusion coefficients

$$DL := 2.07 \cdot 10^{-9} \cdot \frac{m^2}{s}$$

$$DG := 1.58 \cdot 10^{-6} \cdot \frac{m^2}{s}$$

Enter the value of the gas constant

$$R := 8.314 \frac{J}{mol \cdot K}$$

Enter foaming factor
$$FF := 0.9$$

Enter the local slope of the equilibrium curve

$$me := 0.42$$

Enter water density at T

$$\rho W := 970 \frac{kg}{m^3}$$

Enter molecular weights of gas and liquid

$$ML := 32 \qquad MG := 34.2$$

Enter data related to the tray design:

Enter hole diameter and pitch $\qquad do := 4.5 \text{ mm} \qquad p := 12 \text{ mm}$

Principles and Applications of Mass Transfer: The Design of Separation Processes for Chemical and Biochemical Engineering, Fourth Edition. Jaime Benitez.
© 2023 John Wiley & Sons, Inc. Published 2023 by John Wiley & Sons, Inc.

574   **Appendix E-1 Sieve-Tray Design (Mathcad)**

Enter plate thickness: Enter weir height:

$$l := 2 \ \text{mm}$$ $$hw := 5 \ \text{cm}$$

Enter design fractional approach to flooding

$$f := 0.80$$

Calculate the flow parameter

$$X := \frac{mL}{mG} \cdot \sqrt{\frac{\rho G}{\rho L}} = 0.027$$

Specify the ratio of downcomer area to total area

$$AdAt := \begin{Vmatrix} \text{if } X \leq 0.1 \\ \quad \begin{Vmatrix} 0.1 \end{Vmatrix} \\ \text{else if } X \geq 0.1 \\ \quad \begin{Vmatrix} 0.2 \end{Vmatrix} \\ \text{else} \\ \quad \begin{Vmatrix} 0.1 + \dfrac{X - 0.1}{9} \end{Vmatrix} \end{Vmatrix} = 0.1$$

If X is smaller than 0.1, use $X = 0.1$ in equation (4-31)

$$X := \begin{Vmatrix} \text{if } X \geq 0.1 \\ \quad \begin{Vmatrix} X \end{Vmatrix} \\ \text{else} \\ \quad \begin{Vmatrix} 0.1 \end{Vmatrix} \end{Vmatrix} = 0.1$$

Calculate the ratio of hole-to-active area:

$$AhAa := 0.907 \cdot \left(\frac{do}{p} \right)^{2} = 0.128$$

Calculate *FHA*:

$$FHA := \begin{Vmatrix} \text{if } AhAa < 0.1 \\ \quad \begin{Vmatrix} 5 \cdot AhAa + 0.5 \end{Vmatrix} \\ \text{else} \\ \quad \begin{Vmatrix} 1.0 \end{Vmatrix} \end{Vmatrix} = 1$$

Appendix E-1 Sieve-Tray Design (Mathcad)   575

Calculate *FST*:

$$FST := \left(\frac{\sigma}{20 \dfrac{\text{dyne}}{\text{cm}}}\right)^{0.2} = 1.01 \qquad\qquad C1 := FF \cdot FST \cdot FHA = 0.909$$

Enter parameters:

$$\alpha 1 := 0.0744 \text{ m}^{-1} \qquad \alpha 2 := 0.01173$$

$$\beta 1 := 0.0304 \text{ m}^{-1} \qquad \beta 2 := 0.015$$

Iterate to find diameter and tray spacing (*t*):

$$\alpha(t) := \alpha 1 \cdot t + \alpha 2 \qquad \beta(t) := \beta 1 \cdot t + \beta 2$$

$$CF(t) := \alpha(t) \cdot \log\left(\frac{1}{X}\right) + \beta(t) \qquad C(t) := C1 \cdot CF(t) \dfrac{\text{m}}{\text{s}}$$

$$vGF(t) := C(t) \cdot \sqrt{\frac{\rho L - \rho G}{\rho G}} \qquad\qquad QG := \frac{mG}{\rho G} = 0.419 \dfrac{\text{m}^3}{\text{s}}$$

$$g1(D) := 0.5 \text{ m} \cdot \Phi(1 \text{ m} - D) + 0.6 \text{ m} \cdot (\Phi(3 \text{ m} - D) - \Phi(1 \text{ m} - D))$$

$$g2(D) := 0.75 \text{ m} \cdot (\Phi(4 \text{ m} - D) - \Phi(3 \text{ m} - D)) + 0.9 \text{ m} \cdot (\Phi(8 \text{ m} - D) - \Phi(4 \text{ m} - D))$$

Guess Values

$$D := 2 \text{ m} \qquad t := 0.5 \text{ m}$$

Constraints

$$D = \sqrt{\frac{4 \cdot QG}{f \cdot vGF(t) \cdot (1 - AdAt) \cdot \pi}}$$

$$t = g1(D) + g2(D)$$

Solver

$$\begin{bmatrix} D \\ t \end{bmatrix} := \text{Find}(D, t) = \begin{bmatrix} 0.631 \\ 0.5 \end{bmatrix} \text{ m}$$

Calculate some further details of the tray design:

$$At := \frac{\pi \cdot D^2}{4} = 0.313 \text{ m}^2 \qquad \text{Total area}$$

Appendix E-1 Sieve-Tray Design (Mathcad)

$Ad := AdAt \cdot At = 0.031 \text{ m}^2$ Downcomer area

$Aa := At - 2 \cdot Ad = 0.25 \text{ m}^2$ Active area

$Ah := AhAa \cdot Aa = 0.032 \text{ m}^2$ Hole area

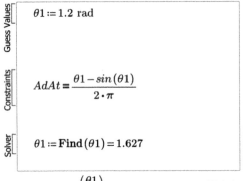

Guess Values: $\theta1 := 1.2 \text{ rad}$

Constraints: $AdAt = \dfrac{\theta1 - \sin(\theta1)}{2 \cdot \pi}$

Solver: $\theta1 := \text{Find}(\theta1) = 1.627$

$Lw := D \cdot \sin\left(\dfrac{\theta1}{2}\right) = 0.458 \text{ m}$ Weir length

$rw := \dfrac{D}{2} \cdot \cos\left(\dfrac{\theta1}{2}\right) = 0.217 \text{ m}$ Distance from tower center to weir

Estimate the gas pressure drop through the tray:

Dry tray head loss, *hd*:

$vo := \dfrac{QG}{Ah} = 13.129 \ \dfrac{\text{m}}{\text{s}}$

$Co := 0.85032 - 0.04231 \cdot \left(\dfrac{do}{l}\right) + 0.0017954 \left(\dfrac{do}{l}\right)^2 = 0.764$

$hd := \dfrac{0.472}{g} \cdot \left(\dfrac{\rho G}{\rho L}\right) \cdot \left(\dfrac{vo}{Co}\right)^2 = 0.021 \text{ m}$

Equivalent head of clear liquid, hl

$va := \dfrac{QG}{Aa} = 1.675 \ \dfrac{\text{m}}{\text{s}}$ $qL := \dfrac{mL}{\rho L} = (4.387 \cdot 10^{-4}) \ \dfrac{\text{m}^3}{\text{s}}$

Calculate capacity parameter   Calculate froth density

$Ks := va \cdot \sqrt{\dfrac{\rho G}{\rho L - \rho G}} = 0.065 \ \dfrac{\text{m}}{\text{s}}$ $\phi e := \exp\left(-12.55 \cdot \left(\dfrac{Ks}{\text{m} \cdot \text{s}^{-1}}\right)^{0.91}\right) = 0.354$

Appendix E-1 Sieve-Tray Design (Mathcad) 577

$$CL1 := 50.12 \; cm \cdot s^{\frac{2}{3}} \cdot m^{\frac{-4}{3}}$$

$$CL2 := 43.89 \; cm \cdot s^{\frac{2}{3}} \cdot m^{\frac{-4}{3}}$$

$$CL := CL1 + CL2 \cdot \exp\left(-1.378 \cdot cm^{-1} \cdot hw\right)$$

$$hl := \phi e \cdot \left[hw + CL \cdot \left(\frac{qL}{Lw \cdot \phi e} \right)^{\frac{2}{3}} \right] = 0.021 \; m$$

Head loss due to surface tension:

$$h\sigma := \frac{6 \cdot \sigma}{g \cdot \rho L \cdot do} = 0.004 \; m$$

Total head loss:

$$ht := hd + hl + h\sigma = 0.046 \; m$$

Convert head loss to pressure drop:

$$\Delta P := ht \cdot g \cdot \rho L = 356.295 \; Pa$$

Check tray design for excessive weeping; calculate orifice Froude number, *Fro*. If *Fro* > 0.5 there is no weeping problem.

$$Fro := \sqrt{\frac{\rho G \cdot vo^2}{\rho L \cdot g \cdot hl}} = 1.114$$

Calculate fractional entrainment, *E*:

$$\kappa := 0.5 \cdot \left[1 - \tanh\left(1.3 \cdot \ln\left(\frac{hl}{do} \right) - 0.15 \right) \right] = 0.024$$

$$h2\phi := \frac{hl}{\phi e} + 7.79 \cdot \left[1 + 6.9 \cdot \left(\frac{do}{hl} \right)^{1.85} \right] \cdot \frac{Ks^2}{\phi e \cdot g \cdot AhAa} = 0.163 \; m$$

$$E := 0.00335 \cdot \left(\frac{h2\phi}{t} \right)^{1.1} \cdot \left(\frac{\rho L}{\rho G} \right)^{0.5} \cdot \left(\frac{hl}{h2\phi} \right)^{\kappa} = 0.024$$

Calculate point efficiency, *EOG*:

$$ReFe := \frac{\rho G \cdot vo \cdot hl}{\mu G \cdot \phi e} = 8.814 \cdot 10^4$$

$$cG := \frac{\rho G}{MG} = 0.035 \; \frac{kg}{m^3}$$

$$cL := \frac{\rho L}{ML} = 24.719 \; \frac{kg}{m^3}$$

$$a1 := 0.4136 \qquad a2 := 0.6074 \qquad a3 := -0.3195$$

$$EOG := 1 - \exp\left(\frac{-0.0029}{1 + me \cdot \dfrac{cG}{cL} \cdot \sqrt{\dfrac{DG \cdot (1 - \phi e)}{DL \cdot AhAa}}} \cdot ReFe^{a1} \cdot \left(\frac{hl}{do}\right)^{a2} \cdot AhAa^{a3}\right) = 0.784$$

Calculate Murphree tray efficiency:

$$DEG := 0.01 \ \frac{m^2}{s} \qquad\qquad \frac{h2\phi}{t} = 0.325$$

$$PeG := \frac{4 \cdot QG \cdot rw^2}{DEG \cdot Aa \cdot (t - h2\phi)} = 93.218$$

$$DEL := 0.1 \cdot \sqrt{g \cdot h2\phi^3} = 0.021 \ \frac{m^2}{s} \qquad\qquad PeL := \frac{4 \cdot qL \cdot rw^2}{DEL \cdot Aa \cdot hl} = 0.76$$

$$N := \frac{PeL + 2}{2} = 1.38 \qquad\qquad \lambda := me \cdot \frac{mG}{MG} \cdot \frac{ML}{mL} = 0.559$$

For perfectly mixed vapor:

$$EMGmixed := \frac{\left(1 + \dfrac{\lambda \cdot EOG}{N}\right)^N - 1}{\lambda} = 0.829$$

For unmixed vapor:

$$EMGunmixed := EMGmixed \cdot \left(1 - 0.0335 \cdot \lambda^{1.07272} \cdot EOG^{2.51844} \cdot PeL^{0.17524}\right) = 0.821$$

$$EMG := \left\|\begin{array}{l} \text{if } 0 < PeG < 50 \\ \quad \left\| EMGmixed \right. \\ \text{else} \\ \quad \left\| EMGunmixed \right. \end{array}\right\| = 0.821$$

Correct tray efficiency for entrainment:

$$EMGE := EMG \cdot \left(1 - 0.8 \cdot EOG \cdot \lambda^{1.543} \cdot \frac{E}{me}\right) = 0.809$$

Appendix E-2
Sieve-Tray Design (Python)

```python
import numpy as np
import scipy as sp
from scipy.optimize import fsolve
import math

# Data presented are from Example 4-10
# Enter data related to the gas and liquid streams:
#  Liquid and gas flow rates, in kg/s
mL = 0.347; mG = 0.494
#  Liquid and gas densities, in kg/m^3
rhoL = 791; rhoG = 1.18
 # Liquid surface tension,dyne/cm; gas viscosity,Pa*s
sig = 21; muG = 1.05e-05
# Temperature, in K; total pressure, in Pa
T = 353;  P = 1.013e05
# Liquid and gas diffusion coefficients, in m^2/s
DL = 2.07e-09; DG = 1.58e-06
# Ideal gas constant,J/mol*K;slope of equilibrium curve, dimensionless
R = 8.314; me = 0.42
#  Foaming factor, dimensionless;water density at T, kg/m^3
FF = 0.9; rhoW = 970
# Liquid and gas molecular weights
ML = 32; MG = 34.2
# Enter data related to the tray design:
#Orifice diameter, in mm; pitch, in mm
do = 4.5; p = 12
# Plate thickness, in mm; weir height, in mm
l = 2; hw = 5
# Design fractional approach to flooding
f = 0.8

# Calculate the flow parameter, X:
X = (mL/mG)*np.sqrt(rhoG/rhoL)
# Specify the ratio of downcomer area to total area, AdAt:
def AdAt(X):
    if X <= 0.1:
        return 0.1
    elif X >= 1.0:
        return 0.2
```

Principles and Applications of Mass Transfer: The Design of Separation Processes for Chemical and Biochemical Engineering, Fourth Edition. Jaime Benitez.
© 2023 John Wiley & Sons, Inc. Published 2023 by John Wiley & Sons, Inc.

579

Appendix E-2 Sieve-Tray Design (Python)

```python
    else:
        return 0.1 + (X - 0.1)/9
# Calculate the ratio of hole-to-active area, AhAa:
AhAa = 0.907*(do/p)**2
# Calculate FHA:
def FHA(AhAa):
    if AhAa < 0.1:
        return 5*AhAa + 0.5
    else:
        return 1.0
# Calculate FST and C1:
FST = (sig/20)**0.2; C1 = FF*FST*FHA(AhAa)
# Modify the flow parameter before using it to calculate CF:
if X >=0.1:
    X = X
else:
    X = 0.1
# Relationship between tray spacing, t, and diameter, D:
def t(D):
    if D <= 1.0:
        return 0.50
    elif D <= 3.0:
        return 0.60
    elif D <= 4.0:
        return 0.75
    else:
        return 0.9
# Iterate to find diameter (D) and tray spacing (t):
def design(D):
    alpha = 0.0744*t(D) + 0.01173
    beta = 0.0304*t(D) + 0.015
    CF = alpha*np.log10(1/X) + beta
    C = C1*CF
    QG = mG/rhoG
    vGF = C*np.sqrt((rhoL - rhoG)/rhoG)
    return D - np.sqrt(4*QG/(f*vGF*(1 - AdAt(X))*np.pi))

D = fsolve(design, 2)
print('The tower diameter, in m, is D = %5.3f' %D)
print('The corresponding tray spacing, in m, is t = %5.3f' %t(D))

# Calculate further details of the tray design:
At = np.pi*D**2/4 # Total area:
Ad = AdAt(X)*At   #Downcomer area
Aa = At - 2*Ad   #Active area
Ah = AhAa*Aa   # Hole area
def f(theta):
    return 2*3.1416*AdAt(X)+math.sin(theta)-theta
```

Appendix E-2 Sieve-Tray Design (Python) 581

```python
theta = fsolve(f,1.2)
Lw = D*math.sin(theta/2)  #Weir length
rw = (D/2)*math.cos(theta/2)
# Estimate the gas pressure drop through the tray:
g = 9.8  #Gravity in m^2/s
QG = mG/rhoG; vo = QG/Ah
Co = 0.85032-0.042318*(do/l)+0.0017954*(do/l)**2
hd = (0.472/g)*(rhoG/rhoL)*(vo/Co)**2  # Dry tray head loss, m
va = QG/Aa; qL = mL/rhoL
Ks = va*(rhoG/(rhoL-rhoG))**0.5
phie = np.exp(-12.55*(Ks**0.91))  #Froth density
CL = 50.12+43.89*np.exp(-1.378*hw)
hl = phie*(hw+CL*(qL/(Lw*phie))**0.667)/100 #Head clear liquid, m
hsig = 6*sig/(g*rhoL*do) #Head loss due to surface tension, m
DelP = (hd+hl+hsig)*g*rhoL #Pressure drop, Pa
print('Gas pressure drop, in Pa = %5.3f' %DelP)


#Calculate the orifice Froude number:
Fro = (rhoG/(rhoL*g*hl))**0.5*vo
print('Orifice Froude number, Fro = %5.3f' %Fro)


# Calculate fractional entrainment:
kappa = 0.5*(1-math.tanh(1.3*np.log(1000*hl/do)-0.15))
h2phi = hl/phie+7.79*(1+6.9*(0.001*do/hl)**1.85)*(Ks**2/(phie*g*AhAa))
E = 0.00335*(h2phi/t(D))**1.1*(rhoL/rhoG)**0.5*(hl/h2phi)**kappa
print('Fractional entrainment, E = %5.3f'%E)


# Calculate point efficiency:
ReFe = (rhoG*vo*hl)/(muG*phie)
cG = rhoG/MG; cL = rhoL/ML
al1 = 0.4136; al2 = 0.6074; al3 = -0.3195
e1 = 1 + me*(cG/cL)*np.sqrt(DG*(1-phie)/(DL*AhAa))
e2 = -0.0029*(ReFe)**al1*(1000*hl/do)**al2*AhAa**al3
EOG = 1-np.exp(e2/e1)
print('Point efficiency, EOG = %5.3f' %EOG)


# Calculate Murphree tray efficiency:
DEG = 0.01; DEL = 0.1*np.sqrt(g*h2phi**3)
PeG = 4*QG*rw**2/(DEG*Aa*(t(D)-h2phi))
PeL = 4*qL*rw**2/(DEL*Aa*hl); N = (PeL + 2)/2
lamb = me*mG*ML/(MG*mL); EMGmix =((1+lamb*EOG/N)**N-1)/lamb
EMGunm = EMGmix*(1-0.0335*lamb**1.07272*EOG**2.51844*PeL**0.17524)
if(PeG<50):
    EMG = EMGmix
else:
    EMG = EMGunm
EMGE = EMG*(1-0.8*EOG*lamb**1.543*E/me)
print('Tray efficiency corrected for entrainment, EMGE = %5.3f' %EMGE)
```

Program results:
The tower diameter is D = 0.631 m
The corresponding tray spacing, t = 0.5 m
Gas pressure drop is ΔP = 356.3 Pa
Orifice Froude number, Fr_o = 1.114
Fractional entrainment, E = 0.024
Point efficiency, E_{OG} = 0.784
Tray efficiency corrected for entrainment, E_{MGE} = 0.809

Appendix F-1
McCabe–Thiele Method: Saturated Liquid Feed (Mathcad)

System: methanol (1) and water (2) at 1 atm. Use Antoine equation for vapor pressure and NRTL for activity coefficients in the modified Raoult' Law for VLE.

$A_1 := 16.5938 \qquad B_1 := 3644.3 \text{ K} \qquad C_1 := 239.76 \text{ K}$

$A_2 := 16.2620 \qquad B_2 := 3799.89 \text{ K} \qquad C_2 := 226.35 \text{ K}$

$$P1sat(T) := \exp\left(A_1 - \frac{B_1}{T - 273.15 \text{ K} + C_1}\right) \cdot \text{kPa}$$

$$P2sat(T) := \exp\left(A_2 - \frac{B_2}{T - 273.15 \text{ K} + C_2}\right) \cdot \text{kPa}$$

$$R := 1.987 \frac{\text{cal}}{\text{mol} \cdot \text{K}}$$

NRTL Equations:

$$b_{12} := -253.88 \frac{\text{cal}}{\text{mol}} \qquad b_{21} := 845.21 \frac{\text{cal}}{\text{mol}} \qquad \alpha := 0.2994$$

$$\tau_{12}(T) := \frac{b_{12}}{R \cdot T} \qquad \tau_{21}(T) := \frac{b_{21}}{R \cdot T}$$

$$G_{12}(T) := \exp\left(-\alpha \cdot \tau_{12}(T)\right) \qquad\qquad G_{21}(T) := \exp\left(-\alpha \cdot \tau_{21}(T)\right)$$

$$\gamma_1(x_1, x_2, T) := \exp\left[x_2^2 \cdot \left[\tau_{21}(T) \cdot \left(\frac{G_{21}(T)}{x_1 + x_2 \cdot G_{21}(T)}\right)^2 + \frac{G_{12}(T) \cdot \tau_{12}(T)}{\left(x_2 + x_1 \cdot G_{12}(T)\right)^2}\right]\right]$$

$$\gamma_2(x_1, x_2, T) := \exp\left[x_1^2 \cdot \left[\tau_{12}(T) \cdot \left(\frac{G_{12}(T)}{x_2 + x_1 \cdot G_{12}(T)}\right)^2 + \frac{G_{21}(T) \cdot \tau_{21}(T)}{\left(x_1 + x_2 \cdot G_{21}(T)\right)^2}\right]\right]$$

Principles and Applications of Mass Transfer: The Design of Separation Processes for Chemical and Biochemical Engineering, Fourth Edition. Jaime Benitez.
© 2023 John Wiley & Sons, Inc. Published 2023 by John Wiley & Sons, Inc.

Appendix F-1 McCabe–Thiele Method: Saturated Liquid Feed

Generate VLE Data:

$P := 1 \text{ atm}$ **Guess T:** $T := (90 + 273.15) \cdot K$

$Teq(x1) := \text{root}\,(P - \gamma_1(x1, (1-x1), T) \cdot x1 \cdot P1sat(T) - \gamma_2(x1, (1-x1), T) \cdot (1-x1) \cdot P2sat(T), T)$

Initial guesses for x and y: $x := 0$ $y := 0.1$

$yeq(x) := \dfrac{\gamma_1(x, (1-x), Teq(x)) \cdot x \cdot P1sat(Teq(x))}{P}$ $xeq(y) := \text{root}\,(yeq(x) - y, x)$

Plot VLE diagram: $N := 20$ $k := 0..20$ $xe_k := \dfrac{k}{N}$

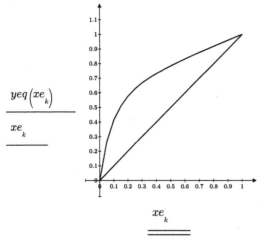

Enter data from Example 6.5

$F := 60.22 \dfrac{\text{mol}}{\text{s}}$ $x_F := 0.36$ $x_D := 0.999$ $x_W := 0.001$

Calculate *Rmin* for intersection of the top operating line and vertical feed line. This assumes there is no pinch point, as the VLE diagram shows for this system.

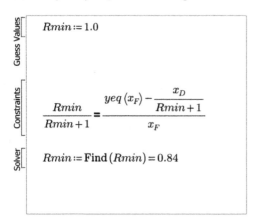

Guess Values: $Rmin := 1.0$

Constraints: $\dfrac{Rmin}{Rmin+1} = \dfrac{yeq(x_F) - \dfrac{x_D}{Rmin+1}}{x_F}$

Solver: $Rmin := \text{Find}\,(Rmin) = 0.84$

Appendix F-1 McCabe–Thiele Method: Saturated Liquid Feed 585

For a reflux ratio that is 50% above the minimum:
$$RR := 1.5 \cdot Rmin = 1.26$$
Solve overall balances:

Guess Values

$$D := \frac{F}{2} \qquad W := D$$

Constraints

$$F = D + W$$

$$x_F \cdot F = x_D \cdot D + x_W \cdot W$$

Solver

$$\begin{bmatrix} D \\ W \end{bmatrix} := \text{Find}(D, W) = \begin{bmatrix} 21.662 \\ 38.558 \end{bmatrix} \frac{\text{mol}}{\text{s}}$$

Stage balances in rectifying and stripping sections: liquid and molar flow rates are L, V in rectifying section; l, v in stripping section. Assume constant molar overflow.

$$L := RR \cdot D = 27.288 \frac{\text{mol}}{\text{s}} \qquad V := (RR + 1) \cdot D = 48.951 \frac{\text{mol}}{\text{s}}$$

For saturated liquid feed: $\qquad l := L + F = 87.508 \frac{\text{mol}}{\text{s}} \qquad v := V = 48.951 \frac{\text{mol}}{\text{s}}$

Rectifying section operating line: $\qquad yrop(x) := \dfrac{RR}{RR + 1} \cdot x + \dfrac{x_D}{RR + 1}$

Calculate the number of stages in rectifying section up to feed line:

$$rectify(xD, xF) := \begin{Vmatrix} i \leftarrow 0 \\ X_0 \leftarrow xD \\ Y_1 \leftarrow xD \\ \text{while } X_i > xF \\ \quad \begin{Vmatrix} i \leftarrow i + 1 \\ X_i \leftarrow xeq(Y_i) \\ Y_{i+1} \leftarrow yrop(X_i) \end{Vmatrix} \\ \begin{bmatrix} X \\ Y \\ i \end{bmatrix} \end{Vmatrix}$$

Appendix F-1 McCabe–Thiele Method: Saturated Liquid Feed

$$\begin{bmatrix} x \\ y \\ f \end{bmatrix} := rectify\,(x_D, x_F) \qquad f = 17 \qquad \text{Feed stage}$$

Stripping section operating line:
$$ysop\,(x) := \frac{x \cdot l}{v} + x_W \cdot \left(1 - \frac{l}{v}\right)$$

Calculate the number of stages (M) in stripping section, including the reboiler.

$$strip\,(j, xj, yjp1, xW) := \begin{Vmatrix} i \leftarrow j \\ X_j \leftarrow xj \\ Y_{j+1} \leftarrow ysop\,(X_j) \\ \text{while } X_i \geq xW \\ \quad \begin{Vmatrix} i \leftarrow i+1 \\ X_i \leftarrow xeq\,(Y_i) \\ Y_{i+1} \leftarrow ysop\,(X_i) \end{Vmatrix} \\ \begin{bmatrix} X \\ Y \\ i \end{bmatrix} \end{Vmatrix}$$

$$\begin{bmatrix} xs \\ ys \\ M \end{bmatrix} := strip\,\left(f, x_f, y_{f+1}, x_W\right) \qquad M = 22$$

Appendix F-2
McCabe–Thiele Method: Saturated Liquid Feed (Python)

```python
import numpy as np
import scipy as sp
from scipy.optimize import fsolve
import matplotlib.pyplot as plt
import sys
import warnings
if not sys.warnoptions:
    warnings.simplefilter("ignore")
# McCabe-Thiele Method: Saturated Liquid Feed
# Data are from Example 6.5: Methanol (1)-Water (2)

# Generate the VLE diagram:
# Antoine equation for vapor pressure
# Psat in kPa, T in K
A1 = 16.5938; B1 = 3644.3; C1 = 239.76
A2 = 16.2620; B2 = 3799.9; C2 = 226.35
def P1sat(T):
    return np.exp(A1-B1/(T-273.15+C1))
def P2sat(T):
    return np.exp(A2-B2/(T-273.15+C2))

# NRTL equations for activity coefficients:
b12 = -253.88; b21 = 845.21; alpha = 0.2994; R = 1.987
def tau12(T):
    return b12/(R*T)
def tau21(T):
    return b21/(R*T)
def G12(T):
    return np.exp(-alpha*tau12(T))
def G21(T):
    return np.exp(-alpha*tau21(T))
def gamma1(x1,x2,T):
    g11 = tau21(T)*(G21(T)/(x1+x2*G21(T)))**2
    g12 = tau12(T)*G12(T)/(x2+x1*G12(T))**2
    return np.exp(x2**2*(g11+g12))
```

Principles and Applications of Mass Transfer: The Design of Separation Processes for Chemical and Biochemical Engineering, Fourth Edition. Jaime Benitez.
© 2023 John Wiley & Sons, Inc. Published 2023 by John Wiley & Sons, Inc.

Appendix F-2 McCabe–Thiele Method: Saturated Liquid Feed

```python
def gamma2(x1,x2,T):
    g21 = tau12(T)*(G12(T)/(x2+x1*G12(T)))**2
    g22 = tau21(T)*G21(T)/(x1+x2*G21(T))**2
    return np.exp(x1**2*(g21+g22))
P = 101.3  # Total pressure in kPa
def Teq(x1):
    x2 = 1 - x1
    def g(T):
        return P-x1*gamma1(x1,x2,T)*P1sat(T)-x2*gamma2(x1,x2,T)*P2sat(T)
    return fsolve(g,350)
def y1eq(x1):
    x2 = 1-x1
    T = Teq(x1)
    return x1*gamma1(x1,x2,T)*P1sat(T)/P
def x1eq(y):
    def h(x):
        return y1eq(x)-y
    return fsolve(h,0.5)
xeq = np.linspace(0,1,101)
yeq = np.zeros(101, dtype = float)
for i in range(0,101,1):
    yeq[i]= y1eq(xeq[i])
    i = i+1
# Plot the VLE diagram
plt.plot(xeq,yeq)
plt.plot([0,1],[0,1],"k--")
plt.title('Equilibriun Diagram')

# Calculate Rmin by the following procedure. It applies only if there is no pinch point
# in the VLE diagram. This condition applies for the methanol-water system at q = 1.0.
# Enter data from Example 6.5:
xF = 0.36; xD = 0.999; xW = 0.001
def g(Rm):
    return Rm/(Rm+1)-(y1eq(xF)-xD/(Rm+1))/xF
Rmin = fsolve(g,1.0)
print('Minimum reflux ratio, Rmin = %5.3f' %Rmin)
RR = 1.5*Rmin
print('Reflux ratio, 50 percent above minimum, R = %5.3f'%RR)
## Solve overall balances:
F = 60.22  # Feed molar flow rate, mol/s
def bal(p):
    D = p[0]; W = p[1]
    return F-D-W, F*xF-D*xD-W*xW
p = np.zeros(2,dtype=float)
p[0] = F/2; p[1] = F/2
p = fsolve(bal,p)
D = p[0]; W = p[1]
print('Distillate flow rate, in mol/s, D = %5.3f'%D)
print('Bottoms flow rate, in mol/s, W = %5.3f'%W)
# Calculate molar flow rates in two sections of tower:
L = RR*D; V = (RR+1)*D #Rectifying section
l = L + F; v = V #Stripping section; saturated liquid feed
```

Appendix F-2 McCabe–Thiele Method: Saturated Liquid Feed

```
# Rectifying and stripping section operating lines:
def yrop(x):
    return RR*x/(RR+1)+ xD/(RR+1)
def ysop(x):
    return x*1/v+xW*(1-1/v)
# Calculate the number of stages in rectifying section:
X = np.zeros(100,dtype = float)
Y = np.zeros(100,dtype = float)
i = 0
X[0] = xD
Y[1] = xD
while(X[i]>xF):
    X[i+1]= xIeq(Y[i+1])
    Y[i+2] = yrop(X[i+1])
    i = i+1
print ('Feed stage =',i)
# Calculate the number of stages in s])tripping section:
Y[i+1] = ysop(X[i-1])
while(X[i]>= xW):
    X[i+1] = xIeq(Y[i+1])
    Y[i+2] = ysop(X[i+1])
    i = i+1
print('Total number of stages =',i)
```

Program results:

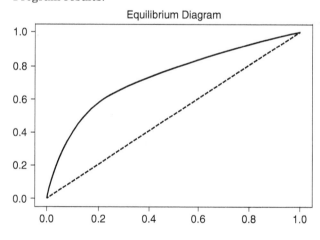

Minimum reflux ratio, $R_{min} = 0.840$
Reflux ratio, 50% above minimum, $R = 1.260$
Distillate flow rate, $D = 21.66$ mol/s
Bottoms flow rate, $W = 38.56$ mol/s
Feed stage = 17
Total number of stages = 22

Appendix G-1
Single-Stage Extraction
(Mathcad)

Single-Stage Liquid Extraction

Data presented are from Example 7-2: acetone-water-chloroform system at 298 K.

$$vx_C := \begin{bmatrix} 0.158 \\ 0.256 \\ 0.360 \\ 0.493 \\ 0.557 \\ 0.596 \end{bmatrix} \quad vx_B := \begin{bmatrix} 0.0123 \\ 0.0129 \\ 0.0171 \\ 0.0510 \\ 0.0980 \\ 0.1690 \end{bmatrix} \quad vy_C := \begin{bmatrix} 0.287 \\ 0.421 \\ 0.527 \\ 0.613 \end{bmatrix} \quad vy_B := \begin{bmatrix} 0.700 \\ 0.557 \\ 0.429 \\ 0.284 \end{bmatrix}$$

$$vsraf := \text{cspline}\,(vx_C, vx_B) \qquad vsext := \text{cspline}\,(vy_C, vy_B)$$

$$x_B\,(x_C) := \text{interp}\,(vsraf, vx_C, vx_B, x_C)$$

$$vxtie_C := \begin{bmatrix} 0.158 \\ 0.256 \\ 0.360 \\ 0.493 \\ 0.557 \\ 0.596 \end{bmatrix} \quad vytie_C := \begin{bmatrix} 0.287 \\ 0.421 \\ 0.527 \\ 0.613 \\ 0.610 \\ 0.596 \end{bmatrix}$$

$$vstie := \text{cspline}\,(vxtie_C, vytie_C)$$

$$y_C\,(x_C) := \text{interp}\,(vstie, vxtie_C, vytie_C, x_C)$$

$$y_B\,(x_C) := \text{interp}\,(vsext, vy_C, vy_B, y_C\,(x_C))$$

$$F := 50\,\frac{\text{kg}}{\text{hr}} \qquad S := 50\,\frac{\text{kg}}{\text{hr}} \qquad x_{CF} := 0.60$$

$$x_{BF} := 0.0 \qquad y_{BS} := 1.0 \qquad y_{CS} := 0.0$$

Principles and Applications of Mass Transfer: The Design of Separation Processes for Chemical and Biochemical Engineering, Fourth Edition. Jaime Benitez.
© 2023 John Wiley & Sons, Inc. Published 2023 by John Wiley & Sons, Inc.

592   **Appendix G-1 Single-Stage Extraction (Mathcad)**

Guess Values

$$E := \frac{F+S}{2} \qquad R := E \qquad x_C := 0.1$$

Constraints

$$F + S = E + R$$
$$F \cdot x_{CF} + S \cdot y_{CS} = E \cdot y_C(x_C) + R \cdot x_C$$

$$F \cdot x_{BF} + S \cdot y_{BS} = E \cdot y_B(x_C) + R \cdot x_B(x_C)$$

Solver

$$\begin{bmatrix} E \\ R \\ x_C \end{bmatrix} := \text{Find}(E, R, x_C)$$

$$E = 76.71 \frac{\text{kg}}{\text{hr}} \qquad\qquad R = 23.29 \frac{\text{kg}}{\text{hr}}$$

$$x_C = 0.189 \qquad\qquad x_B(x_C) = 0.013$$

$$y_C(x_C) = 0.334 \qquad\qquad y_B(x_C) = 0.648$$

Appendix G-2
Single-Stage Extraction
(Python)

```python
# Single-Stage Extraction
# Data from Example 7-2

import numpy as np
from scipy.optimize import fsolve
from scipy.interpolate import CubicSpline

vxC = [0.158, 0.256, 0.360, 0.493, 0.557, 0.596]
vxB = [0.0123, 0.0129, 0.0171, 0.0510, 0.0980, 0.1690]
vyC = [0.287, 0.421, 0.527, 0.613]
vyB = [0.700, 0.557, 0.429, 0.284]
vytieC = [0.287, 0.421, 0.527, 0.613, 0.610, 0.596]
F = 50; S = 50   #kg/h
xCF = 0.60; xBF = 0.0; yBS = 1.0; yCS = 0.0


raf = CubicSpline(vxC, vxB)
ext = CubicSpline(vyC, vyB)
tie = CubicSpline(vxC, vytieC)

def xBR(x):
    return raf(x)
def yCE(x):
    return tie(x)
def yBE(x):
    return ext(tie(x))
def SSEXT(p):
    E = p[0]
    R = p[1]
    xCR = p[2]
    return F+S-E-R, F*xCF+S*yCS-E*yCE(xCR)-R*xCR,\
            F*xBF+S*yBS-E*yBE(xCR)-R*xBR(xCR)
p = np.zeros(3,dtype = float)

# Initial estimates:
p[0] = (F+S)/2; p[1] = (F+S)/2; p[2] = 0.1
p = fsolve(SSEXT, p)
a=xBR(p[2]); b = yCE(p[2]); c=yBE(p[2])
print ('Extract flow rate (kg/h), E = %5.2f' %p[0])
```

Principles and Applications of Mass Transfer: The Design of Separation Processes for Chemical and Biochemical Engineering, Fourth Edition. Jaime Benitez.
© 2023 John Wiley & Sons, Inc. Published 2023 by John Wiley & Sons, Inc.

Appendix G-2 Single-Stage Extraction (Mathcad)

```
print ('Raffinate flow rate (kg/h), R = %5.2f' %p[1])
print ('Acetone fraction in the raffinate, xCR = %4.3f' %p[2])
print ('Chloroform fraction in the raffinate, xBR = %4.3f' %a)
print ('Acetone fraction in the extract, yCE = %4.3f' %b)
print ('Chloroform fraction in the extract, yBE = %4.3f' %c)
```

Program results:

Extract flow rate, E = 76.71 kg/h

Raffinate flow rate, R = 23.29 kg/h

Acetone fraction in the raffinate, x_{CR} = 0.189

Chloroform fraction in the raffinate, x_{BR} = 0.013

Acetone fraction in the extract, y_{CE} = 0.334

Chloroform fraction in the extract, y_{BE} = 0.648

Appendix G-3
Multistage Crosscurrent
Extraction (Mathcad)

Data presented are from Example 7-3: acetone-water-chloroform system at 298 K.

$$vx_C := \begin{bmatrix} 0.158 \\ 0.256 \\ 0.360 \\ 0.493 \\ 0.557 \\ 0.596 \end{bmatrix} \qquad vx_B := \begin{bmatrix} 0.0123 \\ 0.0129 \\ 0.0171 \\ 0.0510 \\ 0.0980 \\ 0.1690 \end{bmatrix} \qquad vy_C := \begin{bmatrix} 0.287 \\ 0.421 \\ 0.527 \\ 0.613 \end{bmatrix} \qquad vy_B := \begin{bmatrix} 0.700 \\ 0.557 \\ 0.429 \\ 0.284 \end{bmatrix}$$

$$vsraf := \text{cspline}\,(vx_C, vx_B) \qquad\qquad vsext := \text{cspline}\,(vy_C, vy_B)$$

$$x_B\,(x_C) := \text{interp}\,(vsraf, vx_C, vx_B, x_C)$$

$$vxtie_C := \begin{bmatrix} 0.158 \\ 0.256 \\ 0.360 \\ 0.493 \\ 0.557 \\ 0.596 \end{bmatrix} \qquad vytie_C := \begin{bmatrix} 0.287 \\ 0.421 \\ 0.527 \\ 0.613 \\ 0.610 \\ 0.596 \end{bmatrix}$$

$$vstie := \text{cspline}\,(vxtie_C, vytie_C) \qquad y_C\,(x_C) := \text{interp}\,(vstie, vxtie_C, vytie_C, x_C)$$

$$y_B\,(x_C) := \text{interp}\,(vsext, vy_C, vy_B, y_C\,(x_C))$$

Stage 1:

$$F1 := 50\,\frac{\text{kg}}{\text{hr}} \qquad S1 := 8\,\frac{\text{kg}}{\text{hr}} \qquad x_{CF1} := 0.60 \qquad y_{CS1} := 0.0$$

$$x_{BF1} := 0.0 \qquad y_{BS1} := 1.0$$

Principles and Applications of Mass Transfer: The Design of Separation Processes for Chemical and Biochemical Engineering, Fourth Edition. Jaime Benitez.
© 2023 John Wiley & Sons, Inc. Published 2023 by John Wiley & Sons, Inc.

596   **Appendix G-3 Multistage Crosscurrent Extraction (Mathcad)**

Guess Values

$$E1 := \frac{F1 + S1}{2} \qquad R1 := E1 \qquad x_{C1} := 0.1$$

Constraints

$$F1 + S1 = E1 + R1$$
$$F1 \cdot x_{CF1} + S1 \cdot y_{CS1} = E1 \cdot y_C(x_{C1}) + R1 \cdot x_{C1}$$

$$F1 \cdot x_{BF1} + S1 \cdot y_{BS1} = E1 \cdot y_B(x_{C1}) + R1 \cdot x_B(x_{C1})$$

Solver

$$\begin{bmatrix} E1 \\ R1 \\ x_{C1} \end{bmatrix} := \text{Find}(E1, R1, x_{C1})$$

$$E1 = 21.643 \frac{\text{kg}}{\text{hr}}$$

$$R1 = 36.357 \frac{\text{kg}}{\text{hr}}$$

$$y_{C1} := y_C(x_{C1}) = 0.604 \qquad y_{B1} := y_B(x_{C1}) = 0.302 \qquad x_{B1} := x_B(x_{C1}) = 0.041$$

Stage 2:

$$F2 := R1 = 36.357 \frac{\text{kg}}{\text{hr}} \qquad S2 := 8 \frac{\text{kg}}{\text{hr}} \qquad x_{CF2} := x_{C1} = 0.466$$

$$y_{BS2} := 1.0 \qquad y_{CS2} := 0.0 \qquad x_{BF2} := x_B(x_{C1}) = 0.041$$

Guess Values

$$R2 := 20 \frac{\text{kg}}{\text{hr}} \qquad E2 := S2 + F2 - R2$$

$$x_{C2} := 0.18$$

Constraints

$$F2 + S2 = E2 + R2$$

$$F2 \cdot x_{CF2} + S2 \cdot y_{CS2} = E2 \cdot y_C(x_{C2}) + R2 \cdot x_{C2}$$

$$F2 \cdot x_{BF2} + S2 \cdot y_{BS2} = E2 \cdot y_B(x_{C2}) + R2 \cdot x_B(x_{C2})$$

Solver

$$\begin{bmatrix} E2 \\ R2 \\ x_{C2} \end{bmatrix} := \text{Find}(E2, R2, x_{C2})$$

$$E2 = 18.594 \frac{\text{kg}}{\text{hr}} \qquad R2 = 25.764 \frac{\text{kg}}{\text{hr}}$$

Appendix G-3 Multistage Crosscurrent Extraction (Mathcad)

$x_{C2} = 0.311$ $y_{C2} := y_C(x_{C2}) = 0.48$ $x_{B2} := x_B(x_{C2}) = 0.014$

$y_{B2} := y_B(x_{C2}) = 0.491$

Stage 3:

$F3 := R2 = 25.764 \dfrac{kg}{hr}$ $S3 := 8 \dfrac{kg}{hr}$ $x_{CF3} := x_{C2} = 0.311$

$y_{BS3} := 1.0$ $y_{CS3} := 0.0$ $x_{BF3} := x_B(x_{C2}) = 0.014$

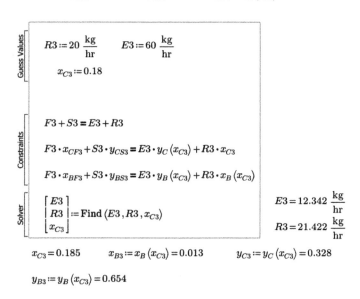

Guess Values:
$R3 := 20 \dfrac{kg}{hr}$ $E3 := 60 \dfrac{kg}{hr}$
$x_{C3} := 0.18$

Constraints:
$F3 + S3 = E3 + R3$

$F3 \cdot x_{CF3} + S3 \cdot y_{CS3} = E3 \cdot y_C(x_{C3}) + R3 \cdot x_{C3}$

$F3 \cdot x_{BF3} + S3 \cdot y_{BS3} = E3 \cdot y_B(x_{C3}) + R3 \cdot x_B(x_{C3})$

Solver:
$\begin{bmatrix} E3 \\ R3 \\ x_{C3} \end{bmatrix} := \text{Find}(E3, R3, x_{C3})$

$E3 = 12.342 \dfrac{kg}{hr}$

$R3 = 21.422 \dfrac{kg}{hr}$

$x_{C3} = 0.185$ $x_{B3} := x_B(x_{C3}) = 0.013$ $y_{C3} := y_C(x_{C3}) = 0.328$

$y_{B3} := y_B(x_{C3}) = 0.654$

Composited extract:

$E := E1 + E2 + E3 = 52.578 \dfrac{kg}{hr}$

$y_C := \dfrac{E1 \cdot y_{C1} + E2 \cdot y_{C2} + E3 \cdot y_{C3}}{E} = 0.495$ $y_B := \dfrac{E1 \cdot y_{B1} + E2 \cdot y_{B2} + E3 \cdot y_{B3}}{E} = 0.451$

Appendix G-4
Multistage Crosscurrent
Extraction (Python)

```python
# Multi-Stage Crosscurrent Extraction
# Data from Example 7-3

import numpy as np
from scipy.optimize import fsolve
from scipy.interpolate import CubicSpline


vxC = [0.158, 0.256, 0.360, 0.493, 0.557, 0.596]
vxB = [0.0123, 0.0129, 0.0171, 0.0510, 0.0980, 0.1690]
vyC = [0.287, 0.421, 0.527, 0.613]
vyB = [0.700, 0.557, 0.429, 0.284]
vytieC = [0.287, 0.421, 0.527, 0.613, 0.610, 0.596]

raf = CubicSpline(vxC, vxB)
ext = CubicSpline(vyC, vyB)
tie = CubicSpline(vxC, vytieC)
def xBR(x):
    return raf(x)
def yCE(x):
    return tie(x)
def yBE(x):
    return ext(tie(x))

# STAGE 1
F1 = 50; S = 8  #kg/h
xCF1 = 0.60; xBF1 = 0.0; yBS = 1.0; yCS = 0.0

def SSEXT(p):
    E1 = p[0]
    R1 = p[1]
    xCR1 = p[2]
    return F1+S-E1-R1, F1*xCF1+S*yCS-E1*yCE(xCR1)-R1*xCR1,\
            F1*xBF1+S*yBS-E1*yBE(xCR1)-R1*xBR(xCR1)
p = np.zeros(3,dtype = float)

# Initial estimates:
p[0] = (F1+S)/2; p[1] = (F1+S)/2; p[2] = 0.1
p = fsolve(SSEXT, p)
print ('STAGE # 1: RESULTS')
```

Principles and Applications of Mass Transfer: The Design of Separation Processes for Chemical and Biochemical Engineering, Fourth Edition. Jaime Benitez.
© 2023 John Wiley & Sons, Inc. Published 2023 by John Wiley & Sons, Inc.

Appendix G-4 Multistage Crosscurrent Extraction (Python)   599

```python
print ('Extract flow rate (kg/h), E1 = %5.2f' %p[0])
print ('Raffinate flow rate (kg/h), R1 = %5.2f' %p[1])
print ('Acetone fraction in the raffinate, xCR1 = %4.3f' %p[2])
print ('')
R1 = p[1]; E1 = p[0]; xCR1 = p[2]
yCE1 = yCE(p[2]); yBE1 = yBE(p[2])


# STAGE 2
F2 = R1; S = 8  #kg/h
xCF2 = xCR1; xBF2 = xBR(xCR1); yBS = 1.0; yCS = 0.0

def SSEXT(p):
    E2 = p[0]
    R2 = p[1]
    xCR2 = p[2]
    return F2+S-E2-R2, F2*xCF2+S*yCS-E2*yCE(xCR2)-R2*xCR2,\
           F2*xBF2+S*yBS-E2*yBE(xCR2)-R2*xBR(xCR2)
p = np.zeros(3,dtype = float)

# Initial estimates:
p[0] = (F2+S)/2; p[1] = (F2+S)/2; p[2] = 0.1
p = fsolve(SSEXT, p)
print ('STAGE # 2: RESULTS')
print ('Extract flow rate (kg/h), E2 = %5.2f' %p[0])
print ('Raffinate flow rate (kg/h), R2 = %5.2f' %p[1])
print ('Acetone fraction in the raffinate, xCR2 = %4.3f' %p[2])
print ('')
R2 = p[1]; E2 = p[0]; xCR2 = p[2]
yCE2 = yCE(p[2]); yBE2 = yBE(p[2])


# STAGE 3
F3 = R2; S = 8  #kg/h
xCF3 = xCR2; xBF3 = xBR(xCR2); yBS = 1.0; yCS = 0.0

def SSEXT(p):
    E3 = p[0]
    R3 = p[1]
    xCR3 = p[2]
    return F3+S-E3-R3, F3*xCF3+S*yCS-E3*yCE(xCR3)-R3*xCR3,\
           F3*xBF3+S*yBS-E3*yBE(xCR3)-R3*xBR(xCR3)
p = np.zeros(3,dtype = float)

# Initial estimates:
p[0] = (F3+S)/2; p[1] = (F3+S)/2; p[2] = 0.1
p = fsolve(SSEXT, p)
print ('STAGE # 3: RESULTS')
print ('Extract flow rate (kg/h), E3 = %5.2f' %p[0])
print ('Raffinate flow rate (kg/h), R3 = %5.2f' %p[1])
```

Appendix G-4 Multistage Crosscurrent Extraction (Python)

```
print ('Acetone fraction in the raffinate, xCR3 = %4.3f' %p[2])
print ('')
yCE3 = yCE(p[2]); yBE3 = yBE(p[2])
E3 = p[0]

#Composited extract
E = E1 + E2 + E3
yC = (E1*yCE1 + E2*yCE2 + E3*yCE3)/E
yB = (E1*yBE1 + E2*yBE2 + E3*yBE3)/E
print ('COMPOSITED EXTRACT')
print ('Flow rate (kg/h), E = %5.2f' %E)
print ('Acetone fraction = %4.3f' %yC)
print ('Chloroform fraction = %4.3f' %yB)
```

Program results:

STAGE # 1

Extract flow rate, E_1 = 21.64 kg/h
Raffinate flow rate, R_1 = 36.36 kg/h
Acetone fraction in the raffinate, x_{CR1} = 0.466

STAGE # 2

Extract flow rate, E_2 = 18.59 kg/h
Raffinate flow rate, R_2 = 25.76 kg/h
Acetone fraction in the raffinate, x_{CR2} = 0.311

STAGE # 3

Extract flow rate, E_3 = 12.34 kg/h
Raffinate flow rate, R_3 = 21.42 kg/h
Acetone fraction in the raffinate, x_{CR3} = 0.185

COMPOSITED EXTRACT

Flow rate, E = 52.58 kg/h
Acetone fraction = 0.495
Chloroform fraction = 0.451

Appendix H
Constants and Unit Conversions

Table H.1 Atomic Mass of Selected Elements

Element	Symbol	Mass	Element	Symbol	Mass
Aluminum	Al	26.98	Lead	Pb	207.19
Arsenic	As	74.92	Lithium	Li	6.94
Barium	Ba	37.30	Magnesium	Mg	24.31
Beryllium	Be	9.01	Manganese	Mn	54.94
Boron	B	10.81	Mercury	Hg	200.59
Bromine	Br	79.90	Nickel	Ni	58.71
Cadmium	Cd	112.40	Nitrogen	N	14.01
Calcium	Ca	40.88	Oxygen	O	16.00
Carbon	C	12.01	Phosphorus	P	30.97
Chlorine	Cl	35.45	Potassium	K	39.10
Chromium	Cr	52.00	Silicon	Si	28.09
Cobalt	Co	58.93	Silver	Ag	107.87
Copper	Cu	63.55	Sodium	Na	22.99
Fluorine	F	19.00	Sulfur	S	32.06
Germanium	Ge	72.59	Tin	Sn	118.69
Gold	Au	196.97	Titanium	Ti	47.90
Helium	He	4.00	Uranium	U	238.03
Hydrogen	H	1.01	Vanadium	V	50.94
Iodine	I	126.90	Zinc	Zn	65.37
Iron	Fe	55.85	Zirconium	Zr	91.22

Principles and Applications of Mass Transfer: The Design of Separation Processes for Chemical and Biochemical Engineering, Fourth Edition. Jaime Benitez.
© 2023 John Wiley & Sons, Inc. Published 2023 by John Wiley & Sons, Inc.

Appendix H Constants and Unit Conversions

Table H.2 Ideal Gas Constant, R

1.987 cal/(mol × K)

8.314 kPa × m^3/(kmol × K)

8.314 Pa × m^3/(mol × K)

82.06 atm × cm^3/(mol × K)

0.08206 atm × L/(mol × K)

Table H.3 Pressure Units

1 bar = 10^5 Pa = 100 kPa

= 0.9869 atm

= 750.06 mm Hg

= 750.06 torr

= 10^6 dyn/cm^2

= 14.5038 psia

Table H.4 Viscosity Units

1 cP = 0.01 P

= 0.01 (dyne × s)/cm^2

= 0.001 kg/(m × s)

= 0.001 Pa × s

= 10^4 μP

$$1 \text{ Å} = 10^{-10} \text{ m}$$

Boltzmann's constant = 1.3806×10^{-23} J/K

Faraday's constant = 96,500 C/g-equiv

Index

Note: Page numbers followed by "*f*" refers to figures and "*t*" refers to tables.

A

Absolute humidity: 449
Absorption
  of an air pollutant: 301
  with chemical reaction: 214–215
  definition: 2, 277
  factor: 201, 248, 278
Acetic acid: 161
Acid ionization constant: 161
Activated carbon: 210–211, 526, 533
  regenerated by countercurrent contact: 211
Activated carbon adsorption: 71
  Freundlich isotherm: 71
  material balances: 71
Activity coefficients: 157, 176
Adiabatic absorber: 299*f*
Adiabatic gas–liquid contact operations: 452*f*, 457
  cooling a hot gas: 457
  cooling a liquid: 457
  fundamental relationships: 458–460
  humidifying or dehumidifying a gas: 457
Adiabatic gas–liquid countercurrent contact: 459*f*
Adiabatic operation of a packed-bed absorber: 296
Adiabatic saturation
  calculations with Mathcad: 452*f*
  curves: 452

Adsorption
  definition: 2
  process: 478
  on silical gel: 190
Adsorption isotherm: 489, 490*f*
  Freundlich isotherm: 491
  Langmuir isotherm: 491
  linear isotherm: 489
Affinity adsorption: 524–526
Agar gel, urea diffusion in: 60–62
Agar gel slab, urea concentration profile in: 61*f*
Agitator power: 421
Air humidification: 124
Alanine: 162
Alcohol dehydrogenase (ADH): 445
Algebraic solution: 159–160
Amino acid: 162–163, 205–206
Amino acid valine: 165
Ammonia molecules: 156
Analogies among transfer phenomena: 66–67
Aniline-water solution, stripping: 271
Annular packed beds: 265
Antibiotics: 431, 522
Antibodies: 524
Antibody based therapies: 434
Antigens: 524
Antoine equation: 158, 177, 294
Aqueous–aqueous phase separation: 434

Principles and Applications of Mass Transfer: The Design of Separation Processes for Chemical and Biochemical Engineering, Fourth Edition. Jaime Benitez.
© 2023 John Wiley & Sons, Inc. Published 2023 by John Wiley & Sons, Inc.

604 Index

Aqueous two-phase extraction (ATPE): 433, 434, 434*f*, 435–436
  of alcohol dehydrogenase: 445
  of lysozyme: 445
  of thyroglobulin: 436
  tie-line calculations for: 444–445
Artificial kidneys: 137
Aspartic acid: 162
Aspen rate-based distillation: 371
Association factor: 23
Asymmetric membranes: 512
Atomic mass of selected elements: 601*t*
Atomic volume contributions: 19*t*
ATPE. *See* Aqueous two-phase extraction (ATPE)
Azeotrope: 386, 440
  water–dioxane solutions: 440
Azeotropic mixtures of ethanol and water: 344

B

Barrer: 481
Batch binary distillation: 373*f*
Batch distillation: 371
  binary batch distillation with constant reflux: 372
    McCabe–Thiele diagram: 372
  constant distillate composition: 375–376
    McCabe–Thiele diagram: 378
  multicomponent batch distillation: 377–378
    BatchFrac: 377
Batch processes: 195, 212
Benzene-recovery process: 185*f*
Berl saddles: 226
Bernoulli's equation: 246
Bessel function of the first kind of order zero: 230
BFW deaerator. *See* Boiler feed water (BFW) deaerator

Billet and Schultes correlation: 237
Binary batch rectification with constant distillate composition: 388
Binary diffusion coefficients: 551–553
Binary distillation: 322
  in packed tower: 351*f*
Binary system: 158
Bio-artificial kidney: 8
  cellophane membrane: 79
  dialysis machine: 8
  semipermeable membranes: 8
  urine: 8
Biochemical equilibrium concepts: 160–165
Biomolecules: 434
Bioproducts, liquid-liquid extraction of: 430–437
Bioseparation processes: 510
Bioseparations: 522–523
Blasius problem: 105*f*
Blasius solution: 106
Blood: 8, 12
  blood cells: 8
  creatinine: 8
  hemoglobin: 12
  plasma: 8
  urea: 8
  uric acid: 8
  waste products: 8
Blood oxygenator: 94
Bohart–Adams model: 534
Boiler feed water (BFW) deaerator: 132, 303
Boilup ratio: 327
Boltzmann constant: 17, 52
Bootstrap condition: 29
Boundary layer theory: 101
  Blasius solution: 106
  concentration boundary layer: 102

Index

605

on flat plate in steady flow: 102
  stream function: 103
Boundary-value problem: 42
Breakthrough curve: 498
Brock and Bird correlation: 24
Bubble-cap trays: 251, 303
Bubble column: 243
Buckingham method: 97
  core group: 98
  dimensional matrix: 98
  rank of the dimensional matrix: 98
Bulk motion contribution: 12

C

Capillary electrophoresis (CE): 511
Capillary tubes: 511
Caprylic acid: 165
Cascades: 196, 404
  McCabe–Thiele diagram: 418
Catalytic cracking of ammonia: 30*f*
Chemical phase equilibria: 156–160
Chemical potential gradient: 14
ChemSep: 371
Chilton–Colburn analogy: 113
Chlorine from water, stripping: 268
Chromatogram: 509
Chromatographic separation of
sugars: 508–510
Chromatography: 478
Cocurrent contact: 194
Cocurrent flow: 484
Cocurrent gas absorption: 306
Concentration boundary layer:
102
Concentration gradient
contribution: 12
Concentration polarization
factor: 517
Coning: 251
Constant distillate composition,
binary batch rectification with: 388
Constant molar overflow: 327

Constant relative volatility:
381, 382
Constants and unit conversions:
601–602
Continuity, equation of: 34
Continuous countercurrent adiabatic
gas–liquid contact: 458*f*
Continuous countercurrent
extraction with reflux: 412–415
Continuous distillation: 322
Continuous flash vaporization
process: 316*f*
Continuous-flow-stirred-tank-reactor:
423
Continuous-phase mass-transfer
coefficient: 425
Continuous rectification: 322
  binary distillation: 322
  boilup ratio: 327
  constant molar overflow: 327
  constant relative volatility: 382
  continuous distillation: 322
  distillation with two feeds: 385
  feed preheater: 336
  feed stage: 328
  Fenske equation: 357
  heavy key: 323
  light key: 323
  McCabe–Thiele
  assumptions: 325
  McCabe–Thiele method: 324
  partial reboiler: 327
  q-line: 329
  rectifying section: 325–326
  reflux ratio: 325
  stripping section: 325
  total condenser: 325
  use of open steam: 342
Convective mass transfer: 3, 91
Conventional filtration: 522
Conventional purification
strategies: 523
Cooling towers: 70, 270, 461

606 Index

Countercurrent cascade of ideal
stages: 197*f*, 198*f*
Countercurrent extraction: 404,
438, 440
  with extract reflux: 415–418
  minimum solvent for: 407*f*
Countercurrent flow: 180, 484
Countercurrent multistage
cascades: 197
Countercurrent multistage
extraction: 404*f*, 405*f*
  with reflux: 412*f*
Countercurrent packed
towers: 285
Covalent surface modification: 522
Crosscurrent extraction: 403, 438
Cross-flow cascade: 213
  of three ideal stages: 214*f*
Cross-flow filtration: 522
Crossflow gas permeation: 484
Cross-flow model for membrane
module: 485*f*
Cross-flow tray tower: 250*f*
Crystal dissolution process: 49*f*
Cut: 386, 483

D

Dalton's law: 5
Damkohler number for first-order
reaction: 62
Dead-end filtration: 521
Debye-Huckel equation: 163
Decanter: 419
Dehumidification: 2
DePriester charts: 380
Depth filters: 521
Desorption, definition: 2
Dew point: 452
Dextran: 433
Diafiltration of proteins: 533
Dialysate: 513
Dialysis: 513–514

definition: 2
membrane-separation process: 513
process: 515*f*
Diethylene glycol, equilibrium data
for: 416*t*
Difference point: 413
Differential distillation
  calculations: 323*f*
  with constant relative
  volatility: 381
  definition: 320
  Rayleigh equation: 320
  steam distillation: 382
Diffusate: 513
Diffusion in electrolyte solutions: 74
  limiting ionic conductances: 75*t*
Diffusion in solids: 50
  combined molecular and Knudsen
  diffusion: 54
  dusty gas model for
  multicomponent diffusion: 57
  hydrodynamic flow in porous
  solids: 55
  Knudsen diffusion in porous
  solids: 52
  molecular diffusion in porous
  solids: 52
Diffusion of A through nondiffusing B:
37, 92
Diffusivity or diffusion
coefficient: 16
  Brokaw equation: 72
  in concentrated liquid solutions: 26
  diffusivity of polar gases: 72–73
  effective diffusivity in
  multicomponent mixtures: 30
  Hayduk and Minhas: 23
  MS diffusivity: 12
  Stokes–Einstein equation: 22
  Vignes formula: 26
  Wilke–Chang equation: 22
  Wilke–Lee equation: 17

Index

607

Dimensional analysis: 96
  Buckingham method: 97
    core group: 98
    dimensional matrix: 98
    rank of the dimensional
    matrix: 98
Dimensionless groups for
mass and heat transfer: 109t
Dispersed-phase mass-transfer
coefficient: 426
Distillation, definition: 1, 315
Distillation–membrane
hybrid: 386, 386f
Distribution-law equation:
157
"Distribution-law" equation:
157
Dittus–Boelter equation: 125
Dowex: 496
Downcomer: 250
Dry-bulb temperature: 452
Drying of soap: 212
Dry tray head loss: 255
Dusty ideal gas model: 57
DVB cross-linked polystyrene
resins: 496t

E

Eddies, turbulent: 91
Effective diffusivity: 30
  multicomponent liquid
  mixtures: 32
  stagnant mixtures: 31
Electrophoresis: 163, 510–511
Entrainment: 249
Eotvos number: 426
Equation of continuity: 34
Equilateral-triangular
diagrams: 394, 395f
Equilibrium considerations: 155
  biochemical equilibrium
  concepts: 160–165

chemical phase equilibria:
156–160
Equilibrium distribution
  of ammonia: 157f
  diagram: 168, 415
Equilibrium-distribution
curve: 156, 158
Equilibrium-distribution
diagram: 416f
Equilibrium stage: 196–198
  and feed-stage location: 336f
  model: 368–370, 369f
Equilibrium thermodynamics: 160
Equilibrium tie-line data
  for system cottonseed oil–propane–
  oleic acid: 442t
  for system n-heptane–aniline–
  MCH: 443t
Equimolar counterdiffusion: 35, 95
Equivalent head of clear liquid: 256
Ergun equation: 130
Essential amino acids: 162
Ethylbenzene, equilibrium
data for: 416t
Extract, definition: 2, 394
Extraction equilibrium: 394
  conjugate or auxiliary line: 398
  distribution coefficient: 396
  plait point: 396
  tie line: 396
  type II system: 397
  type I system: 395
Extraction factor: 410

F

Feed preheater: 336, 337f
Feed stage: 328, 331f
Fenske equation: 357–359
Fenske-Underwood-Gilliland (FUG)
method: 357
  Fenske equation: 357–359
  Gilliland correlation: 366–367

Underwood equations: 361–362
Fick's diffusion coefficient: 16
Fick's first law: 10, 15–16
Fick's second law: 58, 59, 82
Film condensation: 144
Fixed-bed sorption: 498
Flash vaporization
    calculations with Mathcad: 319f
    definition: 316
    partial condensation: 316
    process: 317f
    Rachford–Rice method: 383
Flooding point: 231
Flory-Huggins solution
thermodynamics: 436
Flux: 10
    bulk motion contribution: 12
    concentration gradient
    contribution: 12
    mass diffusion flux: 11
    mass flux: 11
    molar diffusion flux: 11
    molar flux: 11, 35
    molar fraction: 33
Fourth-order Runge–Kutta
method: 43
Fractional approach to
flooding: 240, 253
Fractional extraction: 427, 444
Freundlich isotherm: 491
Froude number: 229
Fruit juice concentration by osmosis:
532
FUG method. *See* Fenske-Underwood-
Gilliland (FUG) method

G

Galileo number, definition: 119
γ-Valerolactone (GVL): 431–432
Gas absorber: 6f
Gas absorption, definition: 2

Gas-absorption operation: 156
Gas holdup: 244, 247
Gas-liquid operations: 225
Gas permeation: 483
Gas-phase mass diffusivities:
84–85
Gas-pressure drop: 237, 255–256
Gas-vapor mixtures, enthalpy
of: 450
Gel electrophoresis: 510
Germanium tetrachloride: 307
Gibbs' phase rule: 158
Gilliland correlation: 366–367
Graham's law of effusion: 53
Grashof number: 101
"Green solvent,"431

H

Hayduk and Minhas correlations:
23
    aqueous solutions: 23
    nonaqueous (nonelectrolyte)
    solutions: 24
Hayduk and Minhas equation: 85
Head loss due to surface
tension: 256
Heat of solution: 293
Heat of vaporization: 209
    Watson's method: 209
Heat transfer, dimensionless groups
for mass and: 109t
Heavy key: 323, 355
Heavy nonkey: 355
Height equivalent to a theoretical
stage (HETS): 429
Height of a gas-phase transfer
unit: 286
Height of a liquid-phase transfer
unit: 286
height of mass-transfer unit
(HTU): 290, 429

Index

Henry's law: 157, 158–159, 193*f*, 205, 209, 278, 290, 481, 490

HETS. *See* Height equivalent to a theoretical stage (HETS)

Hiflow rings: 234, 287

High-performance TFF (HPTFF): 522

High-viscosity biochemical reactors: 147–148

Hirschfelder equation: 17

Hollow-fiber boiler feed water (BFW) deaerator: 132

Hollow-fiber membrane: 130, 208, 484, 517

HTU. *See* height of mass-transfer unit (HTU)

Humid heat: 453

Humidification operations
  definition: 2, 447
  saturated gas–vapor mixtures: 448
    absolute humidity: 449
    enthalpy of gas-vapor mixtures: 450
    molal absolute humidity: 449
  unsaturated gas–vapor mixtures: 451
    adiabatic saturation curves: 452
    dew point: 452
    dry-bulb temperature: 452
    humid heat: 453
    Lewis relation: 456
    partial saturation: 451
    psychrometric ratio: 457
    relative saturation: 468
    wet-bulb depression: 455
    wet-bulb temperature: 454

Hybrid extraction-distillation process for purification: 431–432

Hybridoma cells: 524

Hydraulic characteristics of random packings: 235–236*t*

Hydrodynamic flow
  of gas: 55
    viscous flow parameter: 56
  in porous solids: 55

Hy-Pak: 228

I

Ideal breakthrough time: 507

Ideal gas constant: 602*t*

Ideal gas law: 159

Ideal separation factor: 482

Ideal stages: 198–200
  for absorber: 199*f*, 202*f*
  countercurrent cascade of: 197*f*, 198*f*
  for fractionator: 345*f*
  for stripper: 200*f*, 203*f*

Impeller Reynolds number: 421

Initial sharp concentration peak, symmetric molecular diffusion from: 82

Initial-value problem: 43

In situ solvent extraction: 432

Insoluble liquids, extraction: 411–412

Intalox saddles: 228

Interstitial sites: 51

Ion-exchange equilibria: 494–496
  molar selectivity coefficient: 495

Ion-exchange process: 478

Ionic conductances, limiting: 75*t*

Isoelectric focusing (IEF): 511

Isoelectric point: 162

Isoelectric point of amino acid: 162–163, 205–206

Isotachophoresis (ITP): 511

J

Janecke diagram: 414f, 415, 417f, 441
   related McCabe-Thiele-type construction: 413
j-factor for mass transfer: 113

K

Karr column: 429
Kinematic viscosity: 16
Knudsen diffusion: 52
   coefficients in porous solids: 50
   Graham's law of effusion: 53
   Knudsen diffusivity: 499
   Knudsen flux: 53
   Knudsen number: 52
   in porous solids: 52
Kremser equations: 201, 280, 410
Krogh cylinder-type model: 64f
Krogh diffusion coefficient: 64
Krogh model: 62, 80, 145
   intravascular resistance to oxygen diffusion: 145
   Krogh diffusion coefficient: 64
   parameter values in: 65t
Kronecker delta: 14
Langmuir and Freundlich isotherms: 493f

L

Langmuir isotherm: 491
Latent heat of
vaporization: 78, 468
   Pitzer acentric factor correlation: 78
Length of unused bed (LUB): 505–506
Lennard–Jones constants: 555–556
Lennard–Jones parameters: 17
   collision diameter: 17
   diffusion collision integral: 17
   energy of molecular interaction: 17
Lever-arm rule: 395, 434
Lewis number, definition: 67
Lewis relation: 456
Light key: 323, 355
Light nonkey: 355
Linear isotherm: 489
Liquefied natural gas: 68
Liquid extraction: 2, 213
Liquid–gas humidification and cooling: 471f
Liquid holdup: 227–228
Liquid–liquid extraction
   of bioproducts: 430–437
   definition: 393
Loading point: 231
Logarithmic mean partial pressure: 39
LUB. *See* Length of unused bed (LUB)
Lymphocytes: 524
Lysine: 162, 163
Lysozyme, aqueous two-phase extraction of: 445

M

MAb experimental data: 526t
Mass and heat transfer, dimensionless groups for: 109t
Mass concentration: 4
Mass diffusion flux: 11
Mass diffusivities: 16
   in gases: 551t
   in liquids at infinite dilution: 552t
   in solid state: 553t
Mass flux: 11
Mass fraction: 4
Mass ratios: 190
Mass transfer, definition: 1
Mass-transfer coefficient: 91, 239–240
   in an annular space: 137–138
   definition: 91

Index **611**

for equimolar counterdiffusion: 95
flat plate: 116
Gilliland and Sherwood: 123
hollow-fiber membrane
module: 130
k-type for nondiffusing B: 93
Linton and Sherwood: 123
multicomponent correlations:
133–134
packed and fluidized beds: 128
single spheres: 118
at smooth air–water interface:
148–149
solid cylinder: 122
turbulent flow in circular pipes:
122–123
volumetric mass-transfer
coefficients: 142–143
wetted-wall towers: 123f, 125
Mass-transfer operations,
definition: 1
Mass-transfer resistances: 171–172
Mass-transfer zone: 498
Material balances: 180
cocurrent contact: 194
countercurrent contact: 180
Mathcad numerical solution: 466f,
467f
Mathcad solution: 20, 21f, 26, 27f,
45f, 46f, 47f, 61, 104, 125, 160f, 174f,
178, 193, 200, 288, 318, 320, 375f,
454, 465, 484, 485, 485f, 488f, 491,
492f, 501, 502f, 503f, 504f, 505f, 509f,
510f, 526, 527f, 528f
adiabatic saturation calculations
with: 452f
packed–column design: 563–568
saturated liquid feed: 583–586
Sieve-tray design: 573–578
single-stage extraction: 591–592
Maxwell–Stefan (MS) diffusivity: 12
Maxwell–Stefan equations: 12

Mathcad: 557–558
Python: 559–561
McCabe–Thiele assumptions: 325
McCabe–Thiele construction: 341,
418f
McCabe–Thiele diagram: 374f, 378f,
407f, 410f
McCabe–Thiele method: 96, 324
feed stage: 328–329
large number of stages: 339
minimum number of stages: 332
minimum reflux ratio: 332
optimum reflux ratio: 333f
rectifying section: 325–326
saturated liquid feed: 583–589
stripping section: 326–328
total reflux: 332f
tray efficiencies: 343–344
use of open steam: 342
McCabe–Thiele operating line
for rectifying section: 326f
for stripping section: 327f
Mean free path estimation: 52
Membrane cut: 386, 483
Membrane modules
characteristics of: 513t
cross-flow model for: 485f
Membrane separations, definition: 2
Mercury removal from flue gases by
sorbent injection: 74
MESH equations: 368, 369
Methyl isobutyl ketone (MIBK): 430
Michaelis–Menten kinetics: 64
Microfiltration: 477, 518–520
Minimum impeller rate of rotation:
421
Minimum LS/VS ratio for mass
transfer: 183
Minimum reflux ratio: 332, 415
construction: 333f
minimum stages and: 335f
Minimum solvent/feed ratio: 406

Minimum stages and minimum reflux ratio: 335*f*
Mixer–settlers: 419–420, 443
Modified Bessel function of the first kind and zero order: 500
Modified Raoult's law: 156, 176, 263
Module flow patterns in membranes: 484
  cocurrent flow: 484
  countercurrent flow: 484
  crossflow: 484
Molal absolute humidity: 449
Molar concentration: 5
Molar diffusion flux: 11
Molar flux: 11, 35
  fraction: 37
Molar fraction: 33
Molar velocity: 122, 248
Molar volume: 18
  atomic volume contributions: 19*t*
  liquid at its normal boiling point: 18
  Tyn and Calus method: 18
Molecular diffusion
  definition: 3
  in porous solids: 52
Molecular weight cut-off (MWCO): 519, 523
Mol fractions: 5
Mol ratios: 181
Momentum diffusivity: 66
Monoclonal antibodies: 524–526
MS diffusivity. *See* Maxwell–Stefan (MS) diffusivity
Multicomponent distillation, definition: 354
  Fenske-Underwood-Gilliland (FUG) method: 357
    Fenske equation: 357–359
    Gilliland correlation: 366–367
    Underwood equations: 361–362

key components: 355
  heavy key: 355
  heavy nonkey: 355
  light key: 355
  light nonkey: 355
  sandwich component: 364
  Shiras criterion: 365
rigurous calculation procedures: 368
  Aspen rate-based distillation: 371
  ChemSep: 371
  equilibrium stage model: 368–370
  MESH equations: 368
  nonequilibrium, rate-based model: 370
Multistage crosscurrent extraction: 403, 403*f*
  Mathcad: 594–597
  Python: 598–600
Murphree efficiency: 260
  for absorber: 279*f*
  corrected for the effect of entrainment: 261
Murphree-stage efficiency: 278, 441
  perfectly mixed model: 423
m-value correlations: 380

N

Nanofiltration (NF): 522
Natural convection: 100–101
Natural-draft cooling towers: 465*f*
Neutral side chain: 205–206
Nonessential amino acids: 162
Nonionic polymers: 433
Nonlinear algebraic equations: 160
Nonporous crystalline solids: 51
Nonrandom-two-liquid (NRTL) equation: 340, 341
NORPAC rings: 305

Index **613**

NRTL equation. *See* Nonrandom-two-liquid (NRTL) equation
NTU. *See* Number of gas transfer units (NTU)
Nuclear power plant: 465, 470
  cooling towers: 470
Number of gas-phase transfer units: 286
Number of gas transfer units (NTU): 286
Number of liquid-phase transfer units: 287

O

O'Connell correlation: 343
Oleic acid: 441
  lowers total cholesterol level: 441
  omega-nine fatty acid: 441
Omega-nine fatty acid: 441
Operating line equation: 182, 192, 315, 325
Optical fibers: 307
Optimum reflux ratio: 333f
Organic/aqueous extraction: 432
Osmosis: 515
Osmotic pressure: 515
  of whey protein: 520
Overall efficiencies
  of absorbers and strippers using bubble-cap trays: 303
  O'Connell correlation: 343
Overall height of a gas-phase transfer unit: 290
Overall mass-transfer coefficients: 168, 290, 461
Overall number of gas-phase transfer units: 290
Oxygen: 12
  transfer efficiency: 270

transport in skeletal muscles (Krogh model): 64
transport of oxygen in the body: 12

P

Packed–column design
  Mathcad: 563–568
  Python: 569–572
Packed height: 285, 290
Packed-tower fractionator: 350
  McCabe–Thiele method: 350
Packed towers: 226
  Berl saddles: 226
  Billet and Schultes correlation: 237
  first-generation packing: 226
  flow parameter: 232
  fourth-generation packing: 267
  gas-pressure drop: 237
  Hiflow rings: 234
  Hy-Pak: 228
  IMTP: 228
  mass-transfer coefficients: 239–240
  packing: 226
  Pall rings: 228
  Random packings: 226
  Raschig rings: 226, 266
  second-generation: 227
  Snowflakes: 228
  specific surface area of packing: 232
  structured packings: 228
  Super Intalox saddle: 228
  third-generation: 227
  void fraction distribution: 230
Pall rings: 227, 228
Parameter values in Krogh model: 65t
Partial reboiler: 327, 338f
Partial saturation: 451
Partition coefficient: 479

of bioproduct: 444
Peclet number for mass transfer: 99
PEMA membrane. *See*
Polyethylmethacrylate (PEMA)
membrane
Penicillin: 430
Penicillin F extraction: 440
Permeability: 479
Permeance: 479
Permeate: 477
Permselectivity: 478
Pervaporation: 532
Phase-distribution measurements:
435–436, 444
Phosphate-buffered saline (PBS):
163–165
Photochemical oxidants: 526
Pitzer acentric factor correlation: 78
Plate-and-frame membrane
modules: 512
Podbielniak (POD) extractor: 429
Point efficiency: 258, 259f
Poiseuille's law: 56
Polyacrylamide: 433
Polyacrylamide gel electrophoresis
(PAGE): 511
Polyethylene glycol (PEG): 433
Polyethylmethacrylate (PEMA)
membrane: 483
Ponchon-Savarit method for
extraction: 413
Porous adsorbents: 477
Porous solids: 51
  bootstrap relation: 54
  coefficients in: 50
  hydrodynamic flow in: 55
  hydrodynamic flow of gas: 55
    viscous flow parameter: 56
  Knudsen diffusion in: 52
  molecular diffusion in: 52
  polymeric: 50
  porcelain plate: 80

Power number: 421
Prandtl number, definition: 67
Preloading region: 232
Pressure units: 602t
Pseudoequilibrium
curve: 283, 306, 348
Psychrometric ratio: 457
Pulsed-field gel electrophoresis
(PFGE): 511
Python solution: 84–90, 536–549
  Maxwell–Stefan equations:
  559–561
  multistage crosscurrent
  extraction: 598–600
  packed–column design: 569–572
  saturated liquid feed: 587–589
  Sieve-tray design: 579–582
  single-stage extraction: 593–594

R

Rachford–Rice method: 383
Rackett equation: 353
Raffinate, definition: 2, 394
Random packings: 226
  hydraulic characteristics of:
  235–236t
  parameters of: 241t
Raoult's law: 156, 158, 186, 200, 283,
294, 318
Raschig rings: 226, 266
Raschig super-ring: 267f
Rayleigh equation: 320
Real stages for fractionator: 349f
Reflux ratio: 325
Regular solution: 15
Rejection in UF membranes: 519
Relative saturation: 468
Retentate: 477
Reverse osmosis: 477, 515
Reynolds number, definition: 100
Right-triangular diagrams: 397
Rigurous calculation procedures: 368

Index

615

Aspen rate-based distillation: 371
ChemSep: 371
equilibrium stage model: 368–370
MESH equations: 368
nonequilibrium, rate-based model: 370
Rotating disk contactor (RDC) for liquid extraction: 428

S

Salt rejection: 517
Sandwich component: 364
Saturated air–water mixtures: 451f
Saturated gas–vapor mixtures: 448
  absolute humidity: 449
  enthalpy of gas-vapor mixtures: 450
  molal absolute humidity: 449
Saturated liquid feed
  Mathcad: 583–586
  Python: 587–589
Sauter mean diameter: 425
Scheibel column for liquid extraction: 428
Schmidt number, definition: 67
Screen filters for microfiltration: 521
Scrubbers: 278
Sealed edges, diffusion from slab with: 59–60
Semipermeable membranes: 522
Separation factor: 386, 483
Sherwood number, definition: 99
Shiras criterion: 365
Shock-wave front theory: 506
Shooting method: 43, 104
Sieve-tray design
  Mathcad: 573–578
  Python: 579–582
Sieve trays: 251f
  towers: 251
Silica gel: 79, 524

Simple differential distillation process: 321f
Simultaneous heat and mass transfer: 123–125
  air humidification: 124
  film condensation: 144
  surface evaporation: 146
  transpiration cooling: 144
Single-stage extraction: 400, 401f, 402f
  Mathcad: 591–592
  Python: 593–594
Slab of clay, transient diffusion in: 82
Smooth air–water interface: 148–149
Soap dryer, water balance: 70
Solid sphere, transient diffusion in: 81–82
Solution-diffusion model: 478
  for gas mixtures: 48
  for liquid mixtures: 479–480
"Solve block" capabilities of Mathcad: 160
Solvent in liquid extraction, definition: 394
Sparged and agitated vessels: 243
Spargers: 243, 269
Spiral-wound membrane module: 512
Spray chambers: 226
Stage efficiency: 258
  for liquid-liquid extraction: 423, 429
Stages: 196
  equilibrium stages: 196–198
  ideal stage: 198–200
Steady-state cocurrent process: 194f, 195f
Steady-state countercurrent process: 180f, 182f
Steam distillation: 382
Steam-stripping of benzene: 304
Stefan tube: 40, 41f, 86–88
Steric structure: 525

616   Index

Steroids: 522
Stoichiometric front: 507
Stokes–Einstein equation: 22
Stripping: 2, 277
   stripping an aniline-water
   solution: 271
   stripping chlorine from
   water: 268
   stripping factor: 213, 278
   stripping methanol: 272
   stripping of wastewater: 268
Styrene, equilibrium data for: 416t
Super Intalox saddle: 228
Surface evaporation: 146
Surface tension: 24
   Brock and Bird correlation: 24
Symmetric molecular diffusion from
initial sharp concentration
peak: 82

T

Tangential-flow filtration: 522
Ternary condensation: 145–146
Ternary distillation in a wetted-wall
column: 134–135
Thermal diffusivity: 16
Thermal regeneration of a fixed-bed
adsorber: 533
Thermodynamic factor matrix: 14
Thomas solution: 501, 533
   Bohart–Adams mistakenly
   referred to as the Thomas solution:
   534
Three ideal stages, cross-flow
cascade of: 214f
Thyroglobulin: 436
Tie-line calculations for
ATPE system: 444–445
Tortuosity: 52
Total annual cost (TAC): 431
Total condenser: 325

Total head loss/tray: 255
Total reflux and minimum
stages: 332f
Transient diffusion
   in slab of clay: 82
   in solid sphere: 81–82
Transient molecular diffusion in
solids: 58–62
Transpiration cooling: 144
Tray-deck mass-transfer
devices: 252f
Tray diameter: 252
Tray efficiency: 258–260
Tray geometry: 253f
Tray pressure drop: 255
   dry tray head loss: 255
   equivalent head of clear
   liquid: 256
   head loss due to surface
   tension: 256
   total head loss/tray: 255
   tray spacing: 254t
   weeping: 251, 257
   weir: 253
Tray spacing: 254t
Tray towers: 243, 249–450
   bubble-cap trays: 251
   coning: 251
   downcomer: 250
   dumping: 251
   entrainment: 257
   flooding: 251
   Murphree efficiency: 260
     corrected for the effect of
     entrainment: 261
   point efficiency: 258
   priming: 250
   sieve tray: 251
   stage efficiency: 258
   tray diameter: 252
   tray pressure drop: 255

Index 617

dry tray head loss: 255
equivalent head of clear liquid: 256
head loss due to surface tension: 256
total head loss/tray: 255
tray spacing: 254t
weeping: 251, 257
weir: 253
Trihalomethanes: 246
Tubular membranes: 512
Two-resistance concept: 167f
Two resistance theory: 168f
Two-resistance theory: 166
Tyn and Calus method: 18
Type I equilibrium diagram: 396f
Type II equilibrium diagram: 396f
Typical hollow-fiber membrane module: 131f

U

Ultrafiltration: 477, 518–520
Underwood equations: 361–362
Unsaturated gas–vapor mixtures: 451
adiabatic saturation curves: 452
dew point: 452
dry-bulb temperature: 452
humid heat: 453
Lewis relation: 456
partial saturation: 451
psychrometric ratio: 457
relative saturation: 468
wet-bulb depression: 455
wet-bulb temperature: 454
Urea concentration profile in agar gel slab: 61f
Urea diffusion in agar gel: 60–62

V

Valine: 165
van't Hoff equation: 516, 529
Vapor–liquid equilibrium (VLE) relationship: 317, 340t, 345t
Velocity: 10
diffusion velocity: 10
mass-average velocity: 10
molar-average velocity: 10
Vignes formula: 26
Viscosity: 111, 348
ideal binary solution: 348
method of Lucas for gas mixtures: 111
Viscosity units: 602t
VLE relationship. *See* Vapor–liquid equilibrium (VLE) relationship
VOC. *See* Volatile organic compounds (VOC)
Void fraction: 229–230
Volatile organic compounds (VOC): 526–528
Volumetric mass-transfer coefficients: 142–143
in high-viscosity biochemical reactors: 147–148

W

Wagner's equation for vapor pressure: 318
Washing process: 7f
Wastewater aeration: 269, 270
Water–chlorobenzene–pyridine equilibrium data: 439t
Water–chloroform–acetone: 399f
equilibrium data: 398t
Water cooling with air: 460
Water-immiscible organic solvents: 433

618   **Index**

Water–isopropyl ether–acetic acid equilibrium data: 408*t*
Water solubility of ionized species: 164
Water-transport number: 514
Watson's method: 209
Weber number: 425
Weeping: 251, 257
Wet-bulb depression: 455
Wet-bulb temperature: 454

Wetted-wall towers: 122, 123*f*, 125, 166
Wilke–Chang equation: 22
Wilke–Lee equation: 17, 18, 84–85
Wilson equation: 176, 309

Z

Zonal electrophoresis method: 510
Zwitterion: 162
Zwitterionic amino acid: 165